RURAL ENERGY TO MEET DEVELOPMENT NEEDS: ASIAN VILLAGE APPROACHES

RURAL ENERGY TO MEET DEVELOPMENT NEEDS: ASIAN VILLAGE APPROACHES

Edited by
M. Nurul Islam,
Richard Morse, and
M. Hadi Soesastro

East-West Center
Resource Systems Institute
Honolulu, Hawaii, U.S.A.

Westview Press / Boulder and London

Westview Special Studies in Social, Political, and Economic Development

Research on which this book is based was supported in part by the United States Agency for International Development (USAID), Grant No. AID/Asia-G-1393.

All rights reserved. No part of this publication may be reproduced or transmitted in any form or by any means, electronic or mechanical, including photocopy, recording, or any information storage and retrieval system, without permission in writing from the publisher.

Copyright © 1984 by the East-West Center

Published in 1984 in the United States of America by Westview Press, Inc., 5500 Central Avenue, Boulder, Colorado 80301; Frederick A. Praeger, Publisher

Library of Congress Catalog Card Number: 84-51046
ISBN: 0-86531-770-4

Composition for this book was provided by the editors
Printed and bound in the United States of America

10   9   8   7   6   5   4   3   2   1

CONTENTS

Foreword xv
Acknowledgements xvii
Authors and Editors. xix
Numerical Conversions xxiv

1
ENERGY FOR RURAL DEVELOPMENT: PRINCIPALS, ISSUES, AND METHODS
Richard Morse, M. Hadi Soesastro, and Charles Schlegel

INTRODUCTION 1
   Policy Concerns. 1
   A New Research Community 2
   Our Audience 2
WHO ARE THE PRINCIPALS? 3
   Village Decision Makers 3
   District Administrators, Planners, and Specialists 6
   National Decision Makers 8
   Scientists and Technology Developers 9
OBJECTIVES AND ISSUES 9
   Local Assessment and Organization of Energy
      Systems to Meet Rural Development Needs 10
   Articulating National and Regional with Local
      Development Policies 11
WHAT DO WE KNOW? 13
   Local Knowledge Advantages 14
   Differences and Similarities 15
THE RURAL ENERGY PROBLEM 18
   Rural Poverty and Resource Access 18
   Energy Efficiency, Values, and Intermediate Conversions 21
   Regional Transitions: From Subsistence to
      Commercial Exchange 23
   Converting the Problem to Opportunity 24
REVIEW OF METHODS 24
   Structuring and Interpreting of Secondary Data 24
   Structured Surveys 25
   Participant Observation 25
   Participatory and Action Research 25
   Field and Laboratory Experiments 27
STRENGTH THROUGH DIVERSITY: COUNTRY
   AND TECHNOLOGY EXPERIENCES 27

ENERGY AND RURAL DEVELOPMENT:
TOOWARD A SYNTHESIS 31
    Development Research Goals 32
    A New Energy for Rural Development Research Agenda 33
APPENDIXES 36
REFERENCES 41

2
ENERGY AND RURAL DEVELOPMENT: CRITICAL ASSESSMENT OF THE BANGLADESH SITUATION 43
M. Nurul Islam

INTRODUCTION 43
RURAL DEVELOPMENT POLICY 43
ENERGY IMPLICATIONS FOR RURAL DEVELOPMENT 45
    Interdependence of Subsistence and Development Needs 46
    Commercial Energy Sources 46
    Rural Electrification 47
    Traditional Fuels 48
    Policy Issues 53
ENERGY STUDIES 54
    Bangladesh Energy Study 55
    Energy Use Studies 55
    Two Surveys of Village Resources and Energy Potentials 56
    Wood and Bamboo Consumption Survey 65
    Rural Energy Survey 68
ASSESSMENT OF ENERGY STUDIES 70
    Study Objectives and Scope 70
    Factors Influencing Energy Use and Sample Design 75
    Output Design 76
    Questionnaire Design 76
    Data Collection 76
    Data Processing, Interpretation, and Analysis 77
    Report Preparation and Distribution 77
    Justification of Future Rural Energy Studies 78
    Future Studies 79
CRITICAL ISSUES AND DIRECTIONS FOR
    FUTURE RESEARCH 79
    Need and Resource Assessment for
        Technology Development 79

Enhancing Research and Evaluation Capacities 83
ACKNOWLEDGEMENT 85
NOTES 86
REFERENCES 86

3
POLICY ANALYSIS OF RURAL HOUSEHOLD ENERGY NEEDS IN WEST JAVA 89
M. Hadi Soesastro

BACKGROUND 89
   Rural Development Policy 89
   Energy and Rural Development 90
HOUSEHOLD ENERGY SURVEY OBJECTIVE
   AND METHODS 95
   Scope of Survey 96
   Sample Design 96
   Data Collection Procedure 99
PRESENTATION OF RESULTS 102
   Structure of Rural Household Energy Consumption 102
   Factors Affecting Household Energy Sources 102
   Pattern of Energy Input for Cooking 107
   Pattern of Energy Input for Lighting 111
   Fuel Supply and Availability 112
   Energy "Problems" of Rural Households 114
   Use of Charcoal 119
   Caloric and Moisture Content of Tree Species 119
   Efficiency in the Use of Energy 119
REFLECTIONS ON METHODOLOGY 121
   Deficiencies in Sample Design 122
   Deficiencies in the Survey Instrument 125
   Constraints in Implementation 127
POLICY IMPLICATIONS AND SUGGESTIONS 128
   Variations within Household Groups 128
   Kerosene Consumption 128
   Biomass Consumption 130
   The Study's Use 131
NOTES 132
REFERENCES 132

4
PHILIPPINE RURAL ENERGY RESOURCE AND CONSUMPTION SURVEY ... 135
Planning Service, Ministry of Energy, The Philippines

    INTRODUCTION: THE GENESIS OF THE SURVEY ... 135
    OBJECTIVES OF THE SURVEY ... 135
    CHARACTERISTICS OF THE STUDY AREA ... 136
    SURVEY DESIGN AND PROCEDURE ... 139
        Sampling Design ... 140
        Data Gathering and Analysis ... 141
        Some Limitations ... 143
    RESEARCH METHODS AND ENUMERATION OF FINDINGS ... 144
        First-Order Aggregations ... 145
        Variables Influencing Rural Energy Consumption ... 153
        Regional Energy Resource Endowments and Other Factors ... 155
    CONCLUSIONS AND FUTURE STUDIES ... 159
    APPENDIXES ... 162
    NOTES ... 167
    REFERENCES ... 167

5
ANALYSIS OF RURAL ENERGY DEVELOPMENT IN THAILAND ... 169
Surapong Chirarattananon

    INTRODUCTION ... 169
    THE RURAL ECONOMY AND NATIONAL DEVELOPMENT ... 171
    THE FIFTH ECONOMIC AND SOCIAL DEVELOPMENT PLAN ... 176
    ENERGY ISSUES AND ENERGY POLICY ... 179
    AGRICULTURAL AND RURAL DEVELOPMENT POLICY ... 181
        Agricultural Policy and Energy Implications ... 182
        Rural Development Policy ... 183
    RURAL ENERGY: SOURCES AND CONSUMPTION ... 185
        The 1980 NEA Rural Energy Survey ... 185
        Rural Energy Sources ... 188

    Household Energy Demand 198
    Income Level and Household Energy Consumption 208
    Energy Use in Agriculture 210
    Rural Industry and Energy Use 220
    Energy in Transportation 224
CONCLUSIONS 225
NOTES 226
APPENDIXES 227
REFERENCES 238

6
INTEGRATED RURAL DEVELOPMENT PLANNING AND ENERGY PRIORITIES: PARTICIPATORY SURVEYS IN INDIA MICRO REGIONS 241
T.M. Vinod Kumar

GENESIS OF MICROREGION PLANNING STUDIES 241
INTEGRATED RURAL DEVELOPMENT 242
STUDY APPROACH AND EXPERIENCE 243
NATURAL ENVIRONMENTS 247
    Land and Water Resources 248
    Animal Resources 249
PERSON, SOCIETY, AND RESOURCE DISTRIBUTION 251
    Demographic Characteristics 251
    Productive Assets and Technology 251
ENERGY USE AND CONSTRAINTS 258
    Cooking Fuel Sources 258
    Power for Irrigation 260
    Electricity 261
    Kerosene 263
    Energy for Transportation 264
    Energy Source by Activity 264
REGIONAL PRIORITIES AND CANDIDATE
    ENERGY TECHNOLOGIES 265
ASSESSING ENERGY OPPORTUNITIES 270
ORGANIZATIONAL STRATEGIES 271
LESSONS FOR PLANNING RESEARCH 273
NOTES 276
REFERENCES 276

7
ORGANIZING FOR ENERGY NEED ASSESSMENT AND INNOVATION: ACTION RESEARCH IN NEPAL 279
Deepak Bajracharya

A CONCEPTUAL FRAMEWORK FOR RURAL
  ENERGY DEVELOPMENT 279
BACKGROUND FOR ACTION RESEARCH 285
  Community Characteristics 286
  The Energy Project Objectives and Scope 287
RESPONSES OF THE KUMHALS 289
  Need and Technology Identification 289
  Negotiations During Planning 290
  Promises for Local Contributions 296
  Satisfying Regulations of External Agencies 301
  Management by Local Control 303
  Sharing Benefits and Responsibilities 304
  The Outcome and Some Reflections 305
RESPONSES FROM THE MID-ELEVATION
  COMMUNITIES 307
  Technology Identification 307
  Planning for Implementation 308
  Results and Problem Feedback 310
  Reflections 314
RESPONSES OF THE GURUNGS 315
  Technology Identification 315
  Planning for Implementation 316
THE CHHOPRAK EXPERIENCES IN RETROSPECT 318
  Lessons Learned 318
  Research, Development, and Diffusion for Rural Areas 320
  Suggestions with Particular Reference to Nepal 324
ACKNOWLEDGEMENTS 329
NOTES 329
REFERENCES 333

8
FIELD-BASED ASSESSMENT AND DEVELOPMENT
OF IMPROVED STOVES 337
M. Nurul Islam

OBJECTIVES AND PARTICIPANTS 337
THE COOKING PROCESS 339

Stove Users and Cooking Stages 339
       Food Characteristics 341
       Cooking Pans 341
       Cooking Methods and Fuel Consumption 342
       Cooking Time and Fuel Consumption 343
    STOVES IN USE 343
       In-Situ Built Wood-Burning Stoves 344
       Mass Fabricated Wood-Burning Stoves 344
       Stoves and Housing Conditions 344
       Design Features of Commercial Stoves 346
       Variation in Stove Design Characteristics 346
    FUELS 349
       Kindling Fuels 349
       Biomass Cooking Fuels 351
       Processed Fuels 351
       Smoke and Flue Gases 352
       Fuel Efficiency 353
    CHANGING DESIGN PERSPECTIVES 354
    TESTING STOVE PERFORMANCE 358
       Selecting Stoves for Testing 358
       Heating Water Tests 359
       Cooking Food Tests 361
       Steps to Standardize Stove Performance Tests 368
    FIELD TESTING: A PIVOTAL PROGRAM ELEMENT 370
    INSTITUTIONAL INVOLVEMENT AND
       RELATED POLICY ISSUES 372
    ACKNOWLEDGEMENTS 374
    NOTE 374
    REFERENCES 375

9
SUPPLYING FIREWOOD FOR HOUSEHOLD ENERGY 381
Rick Van Den Beldt

    INTRODUCTION 381
    FIREWOOD CONSUMPTION ISSUES 381
       Rural Firewood Sources 382
       Causes of Deforestation 383
       Industrial Use versus Subsistence Use 383
       Other Uses of Wood and Trees 384

HOUSEHOLD FUELWOOD PROGRAMS 385
   Identifying Demand, Source, and Production Strategy 386
   Mixing Government and Community Goals 387
SPECIES SELECTION, SILVICULTURAL TECHNIQUES,
  AND EXPERIMENTATION 389
   Environmental Assessment 390
   Species Selection 391
   Silvicultural Techniques 395
   Experimentation 400
CASE STUDIES 403
   Center for Policy Research: A Bangladesh Example
    of Firewood Research 403
   Perum Perhutani: Meeting the People to Save
    the Forest 406
ACKNOWLEDGEMENTS 410
REFERENCES 410

10
FINANCIAL AND RESOURCE ANALYSES OF ANAEROBIC DIGESTION (BIOGAS) SYSTEMS 415
Michael T. Santerre

INTRODUCTION 415
PRODUCTS AND BENEFITS 417
CONSTRUCTION AND OPERATION
  RESOURCE REQUIREMENTS 421
FACTORS INFLUENCING THE VIABILITY
  AND BENEFITS OF BIOGAS SYSTEMS 426
   Biogas Development in China 427
   Biogas Development in India 431
FINANCIAL ANALYSES OF HOUSEHOLD
  AND COMMUNITY BIOGAS SYSTEMS 433
   Case Study of a Financial Analysis of Biogas Systems 433
   Analyses of Five Hypothetical Rural Households 436
   Financial Analysis of Community Biogas Systems 452
DISCUSSION AND CONCLUSIONS 456
NOTES 461
REFERENCES 465

11
ORGANIZING CURRENT INFORMATION FOR RURAL ENERGY AND DEVELOPMENT PLANNING 473
Richard Morse, S.C. Agrawal, Carol J. Pierce Colfer, Elizabeth Foster, Ramabhadran Govindarajan, and Jamuna Ramakrishna

RURAL RESIDENTS' INFORMATION NEEDS FOR ENERGY
    PLANNING AND DEVELOPMENT 474
    Decision Criteria of Farming Community Members 475
    Indicators for Analyzing Rural Energy Needs and Potential ... 476
THE COOKING SECTOR 477
    Nutrition and Cooking Fuel as Basic Needs 477
    Estimating Cooking Fuel Requirements and Efficiencies 479
    Analyzing Fuel Consumption by Source and Access 483
    Preferences, Income, and Fuel Availability 491
WOMEN'S AND MEN'S VALUATION
    AND ALLOCATION OF TIME 495
    Farm Household and Labor Household Time Economy 499
IDENTIFYING FARMING SYSTEM ENERGY
    OPPORTUNITIES AND CONSTRAINTS 502
    Local Systems Assessment:
        Indicators of Scarcity and Potential 503
    Valuing Agricultural and Animal Residues
        as Energy Resources 509
POLICY IMPLICATIONS OF SUBREGIONAL
    AND REGIONAL VARIATIONS 510
REFERENCES 514

12
CONVERTING RURAL ENERGY NEEDS TO OPPORTUNITIES 519
Richard Morse, Deepak Bajracharya, Carol J. Pierce Colfer, Barry Gills, and Martin Wulfe

RESEARCH FINDINGS AND LIMITATIONS 519
PEOPLE'S PARTICIPATION IN
    COOPERATIVE RESEARCH 520
    Confidence Through Local Control 520
    Making Research Relevant to Rural Reality 521
    Research and Training Commitment 523
    A New Development Policy-Research Agenda 523

RURAL ENERGY AND FARMING SYSTEMS
RESEARCH AND DEVELOPMENT 524
  Identification of Energy Problems and
    Opportunities in Local Farming Systems 524
  Technology Appraisals and Organization 527
  Institutional and Resource Adaptations 528
  Policy Innovations in Energy and Farming Systems 530
  Evaluation, Iteration, Generalization 532
REGIONALIZING RURAL ENERGY
TRANSFORMATIONS 532
  Identification of Geographic Areas for Extending
    Rural Energy Successes 532
  Converting Local Information Advantages to
    Asset Creation Through New Energy Systems 534
  Comparative Assessment and Generalization
    of Renewable Energy Systems......................... 536
RESEARCH ON NATIONAL AND INTERNATIONAL
POLICY ISSUES 537
  Data Development Guidelines 537
  Priority Issues 538
  Future Research Issues 540
POLICY EMPHASES 541
NOTE 543
REFERENCES 543

INDEX 549

FOREWORD

This volume had its origin at a conference held in 1978 at the East-West Center that considered the short- and long-term energy problems of the Asia-Pacific region. That group of national energy policymakers, scientists, and technologists agreed that providing adequate energy for the rural areas of the developing countries looms large as one of the more critical problems of the region.

Encouraged by this consensus, the East-West Resource Systems Institute obtained a grant from the Agency for International Development for the purpose of initiating a collaborative, multicountry study of rural energy problems. The National Research Council of Thailand and the East-West Center agreed to work closely together as twin foci for the coordination of the effort.

Most persons who have given thought to the problems of social and economic development of the less-developed countries agree that special attention must be accorded the rural areas. In China and in most countries of South and Southeast Asia, three-quarters of the people live in rural areas. In addition to housing the vast majority of the people, the rural areas also harbor the greater part of hunger and poverty in the developing world.

The millions of landless rural poor more often than not lack the necessary resources either to grow food or to buy it. Indeed, rural poverty is perhaps the greatest single cause of hunger in the world today. The rural poor are confronted by the cruel choice of remaining where they are—thus facing continuing hunger and poverty—or of migrating to the cities, where they might well face even greater perils.

We know that rural dwellers can potentially greatly improve their lot. Food production can be increased; the development of small industries can provide jobs; small amounts of electricity can provide important amenities such as lighting and improved communications with the outside world. The availability of fuels for cooking and heating can be improved. Such improvements, however, will require the injection of increased quantities of energy into the rural system.

How can these increased demands for energy best be met? Will rural areas, like the urban ones, become increasingly dependent upon fossil fuels, which for the most part must be obtained from distant sources? Or is sufficient energy available through the intelligent development and management of indigenous energy sources to fill the gap?

The authors of this work discuss some of these difficult questions and describe how individual nations are now grappling with this complex problem. They analyze experiences and lessons from their various cultural perspectives and intellectual disciplines to clarify the unique role of energy in rural household and agricultural activities and to suggest more effective ways for linking energy development with programs to increase agricultural productivity. From those perceptions and experiences emerges a new set of priorities for energy and rural development research and an integrated set of methods for strengthening rural people's participation in energy program development. Above all, this work reveals how villages, regions, and nations can effectively organize to meet their rural energy needs.

Harrison Brown and Sanga Sabhasri

ACKNOWLEDGEMENTS

Many women and men in numerous rural areas of the countries involved in this book are originators of its central ideas and facts, as we hope will be evident from the voices and issues we report. The perception that rural residents' varied energy situations could become a workable theme for cooperative research is owed to a relatively few, tenacious research organizers. Harrison Brown and Sanga Sabhasri are central among these. At a critical point they decided to allot available resources to the preparation of a volume that would bring together the results of the first phase of an insightful intercountry project. Others central in the project's initiation and completion are Artono Arismunandar, Prapath Premmani, J. Gururaja, Shahzad Bahadur, Kulthorn Silapabanleng, and Kedar Lal Shrestha. These decision makers' commitment to seek the reality of village residents' energy problems and potential as a guide to policy formation and to design of renewable energy technologies is the foundation of this volume.

We give particular thanks to Kirk R. Smith for sustained support in the various phases of the project, especially in his penetrating search for clarification of energy's functions and role in society. He contributed significant ideas and information at innumerable authors' meetings and was active in the review of successive drafts to clarify content as well as exposition.

At an early point in conceptualizing the issues, Bruce Koppel's insistence on awareness of the social contexts that shape access to energy and influence its technical organization was rewarding. Colleagues who persisted in efforts to clarify these issues as we and our coauthors proceeded in the research include Iqbal Mahmud, Edilberto Reyes, Vijay Vyas, Varun Vidyarthi, Ramashray Ray, Tushar Moulik, Mansur Hoda, and Ashok Subramanian, among many others. Both intellectual and resource support were given by Norman Brown and Robert Ichord at early and continuing stages of the work.

The book's external reviewers — Ramesh Bhatia, Marilyn Hoskins, Michael Howes, and Amara Pongsapich — immeasurably improved its final content and presentation through their critique and synthesis of the review draft. Marcia Gowen, Fred Hitzhusen, and Mark Phillips also incisively contributed to the structure of several chapters at this final stage.

A book of such wide compass, involving diverse countries and rural areas, early showed signs of growing overlong. Cynthia Shklov, as project editor, found effective ways to reduce and reorganize the material in forms ultimately satisfying to each author. She also found phrasing to capture the styles of the various writers. Those who have undertaken to edit multiauthor works will join in our appreciation of the magnitude of this task. More fundamentally, Cynthia Shklov perceived the human intent and technical structure of the research and, therefore, could be considered in essence a coauthor.

Management support to the project was provided at various stages by John Bardach and Robert Randolph; organizational and information infrastructure by Fredrich Burian; researcher exchange facilitation by Mendl Djunaidy and Kajorn Howard; bibliographic guidance and support by Ann Jacobs and Victoria Rumenapp; and communications support by Prapasri Thanasukarn and Wanasri Samanasena. Karen Ho carefully produced the entire volume through the instrument of a word processor. Doreen McConnel, Jennifer Cramer, Dorothy Izumi, and Jane Smith patiently input earlier revisions. Sheryl Bryson contributed overall editorial guidance in final publication arrangements.

We appreciatively acknowledge these many and valuable aids, while retaining responsibility ourselves for errors and failures to communicate in the pages that follow.

And, returning to our perspectives as editors, we gratefully thank our wives — Mushu, Romola, and Janti — and families for the inspiration and sustained support they provided in this cooperative undertaking.

M. Nurul Islam
Richard Morse
M. Hadi Soesastro

AUTHORS AND EDITORS

M. Nurul Islam is Professor of Chemical Engineering and Director, Institute of Appropriate Technology, Bangladesh University of Engineering and Technology (BUET), Dhaka. He received his B.Sc. Engineering (Chemical) degree at BUET and obtained his Ph.D. at the University of Newcastle on Tyne. After completing the studies he reports in this volume, he was a member of the U.S. National Academy of Sciences panel on Renewable Energy Technology Diffusion in Developing Countries. He was a Visiting Fellow at the Institute of Development Studies and the Science Policy Research Unit, University of Sussex. His special areas of current research interest are the diffusion of technology in rural areas and training for rural energy planning.

Richard Morse is Research Associate, Resource Systems Institute (RSI), East-West Center, and coordinates the Energy for Rural Development (ERD) project. He holds the A.B. of Dartmouth College and the M.A. in Economics of Harvard University, where he completed the general examination for the doctorate in Economics. His research, consultancy, and entrepreneurial activities have centered in Southern Asia, especially in India and Burma. He took part with Sushil Agrawal in the Uttar Pradesh rural energy study cited herein. With Eugene Staley, he co-authored *Modern Small Industry for Developing Countries* (McGraw-Hill 1965). He was study director for the feasibility report on the International Industrialization Institute (U.S. National Academy of Sciences and National Academy of Engineering 1973).

M. Hadi Soesastro is currently Head of the Department of Economic Affairs at the Centre for Strategic and International Studies (CSIS), Jakarta, which he joined in 1971. He graduated from the Faculty of Aeronautical Engineering, Rheinish-Westfaelische Technische Hochschule in Aachen, West Germany, in 1971 and from the Rand Graduate Institute for Policy Sciences, Santa Monica, California, in 1978. He was Director of Studies, CSIS, in conducting the research he reports herein. He is editor of the 1983 volume *Energi dan Pemerataan (Energy and Equity)*.

Sushil C. Agrawal is Research Officer, Planning Research and Action Division, State Planning Institute, Uttar Pradesh. He obtained the M.S. in Social Work and Ph.D. in Sociology from Lucknow University. He has been involved the past 14 years in exploration of various problems of rural development. He directed the Uttar Pradesh study on energy flow systems in five rural communities and was a Research Fellow of the East-

West Center in analyzing and writing the report of this study. His research currently centers on socioeconomic and organizational problems of renewable energy development.

Deepak Bajracharya is Research Fellow, RSI, East-West Center, and coordinates the curriculum development and training program for Participatory Rural Energy Development. He obtained his B.S. degree in Civil Engineering at Lafayette College, the M.S. in Environmental Engineering at Stanford University, and the Ph.D. in Science Policy at the University of Sussex. He worked as an environmental engineer in the United States and Iran on industrial pollution, biogas development, and public health training. He was a member of the UNDP review team on the Regional Energy Development Programme for Asia, 1984–86, and Consultant, FAO-UNDP Seminar on Rural Energy Planning, Beijing, April 1983. In 1983, he received the East-West Center Makana Award for distinguished project contributions through basic research on food-fuel competition and through action research in community organization of rural energy technologies. His current research focus is on nutrition, energy, and work.

Surapong Chirarattananon is Acting Dean and, since 1982, Associated Professor, School of Energy and Materials, King Mongkut's Institute of Technology, Bangkok. He received his Ph.D. in Electrical Engineering from the University of Newcastle, Australia. In 1973 he joined the Faculty of Engineering, Khonkaen University. "Being a regional university," he notes, "it offers an opportunity to learn and feel the way and life of rural people." He participated in the ERD research implementation workshop at Chiang Mai, Thailand, in 1980. His energy research interests include modeling and planning, solar energy, and energy conservation.

Carol J. Pierce Colfer is Farming Systems Researcher, Hawaii Institute of Tropical Agriculture and Human Resources, University of Hawaii, and was Women in Development Specialist while contributing to this study. She is an Adjunct Research Associate with RSI. She obtained her B.A. in Anthropology at Portland State University, the M.A. and Ph.D. in Anthropology at the University of Washington, and the M.P.H. in International Health at the University of Hawaii. A linguist, she has lived in rural communities in Iran, the state of Washington, Bali, and Kalimantan to conduct development-oriented, ecological research. She is now resident social scientist with the University of Hawaii TROPSOILS team in Sumatra working with the Agency for Agricultural Research and Development and the Center for Soils Research, Indonesia, to study and support the initiatives and adaptive adjustments of transmigrant and local farmers in improving farm management and soil conservation practices in resettlement areas.

Elizabeth M. Foster obtained the B.A. in African and Afro-American History and Economics at Stanford University. She was Summer Energy Intern with RSI, working on resource valuation and time allocation, and was intern with the United Nations Development Programme in Kenya. She is currently a Master's degree candidate in International Economics, minor in Energy, at the Food Research Institute, Stanford University.

Barry Keith Gills, with a background in religions and humanities of Asia, obtained the M.A. in Political Science at the University of Hawaii. As an East-West Center awardee he participated in organizing methods reviews for the ERD project. His long-run interests are in ecology and the role of energy in international peace and development. He is continuing graduate research in the M.Phil. program at the London School of Economics and Political Science, in International Relations.

Ramabhadran Govindarajan holds the B.E. in Industrial Engineering of the University of Madras. He conducted research on rural energy systems with the Centre for the Application of Science and Technology in Rural Areas (ASTRA), Indian Institute of Science. He was a Master's student in Agricultural and Resource Economics at the University of Hawaii while working with the ERD project on analyses of energy and agriculture and of household fuel consumption.

T.M. Vinod Kumar, a Civil Engineer and City and Regional Planner, is a faculty member in the Department of Urban and Regional Planning, School of Planning and Architecture, New Delhi. He obtained his B.Sc. in Engineering at Kerala University and his postgraduate degree in Town and Country Planning at the School of Planning and Architecture. He has been project manager and consulting engineer-planner for projects in India, Malaysia, and Indonesia including assignments with the Ford Foundation, World Bank, Danish International Development Authority, and Bread for the World. He was associated with the Centre for the Study of Developing Societies, New Delhi, in the research he reports in Chapter 6.

Gary S. Makasiar, Chief, Planning Services, Ministry of Energy, Philippines, directed the rural energy survey reported in Chapter 4. He holds an A.B. in Mathematics, the M.A. in Economics, the M.B.A. and the M.P.A. In 1973 he joined the Office of the President as Assistant Director in the Policy Analysis and Research Office. Since 1976 he has concentrated on energy policy studies. These include the first construction of time series data on energy use, through nationwide sectoral surveys, and a comprehensive 10-year energy program strategy for the country.

Jamuna Ramakrishna obtained her B.A. in Environmental Studies and M.A. in Geography at the University of Hawaii. She has been Teaching Assistant in the Department of Geography and Graduate Assistant and Research Intern in the Energy Program of RSI. She wrote the annotated bibliography on anaerobic digestion as a component of the computerized ERD reference collection. Presently a Geography Ph.D. candidate in Resource Management, her research field is the environmental aspects of energy use.

Michael T. Santerre is presently working in the capacity of Resident Manager of a shrimp aquaculture company in Sri Lanka. As Research Fellow with RSI from 1978–1982, he developed in collaboration with Kirk R. Smith the FLERT (Fuel-Linked Energy Resources and Tasks) methodology for evaluating and comparing small-scale renewable energy technologies. He holds the B.S. in Education of Massachusetts State College at Fitchburg, majoring in Biology, and the M.S. in Oceanography of the University of Hawaii. As Research Associate with the Hawaii Institute of Marine Biology and the Hawaii Natural Energy Institute he designed and conducted experiments on the environmental physiology and aquaculture of fish and algae, among other marine research activities.

Charles C. Schlegel was a Peace Corps Volunteer in Sabah, Malaysia, in 1965–66. He majored in International Affairs at the University of Colorado and obtained the Ph.D. in Development Sociology at Cornell University. A Research Fellow with the Food Systems and Energy Systems programs of RSI from 1977–1982, he is now Project Officer for Research, Evaluation, and Monitoring with the United Nations Children's Fund (UNICEF) in Indonesia.

Rick Van Den Beldt is Research Associate, Hawaii Natural Energy Institute. He obtained the B.S. in Forestry at Michigan State University and the M.S. and Ph.D. degrees in Agronomy and Soil Science at the University of Hawaii. He was a Peace Corps volunteer in the Philippines, working in village agroforestry projects in several parts of the country, and was Project Officer and Forestry Consultant with the Development Academy of the Philippines. He now coordinates a program of international experimental research and field trials on fast-growing tree species and is Executive Secretary, Nitrogen-Fixing Tree Association.

Martin Wulfe is Consultant in Energy Development and Renewable Energy Technologies. He received the Bachelor of Mechanical Engineering degree from Pratt Institute and the Master of Regional Planning from Cornell University, where he specialized in energy policy planning. He has been a staff member of the Massachusetts Institute of Technology Energy Laboratory. For a year, he lived in West Sumatra teaching village energy planning and doing research on household and agricultural energy, prior to working as a Research Fellow with RSI. He recently completed an energy planning study in two countries of Africa.

NUMERICAL CONVERSIONS

Rural residents obtain energy from many sources — the sun, fuelwood, farm residues, fossil fuels, hydroelectric plants — and are users of various technologies to transform energy into fuels or carriers suited to the task at hand — heating, lighting, mechanical work, cooling. Many rural energy sources and fuels are in nonstandardized forms. There are wide variations in the efficiencies with which they are converted to useful work. A continuing challenge to policymakers and researchers, therefore, is to arrive at reliable and comparable estimates of rural energy production, consumption, and effective use. In this book each author states the units in which the original data were collected and the numerical values used to convert these data to calories or joules (J), the standard international energy unit. For ready reference, the principal energy units, conversion factors, and abbreviations used in the book are listed below.

Energy
J = joule = 0.239 calories
cal = calorie (raises 1 gram of water 1° C) = 4.184 J
kcal (food Calorie) = 1,000 cal
kWh = kilowatt-hour = 3.6×10^6 J

Power (rate of flow of energy)
W = watt = J/second
kW = kilowatt = 1.34 horsepower

hp = horsepower = 746 W = 33,000 ft-lb/minute

Metric prefixes
k = kilo = 10^3
T = tera = 10^{12}
M = mega = 10^6
P = peta = 10^{15}
G = giga = 10^9

1
Energy for Rural Development: Principals, Issues, and Methods

Richard Morse, M. Hadi Soesastro, and Charles Schlegel

INTRODUCTION

This book, coauthored by persons in seven Asian countries in cooperation with the East-West Center, reports methods and results of a cross-section of policy, technology, and development research linking energy and rural development issues. The aim of the book is to distill experiences and lessons from these studies, forming a new agenda of research and methods designed to strengthen the ability of rural people to identify and organize energy resources and technologies to meet their own development needs.

Focusing initially on energy needed to meet basic food and nutrition requirements, the book also develops approaches for assessing energy requirements to increase agricultural productivity and to support other income- and employment-generating activities in rural areas. Individual chapters present and assess methods used in studying those and related policy issues. A new set of priorities for energy and rural development research is identified from this experience, and an integrated set of methods building on rural people's own knowledge and motivations is suggested to address these priority issues.

Policy Concerns

In many Asian nations, rising policy concern for increasing food production and meeting the basic needs of most of the people shifted emphasis in the mid-1970s to incorporate concern for energy constraints as well. Two issues linking nonrenewable and renewable energy resources took on new importance. Suddenly, the world and the region entered a period of rising prices and unstable supplies of petroleum-based fuels, with uncertainties also affecting fertilizer and pesticides. Fossil fuel uncertainties threatened also to accelerate the more gradual but nevertheless widespread environmental depletion associated partly with rural people's heavy reliance on firewood and agricultural residues for cooking fuel. More costly, interrupted supplies of kerosene and diesel oil, it was believed, could lead to a reverse substitution trend of biomass for fossil fuels, increasing environmental damage.

Persons responsible for priority allocation of national resources started turning their attention at this time to determining rural energy needs. Their rationale was twofold. Most people in developing economies live in rural areas. And energy is critical to agricultural production while at the same time it greatly influences the quality of rural life. Accordingly, policies for rural energy development have two main objectives: (1) stimulation of agricultural productivity and food production; and (2) rural development of farm- and forest-based energy resources that can help improve the quality of life and also preserve the local and regional environment and reduce dependence on fossil fuels.

Policymakers, planners, and investors soon became aware of serious information gaps on certain issues. Missing were basic data for defining rural energy problems and for designing policies and programs to meet those problems in accord with national development goals. If agricultural productivity and rural development are to be stimulated through energy inputs, detailed information on energy use patterns and on energy flows is needed. If agriculture is to contribute resources for cooking and other household needs and for other economic sectors through bioconversion, specific information is needed to evaluate alternative technologies, their viability and acceptability, and their environmental impacts.

A New Research Community

Consequently, governments, science academies, technology institutes, universities, and international development agencies have sponsored a variety of energy for rural development research activities within the past decade. Researchers in diverse fields — engineers, agricultural scientists, economists, anthropologists, systems analysts — have brought their specialized skills and concepts to this new realm of inquiry. A sizable new literature has emerged, with many items in unstandardized form.

This book helps connect these policy and research concerns. The book's roots are in an interdisciplinary program involving both survey and action research. The program engages researchers not only in Asian universities and technology institutes but also in government policy, planning, and implementation agencies in cooperation with East-West Center researchers. From the intercountry experience, we identify the principal interest groups involved, examine the issues, and bring together evidence from a review of studies to help clarify the issues and contribute to improved methods for ongoing research.

Our Audience

Researchers and research sponsors are the book's primary audience. If the findings of research are to be in terms useful for policy and action, those conducting the research must be aware of the positions and perspectives of the decision makers concerned. Researchers and research users therefore need close interaction. Yet findings that connect the interests of different users must also be discovered. In this book we not only seek to define standards of research scope and relevance for particular problems; we also try to help decision makers and researchers reach some common ground on procedures and methods for developing pertinent data and for evolving reliable solutions.

The book, then, marks a transition. We hope it will contribute to a new research agenda emphasizing result-oriented energy and rural development methods. Drawing lessons from experiences reviewed here, methods for this new agenda would be built upon the needs and organizing skills of rural people, who produce and use energy in their daily lives. The goal is a new synthesis of rural and national development policies, fostering socially and environmentally appropriate development of rural energy systems.

WHO ARE THE PRINCIPALS?

Decision makers in many different settings sponsor work by this new research community and are users of its findings. The most numerous potential users, of course, are rural residents themselves. They can use, and often request, environment-specific information necessary for appropriate local decisions on changing energy sources and technologies for a wide variety of tasks in individual homes and enterprises, notably for agriculture and cooking.

Village Decision Makers

Prior to choice of technology is of course the clarification of needs. Expression by rural citizens of their own felt needs, as a basis for probing into specific requirements, is exemplified by this dialogue at an early stage in the action research reported by Deepak Bajracharya in Chapter 7:

1st man:	"We should get either electricity or water."
2nd man:	"What we need is mainly water for the winter to plant wheat, cultivate potatoes, cauliflower, and so forth."
Deepak:	"Yes, I see . . ."
1st man:	"Water is really what we need."
2nd man:	"Well, what do our daughters and daughters-in-law want to ask for?"

4 Rural Energy to Meet Development Needs

Arriving for Kamalbari meeting, Chorkate, Nepal. (D. Bajracharya)

Deepak (turning to the women):	"It's your turn now to say what you want."
1st woman:	"All right. How about the elderly, experienced ladies first."
2nd woman:	"What do you want to ask for?"
1st man:	"Looks like they don't want to speak out."
3rd man:	"If I can speak for you and your convenience, I would ask for the mill."
1st man:	"Yes, I think we need a mill which has an oil expeller and a grinder for maize, millet, and wheat."

Deepak:	"What do you say?"
1st woman:	"Yes. To improve the future of our children, we need the mill."
Deepak:	"I don't understand. How is it that the children's future is going to be improved?"
1st woman:	"Presently, the milling and grinding take a lot of time. We can't always manage everything on time. Once we have the mill we can cook early, feed the children, and send them to school."
Deepak:	"I see. How long do you spend in milling and grinding?"
1st man:	"If the person is quick, she gets up as soon as the cock crows the first time, and she starts unhusking the grain...."
2nd woman:	"... at least until 8:00 or 9:00 o'clock."
2nd man:	"You can imagine. How can she manage to cook at 8:00 o'clock and send the children to school on time?"

—Chorkate, Nepal, February 1982.
(Deepak Bajracharya and Chandra Gurung 1984)

The felt need of Chorkate residents to reduce women's drudgery and free their time for more valued uses drove subsequent discussion to underlying needs: for adaptation and innovation in existing social institutions and for a new quality of energy technology to meet the direct needs. The facilitator's role brought new perspectives connecting these underlying needs with the potential of biogas technology and of new organizational forms.

Women and men reached a consensus, in the Chorkate dialogue, on priority of food processing over food production. It was not entirely a dichotomy, since increased oil yield was an expected benefit from the expeller. Complementarity or choice between the food production sector and the food preparation and cooking sector are frequent issues in such village energy decisions:

"When the village assembly was first asked to make its choice of trees to be planted, the men immediately answered: 'Fruit trees.' Fortunately, the women of Chamoli were braver than those of Etawah and argued that these trees will not get them anything. 'The men will take the fruits and sell them by the roadside. The cash will go to buy liquor and tobacco. No, we want fuel and fodder trees,' they argued." (Anil Agarwal 1982)

In the Chorkate meeting, women became involved because their husbands first spoke for them. In Chamoli, the women, as chief fuel gatherers, spoke for themselves. "Finally, both types of trees were planted. An observer notes that otherwise the men were beginning to lose interest in the very proceedings of the ecodevelopment camps" (Agarwal 1982).

The significance of these village decision sessions can be generalized by stating that the motivations, needs, and constraints of rural people constitute the foundation of rural change and development and set the context for defining rural energy issues. Small- and medium-scale farmers, marginal and subsistence farmers, landless families, large-scale farmers, agro-processors, traders, artisans, teachers, and community leaders — all may see these issues somewhat differently. Information provided by these individuals during survey interviews or action research and tasks done or decisions made during participant observation comprise an important resource base of this volume.

District Administrators, Planners, and Specialists

Village officials and councils — elected, appointed, or hereditary — connect the village or cluster of villages with the wider political and governmental framework. On the scale of a rural market area or administrative unit, development planners as well as extension agents and sector specialists (who often serve as advisers to farm family decision makers) are primary clients for research findings. What information do these local planners and experts require?

Needs for local planning data were specified in a discussion in North India in early 1982 to plan use of central government funds (on the order of $135,000) allotted for an energy planning and demonstration project in some 300 villages. The Indian Administrative Service officer responsible for district development convened the meeting, involving the district economics and statistics officer; specialists in agriculture, livestock, and forestry; and a folk education specialist. After reviewing results of a rapid survey of energy consumption in sample villages, they voiced questions such as follows:

> How should we proceed from the sample survey to quantitative estimates of gaps between energy demand and supply, by source?

Grinding and sifting maize, Gorkha District, Nepal. (D. Bajracharya)

What strategies should be adopted to meet these gaps?

Recognizing that even the accelerated rural electrification program in the state will not provide power to these areas in the next five years, what energy alternatives should be assessed for groundwater irrigation in food-deficit, water-scarce parts of the area?

How should the social or opportunity costs of women's and men's time be evaluated in connection with planting, nurture, and maintenance of fuelwood lots to meet cooking needs?

What teaching and orientation materials can be developed by the area's science college for a workshop of village committee members and extension experts, to stimulate participation by rural communities in the planning process? (Richard Morse, pers. comm., 1982)

Six months later, preliminary answers to these questions had been formed and detailed feasibility exercises and resource assessments were underway (A. Khurana, pers. comm., 1982).

This planning session's initial focus on energy led the discussion to related issues of food shortages and labor time valuation. Mechanical energy came into focus as the limiting factor to new irrigation for food production and human energy in its potential for more productive resource use. In effect, energy served the group as an integrating dimension for local development planning.

National Decision Makers

Besides the village and development district, attention to energy extends out through a host of policy, technology, and action agencies — public and private. Nationally, these include ministries of finance, planning, petroleum and power, energy, science and technology, agriculture, forestry, and rural development. These agencies pose contrasting issues for data development and analysis. Decision makers "in different chairs" see the choices from their own position.

> Planner in energy ministry:
>
> "What five-year targets have you set for these new technologies? We must know this to justify your program."
>
> Energy research and development director:
>
> "I can't target. The technologies are still in development and have not gone through sufficient field trials and demonstrations to serve as a basis for extension."

Two years later the research and development director had assumed a position with a major international development bank. The new dialogue:

> Energy ministry planner:
>
> "We are presenting a plan for a five-year energy sector loan for research, development, and demonstration."
>
> Development bank official:
>
> "What is your payback period?"
>
> (Gary Makasiar, pers. comm., 1982)

Field trials were still just as uncertain, but the decision-perspective of the director had changed.

Scientists and Technology Developers

Engineers, scientists, and systems analysts in industrial companies, fuel research and development institutes, forest products institutes, water resource boards, and technical universities and polytechnics need data on resource, demand, and design determinants for improving existing energy technologies and developing new technologies. Each of these groups brings particular knowledge, interests, and judgments to bear in assessing energy needs and potential. Yet each lacks relevant information, techniques, and perspectives held by other professionals.

Moreover, the form and precision of needed information differ sharply among specialized users at various points in the policy and technology system. Even though fallacies of the homogeneous rural group, the homogeneous development project, and the homogeneous analytical procedure are now widely recognized, the departmental structure of the policy system makes difficult an assembly of information that would permit differential responses to local energy needs. Information users in the energy for rural development community are nearly as diverse as the widely varying ecological and social settings where the primary actors — the rural people — live and work.

OBJECTIVES AND ISSUES

Rural energy data collection necessarily requires a clear statement of the policy objectives to be addressed and the issues to be analyzed. These condition the data required and the methodology to be used (Charles Schlegel and James Tarrant 1980). Understanding that, we have reviewed recent studies and methodology papers to observe how specific policy questions have guided selection and design of research methods (see Appendix 1). Thus the book is a learning tool structured around various research projects sponsored by different users. Several of these studies have had narrowly defined objectives. Others have had multiple aims and scope. Many have been exploratory and eclectic. Noting this diversity, we identify in Appendix 2 the principal categories of research objectives that have motivated recent studies. We systematize below their central objectives and issues.

The principal objectives of rural energy research can be defined from either a local (area-specific) or national position: from micro or macro perspectives. Even though their viewpoints may contradict as well as intersect each other, in this book we intend to show how research from these different starting points can be brought together compatibly to meet rural and national development goals.

Local Assessment and Organization of Energy Systems to Meet Rural Development Needs

Attainment by the great majority of rural people of minimum conditions of well-being clearly will require significant increases and changes in their access to energy. Energy will be needed in suitable forms and amounts — and at the right place and time — to increase agricultural and rural industry production, to process and cook more adequate food for millions of families, and to provide clean water, sanitation, lighting, and heat. The community- and environment-specific nature of these needs is a principal determinant of research methods for their fulfillment.

Assessment of User Needs, Problems, and Priorities. Comprehensive understanding of energy potential in the context of rural community needs requires data on supply-demand balances of locally available energy resources. Innovation through new energy forms and technologies requires study of numerous factors. Introduction of biogas technology, for example, requires data on the attributes of specific locations defining parameters for digester design and construction. Such attributes include neighborhood patterns, family size, and other socioeconomic factors; available materials for construction; availability of raw materials to feed into the digester; seasonal and ambient temperatures; topography, land form, and land availability; and water table and subsoil structures.

The most important information to be developed is in relation to people's felt needs. The felt needs of people often are not expressed due to lack of information, suppressed desires, or lack of purchasing power. Information collected will have to include existing and projected needs for a better life. The cultural and anthropological bases of certain traits and characteristics also must be determined to design a suitable technology for actual rural conditions.

Innovation and Development of Renewable Energy Technologies. In this context, scientists and technologists seek to discover, create, and adapt alternative technologies based on renewable resources to remove energy constraints on development. Research concentrates on organizational and technical barriers to the invention, testing, demonstration, adaptation, and adoption of new technologies. Understanding the implicit as well as explicit policy environment for the nurturing and extension of locally appropriate energy technologies is a related concern. Creation of effective organizational, financial, marketing, educational, and maintenance systems to promote the spread of suitable technologies connects research and development activities to the meeting of local needs.

Reversing Environmental Depletion Associated with Deforestation and Other Biomass Removal. Although the specific relationship between traditional fuel use and deforestation and biomass depletion differs greatly from place to place, evidence of environmental damage in many ecosystems, in some instances with perceived downstream effects, has resulted in wide concern for mapping and assessing the severity of this trend. Besides the intrinsic difficulty of quantifying pervasive but often slow environmental change, causes of the problem are complex and interwoven. These include competition between food and fuel needs for scarce land resources and between wood-based fuels used by industry (especially cottage and small industries crucial to rural employment) compared to rapid depletion by logging for timber and pulp. Burning of animal and vegetative wastes in some areas can have an even more imperceptible environmental impact, to a crisis point when the organic humus content of soils is reduced to a nonproductive, infertile two or three percent. The motivating factor in this dimension of energy for rural development research is the need to grasp these trends and devise effective afforestation and other programs to reverse them.

Articulating National and Regional with Local Development Policies

Accent on policies emerging from local settings is implied in this set of research objectives. Diverse rural community energy needs and resources must be known and appreciated as a basis for formulating regional and national development policies. National energy policy research agendas have been motivated by three principal concerns.

Reducing Reliance on Petroleum-Based Energy. The 1970s' oil price shocks and shortages had varying impacts on farm inputs and outputs and rural consumption patterns that reverberated throughout the economy. Diesel oil for irrigation pumps, other stationary engines, and tractors and transport; kerosene for lighting and, in relatively well-to-do households, for cooking; gasoline; and petroleum-based fertilizers are inputs required in different proportions and at different periods of the year among rural regions. Comparative energy supply options, energy use mixes, and future energy demand potentials are issues arising from these impacts. In seeking ways to mitigate these impacts and uncertainties, policymakers face issues such as shifts in the relative supply structure of fossil and renewable fuels; competition for energy resources (e.g.,

12 Rural Energy to Meet Development Needs

Deforestation stage, Gorkha District, Nepal. (D. Bajracharya)

between household and agricultural uses, urban industrial and rural uses); a potentially growing demand for higher quality fuels during development; and appropriate price policies as signals to preferred paths of conservation and substitution.

Establishment of Baseline Data to Determine National Energy Program Priorities. The principal data and information required for rural energy demand projections and supply planning are those that could explain accurately and significantly patterns of household characteristics (income, family size, occupation, education) as related to distinct patterns of fuel consumption by amount, by types, and by purpose. Information on energy flows and requirements for agriculture, processing, transport, and rural industry at village (local), regional, or national scale is also required. For a more detailed analysis involving development and equity goals, social preferences, or other behavioral aspects, studies would be necessary about variations among villages and regions in potential energy resources, supply constraints and opportunities, relative prices of fuel types, and even cooking or other fuel use habits, as well as efficiencies in the uses of fuel.

Integrating National Energy and Rural Development Policies. Many interacting strands of local, farming system, and sectoral issues can be aggregated from two principal policy perspectives: rural development and energy. Efforts to systematize research in this field have started from both perspectives. Angles of view can be illustrated as follows:

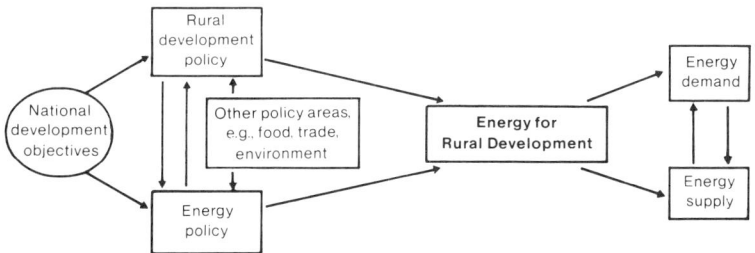

This simplified chart may prevent neglect of any important analytical dimension. Demand management, for instance, is relevant to issues of supply development. One may approach the problem from either perspective, but integration can only be obtained by evaluating both.

The newcomer or "unknown" in the chart is energy for rural development. This is the research problem identified by researchers who met in February 1980 at Chiang Mai, Thailand, as most important: "to assess the energy needs of rural people in the context of energy resources in order to gain better understanding of policy and technology requirements for advancing rural development objectives" (National Research Council and Resource Systems Institute 1980). As countries move from the historic energy changes of the 1970s into the mid-1980s, what is happening? What is changing? As countries reassess the lack of success or the "mistakes" of rural development efforts, which institutional and organizational innovations are being tested? Which are proving effective?

WHAT DO WE KNOW?

Variation in social and farming conditions that determine energy uses and sources is great, and studies to discover regularities are still only infant. Advantages of knowledge are with the rural dweller. Rural producers and users of energy possess knowledge about existing amounts, forms, and attributes of energy, and about needs to be met. The specific knowledge they have is the basis for their assessment of potential changes. Local organization to use and reassess this knowledge in the light of new information on development opportunities is the subject especially of Chapter 7 in this volume. This may entail a mutual learning process, the researcher raising new questions and adding new information. In that process the researcher learns about specific conditions that shape this local knowledge. The process of generalization requires relating these conditions to specific local facts, so that local knowledge can be adapted to other contexts.

Local Knowledge Advantages

The kinds of knowledge that women and men in rural areas have on energy use and resources are listed here, indicating the rural groups holding special advantages.

Knowledge holders and generators	Kinds of knowledge and advantages
Landless families	Collected fuels: wood species, parts, attributes; animal dung, forage and grazing paths; crop residues, green vegetation, depletion. Collection times, distances, units of measure.
Farm laborers	
Artisans: smith, carpenter, potter, weaver, woodsman, leather-worker	Alternative earning opportunities: variability, risks, trade barriers and inequities, wage negotiating factors, implicit valuations.
Service: midwife, tailor, barber	
Marginal and subsistence farm families	Inferior grains and fuels; residue uses and substitution; animal maintenance and services; food and fuel uncertainties; coping strategies; labor and wage terms.

Small- and medium-farm families	Crop rotation; irrigation; storage; animal power, fodder, transport; dairy practice; market variations and barriers; cash management; petroleum products, diesel and kerosene; electric power connections.
Large-farm families, landlords	Cash-crop specialization; fuel purchases; tree and grove management; labor trends, hiring, work organization and supervision; mechanization; marketing; supplier and bank credit; petroleum products, lubricants, gasoline.
Agro-processors, traders	Stocks and storage facilities; truck and rail transport; crop futures, margins, arbitrage; fuelwood marketing; machinery supply, repair and maintenance.
Village and district officials	Selective data on above categories; approximations of range and distributions; secondary data.

Much primary information of this kind is in local terms and units of measure. Many of these terms pertain to barter and in-kind transactions not altogether visible in the price economy. Estimating the magnitude of this unpriced zone and its terms of exchange requires close familiarity with these local knowledge generators. Daily, seasonal, and yearly variations also must be understood.

Much of this knowledge is informal, moreover, and expressed mainly in action and in half-stated conversation — or silences. Observing and sharing these nuances as a member of the community (as a participant observer) is often the only means of discerning the inner facts or values and of appreciating their meaning.

Differences and Similarities

Sharp contrasts in energy availability and use often exist among groups and villages within a single region. Shushil Agarwal (1981) found that not only did the composition of biomass cooking fuels vary markedly in five Uttar Pradesh villages, but the average per capita energy consumed in food preparation differed among villages by a factor of more than two. Bajracharya (1980) observed substantial differences among ethnic groups and settlement clusters in agricultural productivity, per capita food intake, and fuel use patterns within a rural *panchayat* in Nepal. The total area of this panchayat is less than 14 sq km, with a population of about 3,000 persons at the time of the study.

Numerous additional instances could be cited. The point is that in no country does "rural" signify a uniform geographical area inhabited by a homogeneous mass of individuals and families. And just as "rural" lacks a fixed, specific meaning, so a program of "rural development" must, if it is to be successful, be capable of adjustment and modification as variable circumstances dictate. One of the national planner's most perplexing dilemmas, therefore, is to find ways to formulate plans and programs that suit the country in general and yet accommodate different and varying local needs and conditions.

Despite substantial heterogeneity within and among countries, however, some general characterizations about rural energy in developing countries in Asia are possible. Here is a list of a few of the major ones and some of their more obvious implications.

1. In most of the region, the largest single fuel requirement in rural areas is for household use, primarily cooking. The household sector may account for 60–70 percent or more at the present time. Although this proportion will doubtless decline, households will continue to dominate the rural energy picture for some time to come.

2. The most important fuels for household use are the various forms of biomass, including firewood and charcoal, animal dung, and crop residues plus other nonwood vegetation. The energy contained in traditional biomass fuels is most commonly released through direct combustion.

3. Rural energy systems in developing countries are typically small-scale and heavily dependent on local resources. Partly for these reasons, and partly because such a high proportion of total energy is used directly on the farm or in the household instead of in transport and other intermediate stages, most of the decisions relating to the procurement and use of fuel are made directly by individual users, and very often by women.

These general characterizations suggest that some form of biomass system will continue to be the most usual means for supplying local energy requirements, at least until more direct applications of solar power become economically viable. The most important reason for this is that biomass is the most abundant and readily available local energy resource. This fact has direct implications for data gathering required for effective planning and for the selective introduction of energy technologies appropriate to local conditions. National officials usually know with some

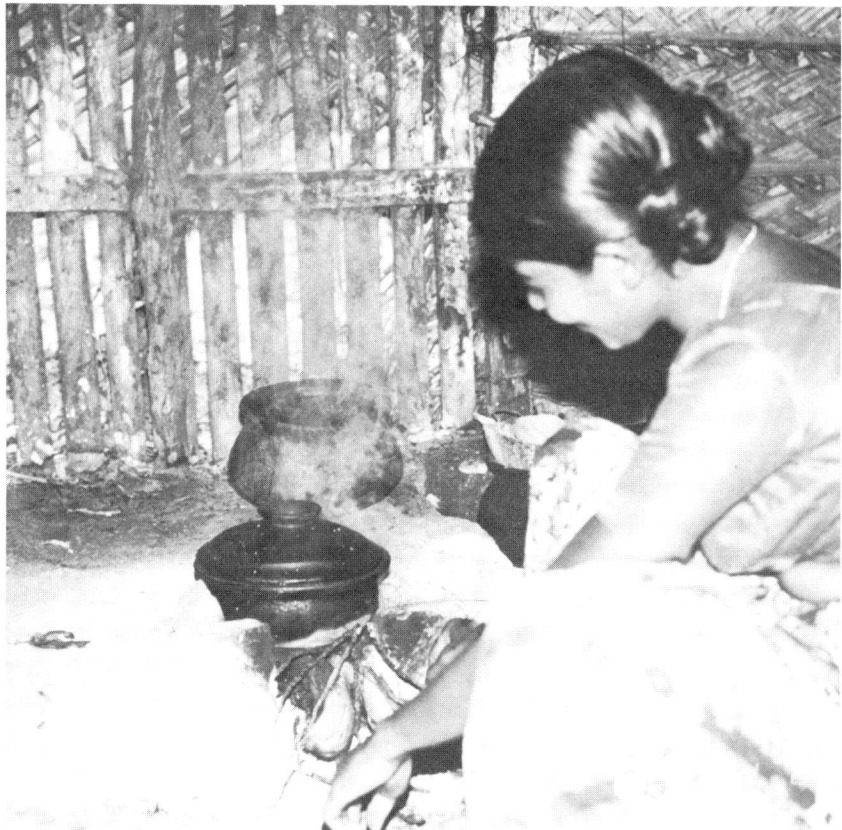

Tending the burning rate, indoor stove, Sri Lanka. (Kirk R. Smith)

accuracy the total amount of petroleum products consumed over one year, but they seldom are able to say where or by whom these products are used, except by tracing this information through records kept at the primary intersections in the distribution system. Hence, data on commercial fuels are progressively less available and less reliable as one moves toward a more disaggregated picture. With biomass fuels, precisely the reverse is true. Aggregate or representative figures ordinarily must be built up or extrapolated from rural energy studies, and most of the results can be applied statistically only to a very restricted geographical area or subpopulation.

In addition, biomass fuels are far more variable in their physical characteristics than kerosene or diesel. The nature of production and conversion processes for these latter fuels assures highly homogeneous products with comparatively constant heating values, presenting no serious problems of accurate measurement. By contrast, biomass fuels vary greatly in physical characteristics, moisture content, and heating values. The mix of biofuels normally differs substantially from household to household and, within households, may vary from season to season or even day to day.

THE RURAL ENERGY PROBLEM

In many areas, development by small-farm and landless families who live in marginal or subsistence conditions is the subject matter of energy for rural development research. Actions by people to change their organization of resources and technology in order to improve their material conditions are the desired results of research, and facts geared to those acts are the research tools. But the establishment of facts in nonmonetized or partly monetized sectors, where both ecology and the dynamics of human work result in differential access to resources and differential efficiency in energy use, is a research process in a partly hidden, changing arena.

The rural energy problem then, centers on rural people's ability to transform institutions and resources, including energy, to overcome poverty and enhance productivity. The problem may be conceived on three related dimensions:

1. Structural relations between rural poverty, resources, and social organization;
2. Differences among resource holders and income groups in efficiency of energy use. Efficiency and productivity differences are influenced by location and ecology, as well; and
3. Contrasts and transitions between subsistence and commercial energy resource management and exchange, linked closely to farming system transitions and often involving resource and income distribution changes.

Transitions from subsistence to commercial exchange are related to infrastructure development within and between regions and to changes in negotiating power between villages and central places within each region. Policies originating from community initiatives intersect here with formation and implementation of regional and national policies.

Rural Poverty and Resource Access

The development process of the subsistence condition differs by country, culture, and region (Shigeru Ishikawa 1981). Ecosystem characteristics determine the potential amounts and forms of energy available per unit area. People's transformations of these natural systems in local farming systems determine how much energy is available per person and to what degree overall availability of energy is problematic.

The farming system, as defined in this book, is a relationship between social and organizational factors, including distribution of resources and benefits, and decisions and actions by men and women to choose, work with, and modify crop, animal, and forest production activities and technologies using the resources of the local environment. The social nature of resource rights and the distribution of land, water, trees, and livestock among families and settlements allocate the supply of energy differentially among the various energy forms. Such allocation

- Constrains or governs family decisions on production and gathering of biomass energy resources, and on purchase of commercial fuels;
- Determines family surpluses and deficits; and
- Influences intrafamily, interfamily, and interneighborhood energy exchanges.

Such allocation also

- Creates apparent surplus or waste energy resources in some neighborhoods and deficits in others;
- Establishes marketable surplus exchanged outside the neighborhood and local area; and
- Determines the proportion of energy transactions occurring through mechanisms of price and market exchange.

Measurement Problems in Subsistence Economies. Measuring scales for subsistence conditions are difficult to define and not uniform. For example, changes in distance walked to collect fuelwood are used in certain energy studies as indicators of scarcity. Estimates of this sort must be done with care, or they may lead to wrong conclusions. A snapshot view rather than a moving picture over time can be one source of such mistakes:

Researcher A (from Asia):
"When some African energy studies state that women carry 20 kilogram headloads 15 miles each day, it is overlooked that the net energy requirements for this work cannot be obtained from the normally available 2,000 kcal of daily food. Data either for the food supply or the fuel collection should be reexamined."

Researcher B (from an East African city):
"When you actually see young girls and women wearing their bodies down and dying at an early age from carrying these loads, you realize that the numbers are correct and that these women are simply using up their energy capital."

(Morse, pers. comm., 1982)

In many areas with rising population-to-land ratios and increasing biomass depletion, transition to negative resource/need balances is the prospect. What kinds of indicators — burning of rice straw; burning of dung — can identify and assess such depletion?

Constraints influencing the choice between providing cooked food to people or to animals are experienced in some areas. In Nabagram villages, most families feed the gruel formed by cooking rice with excess water to cattle and poultry after mixing it with bran. Poor families, however, consume the gruel as part of their own diet; often, they do not have an animal to feed anyway (Nurul Islam 1980). How is a value to be placed on gruel, when it serves such different purposes from one family to another?

Need and Scarcity Assessment. Need is a commonly used and understood word, but difficult to define with precision. People do know their own direct needs best. However, they may not comprehend the conditions influencing their needs or be able to assign values to these conditions.

How people adjust or mitigate their situation when they lack energy resources indicates need. In certain seasons in Bangladesh, some rural families are not able to obtain fuel to cook more than once a day. They dig up roots from dead trees or cut excess amounts of live wood from common forest land. As reported in Chapter 2, these acts have contributed to soil erosion and loss. If families are forced to trespass on others' land and trees to obtain fuel, how should this be measured in terms of need?

In Java, the poorest families consume less fuel because they consume less food. This is clear evidence of what can be called suppressed need for both food and fuel. They find enough energy to cook the food they have. In some villages, minimum fuel needs are met by sharing among households. There, by custom, well-to-do families often permit other families to collect branches and twigs from their land and trees (Hadi Soesastro, pers. comm., 1982).

Scarcity takes many forms. Seasonality is often a major factor. In Serang, Indonesia, access to forests during the rainy season is difficult. Wood supply falls; prices rise to 60–70 percent above the dry season price. Some years back, the rise was steeper. Then a widow set up a new business, buying fuelwood in the dry season, selling it in the wet. Her entry into the market and arbitrage across time reduced the degree of seasonal scarcity and price fluctuations. Even at that, wet season prices curtail access to wood by the poorest groups (Soesastro, pers. comm., 1982).

Identifying and evaluating scarce resources is an important step in assessing local factors that limit increased productivity. It is also an important step in allocating development investments among different areas, according to where and how far input scarcity curtails development or prevents attainment of basic needs.

Energy Efficiency, Values, and Intermediate Conversions

Families who may not be able to obtain sufficient food and fuel for daily needs are, in many rural areas, among those forced to use these resources less efficiently than available technology permits. When women are engaged in farm operations, as during peak sowing, weeding, and harvesting seasons, the time devoted to cooking activities may be more critical than saving fuel. To save cooking time, they use more fuel. Similar tradeoffs between fuel and time are found in urban households. Use of general-purpose stoves that accommodate inferior biomass fuels as well as high-quality fuelwood is another example. Design needs of actual users must be the basis for stove improvement programs, as demonstrated in Chapter 8. But if, at the same time, better firewood supply was assured, designs could lead to great efficiency. People's evaluations of these options and social measures to assure against their risk are essential components for such conserving steps.

African energy studies report that in some places women passed by large trees (which would require chopping), and cut smaller diameter trees farther away. That practice reduces net future growth. In other areas, trees are ring-barked prior to felling. In a direct sense, this is more energy efficient than slash-burn cutting, but nutrients are lost to the soil because the ringed trees lose leaves.

Practices like that provide evidence that energy efficiency differs among ecosystems and between different groups within farming systems. Differences in forms and efficiency of energy use have cause and effect relationships with differences in availability and scarcity of physical resources, landholding structure, farming practices, and task and time allocations among men, women, and children. Many variables influence these energy relationships in crop-animal-forest systems. Changes to new forms of energy conversion will modify those relationships at countless points of end use — in such tasks as cooking — and also at intermediate stages where energy resources are converted into fuels, as well as where fuels are converted to forms or transferred to places for application in end uses (Kirk Smith and Michael Santerre 1980).

For purposes of technology design and selection, energy needs can be divided into these categories: medium-temperature needs (e.g., for cooking); high-temperature uses (for pottery and smith work); mobile mechanical energy (for agricultural operations and transport); stationary mechanical work (for pumping of irrigation and drinking water, food processing, and small industry); and cooling uses, like refrigeration. Changes in conversion technology are motivated by design goals such as increasing energy density (making fuels more homogeneous and easier and cheaper to transport); improving its quality (e.g., by providing heat at higher temperatures); increasing its utility and flexibility (e.g., by using more versatile fuel such as electricity instead of dung); and by making it more convenient and safer (by avoiding pollutant effects, for example).

The scale and location of energy exchanges are also altered by these changes in conversion technology. Consequently, the capacities of different people at different places to initiate or control energy supply and transformation are affected. "System boundaries" for the study of energy in rural development must be defined not simply to anticipate technology change, but more significantly to appreciate the social potential for accomplishing such change and who will benefit and where. Potentially, these gains in efficiency and related productivity offer new added values that rural people, through active planning and organization, can use to improve their quality of life and to create assets for their future.

In attempting to compare efficiencies, care must be taken to treat energy flows from different sources consistently and to express the compared flows in the same units, as "energy use per unit activity" (Resource Systems Institute 1980). These "energy indexing" problems cannot be addressed within the scope of the present book, but should be recognized in qualifying discussions of efficiency. Indexing issues include the definition and measurements of activities: What are the various functions performed by stoves, for example, and how can nonmeasurable functions be compared? Indexing questions also include the identification of consistent system boundaries, especially when comparing energy sources that involve different conversion methods or transport requirements. A related question is where in the fuel cycle should measurement of energy flows occur: at the source, at the point of conversion, or at the point of use? This latter question greatly affects the local-versus-regional incidence of energy transitions and their measurement.

Collecting branches and twigs, Boria village, Gujarat. (Kirk R. Smith)

Regional Transitions: From Subsistence to Commercial Exchange

The patterns of changes in transitions from subsistence farming and fuel gathering to commercial agriculture and fuels have many variants. Demand and supply factors are likely to change at different rates. This complicates inflows and outflows of energy and other commodities in form, quantities, and values. It affects land use and distributive relations. Effects on the poor are often negative, at least initially. The following scenario frequently observed in Java offers signs of what to watch for regarding energy access by different groups.

Access to new roads raises marketed shares of farm produce. Advantages are gained by those who are able to obtain new inputs such as irrigation and fertilizer. Land values rise. Landholders with greater marketable surplus are able to buy more land. Outside speculators also buy land. The poor sell. In five to ten years, land distribution characteristically becomes more unequal. Fuel demand structure also changes. Those with higher incomes are able to buy more kerosene and firewood. Firewood is traded more and more as a commercial energy source. As seen in regions of central Java, urban and industrial access to firewood increases, making rural scarcity more severe. When energy supply worsens, labor's share goes down.

For energy in such transitions to have a positive role, it must contribute to higher productivity and employment. In Java, occupational and skill development and rural industry expansion are key factors at this stage. Is energy a bottleneck? A study of central Java fuel-using industries found the main constraint to be supply of raw materials, not energy directly. Rising firewood prices and lower fuel quality, however, were adversely affecting productivity and quality of output. Insofar as rural industry's demand for crop residues also increased, the ability of the poor to obtain these traditional supplies worsened. The rate and extent of new employment creation, in a national context of the race between income and prices, determines whether the purchasing power of poor groups will rise adequately to offset these effects (Soesastro, pers. comm., 1982).

Converting the Problem to Opportunity

Chorkate residents, after agreeing on their priority need, moved to assess and mobilize resources to create a new cooperative body and a new local energy technology. Policy and resource support were provided by the national government, energized by the community facilitator. The community's structuring of assets and commitments, indeed, led to policy adjustments in the external system, through initiatives of the development agency and bank. That experience exemplifies a paradigm for cooperative rural study and action. Aspects of the rural energy problem just reviewed — resource access, changes in energy efficiency, and transitions from subsistence conditions — add to the dynamic nature of that paradigm. The stage is set in this way for the policy and technology issues addressed in the book, and for a consideration of research methods bearing on these issues.

REVIEW OF METHODS

Methods developed and used to collect information addressing these complex issues are

Structuring and Interpreting of Secondary Data

In most countries, great quantities of information are gathered for administrative units at the scale of a village, cluster of villages, district, and state or province. Such existing data, although often underrated, can usually provide a starting point for identifying critical issues or areas. The compiling and scrutiny of such data in forms relevant for problem identification is a fundamental step to selecting population groups and areas for in-depth research. With little added expense, yet with close attention to problems of definition and measurement, patterns observable in secondary data can be used to test how far and in what respects the findings from local studies can be generalized to wider areas.

Structured Surveys

Formal (structured) interviews using pretested questionnaires can furnish primary data on individual households or villages. Physical measurements of actual fuel consumption by households and biomass potentials of homesteads are often components of such survey schedules. The scope and extent of a survey are determined by the breadth and depth of the research objective. Structured surveys are used mainly to obtain numerical data with measurable accuracy and significance. They are uncertain as a means of collecting nonstandarized data and are very difficult to conduct.

Participant Observation

The researcher or surveyor participates in local activities and in informal discussions with the local residents and observes their daily activities in that locale for a certain period, preferably a full year. This method increases the human contact and level of understanding between surveyor and respondents, thus facilitating the exchange of information regarding nonquantified data, including felt needs, preferences, and priorities. Participant observation may provide a starting point in designing or conducting a structured survey.

Participatory and Action Research

This approach usually begins with participant-observer involvement in regular village meetings and direct interactions with the village community. Active community participation in the research process is often encouraged with this approach, in which the researcher performs as both interviewer and facilitator. Action research in the rural energy context usually is applied to "internalize" and integrate the process of technology development and implementation in a particular village community or area. In other words, action research is meant to promote interactions of the three principal parties, namely, (a) the energy user, (b) the technology designer, and (c) the support agencies (public or private).

Fuelwood market, Vallabh Vidyanagar, Gujarat. (Kirk R. Smith)

Field and Laboratory Experiments

Together with structured surveys, participant observation, or action research, experiments or tests may be conducted to measure actual field conditions, such as efficiencies of cooking stoves or illumination of kerosene lamps. Major parameters affecting fuel use efficiencies can be identified and at least partially replicated in the laboratory.

These methods and approaches have been variously applied and combined in recent studies. Some methods facilitate greater insights into energy problems in particular settings but are not easily generalized for other settings. Others help researchers or decision makers clarify a general situation or issue but often are ineffectual in specific cases. For each research objective, a design of methods that adequately accounts for the main parameters relevant to other objectives could greatly enhance the usefulness of research both locally and nationally. In other words, research designs should allow for integration of micro and macro perspectives. For example: (a) results of large-scale structured surveys can be interpreted more accurately with the help of findings from participant observation or action research with complementary designs; (b) sites for thorough study of particular problems can be selected more precisely by taking into account the results of large-scale structured surveys.

To guide such design and selection of methods, we summarize next the principal lessons of Chapters 2 to 10 on defining research objectives and scope and on matching research methods to these objectives.

STRENGTH THROUGH DIVERSITY: COUNTRY AND TECHNOLOGY EXPERIENCES

We present in Chapters 2 to 6 five country studies that permit a look at national policy settings for fuel substitution and conservation as well as for energy technology development.

The objective of the pioneer study that constitutes the core of Chapter 2 (on Bangladesh) was the assessment of user needs and resources as a basis for technology development. Structured interviews, guided by participant observers, were conducted in all 2,820 households of 23 villages in a contiguous rural area. Energy resources and use in both wet season and dry season were assessed. The chapter assesses biogas potential for domestic and agricultural uses in this rural environment and sets the factual stage for Chapter 8, on user-oriented design and development of improved stoves.

In addition, Chapter 2 reports a companion study in a contrasting rural area and analyzes the quite wide disparity in household fuel use found in these two microstudies. In turn, those studies are compared with two aggregate Bangladesh energy studies and two structured regional surveys centering on biomass. The chapter shows that wide, unexplained divergencies persist in the findings of rural energy studies in Bangladesh to date. These divergencies, unfortunately, are completely ignored in a recent reprise of the earlier study (Russell deLucia, Henry Jacoby, and others 1982) that revives an unduly low estimate of Bangladesh firewood consumption. Chapter 2 assesses methods used in each study, and suggests methods of approach to overcome present deficiencies in basic rural energy data.

The rural energy survey in West Java summarized in Chapter 3 focused on a specific energy policy issue, namely, the effects of kerosene subsidies on equity and on kerosene substitution for firewood in household uses. With this focus, the study also contributed to a larger project on energy and income distribution. To examine the effect of varying rural conditions on the policy issue, the survey was designed to cover a systematic cross-section of rural localities. Questionnaire interviews and fuel measurements were conducted in 533 households distributed among 40 villages and 5 districts. Sample selection factors included district characteristics, village typology, household income, and household size.

A principal methodology conclusion of Chapter 3 is that, for the region concerned, the administrative district is not a suitable sampling unit for studying biomass-fossil fuel interactions. Ecological areas (e.g., a coastal village close to forestland, or a coastal village close to agricultural sites) would represent better those resource, socioeconomic, and spatial factors that are important to determining rural energy sources and use. Other modifications of scope and method are recommended for future energy and rural development policy studies.

The chapters from Bangladesh, Indonesia, the Philippines, and Thailand (2 to 5) facilitate a review of the use of structured surveys in such national policy studies. The strengths of structured surveys in achieving broad coverage and in identifying major regional variations in energy resource and use patterns are brought out. Limitations of the survey method without complementary research, in not serving to establish rural needs and priorities, evaluate noncommercial energy uses and resources, or match technologies to user needs, are also evidenced.

Chapter 4 summarizes a nationwide rural energy baseline study carried out in 1978 by the Ministry of Energy of the Philippines. The main purpose of the survey was to establish a first approximation of "How much of what form of energy was used in which region for what purpose?" The survey, administered in conjunction with a nationwide census of annual household expenditure, also examines whether selected socio-

Principals, Issues, and Methods   29

Regina Santerre adjusting smoke exposure meter, Sri Lanka. (Kirk R. Smith)

economic variables significantly influenced energy consumption patterns.

Another rural energy baseline survey of 2,000 households distributed among 200 villages and 24 provinces of Thailand, designed to represent the entire country, is reported as the core of Chapter 5. The survey was undertaken by the National Energy Administration (NEA) of Thailand in 1980. The discussion of the survey findings is set in the context of Thailand's rapid agricultural development in the past decade. Recognizing land and energy constraints as well as the need for greater progress toward equity in rural development, the chapter summarizes regional energy flow balances, regional energy use in agriculture, and energy use by size of farm in each region.

The Thailand chapter also reports data obtained through local studies in different regions on characteristics of firewood and charcoal use in different stove types. This illustrates how information obtained locally from users of energy resources and technologies can be combined, in hybrid form, with national or regional survey data to make preliminary estimates of fuel substitution potential. The chapter concludes that further microscale studies using regional agroeconomic attributes as a framework for analysis will be required to bring national social and economic development objectives to bear on energy technology development and energy conservation.

The use of survey research in a participatory manner as a vehicle of integrated rural energy planning is exemplified in Chapter 6. Experienced researchers engaged and trained voluntary interviewers who carried out approximately 80 percent of survey work covering all 10,200 households in one development block and 7,900 households in a second block in the state of Bihar in eastern India. The two areas have distinctive ecological and agroclimatic conditions and exhibit different energy problems and priorities. Candidate energy technologies and implementation approaches are presented.

Lessons of the participatory, microregion studies described in Chapter 6 provide a bridge to Chapters 7 to 10, which present approaches to assessing the viability and acceptability of new energy technologies in rural settings. Rural organization for assessment of local needs and selection and implementation of new technologies is the focus of Chapter 7. The next three chapters assess technologies used primarily — but not only — for domestic cooking needs: improved stoves, fuelwood plantings, and biogas systems.

In Chapter 7, the author documents his action research experience involving community participation in needs assessment and technology innovation in a Nepal hill village. The study area, Chhoprak Village Panchayat in Gorkha District, has an ecologically distinguishable population distribution by caste: the Kumhals are at the low elevation; Brahmins, Chhetris, Newars, Sarkis, and Kamis at the middle elevation; and Gurungs at the high elevation. The chapter contrasts community-based action in Chorkate by the Kumhals to assess and organize biogas installations with decisions by individual households in the midelevation community to install improved stoves. Social, technological, and ecological factors in this differential response, as well as the unwillingness of the Gurungs to act responsively to new technology opportunities, are examined.

Field-based assessment and development of improved stoves are reported in Chapter 8. This chapter undertakes to develop methodologies for evaluating stove types and to identify factors important in designing programs for development and extension of improved stoves. Women who make and use their own stoves are the ultimate decision makers. They are the ones who can be expected to guide design and determine acceptance of improved stoves. The chapter describes methods for testing stove performance in actual use conditions as well as in the laboratory.

Chapter 9 discusses the introduction of fast-growing, multiple-use fuelwood species that can be grown both by individual families and on a neighborhood or community scale. Methods for assessing local firewood needs and for matching species selection to land use, soil, and other cultural conditions are illustrated with cases from Indonesia and Bangladesh. Representative yield rates for important species on typical sites and with different silvicultural methods are indicated to facilitate project appraisal. Species providing fodder as well as fuel are examined. Steps to combine continuing experimentation with in-village and on-farm firewood production programs are presented.

Chapter 10 demonstrates assessment methods for both household and community biogas plants, with an analytic framework and numerical case examples that an extension agent could apply in helping a family or group of families gather data for local assessment. The FLERT procedure for resource and technology appraisal of biogas systems (Santerre and Smith 1980) is presented in summary form. The chapter develops scenarios and sensitivity analyses including each principal variable affecting the financial viability and attractiveness of family and community biogas systems.

ENERGY AND RURAL DEVELOPMENT: TOWARDS A SYNTHESIS

Chapters 11 and 12 focus on rural residents' needs for information as inputs to local planning and on the integration of local and regional analyses to address national policy issues and development objectives.

In Chapter 11, the cooking sector is used as the entry point for understanding food-fuel links. Information compiled from previous research clarifies the factors involved in estimating fuel efficiencies and requirements and in assessing cooking fuel sources and access. Work and time allocation by farm family members, especially by women in fuel collection and related food preparation activities, are indicators for assessing opportunity costs and more valued options (including child care) for labor activity. Regarding organization of land and work, the chapter illustrates how to use a set of indicators to identify critical limiting factors and new resource opportunities in the local farming system. If people in local communities contribute their own knowledge in developing and using such indicators to assess and improve resource use, such local assessments can be connected to the shaping of regional and national resource assessments and policies. Regional differences among farming and bioresource systems can also be assessed through such indicators.

Chapter 12 addresses priorities, designs, and methods for continuing research to promote improved energy options meeting rural development needs. Energy needs for the agricultural, agro-processing, rural industry, and transport sectors are addressed. For effective development to occur, technology end uses in these sectors must be meshed with domestic energy needs. Discoveries of new resource mixes and systems demand a new research style capable of contributing to this integration of household and agricultural needs.

Development Research Goals

Lessons drawn from the various cross-country experiences demonstrate the potential for local transformations in human and physical resource use, in a development process that enhances equity while promoting greater productivity. A principal goal of development research is to strengthen people's ability to make these transformations. Development research has the role of helping rural people, especially those with the most acute needs, learn how to build upon and adapt their present knowledge to gain new initiatives in decision making and new advantages in negotiating with outside groups.

Energy and rural development, understood as involving people's decisions on distribution and allocation of work as well as other resources, are integrally connected in the framework of rural energy and farming systems. In the energy and farming system context, farm families and other rural community members need methods and information to

- Improve their knowledge of food-fuel links and of new opportunities for greater productivity in crop, animal, and forest resources;
- Identify crucial factors that limit or block production increases and determine how to overcome them;
- Invent new ways of organizing and allotting work activities, to add value and share benefits more equitably;
- Evaluate alternative energy technologies, both by contributing to their design and by operating, testing, and modifying them as required to fit particular daily and seasonal conditions; and
- Realize economies of scale and of transactions in food and energy resources through appropriate group and community organization.

Cooperative research methods that enable village residents to join with scholars in generating facts relevant to needed innovations, in assessing new resource and technology combinations, and then in organizing and testing them in practice, are the foundation of the new energy for rural development agenda we propose in Chapter 12. This agenda encourages rural people, in cooperation with development institutions, to identify realistically their needs for energy and their means for fulfillment of those needs. Social, allocative, and technological learning and policy guidelines would be fostered by the specific methods chosen to implement the program.

A New Energy for Rural Development Research Agenda

The development research agenda outlined in Chapter 12 provides for integration of various research methods to meet rural energy needs. The research orientation can be described as "village-outward" (see Andrew Vayda, Carol Colfer, and Mohamad Brotokusumo 1980). Tested methods of village and microregion research, combined with regional and sectoral analyses, would be linked with national energy planning and policy formation. Success in three research styles exemplified in chapters of this book instills confidence in the new research community's capacity to achieve such integration. These are (1) action research in specific rural areas, with lively community involvement (Chapter 7); (2) community-based energy needs assessment linked with farming systems research and development, incorporating the *sondeo* or *gaun sallah* method of understanding problems of specific agroclimatic zones and adapting that understanding to the circumstances of other zones (Chapters 2, 6, 7, and 12); and (3) rural energy policy analysis through careful definition of issues and methodical selection of sample study areas and households or farms to represent socioeconomic conditions and ecosystems relevant to the policy issue at hand (Chapter 3).

Participatory Local Research. Village information needs and those of district planning and development agencies would be met in the proposed program through local energy and farming systems research and development. Community groups including farm and labor families would engage with interdisciplinary research teams in appraisals of critical energy constraints as well as new potential linked to farming activities. First attention would be directed to classes of need, including cooking, irrigation, farm power, agro-processing, rural industry, transport, and health. In cooperation with researchers, area residents will organize and monitor technologies identified in this diagnosis stage as ready for use or trial.

For more complex or uncertain systems considered to have promise, local sponsors will conduct feasibility assessments with guidance by research specialists. As more structured research is defined, the research team, with active local participation, would assess larger energy system and resource development opportunities linked with farming system potential. New ways of organizing resources and distributing benefits of added values would be fostered, as suited to these innovations.

Policy issues arising in these local assessments will frequently call for more broadly based research, for which participant observation would establish the basis for structured surveys within the farming system. Examples illustrated in Chapter 12 include the effects of fuel deficits on nutrition and capacity for work and the requirements of small and transitional farms for energy in various forms for more productive crop combinations. These studies, apart from helping participating rural groups progress toward their development goals, would be linked with wider policy analyses.

Regionalization of Renewable Energy Research. Information needs of national and regional decision makers responsible for allocating program resources to different regions and technologies would be served by the program's next phase: enhancing rural advantages in regional energy transitions. Systemic analyses would guide identification of geographic areas for extension and adaptation of successful energy systems. Analyses of energy end use and conversion technologies would define aspects of change in energy density, quality, and utility that influence related economic factors of scale, end use application, transport costs, and locational advantage. Rural communities and enterprises would be assisted, thereby, in building on their information advantages in local farming systems to discover and invest in promising new energy technologies.

National Research Priorities. National development goals and constraints in each country will progressively determine new energy for rural development research priorities. To provide more adequate and systemic data for continuing policy review, we recommend that basic rural energy categories be included in periodic national censuses or statistical surveys. A vital supplement to such data series, however, is required for nonmonetized and subsistence rural energy use and transactions. Special periodic surveys guided by participant observation are recommended to meet this policy information need. Geographic attributes defined through farming system and regional analyses would contribute to sample design.

National policy agencies, it is clear, will build cumulative assessments of rural energy potential and its contribution to rural goals from these basic agenda activities. Specific policy studies, illustrated in Chapter 12, would be selectively conducted to address critical issues as they emerge.

The new energy for rural development agenda addresses the needs of most rural citizens where they live, in terms they can define, and through measures they can actively influence. It is designed to assist researchers in formulating, as well as research users in sponsoring, result-oriented energy assessments. Three principal decisions are central to its implementation. One basic principle to be decided upon is the progressive founding of rural energy research and development in the lives, social organization, and locations of rural people. Next, to support this momentum, priority for research and development funding of institutions located in principal farming regions would be necessary. Local and regional institutions will require flexible budget and organizational support to achieve the combined social science and technology research capabilities projected in Chapter 12. The third policy principle is implicit: encouragement of lively initiatives and autonomy in the issues raised by rural people and in the challenges taken on by research teams, to build on the cultures and inherent resource strengths of these communities.

APPENDIX 1

The principal studies reviewed are
Bhatia, Ramesh.
    1982   Energy Data Systems: The Role of Traditional Sources. Delhi, India: Institute of Economic Growth. Presented at the International Workshop on National Energy Data Systems, Tata Energy Research Institute, December 15–18.

Bhatia, Ramesh, and Mariam Niamir.
    1979   Renewable Energy Sources: The Community Bio-Gas Plant. Presented at the Seminar in the Department of Applied Sciences, November 2, Harvard University, Cambridge.

Bowonder, B., and K. Ravishankar.
    c1982   Energy Use in Rural Areas of India. Hyderabad: Center for Energy, Environment and Technology, Administrative Staff College of India.

Briscoe, John.
    1979   Energy Use and Social Structure in a Bangladesh Village. *Population and Development Review* (December):615–641.

Briscoe, John.
    1979   The Organization of Labour and the Use of Human and Other Organic Resources in Rural Areas of the Indian Subcontinent. Presented at the Conference on Sanitation in Developing Countries Today, sponsored by Oxfam in association with the Ross Institute of Tropical Hygiene, July 5–9, Pembroke College, Oxford, England.

Cecelski, Elizabeth, J. Dunkerly, and W. Ramsey.
    1979   *Household Energy and the Poor in the Third World.* Washington, D.C.: Resources for the Future.

deLucia, Russell J.
    1981   Infrastructure for Rural Energy Systems: A Background Paper. Cambridge, Massachusetts: Meta Systems, Inc. Prepared for the Ad Hoc Expert Group Meeting in preparation for the forthcoming United Nations Conference on New and Renewable Sources of Energy, January 26–30, Cambridge.

Desai, Ashok V.
    1978   India's Energy Consumption: Composition and Trends. *Energy Policy* 6(3):218–230.

Desai, Ashok V.
    1980   India's Energy Economy: Facts and Their Interpretation. Bombay: Centre for Monitoring Indian Economy.

Desai, Ashok V.
   1981   Interfuel Substitution in the Indian Economy. Energy in Developing Countries Series Discussion Paper D-73 B. Washington, D.C.: Resources for the Future.

Desai, Ashok V.
   1982   Rural Energy Surveys in India. A draft report prepared for the International Development Research Centre. New Delhi: National Council of Applied Economic Research.

Energy Research and Development Group.
   1976   *Nepal: The Energy Sector.* Kathmandu: Institute of Science, Tribhuvan University.

Gupta, C.L.
   1981   Energy for Rural Development. *Renewable Energy Sources for Developing Countries.* Reprint, 178–187. England: Heliotechnic Press.

Gupta, C.L., and K. Usha Rao.
   1981   Energy Inputs for Irrigated Farming with Mixed Cropping: A Region. *Urja* (May):106.

Gupta, C.L., K. Usha Rao, and V.A. Vasudevaraju.
   1980   Domestic Energy Consumption in India (Pondicherry Region). *Energy* 5:1213–1222.

Islam, M. Nurul.
   1978   *Study of the Problems and Prospects of Biogas Technology as a Mechanism for Rural Development: Study in a Pilot Area of Bangladesh.* Dacca: Bangladesh University of Engineering and Technology.

Johnston, Peter, and others.
   1977   Rural Energy in Fiji: An Overview. Presented at the UN Economic and Social Commission for Asia and the Pacific Regional Workshop on Biogas and Other Rural Energy Resources, June 20–July 8, Institute of Natural Resources, University of the South Pacific, Suva.

Manibog, Fernando R.
   1979   Patterns of Energy Utilization in a Philippine Village: Sources, End-Uses and Correlation Analyses. A draft report presented to the International Energy Agency, Organization for Economic Cooperation and Development, December, Paris, and to the Rockefeller Foundation, New York.

Meta Systems, Inc.
   1980   State-of-the-Art Review of Economic Evaluation of Non-Conventional Energy Alternatives. Bioresources for Energy Project, U.S. Department of Agriculture Forest Service.

Moulik, T.K., U.K. Srivastava, and P.M. Shingi.
   1978   Bio-Gas System in India: A Socio-Economic Evaluation. Ahmedabad: Indian Institute of Management.

National Council of Applied Economic Research.
   1981   *Rural Energy Consumption in Northern India*. A report sponsored by Environmental Research Committee, Department of Science and Technology. New Delhi: Saraswati Press.

Ravindranath, N.H., and others.
   1978   The Design of a Rural Energy Centre for Pura Village. (Draft). Bangalore: ASTRA.

Ravindranath, N.H., and others.
   1981   An Indian Village Agricultural Ecosystem — Case Study of Ungra Village, Part I: Main Observations. *Biomass* 1(1):61–76.

Reddy, Amulya Kumar N.
   1981   An Indian Village Agricultural Ecosystem — Case Study of Ungra Village, 1981. Part II: Discussion. *Biomass* 1(1):77–88.

Riley, Patricia, and others.
   1981   Profiles of Selected Senegal Villages: Based on Data from the Peace Corps Energy Survey [Senegal]: [Appendix A]. Washington, D.C.: Energy Sector, Office of Program Development, U.S. Peace Corps.

Riley, Patricia, and others.
   1982   Peace Corps Rural Energy Survey, Senegal. Washington, D.C.: Energy Sector, Office of Program Development, U.S. Peace Corps.

Siwatibau, Suliana.
   1978   *A Survey of Domestic Rural Energy Use and Potential in Fiji.* A report to the Fiji government and to the International Development Research Centre, Canada. Suva: The Centre.

Slesser, Malcolm, Chris Lewis, and Ian Hounam.
   1979   Self-Reliant Development — An IFIAS Sponsored Study: 1st Technocratic Report (TC1). Glasgow, Scotland: Strathclyde University, Energy Studies.

Weatherly, W. Paul.
   1980   Environmental Assessment of the Rural Electrification I Project in Indonesia. Jakarta: USAID.

Weragoda, N.V.K.K.
   1977   Report of the Socio-Economic Survey of Pattiyapola Village, Hambantota District, Sri Lanka.

White, Wendy D.
    1980   Annotated Bibliography of Energy Surveys in Developing Countries. Presented at the International Workshop on Energy Survey Methodologies in Developing Countries, January 21–25, Jekyll Island.

Methodology documents reviewed include:

Anonymous.
    1979   Rural Energy-Mix Study: Approach Paper. India: Consulting Engineering Services (India) Private Limited. August.

Arnold, John.
    1982   Report on the Terms of Reference for a Rural Energy Survey in Bangladesh. Bangladesh: Department of State, Agency for International Development (USAID). July 12.

Bangladesh Planning Commission.
    1980   Supply and Demand of Forest Products and Future Development Strategies. Presented at the Energy and Rural Development Research Implementation Workshop, February 5–14, Chiang Mai, Thailand.

Bhatia, Ramesh.
    1980   Energy for Rural Development: An Analytical Framework for Socio-Economic Assessment of Technological and Policy Alternatives. Presented at the Energy and Rural Development Research Implementation Workshop, February 5–14, Chiang Mai.

Bhatia, Ramesh.
    1980   Energy Survey Methodologies: A Framework for Measuring Non-Conventional Energy Sources in Developing Countries. Paper presented at the ESCAP/IEA Workshop on Energy Statistics, October 6–11, Karachi.

Ceylon Electricity Board.
    1980   Rural Energy Consumption and Resources Survey. Presented at the Energy and Rural Development Research Implementation Workshop, February 5–14, Chiang Mai.

Donovan, Hamester, and Rattien, Inc.
    c1979   Africa Energy Survey Methodology. Prepared for the USAID, Bureau for Africa, Office of Development Resources, Washington, D.C.

Forest Research Institute.
    1979   Production and Consumption Surveys of Wood, Fuelwood, Bamboo, and Other Fuel Energy Sources in Different Sectors in Bangladesh: A Proposal. Chittagong: The Institute.

Graham, Thomas.
   1979   Critical Issues for Designing Energy Surveys in Africa. Presented at the Workshop on Fuelwood and Other Renewable Fuels in Africa, November 29–30, Paris.

Isaak, David.
   1980   Report on Workshop on Energy Statistics, Economic and Social Commission for Asia and the Pacific (ESCAP), October, Karachi.

National Academy of Sciences.
   1980   International Workshop on Energy Survey Methodologies for Developing Countries, January 21–25, Jekyll Island, Georgia. Commission on International Relations. Washington, D.C.: National Academy Press.

Openshaw, Keith.
   c1979   Wood Consumption Surveys. University of Dar es Salaam, Division of Forestry. Morogoro. Mimeo.

Peace Corps.
   c1979   Concept Paper for Peace Corps Energy Program. Includes draft of Peace Corps Third World Energy Survey Instrument.

Thomson, James T.
   1980   Firewood Survey: Theory and Methodology. Easton, Pennsylvania: Lafayette College.

APPENDIX 2

Categories of Rural Energy Research Objectives

Establishment of baseline data to determine national energy program priorities

Assessment of rural development implications on energy

Introduction of rural electrification

Assessment of rural energy needs, user problems, and priorities

Introduction of improved energy technologies and energy use practices, including conservation

Assessing deforestation and augmenting biomass energy sources

Identifying potential users of energy technologies

Introduction of new energy technologies

Development of policies for conserving fossil fuels

Estimating substitution potential of renewable energy sources for fossil fuels

Establishment of investment allocations for research and development in renewable energy technology

Development of energy pricing policies

Assessment of factors in successes and failures of renewable energy technologies

Assessing the role of energy in rural employment generation

REFERENCES

Agarwal, Anil.
    1982   Try Asking the Women First. Center for Science and Environment. Report No. 53, April, Delhi.

Agrawal, S.C.
    1981   *Rural Energy Systems in Two Regions of Uttar Pradesh: First Phase Report.* ERD PR I-81-1: IND I-1. Honolulu: Resource Systems Institute, East-West Center.

Bajracharya, D.
   1980   Fuelwood and Food Needs versus Deforestation: An Energy Study of a Hill Village *Panchayat* in Eastern Nepal. In *Energy Analysis in Rural Regions: Studies in Indonesia, Nepal, and the Philippines* by Raymond Atje and others. Honolulu: Resource Systems Institute, East-West Center.

Bajracharya, D., and C. Gurung.
   1984   Dialogue as a Method for Village Level Energy Planning: An Approach to Action Research in a Nepali Village. Honolulu: Resource Systems Institute, East-West Center. Forthcoming.

deLucia, R.J., H.D. Jacoby, and others.
   1982   *Energy Planning in Developing Countries: A Study of Bangladesh.* Cambridge, Massachusetts: Johns Hopkins Press.

Ishikawa, S.
   1981   *Essays on Technology, Employment and Institutions in Economic Development: Comparative Asian Experience.* Tokyo: Kinokuniya Company Ltd.

Islam, M.N.
   1980   Energy Use in Bangladesh. Presented at the International Workshop on Energy Assessment Methodologies, January 21–25, Jekyll Island.

National Research Council of Thailand, and Resource Systems Institute.
   1980   Energy for Rural Development: Implementation Plan for Inter-Country Research Activities 1980–1983. Honolulu: Resource Systems Institute, East-West Center.

Resource Systems Institute, East-West Center.
   1980   Preliminary Research Plan on Energy and Rural Development. Working Document for Participants at the Research Implementation Workshop on Energy and Rural Development, February 5–14, Chiang Mai, Thailand.

Schlegel, C., and J. Tarrant.
   1980   *Thinking About Energy and Rural Development: Methodological Guidelines for Socioeconomic Assessment.* ERD PR-80-3. Honolulu: Resource Systems Institute, East-West Center.

Smith, K.R., and M.T. Santerre.
   1980   *Criteria for Evaluating Small-Scale Rural Energy Technologies: The FLERT Approach (Fuel-Linked Energy Resources and Tasks).* PR-80-4. Honolulu: Resource Systems Institute, East-West Center.

Vayda, A.P., C.J.P. Colfer, and M. Brotokusumo.
   1980   Interactions Between People and Forests in East Kalimantan. *Impact of Science on Society* 30(3):179–190, UNESCO.

2
Energy and Rural Development: Critical Assessment of the Bangladesh Situation

M. Nurul Islam

INTRODUCTION

Bangladesh is a country of 143,998 sq km characterized by a network of rivers and over 230 distributaries that extend about 24,000 km and cover 9,389 sq km. As of July 1980, only 9 percent of the country's 89 million people lived in areas defined by the 1974 census as urban (generally areas with 5,000 persons or more). About 14.4 million households are distributed over 85,650 villages, 65,490 with more than 50 households, and 20,160 with less than 50 households. Agriculture is the occupation of 80 percent of the labor force, and this sector directly contributes 53 percent of the gross domestic product (Government of Bangladesh, Ministry of Planning 1980). The nation's four principal administrative divisions are divided further into 20 districts, 71 subdivisions, 469 thanas, and 4,365 rural unions.

Energy is a vital input for rural development. This chapter proceeds from the assumption that energy planning for rural areas should be considered part of national energy planning and aimed to achieve national rural development goals. The chapter reviews Bangladesh energy policy studies, providing a context for comparison with studies to assess local energy needs and technology potential. In order to examine the role of microregion studies in identifying factors associated with specific variations in rural energy use, the core of the chapter focuses on two in-depth studies in rural areas. This establishes a basis for considering how appropriate blends of macro- and microstudies may be fashioned.

RURAL DEVELOPMENT POLICY

The Five Year Development Plans have had continuing concern with rural development issues. One important social objective of the First Five Year Plan (1973–78) was to achieve equitable distribution of development and to reduce poverty. However, both institutional and resource limitations frustrated even a modest beginning in this direction. After an interim period (1978–80) under a separate plan, the objectives of the Second Five Year Plan (1980–85) were formulated in view of overwhelming problems of massive poverty, unemployment, illiteracy, and malnutrition of the large rural population. Over 80 percent of people in Bangladesh continued to live below the economic poverty line, defined by a minimum caloric requirement of 2,120 kcal. Fifty-four percent were estimated to be below the extreme poverty line, 1,805 kcal (GOB, Ministry of Planning 1980). Inflation has worsened the plight of the poorer groups who work mainly as wage laborers.

The Second Five Year Plan concentrates on several rural development strategies. These will include actions in education and health care, but the foundation for rural development will be agricultural development. The strategy for agricultural development will be to effect a rapid transformation from traditional agricultural practices to modern technology. To attain this goal, adequate resources are necessary as is an equitable share of resources and economic benefits for each group in the rural community (GOB, Planning Commission 1980).

The existing significantly unequal distribution of wealth in the rural areas not only interferes with equitable income distribution but also with equitable opportunities. Because the access to new resources such as agricultural inputs and credit is largely determined by the existing distribution of wealth whose dominant element is land, rural inequality continues due to concentration of land in the hands of the few. Lack of opportunities both inhibits productive energies and handicaps any poverty-oriented strategy. A new more viable distribution system as well as appropriate political decisions and allocation of resources are required.

The Second Five Year Plan also recognizes regional planning as an essential element in comprehensive rural development. A balanced regional growth pattern with meaningful dispersion of industries, cash crops, and other economic activities requires dividing the country into suitable regions, subregions, or supraregions based on economic and physical characteristics. The natural primary unit for regional planning is the village. Between the village and the nation is a set of alternative intermediate tiers. All of the substantive tiers are slated for development over the next five years. The plan calls for disaggregating national production goals into regional targets to ensure agricultural commodity specialization according to specific soil types and other resource endowments. Moreover, regional input and service packages supporting such commodity specialization will be matched to regional plans.

The Integrated Rural Development Programme (IRDP) of Bangladesh is based on the experience of the Comilla district cooperative model developed by the Bangladesh Academy for Rural Development. Initiated in 1961 in Kotowali *thana*, the Comilla model expanded by the late 1960s to all other thanas in the district. In 1970, the program was expanded to the rest of the country as the IRDP, under a separate directorate of the Ministry of Local Government, Rural Development and Co-operatives. When subsequent evaluation found that the cooperatives did not produce sufficient yields, several "crash programs" were initiated in selected thanas with large financial assistance from external development agencies.

A Swedish International Development Agency report presents a comprehensive review of the Comilla model and the modifications made through IRDP and those "crash programs." According to that review, the cooperatives failed to change effectively the pattern of poverty, landlessness, inequality, and unemployment (S.D. Vylder and Asplund, n.d.). The report also found that, at that time, cooperatives almost exclusively benefited a small minority of the peasants and favored some urban-biased interests as well. The landless or near landless (most of the people) were not participating at all. The intensive program for Integrated Rural Development is a costly approach considering either area or population. Even if a particular zone "achieves" rural development using that approach, the least developed areas would have to be selected first to equalize overall development. However, the most developed areas tend to be selected first for quick investment returns.

New area development projects were not planned for the Second Five Year Plan period. Projects in progress were scheduled for completion and evaluation regarding appropriateness and effectiveness (GOB, Planning Commission 1980).

Although the Directorate of IRDP was created to organize social institutions (cooperatives) to enhance agriculture and economic development, a separate corporation of the Ministry of Agriculture (the Bangladesh Agriculture Development Corporation, BADC) is responsible for the delivery and maintenance of agricultural inputs. These inputs include irrigation pumps and fuels, fertilizer, pesticides, and seeds.

ENERGY IMPLICATIONS FOR RURAL DEVELOPMENT

In Bangladesh, the land available for farming cannot produce enough food for the dense population using traditional agricultural technology. Increasing production means using more energy for running irrigation pumps and for producing chemical fertilizer. More energy is needed for processing and transporting agriculture products and for producing finished goods. These development activities, in turn, would provide off-farm employment. These end uses require only a small but vital portion of the total rural energy need, but the energy used must be high quality.

Interdependence of Subsistence and Development Needs

At present, energy in rural Bangladesh is used primarily for cooking and lighting, besides agricultural purposes. Cooking fuel alone constitutes most of the total energy requirement, although of lower quality than energy needed for development activities. Traditionally, locally available fuelwood met the cooking fuel requirement. But the unfavorable land-man ratio and the resulting scarcity of fuelwood together have initiated changes to other cooking fuels. Agriculture residues are used instead in many rural areas now but since they do not burn well and are harder to store, they are considered low grade compared to fuelwood. Even residue fuels are scarce in some critical situations. Further, where residues are used as cooking fuel instead of as fertilizer, the quality of the soil and, therefore, crops suffer. Reportedly, on 60 percent of Bangladesh's arable land, organic matter content has deteriorated to a critical 2 percent level because organic fertilizer is not being used. If that percentage is not raised to at least 3 percent within the next 10 years, chemical fertilizer efficiency will remain static and crop production will not rise (*The Bangladesh Times* 1982).

All agricultural production in Bangladesh depends on human and animal power. Chronic shortage of food and fodder affects the quality of this energy source. Moreover, increasingly more intense cropping for higher output has increased demand for that energy. Because the land-man ratio has reached a critical level, the relationships among food, fuel, and fodder should be considered together in a rural development program. Design of any energy program must take into account the locale, season, and specific resident population. Planning attention is required to maintain and develop energy supplies in both the quantity and quality necessary to meet subsistence and development requirements. This raises complex policy issues regarding both commercial and traditional fuels.

Commercial Energy Sources

Bangladesh has limited commercial energy sources. Potential reserve statistics are natural gas, 10.4×10^{12} SCF; coal, 527×10^6 ton; peat, 133×10^6 ton; and hydroelectricity, 690 MW. No petroleum reserves are known.

Two major rivers flowing from north to south divide the country into an Eastern Zone and a Western Zone. Energy use patterns vary greatly between the two zones. All the known natural gas reserves and hydroelectric resources are located in the Eastern Zone. Indigenous natural gas is available for certain industries there but in the Western Zone industries depend on imported petroleum. In the Eastern Zone electricity is generated by natural gas and hydropower and generation capacity exceeds the demand. In the Western Zone where imported petroleum fuel is used to generate electricity, demand exceeds generating capacity. Reliable supplies of electricity and industrial fuel as well as cheaper unit price of industrial fuel (from natural gas) have contributed to accelerated industrial growth in the Eastern Zone.

In 1972, the price ratio of petroleum fuel per thermal unit to that of natural gas used in industry was 2.3:1. In 1983, the ratio will probably reach 4.6:1. For power generation the same ratio is 11.2:1 (1972) and 18.8:1 (1983). A uniform price of electricity to consumers is maintained throughout the country.

Government policy encourages maximizing use of indigenous natural gas. Present annual consumption is 0.5 percent of total proven gas reserves; carrying capacity of the gas transmission line is 0.9 percent of total proven reserves. Financial constraints to date have limited extension of the natural gas pipeline within the Eastern Zone, but one goal of the current project is the extension of a pipeline to the second largest city, Chittagong.

New electricity generation stations have been located near the Eastern Zone gas reserves while existing power stations near gas reserves have been converted from petroleum to natural gas. When current work to interconnect the zonal grids is completed, electricity generated using natural gas in the east will also be used in the west. Expansion of natural gas lines and of electricity distribution are both capital-intensive ventures. Bangladesh with its weak balance of payments needs at least 10 years lead time to implement this kind of project.

Energy-poor areas inevitably develop more slowly than areas where such capital-intensive projects are more techno-economically feasible. This pattern is reflected in Bangladesh where the Eastern Zone is developing faster than the Western Zone.

Rural Electrification

In 1977, the Rural Electrification Phase I project began by establishing the Rural Electrification Board. The Phase I project was planned to provide electric connections to potential consumers in six selected areas of the country, covering 4,373 sq mi (8 percent of total area of the country). Total households in the project area were 1.1 million. Twenty-two percent of the households in the area were estimated to be able to get electric connection (GOB, Rural Electrification Board 1977).

In the draft Second Five Year Plan, 14.5 percent of the total public sector expenditure was allocated for the development of energy sector.

Allocation for rural electrification alone was 20 percent of the budget of energy sector. One goal of the Second Five Year Plan (ending in 1985) is that all thana headquarters and more than 23,000 villages (35 percent of total villages) will have access to electricity. Twenty percent of households in 23,000 villages are expected to be connected to electricity (GOB, Planning Commission 1980).

Traditional Fuels

Traditional fuels are obtained from three major sources: tree biomass, agricultural residues, and animal residues.

Tree Resources. Tree-covered areas of Bangladesh are either reserve forests (89 percent of tree-covered areas) or homestead woodlots (11 percent). The reserve forests are either state owned and administered by the government's forest department (51 percent of tree-covered areas) or by the Chittagong Hill Tracts district authority (38 percent). Trees in reserve forests are natural growth and used for timber, industrial raw materials, and firewood.

Reserve forests are located in five geographic areas. Their distribution with that of other resources and population of each district is shown in Table 1. As the table indicates, resources are not distributed evenly among population groups. In the north and northwest, eight districts with 37 percent of population have only 0.6 percent of total forest areas. On the other hand, Chittagong Hill Tracts district, with only 0.7 percent of the total population, has 55 percent of the forest areas. The only hydropower station in the country is located in this district. The districts with the least forest-covered area also lack commercial energy resources.

Homestead woodlots play an important role in meeting the need of traditional fuel resources. Within the social context of Bangladesh, the homestead is defined as an area physically occupied by one or more households for habitation. Functional usages of homestead land area are for housing, cooking space, water sources, latrine, postharvest processing of crops, animal shed, planting area for trees and vegetables, and limited grazing space. On homesteads having more than one household, sharing of different facilities and their physical location depends on the distribution of ownership of homestead land.

The homestead woodlots are planted by individual owners in their private plots located within and near the homestead. Generally, multi-purpose trees are grown along with other vegetative crops. Homestead woodlots are distributed evenly throughout the country. Although

Critical Assessment of Bangladesh Situation 49

Transporting fuel logs to town, Rajshahi. (R. Van Den Beldt)

homestead woodlots cover much less area than reserve forests, almost all rural people depend on them for fuel supply. Since the lots are owned by the people and since they are easily accessible, trees in homestead lots are used more efficiently than reserve forest trees. Trunks of trees, branches of trees, and fallen leaves are all used as cooking fuels. Often only branches are lopped off periodically.

Trees from reserve forests are cut at ground level in most cases, and only trunks and major branches are taken away for fuelwood. Within a particular area, the rate of extraction of fuelwood from reserve forests and homesteads is related to urban and rural demands. Generally, all cooking fuel needs of rural areas are met from resources available within the village. If a reserve forest is located adjacent to a village, it may supply some part of the village's fuel. Reserve forests are the major supply source of fuelwood for urban areas. Distances supplied are influenced by transport cost. Manpower and transportation considerations tend to limit extraction of fuelwood to the dry season.

Lacking specific data, a possible scenario for the supply of fuelwood to various urban areas of Bangladesh follows.

TABLE 1.
Distribution of Resources and Population in Different Districts of Bangladesh, 1974-75

Divisions & Districts	Total Area (a)	Not Available for Cultivation (b)	Reserve Forest (c)	Cultivable Wastes (d)	Current Fallows (e)	Net Cropped Area (f)	Total Cropped Area (g)	Irrigated Area (h)	River Covered Area (i)	Chemical Fertilizer (j)	Cattle (k)	Total Population (l)	Urban Population (m)
						Percentage							
DHAKA													
Dhaka	5.2	6.1	1.2	1.3	6.2	6.1	6.1	8.7	5.0	8.5	5.6	10.9	36.7
Kishoregonj	3.9	4.7	0.05	7.2	5.1	4.4	4.7	10.5	3.3	5.8	11.4	10.5	6.4
Mymensingh	5.3	4.3	1.2	2.5	6.2	6.7	7.8	5.3	1.2	6.1	2.2	2.9	1.6
Tangail	2.4	1.2	2.1	3.0	0.7	3.0	3.4	2.4	1.2	2.1	2.2	2.9	1.6
Faridpur	4.9	6.7	-	0.7	4.0	5.8	6.1	2.3	6.1	1.3	4.7	5.7	1.8
CHITTAGONG													
Chittagong	4.9	4.9	9.5	6.0	5.3	3.6	3.7	6.7	6.9	14.0	6.4	6.1	14.4
C. Hill Tracts	9.2	1.0	54.9	1.5	0.3	0.8	0.9	0.8	5.1	0.9	0.7	0.7	0.8
Noakhali	3.4	2.4	0.1	10.7	6.7	4.0	4.0	5.0	13.8	5.5	2.5	4.5	0.8
Comilla	4.7	3.9	0.03	2.1	11.8	5.6	5.3	8.3	3.7	12.9	5.9	8.1	5.7
Sylhet	8.7	12.5	3.6	20.7	7.5	8.5	7.5	18.7	1.4	5.2	9.3	6.6	2.0

RAJSHAHI											
Rajshahi	6.6	7.6	0.1	11.0	4.3	8.1	7.3	7.9	1.5	5.6	5.4	5.9	3.7
Dinajpur	4.7	4.5	0.4	9.4	8.7	5.4	5.3	2.6	0.5	5.1	5.7	3.6	1.7
Rangpur	6.7	7.9	0.1	12.8	5.7	8.0	9.7	4.9	5.6	5.2	9.3	7.6	3.9
Bogra	2.7	3.3	–	0.3	1.3	3.5	3.7	2.9	1.0	2.9	4.1	3.1	1.2
Pabna	3.4	3.5	–	0.1	7.5	4.0	4.1	2.3	4.8	5.7	3.7	3.9	3.2
KHULNA											
Khulna	8.4	6.8	26.1	0.9	6.3	4.7	4.0	2.0	16.7	1.5	5.1	5.0	9.9
Barisal	4.7	6.8	0.2	3.7	1.7	5.6	5.6	4.4	17.6	3.3	4.6	5.5	2.3
Patuakhali	3.0	4.6	0.4	2.2	3.2	3.1	2.6	0.8	8.4	2.9	5.3	2.1	0.6
Jessore	4.6	5.2	–	3.4	1.1	6.1	5.5	1.6	1.3	4.4	4.8	4.6	2.7
Kushtia	2.5	2.1	–	0.3	6.3	3.0	2.7	1.9	1.0	1.0	3.3	2.6	2.4
BANGLADESH	99.9	100.0	99.9	99.8	99.9	100.0	100.0	100.0	100.0	99.9	100.0	99.6	99.8
			Million Acres							Million Tons	Million	
TOTAL	35.3	6.58	5.46	0.67	2.0	20.56	29.92	3.56	2.32	0.28	28.9	76.0	7.0

Source: Figures are from Statistical Year Book of Bangladesh, 1979, B.B.S.

Note: Not available for cultivation: Rivers, tidal creeks, lakes, ponds, roads, homesteads.
Cultivable wastes: Area suitable for cultivation but lying fallow for more than one year. Current fallow: Area brought under cultivation, but not cultivated during the year.
Total area: (a) = (b + c + d + e + f).

1. The main source of fuelwood for urban areas is reserve forests in their particular district. Any additional need is met from homestead woodlots. This is the situation in eight districts (Khulna, Patuakhali, Barisal, Mymensingh, Tangail, Sylhet, Chittagong, and Chittagong Hill Tracts).

2. Districts with disproportionately higher urban population obtain their fuelwood supply from three sources: homestead woodlots, reserve forests located within the particular district, and reserve forests located in other districts. For example, Dhaka district, with 37 percent of total urban population of the country, has only 1.2 percent of total reserve forest area. In addition to sources within the district, the fuelwood supply for urban areas of Dhaka district comes from adjacent Mymensingh and Tangail districts and also from coastal forests located in the Khulna and Patuakhali districts. The transport system originates from Dhaka, the capital city, carrying consumer goods and intermediate goods to peripheral districts. Fuelwood is brought back on the return journey at small marginal cost. The operating distance of road transport is less than the operating distance of water transport.

3. Districts with very few reserve forests for fuelwood supply within their own boundaries depend on homestead woodlots located within the district and reserve forests located in other districts.

The demand of fuelwood in a particular urban location is related to the socioeconomic status of users and to availability and price of commercial cooking fuels. For example, part of the urban population of Dhaka and Narayanganj have piped natural gas for cooking. With a domestic gas connection, the monthly cost of cooking with natural gas was only Tk. 35[1] (US $1.75) as of July 1982. An average family without natural gas for cooking would pay about Tk. 140–150 to cook with kerosene and Tk. 120–160 to cook with fuelwood.

Village homestead woodlots play a very important role in supplying cooking fuel not only for rural population but also for urban locations. Some recent attempts have been made to establish forest nurseries to supply seedling for homestead woodlots. Existing infrastructural facilities are inadequate to meet existing need.

Agricultural and Animal Residues. Use of agricultural and animal residues as cooking fuels differs according to location, season, and socioeconomic status of consumers. Transportation difficulties and storability limit residue consumption as cooking fuel to rural areas in the dry season. Availability of agricultural residue is related to agroclimatic zone and harvesting season of specific crops. Residue from crops harvested in the dry season is in high demand as fuel as well as for other uses. Residue available in wet months but unpreservable cannot be used as fuel. Rice straw and jute sticks both are available in wet months. Rice straw decomposes easily, but jute sticks are strong and can be preserved. Animal residue produced in wet months cannot be used as fuel because it cannot be dried.

Note that the production of traditional fuels such as fuelwood and agricultural and animal residue depends on land. Actual consumption of these types of fuels by individual families depends on ownership of land or relationship between consumer and landowner.

Policy Issues

This profile of the distribution of commercial and traditional energy sources and forests suggests certain policy issues related to energy needs and rural development. In districts critically short of trees, is substitution of fossil fuels for fuelwood as industrial fuel and as urban domestic fuel an appropriate policy option to retard the adverse effect of deforestation on agriculture production in these districts? Coal, fuel oil, bottled petroleum gas (LPG), and compressed natural gas (CNG) might be considered as fuelwood substitutes for appropriate end uses. Alternatively, should tree plantation be undertaken for areas critically short of trees? This policy would require a change in attitude in existing forestry policy. Traditionally, forestry infrastructure has been developed for conservation and commercial extraction of forest products from naturally forested areas. Areas with no natural forests have the minimum infrastructure but are in greatest need.

In 1979, in order to stimulate growth of industries in backward areas, the government decided to pay approximately 60 percent rebate on purchase of fuel oil to industries established or located in seven districts (Dinajpur, Rangpur, Rajshahi, Pabna, Bogra, Barisal, and Patuakhali) of the Western Zone. This subsidy is of course limited only to certain privately owned industries approved by the industries department. The total number of industries so far availing of the rebate is less than twenty. One policy issue is whether all the industries located in these districts should be considered under some energy price incentive scheme for the reason just discussed. Delivery of commercial energy and price incentives, potentially including LPG and CNG, may also be considered as a policy option to establish agro-processing industries in the least developed areas of other districts to provide opportunity for rural employment.

At existing prices of diesel and electricity, it is more expensive to irrigate with a diesel-operated pump than with an electrically operated pump. The existing electrical network covers only a small fraction of the total irrigated area. Similar to the provision of price incentives and industrial fuels to backward areas, would a set of price incentives for diesel fuel used for irrigation be an appropriate policy option? This facility would benefit a large number of users in rural areas and would contribute to agriculture production.

The program for rural electrification raises related policy issues. On the one hand, two major weaknesses of rural electrification can be identified:

1. Twenty percent of the rural households are the potential users of rural electricity. This group also controls most of the rural resources. Therefore, rural electrification may enhance rural development by improving the standard of living of the upper 20 percent, which may remove disparities between the rich urban and rich rural group. But most of the rural people will be deprived of direct benefit of rural electrification.

2. Project areas brought under rural electrification were selected according to techno-economic feasibility. Therefore, it is the more developed area of the country that has been selected for rural electrification. This will result in unequal development of different regions of the country.

On the other hand, rural electrification has important potential for increasing irrigation and farm productivity, thereby increasing both employment and residue as well as food production. If based on generation from natural gas, it may reduce oil imports also. In some areas landless families have organized pump groups, selling water-lifting services to nearby marginal and subsistence farmers. Optimal use of irrigation capacity and equitable distribution of water depend heavily, however, on effective organization by user groups — for which extension services are currently limited.

The various policy options just discussed imply that location-specific policy actions are necessary for achieving the objectives of rural development. One question may be phrased more broadly: is energy pricing policy an appropriate and effective measure to offset the locational supply imbalance of commercial energy sources? It should of course be stressed that any incentive measure with energy price must be linked with the guarantee of its efficient use by the beneficiary group.

ENERGY STUDIES

The objectives, scope, and methodologies of surveys related to energy use in Bangladesh vary widely. The focus of this chapter is on in-depth studies of household energy use and resources in two rural areas, oriented to technology needs and development. A comparative setting is offered by brief reviews of other studies.

Bangladesh Energy Study

The Bangladesh Energy Study (BES) was funded by the Asian Development Bank through UNDP and was carried out by a group of expatriate consultants. That study estimated total energy consumption of Bangladesh in 1973–74 at 287 PJ, or 3.8 GJ/person (GOB 1976). Traditional fuels were estimated to supply 73 percent of total energy (2.8 GJ/person) and commercial fuels supplied 27 percent (1 GJ/person). Only 1.6 percent of the total energy was electric.

Uncertain about their own estimate, the BES stressed the need for detailed studies on rural energy supply and demand. The study highlighted the importance of traditional energy in the national economy but did not recommend investment in this important sector. One of the major uncertainties was in demand and supply estimates for firewood. On the basis of aerial survey data, the BES estimated possible availability of tree biomass at 2.1 million tons per year. Firewood consumption estimates in other studies varied from 4 to 22 million tons per year. The BES argued that only the more wealthy rural households actually use firewood obtained from their homestead. The poor rural households mainly depend on residual biomass. Based on information from aerial photographs, the report found no evidence of overcutting of trees in homestead areas.

Arguments by BES justifying their estimate were not accepted at field level. Overcutting of trees has become a general concern for various groups of people and is attributed to the wide gap between the supply and actual consumption.

Energy Use Studies

Based on secondary data, Rodney Tyers (1978) estimated total rural energy consumption in Bangladesh in 1974–75 at 616 PJ. Estimated annual rural energy use was 8.7 GJ/person, of which traditional fuels accounted for 5.0 GJ. That study emphasized the importance of fuel supply for food processing and preparation and the risk of declining per capita crop residue availability with adoption of high-yield, short-straw crops. The available data were analyzed in terms of food production models, incorporating growth rate of food grain, livestock, crop residues for cooking, and economic and technological trends.

In 1978, an in-depth study of rural energy use was made in a small randomly selected sample of families in Ulipur village, Comilla district (John Briscoe 1979). The study population of 300 people belong to 40 Muslim agriculturist families and 8 Hindu fishing families. Energy use data were collected over 8.5 months by participant observers. The three sources of information were a household questionnaire (demographic, economic, land utilization, crop production, animals, and residue production); fortnightly measurement for each study family (human and animal labor, crop production, crop residue production, cattle feed and dung, and fuel for storage); and quarterly detailed measurement for each family (types and quantities of fuel used for cooking food and for parboiling paddy).

Ownership of fuel-producing assets was found to be highly concentrated: 16 percent of the families own 80 percent of the trees, 55 percent of the cropped land, and 45 percent of the cattle. The annual fuel consumption of the community was estimated as 6.8 GJ/person. About 3 calories of fuel was consumed for each calorie of food grain cooked, indicating a very low efficiency of the cooking process as will be seen in the analysis of cooking fuel requirements in Chapters 8 and 11. Families of different classes used different types of fuels for cooking. The average number of trees per person in the study area was estimated as 6.4. Trees are accessible to their respective owners only. This type of study reveals the influence of sociopolitical factors on energy use patterns under noncommercial transactions.

Based on a study in Dhanishwar, Bangladesh, yearly per capita cooking energy demand for different socioeconomic groups was estimated as: landless, 4.3 GJ; poor, 4.7 GJ; middle, 5.8 GJ; and rich, 7.6 GJ. Supplies of cooking energy for different groups were estimated as: landless, 0.7 GJ; poor, 4.6 GJ; middle, 5.2 GJ; and rich 13.8 GJ. A wide gap between the demand and supply of cooking fuel for landless families in Dhanishwar indicated the scarcity of their cooking fuel. The landless and poor have food deficits, with daily food intake of 1,600 kcal and 1,770 kcal respectively, and obtain food and fuel from the rich in exchange for labor at the time of peak demand (David Hughart 1979).

Two Surveys of Village Resources and Energy Potentials

A 1978 study (Nurul Islam 1980) surveyed the total resources of 23 villages in one rural area of southern Bangladesh to (1) evaluate the current energy situation, (2) assess the prospect of alternative energy technologies, (3) test the survey methodologies for a countrywide rural energy survey, and (4) use the survey area for long-term field testing of various alternative technologies.

When the survey began, no previous study was available in Bangladesh to guide design of methodology. Since they did not have access to published literature of other developing countries, the multidisciplinary research team planned, designed, and executed the survey based on mutual discussions. A detail discussion of methodological aspects follows.

Two alternatives for the survey's geographic coverage were considered: either a group of villages from the same location or a number of villages from different locations. The first alternative was chosen considering the importance of rural institutions in rural development and in diffusion of technology. The area of a *union* (usually about 15 villages) was considered appropriate since village development is controlled and coordinated by thana-level officials through elected Union Council members. Nabagram Union of Barisal district was selected on the following criteria: it had not been surveyed before; it had a traditional energy consumption pattern (no electricity); and it was free from direct influence of urban and/or industrial center. Nabagram Union has 23 villages over an 8 sq mi area. All the households in the area were covered by the survey, identifying the traditional settlement pattern (homestead). The survey area is representative of a tidal zone.

The data for this survey were collected through personal interviews by trained investigators using a carefully designed pretested questionnaire. Two senior research staff supervised the field data collection and acted as participant observers in special cases. Recruiting resident field investigators from the project area was judged desirable because of the nature of the data and the project's long-term objective. Considering the educational background and experience of the investigators, the idea of coding the questionnaires was dropped during pretesting. Data were coded later by an analyst.

A preproject course was arranged for field investigators. Of 42 participants, 20 investigators were selected, one from each large village in the project area. Each investigator initially surveyed his own village and later transferred to other villages as necessary.

Most of the field survey was conducted during the 1978 rainy season (July–October). The rain meant reaching the respondents was more difficult and field research staff sometimes had to depend on country boats for transportation, but the rain also kept the respondents at home. Also, rural people's resources, especially their cooking fuels, are more vulnerable during the rainy season. Consumption of individual fuels was recorded separately for the wet and dry seasons, based on reported and observed information for the wet season and reported information for the dry season. A second part of the survey was conducted in April 1979, during the dry season. Stove dimensions were then measured and utilization of tree resources was assessed. A subsequent series of laboratory and field experiments was carried out to evaluate the fuel consumption data estimated by the survey. In addition, the performance of traditional cooking stoves was evaluated as the basis for development of an improved stove design. For more details on methodology of the study, see Islam 1980.

A similar methodology was used in a survey of four villages of Kulaghat Union in Rangpur district, in northern Bangladesh. A field survey was carried out in March–June 1981 (dry season). The field research supervisor of the Nabagram study was part of an otherwise new team of researchers from the Chemical Engineering Department at Bangladesh University of Engineering and Technology responsible for the study (A.K.M.A. Quader and K.I. Omar 1982).

Table 2 summarizes important observations of those studies.[2] Some of the data from the reports (Islam 1980; Quader, Omar 1982) have been recalculated to reflect the same basis. Neither survey area has an electricity supply or nearby reserve forests, but Nabagram is located in a more forested region than Kulaghat. The average minimum temperature and rainfall in Kulaghat is less than Nabagram. The numbers of households per homestead in Nabagram and Kulaghat are 3.23 and 1.5 respectively. In Nabagram, homesteads are built above flood level on raised earthen mounds — in a more concentrated settlement pattern. Twenty-two percent of the households in Nabagram are landless and 34 percent are landless in Kulaghat. Cattle population per acre and per person is slightly higher in Kulaghat, but almost the same percentage of households have no cattle. Since a family-size biogas plant needs the dung from four or more head of cattle, only 11 percent of the households in Nabagram and 14 percent in Kulaghat can install a family-size plant.

Proportionately, Nabagram has slightly more tree resources but a smaller proportion of bamboo bushes. Fifty-seven percent of Nabagram's trees are fruits and timber; in Kulaghat, 65 percent are betel nut. Nabagram has two ponds per homestead and Kulaghat has 0.1. Many Nabagram ponds have been dug to make the earth mound to situate a homestead above flood level. Kulaghat's fewer ponds are augmented by a higher proportion of tubewells as a source of drinking water. In both the locations, roofing materials of most houses are of biomass sources. Old roofing materials are used as cooking fuels after periodic roof repairs. Most households in both areas have an inadequate local supply of food grains.

Average yearly consumption of biomass fuel for household cooking was 353 kg/person in Nabagram and 634 kg/person in Kulaghat. Household cooking includes cooking food, parboiling paddy, and making ghur

TABLE 2.
Summary of Resources Survey of Two Rural Locations of Bangladesh

Description	Survey Locations	
	Nabagram	Kulaghat
Number of villages surveyed	23	4
Total area of survey villages (acres)	5,120	1,845
Number of homesteads	871	638
Number of households	2,820	954
Population	16,228	5,301
Households per homestead	3.2	1.5
Homesteads with single household	9.0%	67.4%
Households with no cultivable land	22.0%	34.0%
Family size (person/household)	5.8	5.6
Persons per acre	3.2	2.9
Cattle per acre	0.65	0.96
Cattle per person	0.21	0.33
Households having four or more cattle	11.0%	13.7%
Households with no cattle	55	51
Tree Resources Types (trees/person)		
Fruit and timber trees	5.0 (57%)	1.1 (16%)
Palms (betelnut, coconut, date palm)	2.9 (33%)	4.8 (66%)
Firewood and other trees	0.9 (10%)	1.3 (18%)
Total trees	8.8 (100%)	7.2 (100%)
Bamboo (bushes/person)	0.17	0.33
Housing and Services		
Dwelling house per household	0.99	1.7
Kitchen per household	0.83	0.71
Cowshed per household	0.41	0.5
Open yard per household	0.63	0.75
Open yard per homestead	2.04	1.11
Kitchen yard per household	0.73	0.36
Ponds per household	0.64	0.07
Ponds per homestead	2.00	0.08
Tubewell per homestead	0.03	0.05
Persons per tubewell	213	118
Building Materials and Roof of Different Houses		
Pucca	0.6%	0.06%
Galvanized iron sheet	20.0%	12.0%
Tiles	11.0%	0.1%
Biomass (straw, grass, leaves)	68.4%	88.0%
Food Grain Supply Situation		
Households growing more food grain than own need	3.5%	14.0%
Households growing sufficient food grain	13.5%	14.0%
Households having deficit of food grain	66.0%	42.0%
Household not growing food grain	-	30.0%
No response	17.0%	-
Average Annual Fuel Consumption		
Biomass fuels for cooking (kg/person)	353	634
Kerosene for illumination (Imperial gallon/household)	12.7	8.15

Average Annual Fuel Energy Consumption		
Cooking fuel (GJ/person)	4.9	8.1
Illuminating fuel (GJ/person)	0.4	0.24
Total fuel energy (GJ/person)	5.3	8.34
Types of Stoves Located Inside the Kitchen (stoves per household)		
One-mouth insitu	1.17	1.14
Two-mouth insitu	0.25	nil
One-mouth movable	0.11	0.013
Tafal insitu	0.02	0.002
Three-stone temporary	nil	0.03
Kerosene	0.016	nil
Types of Stoves Located in the Kitchen Yard (stoves per household)		
One-mouth insitu	0.92	0.6
Two-mouth insitu	0.92	nil
One-mouth movable	0.016	0.01
Tafal insitu	0.06	0.06
Three-stone temporary	nil	0.4
Cooking Fuels Consumed per Person in Wet Season (GJ/six months)		
Firewood and branches	2.1 (84%)	2.3 (59%)
Leaves	0.05 (2%)	0.2 (5%)
Agriculture residues	0.30 (12%)	1.3 (33%)
Animal residues	0.05 (2%)	0.1 (3%)
Total biomass fuel in wet season	2.5 (100%)	3.9 (100%)
Cooking Fuels Consumed per Person in Dry Season (GJ/six months)		
Firewood and branches	0.87 (36%)	1.9 (45%)
Leaves	0.46 (19%)	0.4 (10%)
Agriculture residues	0.89 (37%)	1.8 (43%)
Animal residues	0.18 (8%)	0.1 (2%)
Total biomass fuel in dry season	2.40 (100%)	4.2 (100%)
Illuminating Devices (per household)		
Open wick lamp	2.46	2.05
Glass chimney fitted lantern	0.56	0.39
Zonal Information		
District	Barisal	Rangpur
Population density of the district (persons/km^2)	578	566
Average minimum temperature (F)	71	68
Rainfall (cm)	216	198

Source: Statistics for Nabagram are from (Islam 1980); for Kulaghat, from (Quader, Omar 1982).

(crude sugar). An average of 15 to 20 percent of biomass fuel consumed in rural areas is required for parboiling of paddy. Except in the districts of Chittagong and Sylhet, parboiling is practiced throughout Bangladesh. Parboiling is a method of precooking rice in the husk with water and heat that hydrates and gelatinizes the starch kernel. This process eases dehusking, reduces grain breakage, and improves storability.

The higher per capita fuel consumption estimated in Kulaghat may be attributed to

1. Seasonal bias in fuel estimation by the users. The Nabagram study was carried out in rainy season and the Kulaghat study was carried out in dry season. Normally cooking fuel is scarce in the wet season and abundant in the dry season.
2. Higher availability of agriculture residues (jute sticks) at the homestead level.
3. Higher consumption of fuel for water heating and body warming due to low winter temperature.
4. Use of less efficient cooking devices.

Most stoves used in both the areas are *insitu* built. Generally, one-mouth stoves are used for daily household cooking; other types, such as two-mouth stoves and three-stone stoves, are used for parboiling and ghur making. More one-mouth stoves are used in Nabagram than in Kulaghat. The numbers of stoves inside the kitchen (wet season) and outside the kitchen (dry season) differ between the two areas. Notably, in Nabagram two-mouth stoves are used outside by nine in every ten households and inside by one in every four households, mainly for parboiling paddy as well as other cooking tasks. There are no two-mouth stoves in Kulaghat. There, two of every five households have three-stone stoves for parboiling paddy. Two-mouth stoves may consume less fuel than three-stone stoves because of the former's partially shielded combustion chamber. This difference might have contributed to lower per capita fuel consumption in Nabagram.

Variations in stove design for domestic cooking may be attributed to specific cooking needs, duration of cooking, types of fuels, and cultural practices. Varieties of traditional wood-burning stoves observed during field visits may be classified as with or without grate; single or multiple mouth; below or above ground level construction; and small or large. Figures 1, 2, and 3 show types of one-mouth, two-mouth, and three-mouth stoves used for domestic cooking in rural Bangladesh. Use of two- and three-mouth stoves is common in areas where long straw (1.5 m or 3 m or longer) available from deep water paddy is used as fuel, such as in Comilla district. In urban areas one-mouth stoves are mainly in use. An assessment of variation in local practices of stove use can be a good basis for future research on diffusion of improved stoves.

The small seasonal variation of fuel consumption observed in the two survey areas may be attributed to seasonal variation of operational efficiency of stoves, or to limitation of methods used in data collection. In

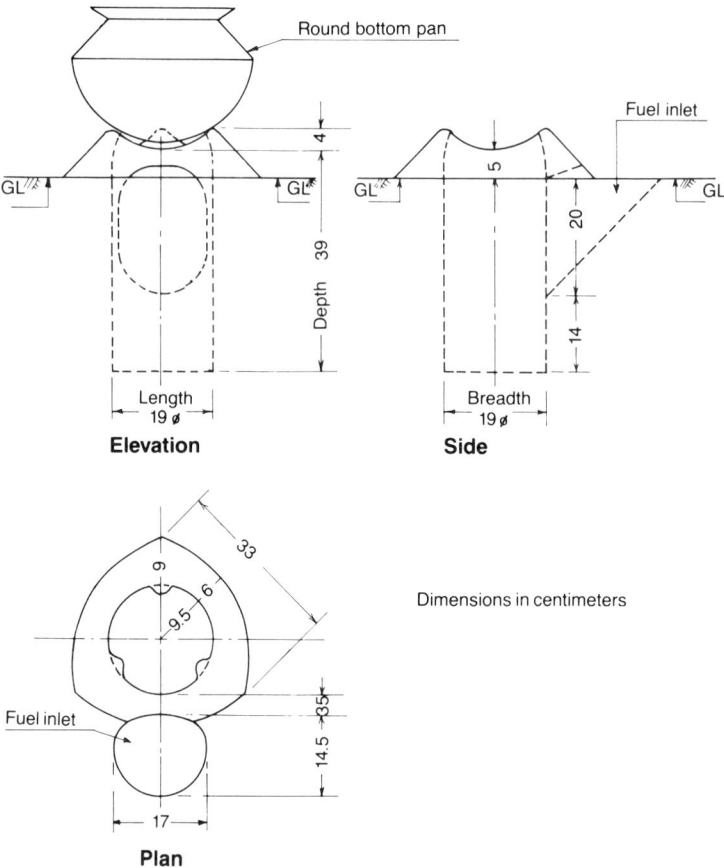

Figure 1. One-Mouth Stove

Nabagram, the average operational hours of stoves are longer in the dry season — sometimes up to 8 to 12 continuous hours due to parboiling paddy. In addition, comparatively dry fuels are used in this season. In the wet season, stoves are used intermittently and comparatively wet fuels are used. As a result, with nearly the same estimated fuel consumption, more cooking might have been performed in the dry season due to an average higher end use efficiency. This attribution needs experimental verification to ascertain the effect of seasonal temperature variation on fuel consumption.

In both studies, at the time of survey the respondents were asked to estimate the amount of fuel consumed in the dry season and in the wet season. Respondents might have equally divided their yearly fuel con-

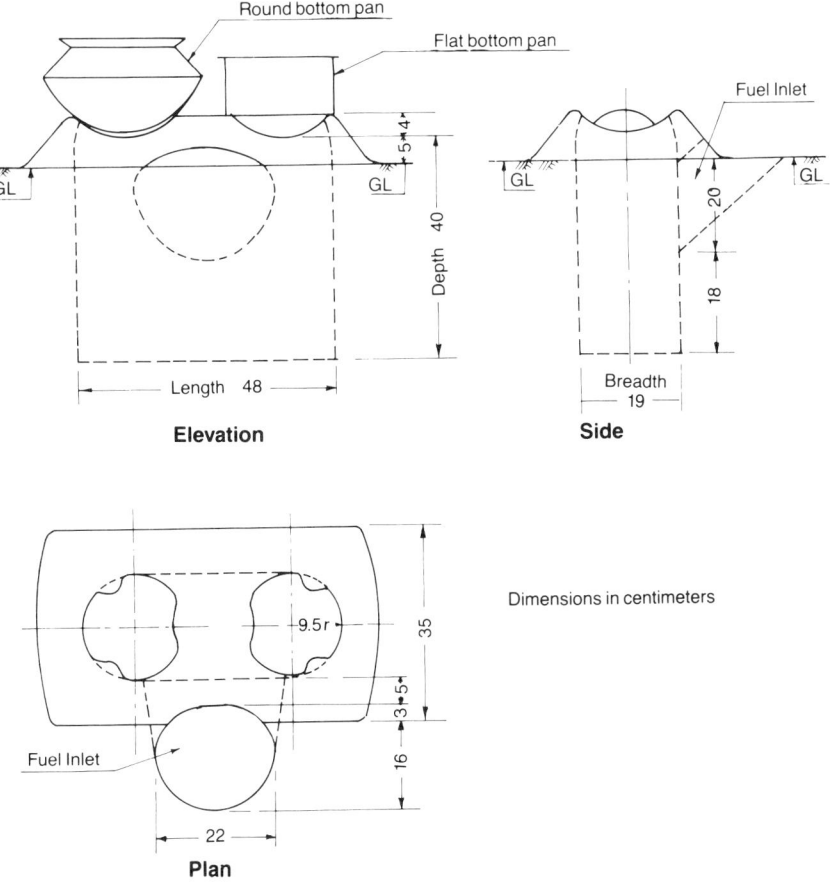

Figure 2. Two-Mouth Stove

sumption at the time of reporting. Reliable estimation of seasonal variation of fuel consumption requires data collection in different seasons.

The pattern of per capita fuel consumption in households of different size in four villages of Nabagram is indicated in Table 3. Less fuel per capita is consumed as the number of persons per household increases, implying an economy of scale in cooking.

Per capita consumption of food and fuel of a particular village of Kulaghat is shown in Table 4, according to landholding size. Per capita consumption of household cooking fuels increases as size of landholding increases from 6.91 GJ/year for landless to 16.9 GJ/year for landholdings of 10 acres (Quader, Omar 1982). Higher per capita fuel consumption by households with larger landholdings may be attributed to higher amount

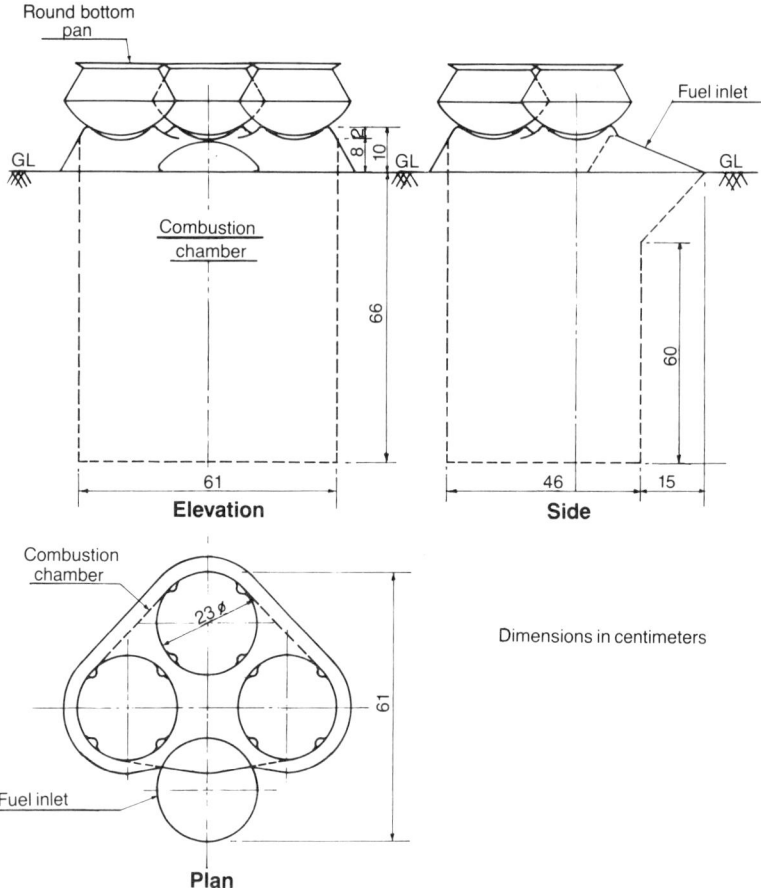

Figure 3. Three-Mouth (Comilla) Stove

of food cooked and to increased parboiling and ghur making.

Caloric food intake per person also increases with increase of landholding size from 2.9 GJ/year (1,910 kcal/day) for landless to 5.8 GJ/year (3,800 kcal/day) for landholdings of 10 acres. Higher per capita food consumption by households with larger landholdings may be attributed to better accessibility as well as to the method of payment to agriculture laborers. In Bangladesh, agriculture laborers traditionally are paid partly cash and partly with food. The number of meals provided with part cash payment may vary from one to three meals a day, depending upon the locale and season. Per capita food consumption has been estimated by dividing the yearly food consumption reported by the

TABLE 3.
Per Capita Consumption of Cooking Fuels in Four Villages of Nabagram

Number of Household Members	Fuel Consumption			
	Village 8	Village 9	Village 20	Village 22
	GJ/year			
1	–	–	11.00	10.90
2	7.56	8.91	7.40	6.77
3	4.46	8.44	5.29	5.93
4	4.37	8.44	5.14	4.44
5	4.04	6.45	4.25	3.49
6	4.05	5.80	4.60	3.28
7	3.25	5.60	4.50	3.65
8	3.14	4.68	4.90	3.55
9	2.18	7.20	3.84	2.75
10	3.29	5.27	–	1.38
11	2.28	–	–	1.80
12	2.33	4.99	–	3.49
13	5.98	3.48	–	–
14	1.73	–	–	–
20	4.98	–	5.35	–
Weighted Average	3.43	5.93	4.80	3.82

Source: Figures are calculated from Islam (1980).

household by the number of permanent household members. Amounts of food consumed by casual laborers thus may contribute to the higher per capita food consumption estimate for larger landholders who employ them, and lower per capita food in their own households.

Another factor may contribute to the lower per capita food consumption indicated for landless laborers. At times of acutely scarce employment within a village, they migrate temporarily to other villages or to an urban area in search of jobs. Yearly food consumption reported by the household would normally include only the amount consumed by them during their stay in the village.

Wood and Bamboo Consumption Survey

The Institute of Statistical Research and Training, Dhaka University, designed and executed a survey on Rural Consumption of Wood and Bamboo in Bangladesh for the Forestry Section of the Planning Commission, in conjunction with the Food and Agriculture Organization (FAO). The basic framework was a stratified, three-stage, self-weighted sample of 6,000 households selected from 430 villages in 47 thanas. Six major zonal stratifications were used. Four of them were based on ecological

TABLE 4.
Distribution of Food and Fuel Consumption in Sakoa Village
of Kulaghat Union by Landholding Size

Description	Landless	0-0.5	0.5-1	1-2	2-3	3-5	5-10	10+
		Farm Size (acres of cultivable land)						
1. Acres/person	0	0.05	0.14	0.30	0.41	0.47	0.87	1.34
2. Cattle/person	0.04	0.06	0.29	0.38	0.31	0.37	0.71	1.04
3. Tree/person	3.1	9.1	14.60	23.40	18.70	18.40	22.30	23.40
	GJ/year per capita							
4. Food consumption (rice and wheat)	2.9	3.39	3.46	3.48	3.64	3.40	4.16	5.79
5. Food cooking fuels								
a. Firewood	1.66	2.11	2.27	4.38	3.17	3.02	4.84	4.54
b. Branches, twigs, leaves	2.72	2.42	2.12	2.42	1.97	1.51	1.66	2.72
c. Agriculture residues	2.26	2.00	1.88	2.14	2.00	1.88	2.26	3.26
d. Total (a+b+c)	6.64	6.53	6.27	8.94	7.14	6.41	8.76	10.52
6. Fuel for parboiling	0.27	0.53	1.07	1.73	2.13	2.27	2.67	4.00
7. Fuel for ghur making	-	-	-	0.15	0.07	0.30	2.16	2.38
8. Total household cooking fuel (5+6+7)	6.91	7.06	7.34	10.82	9.34	8.98	13.59	16.90
9. Kerosene for lighting	0.20	0.20	0.20	0.30	0.28	0.30	0.31	0.48

Source: Figures are calculated from Quader, Omar 1982.

Note: Total households, 250; total population, 1,407.
Caloric value of rice and wheat = 14.8 MJ/kg = 3,540 kcal/kg

characteristics; one, on type of "urban fringe" area; and one, by proximity to reserve forest. The field survey was carried out in January and February 1980 (the dry season).

The methodology of data collection and the survey results have been compiled in a series of reports (FAO 1980, FAO 1981a, FAO 1981b, and FAO 1981c). Two types of household questionnaire were used. The main questionnaire was based on reported and observed information collected by personal interview for the entire sample. A full measurement questionnaire was based on measured information and was applied to a subsample of 1,200 households. (The latter was primarily designed to calibrate the result of the former.) The items included were

1. Quantity of fuel consumed yesterday, last week, and last year.
2. Separate fuel consumption data for fuelwood, branches, other parts of trees, bamboo, agriculture and animal residues, waste papers, charcoal, oil, and gas.

The questionnaires did not mention specific end uses of fuels, such as cooking and lighting.

Salient points of the observed results follow. Average annual fuel consumption was 501 kg/person by the daily recall method and 360 kg/person by the yearly recall method. The higher estimate by the former method probably reflected the fuel use pattern of the dry season, when the survey was made.

According to the data (Table 5), per capita consumption of biomass fuels decreases as landholdings increase: 5.4 GJ/year for 0.1 acre landholding size to 3.8 GJ/year for landholding size of 7 acres and above. The figures are opposite to those of Quader and Omar (1982). Lower per capita fuel consumption by higher landholding group has been attributed to economy of scale of cooking, variation in type of food cooked by different landholding size, and higher technical heat utilization efficiency of the type of fuel used by higher income group households (FAO 1981a). According to Table 5, per capita consumption of the low-grade biomass fuels (residues, twigs, leaves) decreases with the increase of landholding. Per capita fuel consumption and its composition, however, also depend on specific location. Interpretation of data aggregated by landholding size for the entire country is difficult. Fuel consumption data of different agroclimatic zones tabulated separately by landholding size would have been more useful.

The average per capita fuel consumption estimated by the study was 4.44 GJ/year. The amount and type of estimated fuel use varied by region (Table 6). The estimated per capita consumption of Region 2 was the lowest (2.95 GJ/year), which may be attributed to its location in a poorly forested area. Region 4's estimated per capita fuel consumption (3.77 GJ/year) and estimated tree fuel consumption (210 kg/year) (FAO 1981a) were similar to Islam's estimates (1980) of 4.9 GJ/year and 229 kg/year for 23 villages in the same region. In Region 5, higher per capita fuel consumption (4.85 GJ/year) may reflect higher availability of tree fuel (43 percent of total biomass fuel). The highest per capita fuel consumption, 5.23 GJ/year in Region 6, was attributed to the effect of urban-induced high income (FAO 1981a).

The study did not specify end uses of kerosene. Although kerosene is used in rural Bangladesh mainly for lighting, aggregating kerosene with gas and charcoal has made it difficult to analyze their consumption pattern. Even though a systematic statistical procedure was followed in the survey design, this particular set of data lost its useful meaning due to method of presentation.

TABLE 5.
Composition of Biomass Fuels Consumed in Rural Areas of Bangladesh

Landholding	Type of Biomass Fuel						Total Biomass
	Grade 1				Grade 2	Grade 3 Other Parts	
Acres	Firewood	Branches	Bamboo	Total	Residues	of Trees	
	GJ/year per capita						
0-1	0.51	0.65	0.38	1.54	2.24	1.63	5.41
1-2	0.36	0.83	0.48	1.67	1.87	1.54	5.08
2-3	0.51	0.68	0.51	1.70	1.87	1.28	4.85
3-4	0.41	0.76	0.41	1.58	1.66	1.08	4.32
4-5	0.57	0.66	0.36	1.59	1.57	1.07	4.23
5-6	0.76	0.71	0.48	1.95	1.39	1.03	4.37
6-7	0.59	0.68	0.50	1.77	1.58	0.79	4.14
7+	0.82	0.95	0.35	2.12	0.94	0.75	3.81
Weighted Average	0.54	0.74	0.42	1.70	1.80	1.32	4.82

Source: Figures are calculated from (FAO 1981a), Yearly Recall Data.

Note: Firewood is wood from the main trunk of a tree. Branches are large branches cut from a tree. Residues are agricultural and animal residues. Other parts of trees are defined as small branches, twigs, leaves, and bark. Grade 1 fuel quality is superior to Grade 2 and Grade 3. Grading has been set by me based on my experience. (See Note 2 at the end of this chapter for conversion factors used on original fuel consumption data.)

Rural Energy Survey

The Centre for Policy Research (CPR) at Dhaka University published a report on a rural energy survey of 23 villages in nine districts of Bangladesh (Ataur Rahman 1982). The survey data were collected between late 1980 and mid 1981, over three different seasons of the year: winter (December), summer (April), and rainy season (July). In each period the survey took two weeks.

The entire area of Bangladesh was divided into five geographic groups with these characteristics in mind: (a) forest areas, (b) densely populated areas, (c) coastal areas, (d) landholding patterns, (e) development activities, and (f) selected socioeconomic characteristics. Of 1,200 households selected for the survey by a three-stage stratified sampling process, complete data were available for 760 households. Responses from the rest were not obtained due to migration, temporary absence during interviews, and lack of communication with interviewer.

Fuel consumption data were gathered only for household cooking. Biomass fuels in the questionnaire included wood from main trunk, branches, leaves, bamboo, and agricultural residues (paddy straw, bagasse, animal dung, other crop residues). Also included were charcoal, kerosene,

TABLE 6.
Household per Capita Fuel Usage, by Regions of Bangladesh

Type of Fuel	Fuel Consumption Regions					
	1	2	3	4	5	6
			GJ/year			
Firewood	0.79	0.88	0.84	1.20	2.07	1.99
Other Tree Fuel	1.42	0.79	1.22	1.22	1.04	1.0
Residues Including Bamboo	2.84	1.28	2.43	1.35	1.74	2.24
Total	5.05	2.95	4.49	3.77	4.85	5.23

Source: Figures are from FAO 1981a

Note: Region 1 : Rajshahi, Bogra, Rangpur, Dinajpur (northwest)
Region 2 : Kustia, Pabna, Faridpur, Jessore (west)
Region 3 : Mymensingh, Dacca, Comilla (north central)
Region 4 : Barisal, Patuakhali (south)
Region 5 : Khulna, Sylhet (forest fringe)
Region 6 : Dacca, Khulna (urban fringe).

Heating values of fuelwood = 12.87 MJ/kg, twigs and leaves = 12.87 MJ/kg and cowdung = 11.63 MJ/kg used by the study (FAO 1981a) in this table are different from those used in previous tables (see Note 2 at the end of this chapter). Due to unavailability of raw data, it has not been recalculated.

petrol and other oils, and electricity. No mention was made about fuel used for household lighting.

According to the survey, the potential per capita availability of fuel energy from tree biomass was estimated as 4.1 GJ/year. However, the report did not explain the method of estimating standing tree biomass or state the heat values used for calculation. The composition of sources of supplies for cooking fuel was tree biomass, 56 percent; crop residues, 38 percent; animal dung, 5 percent; and kerosene and charcoal, 1 percent (Rahman 1982). The study reported that 61 percent of the families used fuelwood for cooking in the rainy season compared with 39 percent in the winter and 38 percent in the summer. Seventy-six percent of the families used paddy straw for cooking in winter, compared with 28 percent in the rainy season and 36 percent in the summer. Similar seasonal variation was observed for leaves used for cooking.

Total energy used by respondent households reportedly was 1,700 GJ/year. Various end uses were household cooking, 72 percent; fertilizer, 22 percent; and rural industries, 6 percent. Without explaining the basis, the study reported estimated fuel use per capita for cooking as 1.6 GJ/year—a wide discrepancy from estimated fuel consumption reported in the same table (Table 4 in Rahman 1982). Even the higher per capita fuel consumption estimate of the study is much less than any of the previous estimates. Without statistics of the original weight of fuel and the corresponding calorific value, nothing can be inferred about this wide difference in estimated fuel consumption.

Energy consumption per household was reported to increase as the amount of land owned increases. The percentage of fuelwood, straw, and kerosene of total fuel consumed increased with the increase of landholding size. However, use of branches, leaves, and cow dung decreased with the increase of landholding size.

Households in districts with more forested area (Chittagong, Sylhet, and Barisal) consumed more fuel than did forest deficit districts (Pabna and Bogra). Districts with larger supplies of food grains used a higher proportion of straw for cooking (Rahman 1982).

The study made the following recommendations: planned, efficient land use for agricultural yield and energy plantation; plantation of fast-growing trees; diffusion of improved stoves; and massive inputs of new energy sources to improve the quality of life. Long-term recommendations of the study included rural electrification and expanded rural use of natural gas.

ASSESSMENT OF ENERGY STUDIES

Eight studies are available in Bangladesh for deriving estimates of rural energy use. Estimated per capita consumption of biomass fuels for cooking and their composition vary according to the particular study (see Table 7). Definitions of fuel types differed so composition is only roughly estimated. Because of widely varied objectives, scope, and methodologies (Table 8), the studies do not provide a reliable overall database for formulating policy on rural energy. This section aims to assess critically the strengths and weaknesses of previous studies with the objective of providing a guideline for future rural energy studies.

Study Objectives and Scope

On the basis of differences in objectives, previous rural energy studies can be classified into two groups: policy formulation (GOB 1976, FAO 1981a) and research to assess energy needs (Tyers 1978; Briscoe 1979; Islam 1980; Rahman 1982; and Quader, Omar 1982).

The Bangladesh Energy Study (BES) attempted to formulate an overall national energy policy (GOB 1976). Due to easy availability of data on commercial energy matched with the experience of experts, a detailed analysis was made of the commercial energy sector. Since data were

TABLE 7.
Composition of Biomass Fuels Used For Cooking in Bangladesh

Reference	Composition of Sources of Biomass Fuels				Per Capita Fuel Consumption
	Tree	Agriculture Residues	Animal Residues	Total	
	Percentage (on heating value basis)				GJ/year
(GOB 1976)	12.5	62.5	25	100	2.8
(Tyers 1978)	7	66	27	100	5.0
(Briscoe 1979)	36	61	3	100	6.8
(Hughart 1979)					4.3-7.6
(Islam 1980)	71	24	5	100	4.9
(FAO 1981a)	63	37	+	100	4.4
(Rahman 1982)	57	38	5	100	1.6
(Quader, Omar 1982)	59	38	3	100	8.1

+Included in agricultural residues.

Note: Tree: firewood, branches, twigs and leaves; agriculture residues: rice straw, rice hulls, jute stick, bagasse, and other crop residues; animal residue: dry cow dung.

absent and understanding of rural energy needs was limited, a simplified method based on supply-side estimation was used to approximate traditional fuel use. From a national energy planning perspective, BES provided useful policy guidelines for development of commercial energy, but its contribution to development of traditional energy was not effective. Although approximate, as the first of its kind, the BES estimate of traditional fuel supply was used by later studies as a basis for comparison. Its results subsequently have been shown to be "highly misleading" and its supply-side estimation methods have been termed "not a viable substitute for direct consumption estimation" (James Douglas 1983).

The BES report suggested that a detailed rural energy study be made. This could have been a very useful input in formulating rural energy policy in the Second Five Year Plan, but such a study has not been undertaken yet. Reasons for not undertaking a rural energy study may include lack of financial resources and shortage of trained researchers. Eight years have passed since the first national energy study was done: enough

TABLE 8.
Methodological Comparison of Energy Studies in Bangladesh

Reference	Objective	Scope	Method
GOB 1976	Assessment of energy consumption and sources for policy formation.	Entire country	Supply side estimation of traditional sources from aerial photograph.
Tyers 1978	Research to assess rural energy need.	Rural Bangladesh	Secondary data.
Briscoe 1979	Research to assess rural energy need.	48 households from one village.	Participant observation and field measurement by principal researcher. Survey duration: 8-1/2 months (May 1977 through January 1978).
Hughart 1979	Village energy balance.	Total census of 77 households in one village.	Unknown.
Islam 1980	Research to assess rural energy need and potential for introducing new energy technology.	Total census of 2,820 households in 23 villages in one area (Nabagram Union, Barisal District).	Data collected through questionnaire and participant observation: 2 field supervisors, 20 field investigators. Data analyzed manually by 8 data analysts. Survey duration: July through October (rainy season) 1978 and April (dry season) 1979.
FAO 1981a	Assessment of supply and demand of forest products for policy formulation in forestry.	Main survey covered 6,000 households, from 430 villages of 18 districts. Field measurement, 1,200 households. Three-stage stratified random sample covering whole country.	Data collected through questionnaire by 2 field coordinators, 14 supervisors, and 70 field investigators. Data analyzed by computer. Survey duration: January through February (dry season) 1980.
Rahman 1982	Research to assess rural energy need.	760 households from 23 villages of 9 districts, selected by three-stage stratified random sampling.	Data collected through questionnaire by 25 investigators. Survey duration considered the effect of seasonality: winter (December), summer (April), rainy season (July).
Quader, Omar 1982	Research to assess rural energy need and potentials for introducing new energy terminology.	Total census of 954 households of 4 villages in one area (Kulaghat Union, Rangpur District).	Data collected through questionnaire and participant observations by 2 field supervisors, 12 field investigators. Data analyzed manually by 3 data analysts. Survey duration: (March through July and December 1981).

time to train a group of indigenous experts. Because of weakness in science and technology policy planning, however, the country is in no better position with respect to trained personnel for energy assessment than it was eight years ago. Any research capabilities developed in the energy field in this period have been the result of individual researchers' initiative and not part of national research planning. Preferences of individual researchers have resulted in a general bias towards development of particular energy technologies. Few indigenous researchers have been involved in policy-oriented energy research. The quality of their work and its application in policy formulation have suffered due to lack of coordination.

Decision makers have an inherent preference for energy studies carried out by expatriates. This helps them mobilize scarce financial resources needed for the survey. It also assures further funds for implementing projects identified by the survey. This attitude of decision makers matches the outlook of external funding agencies, who are often more confident about the capability of expatriates than local experts.

The objective of rural energy studies should be not only to estimate energy supply and demand but also to develop a system that is socially and culturally acceptable to meet future needs. In this regard, expatriate experts find limited acceptability of their suggested technological solutions. In the absence of comprehensive rural energy assessment, decisions about supplying energy for rural areas are made on an ad hoc basis. Projects with external funding prospects get priority in decision making.

The BES as well as policy actions regarding national energy planning provide some directions concerning the objectives and scope of future energy studies. Adequate attention should be given to assessment of rural energy. National energy surveys should not be considered a once-in-a-while ad hoc exercise. Time series data gathering is required for long-term energy planning and for assessing the effect of short-term plans. Appropriate institutions must be selected and people trained to conduct energy surveys at regular intervals. In addition, energy demand and supply data may be gathered as a part of other national surveys (e.g., agricultural, forestry, livestock, nutrition, household expenditure, and industries surveys). This would of course require close collaboration among different agencies to make adjustments for the various sampling designs used in collecting primary data.

The Wood and Bamboo Consumption Survey (FAO 1981a), designed to provide a database for planning the Forestry Sector, had limitations discussed earlier in assessing rural energy need. With modified questionnaires, information on certain aspects of rural energy use could be obtained. A separate document (FAO 1980) has critically reviewed the

Industrial firewood, trunks and branches, Faridpur. (R. Van Den Beldt)

methodology of the forestry survey and suggested integration of forestry statistics with a continuing agricultural survey and later with rural household surveys. This would permit analysis of forest data in relation to local agroclimatic, landholding, and user characteristics.

The importance of traditional energy sources in supplying the total energy need was identified by BES. Tyers (1978) pointed out the risk of declining per capita residue availability with adoption of high-yield, short-straw crops. Such observations at the macro level are not adequate for making specific policy decisions.

The scope of study on patterns of energy use in 48 households (Briscoe 1979) provided insight on energy transactions of different socioeconomic groups, although the single study village cannot be assumed to be representative of other villages. This type of study has merit in identifying the sociocultural factors affecting the diffusion of technology in rural areas. Actual measurement of fuel consumption and participant observation by the researcher for the total period of study are the two distinct methodological differences from other studies. Those aspects enriched the quality of work at the expense of higher personnel cost.

Two microregion surveys (Islam 1980; and Quader, Omar 1982)

undertaken by the Chemical Engineering Department, Bangladesh University of Engineering and Technology, focused on the interrelationships between different village resources and energy patterns. These studies covered the first phase of introducing alternative energy technologies. Lack of continuity of research funding, however, handicapped the testing of various energy technologies in selected locations within the survey villages.

Both macro- and microstudies are necessary for national energy planning. Macrostudies are needed to assess regional and national demand and supply to formulate national energy policy. In a rural context, microstudies are essential to understanding energy needs of different socioeconomic groups in relation to their cultural background. Assessing the potential of various energy technologies and designing their delivery system also requires inflow of knowledge from microstudies.

Factors Influencing Energy Use and Sample Design

Household cooking fuel consumption is a central concern of rural energy studies. From available studies, factors affecting per capita fuel consumption include

1. Location of reserve forests in relation to the distribution of urban and rural population.
2. Ownership accessibility of fuel-producing resources, such as trees, agricultural land, and animals.
3. Agriculture landholding and type of crop. Some crops are net producers (jute, cotton), net consumers (tobacco), and both producers and consumers (rice and sugarcane) of household fuels.
4. Household or mill processing of paddy and sugarcane.
5. Local practices of providing food as part payment of agriculture wage.
6. Seasonal migration of rural laborers and families.

When village or household samples are drawn at random without considering these factors, sharply conflicting figures on rural energy consumption may be obtained. Without knowledge of interrelated factors, incorrect inferences may be drawn about the accuracy of data collection. It should also be noted that factors such as choice of cropping system, method of postharvest processing, method of wage payment, and temporary migration may vary from year to year due to individual households' decisions in response to changing circumstances.

The ultimate size of sample used in different studies has been decided mainly by resource and time constraints. Careful attention is needed to identify appropriate sample designs for different types of surveys, to match their specific objectives and determining factors.

Output Design

Considerable savings in data gathering and analysis can be achieved if the output format is decided according to the objectives of survey prior to questionnaire design. Effective output and questionnaire designs depend on the experience of researchers. A critical analysis of energy surveys indicates that either a small fraction of data gathered has been actually presented for discussion or that the tabulated results could not be interpreted due to lack of specific data. This of course indicates that, so far, a standard survey methodology has not been developed.

Questionnaire Design

Some common weaknesses and unresolved issues of energy survey questionnaires used in Bangladesh are

1. Fuel amounts consumed by the household were recorded without reference to specific end uses.
2. Different terminology or classification was used in identifying traditional fuels.
3. Except in one microstudy, the amount of fuel consumed was estimated by memory recall method. Time periods used in different surveys were daily, weekly, six monthly, and yearly. Direct measurement of fuel is reliable, but to what extent it is feasible requires proper assessment.
4. In one macrosurvey, the same questionnaire was used for urban and rural areas, resulting in confusion.
5. Although introduction of efficient stoves to save cooking fuel is considered as a policy option, only two studies have included information about existing stoves.

A review of questionnaires reveals two important points: a single questionnaire is not adequate for data gathering in rural energy surveys, and multidisciplinary expertise is needed in designing questionnaires.

Data Collection

In most studies the field survey was done during the dry season, mainly because the sites were more accessible and funds were more easily available. As discussed earlier, biomass fuels are more abundant in the dry season. Estimates of fuel use by memory recall might reflect the fuel supply situation at that time.

Survey people stayed in the field from two weeks to a few months. The reports do not identify the actual time spent for data gathering. For specific purposes such as actual measurement or estimating seasonal variations, data were gathered more than once. When the nature of data requires multiple contact with the respondents and access to private premises, such as with measurement of stove dimensions (Islam 1980), recruiting field investigators from the survey area is convenient. Presence of an experienced researcher to act as participant observer as well as supervisor can improve the study's overall quality. If qualified people are unavailable, less experienced people may be recruited and trained for supervisory roles in large-scale surveys, with an experienced researcher providing overall observation of the areas surveyed. When the National Statistics Office assumes charge of a field survey, an experienced energy researcher should be involved at various stages of the survey.

Data Processing, Interpretation, and Analysis

Gathering and analysis of rural energy data are not an end in themselves. Data should be presented in such a way that they can be easily used as instruments for decision making. Some lessons from the weaknesses of data analysis of previous studies are as follows. Islam (1980) did not consider socioeconomic stratification in presenting survey data. As a result, the magnitude of problem of different socioeconomic groups could not be determined. For policy planning purposes data presented for specific administrative units (e.g., district, thana) are convenient. If these units are made up of more than one agroclimatic zone, however, methods must be found to obtain and present data separately for the respective zones.

Report Preparation and Distribution

Because of variation in objectives, scope, and availability of resources and expertise, distinct differences emerge between the reports of policy and research studies. Policy studies have published multiple volume reports, with main policy issues in one volume and specific subjects in separate volumes. In the Forestry Study, each volume was prepared by an experienced subject matter specialist.

Research-oriented studies have presented their findings in a single volume. Some of those reports have details on survey methodology, unconverted survey data, conversion factors, survey results, and findings. Others include mainly the survey results and findings. From a research perspective, the former scope of report is more useful than the latter. Those reports have valuable experience for future policy studies but could not make much impact on existing energy policy. Their limited contribution can be attributed to the following factors: limited geographic scope, lack of researcher experience in linking survey findings with national policy, and lack of understanding of technical aspects of energy data. These studies have been the first experience of energy survey for their respective research leaders.

For effective use of reports, careful consideration is needed in choice of the title and format. A study covering different subjects may need separate reports. Policy issues might be discussed in one volume and specific subjects presented separately. On this point, the author's own experience with the distribution of a research report (Islam 1980) is an example. The study was carried out with a grant from the International Development Research Centre (IDRC), Canada, under the project on Socio-Economic Evaluation of Biogas Technology. To comply with the formalities of the research grant, the report was entitled "Study of the Problems and Prospects of Biogas Technology as a Mechanism for Rural Development: Study in a Pilot Area of Bangladesh." To present the actual nature of the study, a second title was also included, "Village Resources Survey for Assessment of Alternative Energy Technology." With the first title, the report mostly drew the attention of biogas researchers. Researchers working on rural energy surveys realized the report's relevance only through the second title. Although a substantial part of the report was devoted to stoves, it had little distribution among stove researchers. At least two separate volumes would have been a much more effective format.

Justification of Future Rural Energy Studies

The availability of eight rural energy surveys in Bangladesh may tend to impress the uncritical analyst that enough information is available to formulate rural energy policy. The various limitations of these studies have been shown, however. This background reveals a definite need for a rural energy study to decide future policy options. Careful consideration should be given to deciding the objectives (short term, medium term, and long term), as well as scope and methodologies for the study.

Future Studies

The Asian Development Bank has agreed to finance an Energy Planning Project to be implemented by UNDP. This involves updating energy data and preparing an Energy Master Plan for Bangladesh to the year 2000. Special attention will be given to energy pricing policy and rural energy development. A Rural Energy Survey sponsored by USAID is under government consideration. The information gathered would provide data for the Energy Planning Project.

CRITICAL ISSUES AND DIRECTIONS FOR FUTURE RESEARCH

For rural development, energy need must be assessed to provide food, fuel, fertilizer, off-farm employment, and transport for rural people and fodder for their domestic animals. Most rural energy studies have concentrated on assessing energy need for domestic cooking and lighting. An important area for future research is to identify energy-related projects to facilitate optimum use of resources to meet the basic requirements of rural development.

Several Bangladesh research institutions are engaged in developing various alternative energy technologies expressly for rural areas. Their research has concentrated mainly on developing prototypes and on performing scientific and engineering tests under laboratory conditions. Limited field testing and demonstration of alternative technologies have been attempted. Lack of understanding of users' needs and absence of appropriate institutions have severely limited success. Based on assessment of various energy studies in Bangladesh, research on new energy technology development should be conducted in close collaboration with need and resource assessment studies.

Need and Resource Assessment for Technology Development

As land is the basic means of rural production, research should first aim to optimize use of land resources. Optimum use of both farmland and homestead land should be considered in an integrated farming system. During energy studies it has been observed that homestead land is used for different productive purposes such as multipurpose trees, seasonal vegetable crops, domestic animals and poultry birds, fish, and processing agriculture crops. Similar observations are reported by Richard Harwood (1979). So far no research for the optimum use of homestead land has been undertaken. Isolated attempts have been made to optimize some subsystems but with only limited success. Various technical innovations such as improved stoves, fast-growing trees, and biogas technology need to be assessed along with existing homestead land use and other subsystems.

In Bangladesh agriculture research has been initiated by monocrop institutions. In recent years, to improve the diffusion of research results to farmers, attempts have been made to assess costs and returns of different crops under selected cropping systems (BRRI 1981, Hoque, Hobbs 1981; Hoque and others 1982). From the energy point of view, analyses of single crops are of limited use. The studies have ignored the contribution of agriculture residues and have mainly compiled output data in terms of money rather than consumption units.

Farmers' decisions on the use of crop residues as fodder, building materials, and fuel should be reflected in energy-oriented analysis of local farming systems. With less competitive price of jute fibers, farmers have been found to cultivate jute to obtain the benefit of jute sticks as fuel. In the early 1970s, intensive cultivation of tobacco in some northern districts reportedly resulted in rapid depletion of tree resources for tobacco curing. Later, tobacco cultivation had to be reduced due to fuelwood scarcity. Farmers have responded to the scarcity of fuelwood by cultivating cotton to use cotton stalks as cooking fuel. Excessive cutting of trees has caused droughts in this area that in turn affected cotton cultivation (*The Sangbad* 1982).

Development of optimum farming and energy systems should start on the farmers' field and homestead land, not in the artificial environment of research institutions (Harwood 1979; Willis Shaner, P.F. Philipp, and W. Schmehl 1982). Once optimum farming systems are identified, arrangements must be made for necessary external inputs in the form of credit, extension services, fertilizer, irrigation, energy for irrigation, pesticides, and marketing opportunities.

The farming system alone would not suffice to provide full employment for the rural people. A broad-based assessment of off-farm employment opportunities would require analysis of the potential of raw materials, energy, local skill, storage, transport, and marketing. Selection of a technology package should be oriented toward maximizing productive employment, recognizing capital limitations and contributing to capital formation. Identification of appropriate research institutions, mobilization of multidisciplinary research teams, and necessary logistic support are the critical issues for the successful implementation of such a program.

Combining of needs assessment with development of a specific technology is illustrated in Chapter 8, where user needs and resource constraints affecting stove improvement are documented from the Nabagram study reported in this chapter. In the next section, need and resource assessment for technology development are illustrated by application of microstudy findings to biogas development.

The two major physical constraints for biogas diffusion are unavailability of sufficient dung and high capital cost. The surveys by Islam (1980) and Quader, Omar (1982) observed that more than 50 percent of the households did not possess any cattle. Less than 15 percent of the households possessed the four heads of cattle necessary to sustain a family-size biogas plant fueled by cattle dung. Only rich households can individually own a family-size biogas plant. The average cost of a family-size biogas plant (floating-gas-holder type) is from US $600 to $750.

Dry dung available from three adult cattle is 6 kg/day — sufficient to meet the cooking fuel need of a family of six persons. If, then, the family could choose whether to invest US $600 to make a biogas plant or to buy three cattle, they would definitely choose the cattle because they would have both an opportunity for income as well as a way to meet the cooking fuel need.

Cooking requires 90 percent of all fuel consumed in rural households in Bangladesh and cooking fuel is scarce. Therefore, promotion of biogas technology is considered a policy option for solving the rural cooking fuel problem — one that would also reduce the deforestation rate. However, ownership of land producing biomass resources is concentrated. Wealthier households able to purchase a biogas plant are not affected by the fuelwood crisis and do not have a "felt need" for use of biogas for cooking. They easily meet their cooking fuel needs without any capital investment. On the other hand, poorer households suffer from scarce supplies of cooking fuels. They cannot afford a biogas plant; they must continue using biomass for cooking.

Savings of fuelwood possible if richer households used biogas would not necessarily be available to the poorer households. Therefore, biogas plants installed to fit the resources and motivations of individual households in meeting their cooking fuel needs would not solve the cooking fuel crisis in rural areas.

If a biogas plant is designed to meet energy needs for irrigation or any other productive purpose, users would appreciate the savings in petroleum fuels. However, in rural locations where electricity is available from a grid for irrigation and industries, users might not have the same felt need for a biogas plant.

Installation of community biogas plants is considered a strategy for reducing capital costs for biogas generation (see Chapter 10). Few examples exist of successfully operated, community-owned rural technology in

Bangladesh. The successful operation of a community biogas plant depends on the strength of social institutions (see Chapter 7). Design of a cooking system that meets demands of different users is a complex task.

Since dung is available close by, the prospect of community biogas plants may be assessed on a homestead basis. But the number of homesteads with an adequate cattle/person ratio to produce sufficient dung to generate biogas for cooking would also be very limited. Homesteads with sufficient cattle could install biogas plants to meet just the energy need for lighting. The biogas requirement for lighting (not including portable lights) in an average homestead with three households is

$$\frac{3 \text{ household}}{\text{homestead}} \times \frac{3 \text{ lamp}}{\text{household}} \times \frac{3 \text{ hour}}{\text{day}} \times \frac{0.07 \text{m}^3}{\text{lamp hour}} = 1.9 \text{ m}^3, \text{ say } 2 \text{ m}^3.$$

The 2 m³ per day necessary for lighting in such a homestead is approximately the amount of gas required for cooking by a single family. Alternatively, small biogas plants integrated with sanitary latrines could be designed to meet the energy need for lighting only. A biogas plant installed for lighting would save approximately 6 gal of kerosene per household per year, resulting in reduction of kerosene imports. This option would become more viable if the price of petroleum again rises.

In Bangladesh, one of the important uses of energy for productive purposes is for pumping irrigation water. The energy need for irrigation is seasonal and is limited to dry months (including the cold months). Generation of biogas is considerably reduced at low temperature. In specific locations, a biogas plant designed to meet the energy need for irrigation should be based on a gas production rate corresponding to the temperature of winter months. For the rest of the year, due to high ambient temperatures, gas production rate would be higher. An appropriate use of generated gas during this period must be found for the optimum use of the plant.

In a local area, a biogas plant can also be designed to meet the energy needs of different end uses as a part of an integrated system. Successful operation of such plant would depend on the strength of social organization, not success of technical controls. The importance of social factors in organizing use of alternative organic materials as feedstock and in using digested dung as high-quality manure is further stressed in Chapter 10.

In Bangladesh, research on biogas technology was initiated in 1972. For work until October 1980, see Islam (1980a). In 1980, the Environmental Pollution Control Directorate (EPCD), a department under the Ministry of Science and Technology, started a project to promote biogas. In the first phase, they plan to install 250 demonstration plants at different locations around the country. In 1980–81, EPCD installed 110 fixed-dome type, family-size biogas plants at no cost to the users. The total cost of each plant was Tk.6,500 (US $325) and was met from government fund. The beneficiaries were selected by district administrative authorities and EPCD was responsible for installation of the plants. Preliminary assessment indicated unsatisfactory performance of most fixed-dome plants.

At the beginning of 1982, EPCD initiated a research project to reduce the cost of biogas plant. They claimed to have developed a floating gas holder plant costing only Tk.3,500 (US $175). In this plant, the digester is made of 5-inch thick brick wall. Ordinary clay is used to join brick, and the inside of the digester is plastered with sand and cement. A gas holder is made of 24-gauge galvanized iron sheet. Construction of these plants is planned for different locations of the country to evaluate field performance.

Based on a preliminary survey, IRDP estimated that in 350 IRDP thanas 7,000 households would be willing to install a biogas plant at a cost of Tk.5,000 (US $250). If the cost were more, the number of prospective users would be less. Reportedly biogas plants would be in considerable demand if they are designed for productive use, such as running shallow tubewells.

Enhancing Research and Evaluation Capacities

Energy needs vary according to particular areas, seasons, and socioeconomic conditions. Observations from microsurveys should be assessed considering energy demand and supply in intermediate zones to conceptualize regional and aggregate national opportunities and policies for rural energy development. The objectives should be not only the development of a particular rural area or zone but also the removal of differences between zones and socioeconomic groups. This strategy for rural energy planning fits within the framework of regional and spatial planning outlined in the Second Five Year Plan.

In the past, technologies selected for rural development assumed a guaranteed supply of energy. Energy need should be a criterion of the technology packages selected for rural development. Technologies selected also should allow for using local resources. Assessment is necessary of the prospect of using biomass to meet interlinked rural energy needs for subsistence as well as development.

This chapter shows that energy studies in Bangladesh have depended on ad hoc institutional arrangements with consultants and university researchers. Considering the importance of energy, collection of energy statistics should be considered part of a national statistical system. No institution exists to research the issue of traditional and conventional energy sources. A definite policy guideline for research and development on rural energy is needed, balanced between laboratory and field research, and between hardware development and policy research to assess applicability.

Education and training on energy issues should be considered an integral part of rural energy delivery systems. Training programs should be designed not only for workers in the energy sector but also for other departments such as Agriculture, Livestock, Forestry, and the IRDP. Energy should be an integral part of curricula at various levels of education. People also should be informed about future changes in energy sources and policies.

The methodology adopted in the Nabagram study for assessing alternative energy technology is satisfactory. A nationwide survey may select sample villages throughout the country based on agroclimatic zones. Within a selected village, survey of total households is appropriate instead of further sampling in order to gain deeper understanding of varying energy use patterns and relationships. Ownership of land is normally used to stratify rural households; however, for the same landholding group, per capita fuel consumption may vary under different household and homestead circumstances.

Another issue that needs serious consideration is identification of an appropriate methodology for evaluation of rural development projects. The success of a rural delivery system is normally evaluated in terms of the number of units installed in the rural area rather than the actual number used and level of operation. Given the inadequacies in the evaluation system, persons engaged in rural development activities have failed to benefit from an appropriate feedback mechanism. As a result, irrelevant programs have been perpetuated by the development authorities without constructive criticism. Appropriate indicators of success in energy-related projects include (1) biogas — the number of units in operation and their impact on solving rural energy problems, (2) forestry — the number of live seedlings after one year, (3) rural electrification — the number of pumps and other rural productive units actually using electricity and the economic limitations of most people not using electricity, and (4) improved stoves — the number of improved stoves actually in use and the project's contribution toward solving the fuelwood problem.

Common observable characteristics of energy surveys in developing countries include supervision by university academics, surveys scheduled during yearly university holidays, selection of survey areas close to the university, accessibility of area by modern transport, and use of final year or graduate students as survey enumerators or field investigators. One reason for this pattern of university surveys is the expected relative savings on survey costs. Ultimately, a survey identifies projects or contributes to decisions for activities much more costly than the survey itself. For the success of the identified projects, the quality of survey should not be sacrificed in order to save survey cost. Although university academics have professional research experience, if the research demands continuous attention their choice as supervisors might not be appropriate. Moreover, if competence of academic researchers is assessed by the total number of research publications, and the researcher maximizes available research time for publications, time available for field visits might not be adequate or appropriate. Two to three weeks is not adequate time for urban-biased university students to familiarize themselves with a particular rural situation. Training of field investigators and enumerators is very important to maintaining the quality of data.

Energy plays a vital role in rural development. To make the development effort meaningful, careful understanding of specific rural energy problems is essential. Steps required for such understanding are involvement of relevant personnel and agencies, choice and combination among action research and survey research methods, data analysis, problem and project identification, design of technology development and diffusion system, implementation mechanism, and evaluation and feedback procedures. No uniform solution is likely to meet the need. Observations of various energy studies carried out in Bangladesh indicate that lack of understanding may result in detours away from the main route of rural development.

ACKNOWLEDGEMENT

Background material for this paper was prepared while the author was at the University of Sussex (Institute of Development Studies and Science Policy Research Unit) with an IDRC-sponsored fellowship.

NOTES

1. TK.20.0 = US $1.0.
2. The lower heating values of different types of traditional fuels reported by the Bangladesh Energy Study (GOB 1976) were used for conversion in Tables 2 to 5. They are as follows:
    Fuelwood with average moisture = 15.12 MJ/kg
    Bagasse (50% moisture) = 7.44 MJ/kg
    Cow dung (50% moisture) = 8.61 MJ/kg
    Vegetable wastes, rice hulls, rice straw, twigs and leaves = 12.6 MJ/kg.
Agriculture residues include rice hulls, rice straw, bagasse, jute sticks, and other crop stalks. The conversion factor used for kerosene was 46,055 kJ/kg.

REFERENCES

Bangladesh Rice Research Institute (BRRI).
  1981 *Proceedings of the Workshop on Rice-Based Cropping Systems Research and Development.* Dhaka.
Briscoe, J.
  1979 Energy Use and Social Structure in a Bangladesh Village. *Population and Development Review* (December): 615–641.
Douglas, J.
  1983 Surveying Wood Fuels in Bangladesh. In Annex II, *Wood Fuel Surveys.* Food and Agriculture Organization, Rome.
Food and Agriculture Organization (FAO).
  1980 *Household Survey of Rural Consumption of Wood and Bamboo in Bangladesh: A Review of Methodology.* Field Document 1. UNDP/FAO/Planning Commission (Government of Bangladesh) Project BGD/78/010. (D. Asrat, Consultant)
FAO.
  1981a *Supply and Demand of Forest Products and Future Development Strategies: Consumption and Supply of Wood and Bamboo in Bangladesh.* Field Document 2. UNDP/FAO/Planning Commission Project. (J.J. Douglas, Project Coordinator)
FAO.
  1981b *The Industrial Forestry Sector of Bangladesh.* Field Document 3. UNDP/FAO/Planning Commission Project. (J.J. Douglas, Khalilur Rahman, and Akramul Aziz)

FAO.
　1981c　*Future Consumption of Wood and Bamboo in Bangladesh*. Field Document 4. UNDP/FAO/Planning Commission Project. (N. Byron, Consultant)

Government of Bangladesh (GOB).
　1976　*Bangladesh Energy Study*. Montreal Engineering Ltd., and others. Administered by the Asian Development Bank, Project of UNDP. (BGD/73/038/01/45)

GOB. Rural Electrification Board.
　1977　Project Proforma, Area Coverage Rural Electrification Phase I.

GOB.
　1979　Project Proforma, A Study on Supply Demand of Forest Products and Their Future Development Strategies. UNDP/FAO/Planning Commission Project. (BGD/78/010)

GOB. Ministry of Planning.
　1979　*Statistical Year Book of Bangladesh, 1979*. Dhaka: BBS Ministry of Planning.

GOB. Ministry of Planning.
　1980　*Statistical Year Book of Bangladesh 1980*. Dhaka: BBS Ministry of Planning.

GOB. Planning Commission.
　1980　*The Second Five Year Plan 1980*. Dhaka, May draft.

Harwood, R.R.
　1979　*Small Farm Development: Understanding and Improving Farming Systems in the Humid Tropics*. Boulder, Colorado: Westview Press.

Hoque, M.Z., and P.R. Hobbs.
　1981　Rainfed Cropping Systems. Report of Research Findings at Bhogra Village 1975–79, Bangladesh Rice Research Institute, Bangladesh.

Hoque, M.Z., N.U. Ahmed, Nur-E-Elahi, and M.R. Siddiqui.
　1982　Recent Findings on Cropping Systems in Deepwater Rice Areas. *Proceedings of the 1981 International Deepwater Rice Workshop*. Manila, Philippines: International Rice Research Institute.

Hughart, D.
　1979　Prospect for Transitional and Non Conventional Energy Sources in Developing Countries. World Bank Staff Working Paper no. 346. Washington, D.C.

International Rice Research Institute (IRRI).
　1981　Recent Findings on Cropping Systems in Deepwater Rice

Areas. *Proceedings of the 1981 International Deepwater Rice Workshop.* Manila, Philippines: IRRI.

Islam, M.N.
1980 *Village Resources Survey for the Assessment of Alternative Energy Technology.* Prepared for the International Development Research Centre, Canada. Dhaka: Bangladesh University of Engineering and Technology, Department of Chemical Engineering.

Islam, M.N.
1980a Design Construction and Operation of a Biogas Plant Suitable for the Rural Area of Bangladesh. Dhaka: Bangladesh University of Engineering and Technology, Department of Chemical Engineering.

Manibog, F.R.
1982 *Bangladesh: Rural and Renewable Energy Issues and Prospects.* Energy Department Paper no. 5. Washington, D.C.: World Bank.

Quader, A.K.M.A., and K.I. Omar.
1982 *Resources and Energy Potentials in Rural Bangladesh: A Case Study of Four Villages.* Prepared for the Commonwealth Science Council, London.

Rahman, A.
1982 *Energy for Rural Bangladesh.* Dhaka: Dhaka University, Centre for Policy Research.

The Bangladesh Times.
1982 Greater Use of Cowdung in Agriculture Land Essential. March 23.

The Sangbad.
1982 Uncontrolled Use of Chemical Fertilizer and Pesticides Are Causing Environmental Pollution. November 12.

Shaner, W.W., P.F. Philipp, and W.R. Schmehl.
1982 *Farming Systems Research and Development: Guidelines for Developing Countries.* A Consortium for International Development Study. Boulder, Colorado: Westview Press.

Tyers, R.
1978 Optimal Resources Allocation in Transitional Agriculture Case Studies in Bangladesh. Ph.D. diss., Harvard University, Division of Applied Science.

Vylder, S.D., and Asplund.
Contradictions and Distortions in a Rural Economy: The Case of Bangladesh. Stockholm, Sweden: SIDA, Policy Evaluation Division.

3
Policy Analysis of Rural Household Energy Needs in West Java

M. Hadi Soesastro

BACKGROUND

The Republic of Indonesia is an archipelago of about 13,000 islands, with a total land area of about 2 million sq km. Administratively, the country is divided into 27 provinces, 295 districts (*kabupaten*) and municipalities, and then into 3,329 counties (*kecamatan*), administering 62,865 villages (*desa*). The 1980 population was about 147 million people, of whom 78 percent live in relatively underdeveloped rural areas. The second Five Year Development Plan (1974/75–1978/79) concentrated on rural development to improve the socioeconomic conditions of that majority population.

Rural Development Policy

A desa, or Indonesian village, is considered an entity of territory and population and an integral part of the country's administrative network. Rural development strives to strengthen those elements in all their aspects of life, political, economic, and sociocultural.

The concept of development in villages is based on strengthening the initiative and self-help of communities, so they can utilize their resources better. Establishment of community institutions has been encouraged to increase the self-governing ability of the people, their awareness of common needs, and their involvements in community-initiated activities. In an Indonesian village, a Village Social Committee (LSD) is responsible for planning and implementing village development activities. The Family Welfare Unit (PKK), in cooperation with the Village Social Committee, provides community-based services.

Villages do depend on each other and to a great extent they rely on the facilities in their surrounding areas, including urban centers, so their development efforts require coordination at various administrative levels. The Local Development Work Unit (UDKP) was established at the kecamatan level to integrate the efforts of the rural community and the government in implementing rural development programs. A county-town (*kota kecamatan*) usually is the nearest growth center for villages in the area.

The main tasks of the Local Development Work Unit are to identify (a) its role and function within a broader area, (b) potentially usable resources in the area, (c) requisite local skills, and (d) training programs for the communities. It also formulates, promotes, and adopts the appropriate technology for rural development. Programs and projects suggested by the Local Development Work Unit are discussed at county meetings chaired by the head of the county (*Camat*) and attended by village chiefs (*Lurah*), representatives of various government agencies, Village Social Committee members, field workers, and influential informal leaders within the respective county. Programs adopted this way are submitted to district and province levels for approval.

Each year since 1972, villages in Indonesia have been classified according to three typologies to help development planners identify them by socioeconomic conditions and potentials as well as by their ability to share responsibility in their own development. The three typologies of villages are *swadaya* (traditional), *swakarya* (transitional), and *swasembada* (modern).

Ten major sets of indicators are used in classifying villages. The three basic indicators are (1) population density; (2) natural environment (form of land, rainfall, soil productivity); and (3) orbitation (distance to capital city of the province or county and condition of roads connecting village with urban centers). Developing indicators are (4) economic structure (percentages of population engaged in agriculture, industry, and services); (5) village product (monetized value); (6) traditional customs (strictly or loosely applied in the community); (7) village and community institutions; (8) level of education (percentages of population with six-year elementary schooling); (9) level of self-help; and (10) infrastructure (roads as well as production, marketing, and social infrastructures). Scores or ranks are calculated for those ten indicators, and the total score determines the typology of a village. Traditional villages are levels I to V with scores from 7 to 11, transitional villages are levels I to V with scores from 12 to 16, and modern villages are levels I to V with scores from 17 to 21.

The national goal of rural development is for all villages to qualify as modern villages by the year 2000. As of 1978, 32.7 percent of the villages in Indonesia were classified as traditional, 56.1 percent as transitional, and only 11.3 percent as modern. The challenge, therefore, is considerable.

Energy and Rural Development

The indicators of village development levels have not included items directly relating to energy. Indeed, in the past, energy was not a major concern in rural development planning. Only recently has a perceived dwindling of biomass resources fostered a growing awareness of rural people's energy problems. The introduction of commercial energy, primarily kerosene, into the rural areas has also brought attention to their problems.

The current debate on the energy issues in rural development reveals two main points of view. One is that the energy needs of the rural population should be satisfied by cheaply developed local resources. Advocates say kerosene should be used only minimally because its wide use increases the economic vulnerabilities of the rural areas. The relatively low, heavily subsidized, kerosene prices tend to stimulate use of kerosene, and abolishment of those subsidies could severely affect the rural population if most of them are using kerosene fuel. Defenders of kerosene use argue that since kerosene is inherently a "better" type of fuel, it should be made available to the rural population for equity reasons. As of 1982, kerosene was being subsidized exactly with this objective in mind.

Apart from the debate on the use of kerosene by the rural population, more fundamental energy in rural development issues have been raised recently. Energy is a basic need (Hadi Soesastro 1980). Because energy is also a means to improve quality of life and to increase productivity and employment, it is an important part of rural development. The objective of various current studies by various institutions in Indonesia is a better understanding of this problem. Meanwhile, basic information and data on energy consumption patterns in rural areas are not available. Information about who uses what types of energy, for what purposes, and by what amount is important for policy planning and analysis.

In addition to biomass and kerosene, electricity has gradually become another important component in the rural household energy budget. The 1980 national census in Indonesia reveals the relative importance of the different types of energy in rural areas (see Table 1). About 73.9 percent of rural households in Indonesia use biomass and about 24.6 percent use kerosene for cooking. In the province of West Java, the percentage of households using biomass for cooking is lower (65.9 percent) and the percentage using kerosene is higher (32.9 percent) than the national average. Kerosene is used by 85 percent of rural households for lighting; about 14 percent use electricity.

As of July 1980, about 7 percent of all villages in Indonesia had electricity (about 525,000 households). In 3,016 villages, electricity was provided by the state electricity company (PLN), and, in 1,439 villages, electricity was generated by local governments, Village Social Committees, or private individuals. According to the Third Five Year Development Plan (1979/80–1983/84), the state electricity company

TABLE 1.
Structure of Energy Consumption of Rural Households (1980)

Types of Energy	Percentage of Households	
	Cooking	Lighting
All Indonesia		
Biomass	73.9	–
Charcoal	0.2	–
Kerosene	24.6	85.0
Gas	0.5	–
Electricity	0.2	13.9
Others	0.4	0.8
Not Stated	0.2	0.3
West Java*		
Biomass	65.9	–
Charcoal	0.1	–
Kerosene	32.9	85.7
Gas	0.3	–
Electricity	0.2	12.7
Others	0.5	1.2
Not Stated	0.2	0.4

* Later sections of this chapter refer to West Java.

will add 3,700 modern villages in its rural electrification project with a total installed capacity of 130 MW, adding about 625,000 households as consumers. The long-term target of the rural electrification project is to electrify 30,000 villages — an estimated 1.2 million households by the year 2000.

Biomass. As shown earlier, biomass still is the predominant rural energy source. Since firewood as a source of energy is regarded as a basic need, the rights of village people to gather firewood for personal use from public land are granted by law, the so-called *sprokkelrecht*. Biomass is used not only for cooking by rural households but quite extensively by rural industries.

A major problem in examining the issues relating to biomass is the lack of accurate data on consumption and production. Comprehensive large-scale surveys have not been conducted yet. Data have been gathered in several uncoordinated undertakings at universities and institutions, but their estimates lack consistency (see Table 2). Several studies indicated a quite substantial shortfall of firewood, based on estimated consumption and production figures, thought to have resulted in increasing deforestation and to have caused vast areas of private farmland to become critical (Satyawati Hadi 1979). Firewood shortage is not only felt in densely populated areas, such as Java, but also in sparsely populated arid regions in eastern Indonesia. The Forest Service recognized this situation in early 1960s, and, with the First Five Year Development Plan (1969/70–1973/74), a national program of greening of the land was carried out. The program consisted of two efforts, namely, (a) the establishment of new forests on government or public forest land (the Reforestation Program), and (b) the encouragement of the planting of perennial species on private and community property (the Greening Movement). Subsequent development plans continued the program. The Third Plan targets are 300,000 ha reforestation and 700,000 ha greening per annum.

Kerosene. To take the place of dwindling firewood supplies, kerosene was distributed to more and more villages. Prices of kerosene were also kept low by subsidies. From 1970 through 1980, the national consumption of kerosene grew by about 11 percent per year.

Gradually kerosene supply became a major national problem, partly because the growing subsidies were affected by the dramatic increases in oil prices and by the high growth rate of consumption and partly because pressures were created on the supply side. An oil-producing country, Indonesia became an importer of kerosene. The rate of growth of kerosene consumption declined slightly to 9.3 percent per year in the period 1977–81 from a higher rate in the early 1970s. However, kerosene imports more than tripled from 1977 to 1981. Subsidies of kerosene increased substantially from Rp 14.60 (US $0.04) per liter in 1977 to Rp 88.20 (US $0.14) per liter in 1981 (Table 3).[1]

In 1981, then, the government subsidized more than twice the amount paid by the consumer, Rp 37.5 per liter (US $0.06). The total subsidies on petroleum products increased to about US $2 billion in FY 1981/1982 — equal to about 20 percent of the total government development expendi-

TABLE 2.
Estimates of Per Capita Biomass Consumption in Rural Indonesia

Region	Year	Consumption (m³/year)	Source/study
All Indonesia	1956	0.50	Team Survey LPHH 1970, in Wiersum 1979
	1970	0.72	FAO Estimate; T. Silitonga 1974, in Wiersum 1979
	1975	0.86	Chandrasekharan 1977, in Wiersum 1979
	1976	0.84	Raymond Atje 1979
Java	1976	0.79	Atje 1979
	1978	1.00	Satyawati Hadi and others 1979
East Java	1971	0.52	Sumarno and Sudiono 1973
	1978	1.27	Hadi and others 1979
Central Java			
Whole province	1978	0.64	Hadi and others 1979
Solo WS (watershed)	1969	0.74	LPHH 1969, in Hadi and others 1979
	1975	0.36 (0.20-0.43)	Wiersum 1976
	1976	1.13	Wangsadidjaja and others 1979, in Wiersum 1979
Wonogiri Area	1978	0.8-0.9	Directorate of Land Use, Dept. of the Interior 1969, in Wiersum 1979
	1977	0.54	Yudodibroto 1978, in Wiersum 1979
West Java			
Whole province	1977	2.08	Herman Haeruman and others 1977
	1978	0.43	Hadi and others 1979
Citanduy WS	1977	2.22	Nasendi 1978
Cimanuk WS	1978	0.49-1.22	Haeruman and others 1977
Citarum WS	1979	2.53	Rusydi and others 1979
Sukabumi Area	1977	0.40	Anonymous 1978, in Wiersum 1979
Bogor Area (Babakan)	1977	0.35	Komarudin 1977
Bandung Area (Salamungkul)	1979	0.96	O. S. Abdoellah 1979
Bali	1978	1.06	Hadi and others 1979

tures or to the amount of all foreign assistance to Indonesia in that fiscal year. In early 1982, petroleum products' prices increased by about 60 percent, but subsidies on kerosene still stood at about Rp 72.00 (US $0.11) per liter.

Raising petroleum prices traditionally has stirred great controversy in Indonesia. Some people feel that higher income households benefit more

TABLE 3.
Kerosene Consumption, Imports, and Subsidies (1977-1981)

Year	Consumption (million liter)	Imports (million barrel)	Kerosene subsidy* (Rp/lt)	Total subsidies on petroleum products as percent of Government development expenditures+
1977	5,840	5.00	14.60	3.0
1978	6,631	4.37	17.90	7.2
1979	7,228	5.52	33.00	13.2
1980	7,787	7.46	62.50	17.6
1981	8,347	16.56	88.20	20.1

* At year end.

+ Total subsidies of all subsidized petroleum fuels (kerosene, automotive diesel oil, industrial diesel oil, and fuel oil) in the respective fiscal year, starting April 1.

from kerosene subsidies and that the urban population benefits rather than the rural population. No accurate information about usage is available. A study by Alan M. Strout in 1978 (based on figures collected in 1976 by the National Socioeconomic Survey, or SUSENAS) estimated that 20 percent of all kerosene consumed nationally is consumed by urban households for cooking, 16 percent by urban households for lighting, 20 percent by rural households for cooking, and 43 percent by rural households for lighting. According to the SUSENAS Survey, per capita consumption of kerosene by urban households was about three times the consumption of kerosene by rural households.

Other people argue that an increase of kerosene prices would cause greater deforestation, and thus severe environmental problems. According to one article, however, demand for firewood does not seem to be influenced by the price of kerosene. Raising the price of kerosene means a short-term increase in the price of firewood that encourages illegal foraging. However, raising the kerosene prices also provides an incentive over time to increase commercial firewood production by spontaneous reforestation (Howard Dick 1980).

HOUSEHOLD ENERGY SURVEY OBJECTIVE AND METHODS

A rural energy survey was undertaken by the Centre for Strategic and International Studies (CSIS) as part of a larger study project on energy and income distribution in Indonesia. The entire study was exclusively financed by CSIS and was administered and carried out by CSIS, assisted by local consultants (forestry experts with substantial experience in designing and conducting fuelwood consumption surveys). The objective of the larger CSIS study was to understand the role of energy in the national efforts to promote a better distribution of income in the development process. Hence, the focus of the study was to examine the energy problems of the poor.

Scope of Survey

The first phase of the study concentrated on rural areas, particularly on specific policy issues such as the effects of reducing kerosene subsidies upon the quality of life of the rural population. The survey consisted of two parts: a rural household energy survey, and a rural (cottage) industry energy consumption survey. (The industry survey was a pilot survey of 70 establishments covering 20 types of industrial activities in four districts.) The discussion here is confined to the household survey, conducted in February and March 1980.

The rural household energy survey was conducted to provide baseline information for analyzing the long-term trends of energy demand and supply in rural areas as influenced by income changes, fuel substitution, or changes in technology. The survey was designed to collect these kinds of information:

1. Changes in the level and structure (composition by fuel types and end uses) of energy consumption with improvement in rural people's standard of living.
2. The structure of biomass energy consumption, or the uses of firewood and agricultural residue.
3. The effects of commercialization of energy, especially of firewood, upon consumption.
4. Efficiency in the uses of energy in rural areas.

Sample Design

The main considerations guiding design of the survey involved collection of information from a relatively large sample of households having specific characteristics reflecting different locations in West Java. The main characteristics were the general ecological and geographical environment of the district and the level of socioeconomic development of the village. West Java was selected because a large-scale survey in this

TABLE 4.
Characteristics of Selected Districts

District	Geography	Ecology	Population density
Bandung	inland good roads 266 villages	high plain plantation cultivated land forest	716
Ciamis	inland/coastal poor roads 197 villages	low plain cultivated land brushwood forest firewood exporter	411
Cirebon	coastal good roads 294 villages	low plain cultivated land critical land firewood importer	1,135
Garut	inland/coastal poor roads 176 villages	high plain forest plantation brushwood firewood exporter	400
Serang	coastal good roads 282 villages	low plain cultivated land critical land	519

Note: Population density is given in number of persons per square kilometer as of 1977/78 information.

province could be administered from Jakarta more easily than one in other provinces and at a more reasonable price and because, according to forestry experts, forestry resources in many areas of West Java were rapidly deteriorating.

This survey's sampling procedure was purposive rather than random. After determining that about 1 percent of the villages in West Java would be treated, districts would be selected first and then counties. Of the 4,039 villages recorded in the 1977-1978 statistics, 40 villages were selected. Five of the 24 districts were selected according to observed ecological and geographic differences (see Table 4 and Fig. 1). And in each district, eight villages were selected by consulting the Village Statistics of the Department of the Interior. First selected was one county with most of the socioeconomic levels to be represented in the survey. Then villages with other socioeconomic levels were selected from adjacent counties.

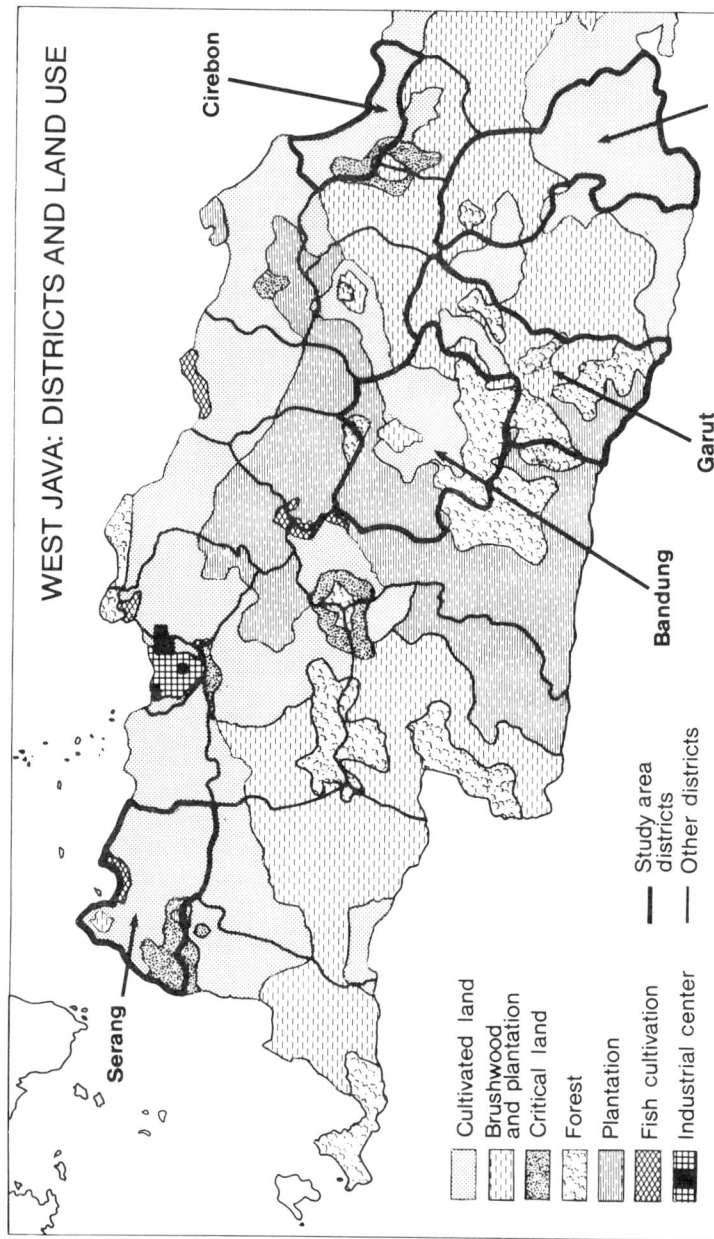

Figure 1. West Java: Districts and Land Use

Of the 4,039 villages in West Java, about 13 percent were categorized as traditional villages, 68 percent as transitional, and 19 percent as modern villages. The eight villages selected in each of the five districts (Table 5) included one traditional village, five transitional villages, and two modern villages. Thus, the sizes of those groups were approximately proportional to the sizes of categories of villages in West Java as a whole.

The number of households was not predetermined; a sample size of 1 percent of total households was selected in each village. By the end of 1979, a pilot survey was made to examine how accessible and feasible the selected villages were for the survey. Information also was collected on economic activities and demographic characteristics of the villages. The number of households in each village was registered. The 1-percent intensity of the sample resulted in a total of 533 households, with a minimum of four and a maximum of 69 households in one village.

Since income was considered an important variable for the study, the sampling included households from different income groups in each village. The village heads were consulted, and in some cases, the list of taxes paid by individual households was used. If tax records were used, households were ranked according to taxes paid, and that list was divided into at least four groups of equal tax brackets. The sample households were selected from each group randomly and, as far as possible, proportionally. It should be noted that the heads of villages (or their secretaries) often intervened in the selection of households. Therefore, a source of possible bias was present in household selection.

Data Collection Procedure

The survey was an interview, combined with field measurements of fuel consumption for cooking and lighting in the households surveyed. Twenty-five surveyors were allocated to the five districts according to the number of households covered, namely, nine for the Bandung district, five each for the Ciamis, Garut, and Cirebon districts, and four for the Serang district. Five supervisors, one for each district, oversaw surveys that ranged from 12 days (Serang) to 17 days (Bandung). The surveyors were students experienced in fieldwork, who participated in a one-day briefing about administering the questionnaire and making field measurements. They had done field surveys under the same supervisors. In turn, the supervisors had helped the design of the survey instrument.

The collected data can be categorized into five major sets of items: household characteristics, fuel use habits, fuel consumption, fuel supply and choice, and biomass types and biomass potentials (see Table 6).

TABLE 5.
Sample Description

District	County	Village	Village Typology		Sample of Households
Bandung	Pangalengan	Margaluyu	Traditional	V	17
		Lamajang	Transitional	I	14
		Sukamanah	Transitional	II	58
		Pangalengan	Modern	II	69
	Rancaekek	Bojongsalam	Transitional	IV	12
		Cangkuang	Transitional	V	16
		Haurpugur	Modern	I	8
	Pameungpeuk	Sangkanhurip	Transitional	III	12
Ciamis	Pangandaran	Babakan	Transitional	V	11
		Sukaresik	Modern	I	10
		Pananjung	Modern	II	19
	Cijulang	Kertayasa	Transitional	I	13
		Legokjawa	Traditional	V	13
		Kertaharja	Transitional	II	13
		Sidangsari	Transitional	III	10
	Ranca	Kaso	Transitional	IV	12
Cirebon	Astanajapura	Waruduwur	Traditional	V	4
		Japura Lor	Transitional	I	9
		Kendal	Transitional	II	5
		Kanci	Transitional	V	14
		Martapada Kulon	Modern	I	10
		Martapada Wetan	Modern	II	9
	Losari	Pasuruan	Transitional	III	9
		Pebedilan Kulon	Transitional	IV	14
Garut	Malangbong	Malangbong	Modern	II	13
		Cihaur Kuning	Transitional	V	16
	Cikajang	Cikajang	Modern	I	17
		Dangiang	Transitional	IV	10
		Tanjuang Raya	Transitional	I	8
		Bojong	Traditional	V	6
		Mekar Jaya	Transitional	III	12
		Cipangramatan	Transitional	II	10
Serang	Kopo	Pasir Buyet	Traditional	V	6
		Cidahu	Transitional	I	5
	Petir	Padasuka	Transitional	II	5
		Seuat	Transitional	III	8
		Panunggalan	Transitional	IV	6
		Petir	Modern	I	8
	Keragilan	Sentul	Transitional	V	10
		Keragilan	Modern	II	12

TABLE 6.
Principal Data Collected

Household Characteristics	Fuel Use Habits	Fuel Consumption (measured)	Specific Fuel Data	Fuelwood Data
1. Household Size and Composition	1. Cooking: daily menu cooking stoves cooking appliances preparation and handling of fuels	1. Cooking: types and amount (days 1-3)	1. Source of Supply: by types prices methods of collection (distance, labor used, frequency, parts of tree)	1. Collection of 2,000 samples for lab tests: species and date collected moisture content and caloric value
2. Education levels	2. Lighting: types of lamp area lighted how long for what purpose	2. Lighting: types and uses (days 1-3)	2. Supply problems: kind of problem household-level solutions frequency	2. Estimate of biomass potential of home gardens by: counting fuel-producing plants identifying species measuring diameter
3. Household income: main income seasonal income transfer income	3. Other: ironing radio, tv, other	3. Other: charcoal batteries electricity	3. Fuel choice: reasons for using or not using reasons for changes rural electricity	
4. Land ownership, area farmed, cattle ownership				
5. Crops produced				
6. Household expenditures (days 1-3)				

PRESENTATION OF RESULTS

Results reported in terms of the *average* consumption by end use and by type of energy, as in most studies on rural household energy consumption, provide a basis for identifying rural household energy needs and thus for formulating broad energy supply policies. Information for policy analysis on price, equity, and development effects, however, must show whether significant differences or variations exist in household or individual energy consumption, and the factors influencing such variations.

Structure of Rural Household Energy Consumption

The overall rural household energy budget established in the present study is much lower (an average 6,500 kcal per capita per day) than the earlier study by Herman Haeruman and others (an average of 12,000 kcal per capita per day) (see Table 7). According to the CSIS study, about 80 percent of all energy input is for cooking and only 17.5 percent for lighting. The statistics from the Haeruman study are 88 percent for cooking and 11 percent for lighting.

The average energy budgets differ primarily because the structures of energy input differ. Commercial energy constituted 40 percent of total energy input in the CSIS study but in the Haeruman study was about 15 percent. Thus the level of energy input increased as the proportion of noncommercial energy (biomass) in the total energy budget increased.

The difference between the two studies is partly due to different conversion factors. The CSIS study used 3,200 kcal/kg for biomass, 8,900 kcal/l for kerosene, and 7,000 kcal/kg for charcoal. On the other hand, the Haeruman study used 3,500 kcal/kg for biomass, 9,000 kcal/l for kerosene, and 6,700 kcal/kg for charcoal.

Studies usually measure the amount of energy input, rather than the amount of useful energy, because of the practical difficulty of measuring energy output. A clear picture of the structure of energy consumption by different household groups is therefore important. Some households only use biomass, some only use kerosene, and others use both for cooking. Since kerosene is a more efficient type of energy, households using only kerosene would need less energy inputs than those using only biomass for cooking.

Factors Affecting Household Energy Sources

Since energy for cooking predominates in the household energy budget, our study first disaggregates sample households into three

TABLE 7.
Rural Household Energy Budget
(per capita per day)

	CSIS (1980)					Haeruman and others* (1977)				
Number of villages	40					10				
Number of households	533					200				
Consumption	kg	l	kWh	kcal	(%)	kg	l	kWh	kcal	(%)
Cooking										
Biomass	1.16			3,710	57.1	2.85			9,990	83.4
Kerosene		.17		1,510	23.3		.05		510	4.3
Subtotal				5,220	80.4				10,500	87.7
Lighting										
Kerosene		.12		1,070	16.4		.14		1,340	11.2
Electricity			.08	70	1.1			–	–	–
Subtotal				1,140	17.5				1,340	11.2
Other										
Charcoal	.02			140	2.1	.02			140	1.1
Total				6,500+	100.0				11,980+	100.0

* Original figures were given in per capita consumption per year. (See *Studi Konsumi Sumber Daya Energi Pedesaan* 1977).

+ Respectively 27.2 MJ and 50.1 MJ.

groups based on type (or types) of fuel used for cooking: biomass only, biomass and kerosene (mixed), and kerosene only. Of the 533 sample households, 232 households (or 43.5 percent) use only biomass, 112 households (or 21.0 percent) use both biomass and kerosene, and 189 households (35.4 percent) use only kerosene for cooking (see Table 8). The structure of energy consumption for cooking for the three groups is examined by income level, size of household, village typology, district, and stated reasons for use.

TABLE 8.
Structure of Energy Consumption for Cooking

Household Categories	Percentage of Households	Percentage of Households Using		
		Biomass Only	Mixed	Kerosene Only
Income (Rp/cap/month)				
Under 5,000	37.7	59.7	18.9	21.4
5,000 - 7,999	28.7	36.0	19.6	44.4
8,000 - 14,999	23.3	30.6	25.8	43.5
15,000 and above	10.3	34.6	21.8	43.6
Household Size				
Less than 5	33.4	52.8	10.7	36.5
5 - 7	45.0	39.2	25.0	35.8
8 and above	21.6	38.3	28.7	33.0
Village Type				
Traditional	8.6	60.9	10.9	28.2
Transitional	58.5	50.3	22.4	27.3
Modern	32.8	26.9	21.0	52.0
District				
Ciamis	18.9	80.2	4.0	15.8
Garut	17.3	69.6	16.3	14.1
Serang	11.3	58.3	20.0	21.7
Cirebon	13.9	29.7	21.6	48.7
Bandung	38.6	14.6	31.5	53.9
All Households	100.0	43.5	21.0	35.4

Income. Households are grouped by average per capita income per month. A poverty level has been set at about 25 kg of rice or Rp 5,000 per capita per month. Households above the poverty level are further disaggregated (rather arbitrarily) into three income classes. Nearly 60 percent of households in the lowest income group use only biomass for cooking; between 31 and 36 percent use only biomass in higher income categories. This is predictable since biomass is available free. Notably, once income reaches the poverty level, the proportion of households using only kerosene for cooking increases from about 21 to 44 percent and stays at that percentage for higher income categories.

Household Size. As the size of the household (size includes nonmembers eating in respective household) increases, the proportions of households using only biomass or only kerosene for cooking both decline. In other words, larger households tend to use a mix of biomass and kerosene for cooking.

Village Type. A majority of households in traditional villages use only biomass for cooking, whereas about 52 percent of households in modern villages use only kerosene for cooking. This is probably because commercial energy is more accessible in modern villages.

District. The five districts surveyed exhibit different ecological and geographic characteristics. The Ciamis and Garut districts both have about 400 persons per sq km; they both have poor roads and both export firewood. However, Ciamis is low plain while Garut is high plain (mountainous). Most rural households in these districts use only biomass and only a few use only kerosene for cooking. The Bandung district is more densely populated (716 persons per sq km) than Ciamis and Garut. It has plantation and forest areas while villages and cities are well connected by good roads. There, about 54 percent of rural households use only kerosene and only 15 percent use only biomass for cooking.

Cirebon is very densely populated, with 1,135 persons per sq km. The Cirebon district has relatively good roads and cultivated lands and imports firewood and agricultural residue. Nearly half of the rural households in Cirebon use only kerosene for cooking.

The Serang district is less densely populated than Bandung and Cirebon and has relatively good roads. About 58 percent of rural households use only biomass, while about 22 percent use only kerosene for cooking. Apparently, as population density increases so does the proportion of households using kerosene for cooking. (The trend also was reflected in the 1977 Haeruman study.) Also, physical infrastructure, apart from local availability of fuels, does strongly affect the structure of energy consumption by households.

Reasons for Fuel Selection. Most respondents who used only kerosene for cooking said they did because it was "clean and practical" (see Table 9). Those respondents who did not use kerosene for cooking said firewood was inexpensive.

The prices of both kerosene and firewood do differ from area to area. In the firewood-producing and exporting district of Ciamis, for example, the price of firewood is lower than in other districts. The price of kerosene is higher there than in other districts largely because of its poor roads and transportation system.[2]

TABLE 9.
Reasons for Using or Not Using Kerosene for Cooking

Households Using Only Kerosene	(n = 103)
1. Clean and practical	61%
2. Efficient in use	33%
3. Inexpensive	30%
4. Firewood is difficult to use; more expensive	18%
Households Not Using Kerosene	(n = 29)
1. Firewood inexpensive	38%
2. Kerosene expensive	7%

TABLE 10.
Structure of Energy Consumption for Lighting

Household Categories	Percentage of Households	Percentage of Households Using		
		Kerosene only	Mixed	Electricity only
Income (Rp/capita/month)				
Under 5,000	37.3	84.1	4.0	11.9
5,000 - 7,999	28.7	76.5	3.3	20.2
8,000 - 14,999	23.3	72.6	8.9	18.5
15,000 and above	10.3	60.0	14.5	25.5
Household Size				
Less than 5	33.4	78.6	5.1	16.3
5 - 7	45.0	78.7	4.6	16.7
8 and above	21.6	69.6	10.4	20.0
Village Type				
Traditional	8.6	100.0	–	–
Transitional	58.5	81.4	5.5	13.1
Modern	32.8	63.4	8.6	28.0
District				
Cirebon	13.9	100.0	–	–
Ciamis	18.9	93.1	5.9	1.0
Serang	11.3	90.0	10.0	–
Garut	17.3	89.1	1.1	9.8
Bandung	38.6	51.0	9.2	39.8
All Households	100.0	76.7	6.0	17.2

Lighting is the second major use of energy in rural households. Of the surveyed households, 409 households (76.7 percent) use only kerosene, 32 households (6.0 percent) use both kerosene and electricity for lighting, while 92 households (17.2 percent) use only electricity (see Table 10). Among the lowest income households, about 84 percent use only kerosene and about 12 percent use only electricity for lighting. In the highest income category, about 25 percent of households use only electricity and about 15 percent use electricity in combination with kerosene lamps. The survey suggests that larger households are more inclined to substitute electricity for kerosene. At the time of the survey some households with generators gave away their excess electricity to some of their poorer neighbors.

The most striking difference in the structure of energy consumption for lighting is that all households in traditional villages use only kerosene for lighting. With improvements in the socioeconomic conditions of villages, the proportion of households using only kerosene declines. In modern villages the proportion is 63 percent, probably because the national rural electrification program starts with villages of higher socioeconomic levels.

About 90 percent or more of the rural households use only kerosene for lighting in four of the districts, excluding Bandung. In Bandung, where infrastructure is relatively better than in the other districts, close to 40 percent of rural households already use only electricity and only 51 percent rely on kerosene for lighting.

Most respondents who used only electricity for lighting said they did because it was "clean and practical" (see Table 11). Those not using electricity said the main reason was the high installment cost.

Pattern of Energy Input for Cooking

To compare the levels of energy consumed for cooking by type (or types) of fuel, it is instructive to compare the level of energy input of households using only biomass with the level of energy input of those using only kerosene for cooking (see Table 12).

As per capita income increases, per capita input of energy for cooking increases, both for biomass and for kerosene. The average per capita kerosene input for cooking for the highest income group is twice as much as for the lowest income group. Use of biomass per capita by the highest income group is twice its use by the lowest income group in terms of kilograms, but only 1.5 times its use by the lowest income group in heat equivalent (kilocalories or joules). Apparently the highest income group uses less "efficient" biomass than does the lowest income group perhaps

TABLE 11.
Reasons for Using or not Using Electricity for Lighting

Households Using Only Electricity	(n = 34)
1. Clean and practical	71%
2. Facilities provided	21%
3. Better life conditions	18%
Households Not Using Electricity	(n = 101)
1. High installment cost	50%
2. Awaiting connection	43%
3. No access	8%

TABLE 12.
Input of Energy for Cooking
(per capita per day)

Household Categories by Income	Households using				
	Biomass only			Kerosene only	
(Rp/cap/month)	kg	kcal/kg*	MJ	l	MJ+
Under 5,000	1.57	3,330	21.9	.27	9.7
5,000 - 7,999	2.04	3,150	26.9	.32	10.0
8,000 - 14,999	1.71	3,160	22.6	.38	13.7
15,000 and above	3.41	2,260	32.2	.53	19.0
All Households	1.86	3,090	24.0	.35	12.7

* Based on results of laboratory examination of biomass samples taken from the respective households.

+ Assumes that 1 liter of kerosene has a heat content of 37.2 MJ.

because poorer households usually collect dry twigs, and the highest income households use freshly cut branches of trees from their own homeyards and gardens.

The pattern of energy inputs for cooking at four different income levels reveals that kerosene generally is twice as efficient as biomass as a fuel for cooking (Fig. 2). In early 1980, people in the lowest income group using commercial firewood for cooking paid about Rp 16 to Rp 20 (about US $0.03) per capita per day (based on an average market price of commercial firewood of Rp 10 to Rp 12.50 per kg in Cirebon). If people in that same group used kerosene, they paid about Rp 10 per capita per

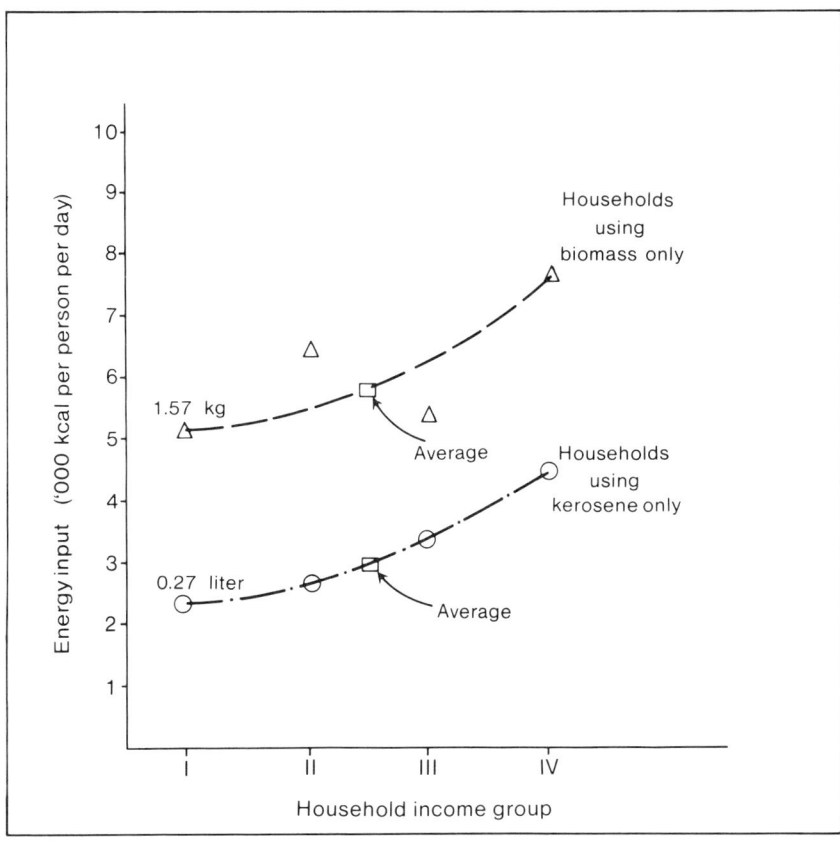

Figure 2. Energy Input for Cooking by Income Group

day in 1980. Prices of kerosene vary by district so comparison is based on the average price in rural markets of Cirebon of Rp 38 per liter. With a new government policy to reduce kerosene subsidies in early 1982, kerosene costs increased for the lowest income group to Rp 22 per capita per day at a market price of about Rp 80 per liter in rural markets, with an official price of Rp 60 per liter. Thus, it became more expensive to use kerosene than commercial firewood at then prevailing market prices. Apparently (without 1982 fuelwood prices available), the price of commercial firewood tends to increase with the price of kerosene.

Households using only biomass for cooking have a lower average per capita income (Rp 6,310 per month) than either households using only kerosene (Rp 9,080 per month) or households using both kerosene and biomass (Rp 8,870 per month). Households using mixed fuels for cooking (Table 13) usually need a higher energy input per person (32.6 MJ/day)

TABLE 13.
Input of Energy for Cooking of Households Using Mixed Fuels
(per capita per day)

Household Categories by Income (Rp/cap/month)	Energy Input					Kerosene with Mixed Fuels as percent of "normal" use *
	Kerosene		Biomass		Total	
	l	MJ	kg	MJ	MJ	
Under 5,000	.16	5.7	1.32	18.4	24.1	58.7
5,000 - 7,999	.25	9.0	1.53	24.5	33.5	78.2
8,000 - 14,999	.28	10.2	2.07	28.1	38.3	74.6
15,000 and above	.40	14.4	2.14	27.6	42.0	75.6
All Households	.24	8.8	1.68	23.8	32.6	69.5

* "Normal" use is defined as use of only kerosene.

than households using only kerosene (12.7 MJ/day) or households using only biomass (24.0 MJ/day). Households using mixed fuels for cooking may exhibit different cooking habits from other groups of households. Only about one-third of all households using mixed fuels for cooking belong to the lowest income group; 53.6 percent are households of 5 to 7 people, 62.5 percent are households in transitional villages, and 58 percent are households in Bandung district.

In households using mixed fuels for cooking, about 27 percent of the energy input for cooking is kerosene. The change in this proportion from 24 percent in the lowest income group to about 34 percent in the highest income group suggests a substitution of kerosene for biomass as income rises. (Based on survey results, income elasticity of kerosene input was 0.4 for households using mixed fuels and 0 for households using only kerosene for cooking.) If households using only kerosene for cooking are named "normal," households using mixed fuels tend to use about 75 percent as much kerosene as the "normal" amount, if the household is above the poverty line. For households below the poverty line, the proportion is about 59 percent.

That per capita energy input for cooking declines as household size increases suggests some economies of scale. Per capita cooking energy input for households with eight or more members is about half that of households with less than five members. If, due to socioeconomic development, the size of rural households tends to become smaller, per capita energy input for cooking can be expected to increase.

Average per capita energy input in transitional villages tends to be higher than in traditional or modern villages, except for the group of

TABLE 14.
Pattern of Energy Input for Cooking (per capita per day)

Households Categories	Only Biomass		Households Using				Only Kerosene	
			Mixed					
			Biomass		Kerosene			
	kg	MJ	kg	MJ	l	MJ	l	MJ
Size								
Less than 5	2.47	31.4	2.99	42.3	0.35	12.7	0.48	17.2
5-7	1.49	19.1	1.49	21.4	0.23	8.5	0.30	10.9
8 and above	1.32	18.7	1.25	17.5	0.19	7.1	0.24	8.8
Village Typology								
Traditional	1.71	22.9	1.34	19.3	0.20	7.3	0.34	12.4
Transitional	2.01	25.0	2.09	29.6	0.27	9.7	0.33	11.9
Modern	1.43	18.1	0.95	13.3	0.20	7.4	0.37	13.4
District								
Bandung	2.45	34.9	2.18	30.8	0.30	11.0	0.35	12.6
Caimis	2.17	25.4	1.11	18.3	0.20	11.4	0.26	9.4
Garut	1.70	22.9	1.27	15.0	0.12	4.4	0.30	10.7
Serang	1.43	20.2	0.98	9.6	0.20	7.2	0.41	14.9
Cirebon	1.00	13.5	0.89	13.4	0.16	5.7	0.39	14.1

households using only kerosene for cooking (see Table 14). Households in modern villages use more kerosene for cooking and the proportion of households using only kerosene for cooking is markedly higher in modern villages. Per capita biomass use is highest in transitional villages perhaps because 36 percent of all those sample households are in the Bandung district, where per capita biomass input is highest. That level is even higher than in the two firewood-exporting districts, Ciamis and Garut, perhaps because the average income of the rural population in Bandung is also higher.

Pattern of Energy Input for Lighting

As income level increases, per capita input of energy for lighting also increases (see Table 15). Among households using only kerosene and those using both kerosene and electricity for lighting, the highest income group uses on average per capita more than twice the amount used by the lowest income group. Among households using only electricity, the increase is even larger.

TABLE 15.
Input of Energy for Lighting (per capita per day)

Households Categories by Income (Rp/cap/month)	Households Using							
	Kerosene only		Mixed			Electricity Only	
	l	MJ	l	kWh	MJ	kWh	MJ
Under 5,000	.11	4.2	.04	0.12	1.9	.24	0.9
5,000–7,999	.16	5.7	.06	0.08	2.5	.31	1.1
8,000–14,999	.17	6.3	.06	0.20	3.2	.47	1.7
15,000 and above	.28	10.0	.09	0.26	4.3	.72	2.6
All Households	.15	5.6	.06	0.18	3.0	.39	1.4

The average per capita income of households using only kerosene for lighting (Rp 7,200 per month) is lower than the income of households using only electricity (Rp 8,990 per month), or of households using both kerosene and electricity (Rp 12,510 per month). The kerosene input of households using both kerosene and electricity is about one-third of the kerosene input of households using only kerosene for lighting at all income levels.

As household size increases, per capita energy input for lighting declines, according to the survey. Village typology does not seem to directly affect levels of energy input for lighting. Traditional villages in general rely totally on kerosene for lighting. Villages surveyed in the Cirebon district have no access to electricity and rely totally on kerosene. Generally, in villages where electricity is not widespread, per capita kerosene input for lighting tends to be highest.

Fuel Supply and Availability

Biomass and kerosene are the main sources of energy for rural households. For the total sample of households, biomass constitutes on average about 60 percent of the total household energy budget, and kerosene the remaining 40 percent. Nationwide, the share of biomass could be much larger.

Biomass is essentially a locally available source of energy. Intra- and interdistrict trade of biomass is undertaken mainly in response to the needs of rural industrial establishments. Commercial firewood is only used by a limited number of households. In the Serang district, for example, higher income households purchase their firewood from industrial establishments supplied regularly by traders.

Farm-forest fringe, Sukabumi, Central Java. (R. Van Den Beldt)

Kerosene is supplied from national depots, administered by Pertamina as a state enterprise, and distributed through dealerships operated by private companies or agents. From the depots, kerosene is transported in fuel trucks by distribution dealers to their subagents out to the districts. Then barges, small boats, or carts distribute it to vendors or local retailers. So, by the time kerosene reaches consumers in the villages, the average prices often become 50 to 100 percent higher than the official retail price. Kerosene is not evenly distributed throughout Indonesia. Villages in remote and mountainous areas usually have no access to kerosene due to lack of infrastructure. In the Ciamis district, where roads are still in a poor condition, some households have to go as far as 10 km to get kerosene.

Of the 533 households surveyed, 344 households (64.5 percent) use biomass for cooking. The sources of their biomass supply are mainly their home yards and their gardens and estates (see Table 16). Forests or brushwood contribute only 5 to 8 percent.

Households that collect their own biomass also reported on the typically used parts of the tree. About 61 percent of the households

TABLE 16.
Sources of Biomass Supply

Source		Percentage of Households
Collecting from		
Home yards only		7.6
Garden/estate only		64.1
Forest/brushwood only		5.2
Home yards and garden/estate		5.0
Home yards and forest/brushwood		1.2
Garden/estate and forest/brushwood		1.5
Other		2.0
	Sub-total	86.6
Purchasing		13.4
	Total	100.0

TABLE 17.
Tree Parts Collected as Fuel

Parts	Percentage of Households
Twigs only	27.4
Branches only	5.1
Trunk only	12.3
Twigs and branches	28.8
Twigs and trunk	12.0
Branches and trunk	12.7
Other (roots, etc.)	1.6
Total	100.0

collecting their own biomass only collect twigs and prune branches. About 37 percent of the households cut the trunk (see Table 17).

Energy "Problems" of Rural Households

The questionnaire included items related to problems of supply and availability as faced or perceived by rural households in the fulfillment of their energy needs, both for cooking and lighting (see Table 6). It can be expected that different types of households face different problems. Their solutions to those problems will also be different, depending upon constraints and opportunities, both at the household level (budget constraints or substitution possibilities) and location-wise (availability of fuels in use in nearby villages or availability of alternative sources of energy).

In terms of fuels for cooking, about 23 percent of the 533 households surveyed face frequent disruptions in the fulfillment of their energy needs. Of those households, 62 percent face problems relating to the use of biomass and 38 percent in relation to kerosene. The nature of the problem may be nonavailability (experienced by 37 percent of those households), scarcity and/or cost consideration (55 percent), or other reasons (7 percent), such as weather condition (biomass becomes too wet). In dealing with those problems, about 50 percent of those households look for substitutes, about 38 percent seek other sources of supply (in nearby villages) or continue to use the fuel at higher prices, and about 12 percent reduce their consumption of energy for cooking.

Table 18 shows the differences in magnitude and nature of the problem by district, by village type, and by household income. These results suggest that rural households in the districts of Bandung and Garut are more prone to facing disruptions (around 30 percent of households). This suggests that while kerosene has penetrated relatively more extensively in Bandung district, its distribution network is still relatively poor. In Garut, a firewood-exporting region, scarcity of biomass has become a problem as a result of diversion from home markets to outside regions, which also tend to increase prices in home markets.

In Bandung, households facing disruptions in the supply of kerosene tend to substitute biomass for kerosene. In Garut, most households look elsewhere for biomass supply or continue to use it at higher prices. In the three other districts, only about 13 to 14 percent of households face energy supply problems. Their responses differ: in Cirebon about 40 percent of households reduce their energy consumption, and in Ciamis and Serang most households look elsewhere for fuel supplies.

About 50 percent of households in traditional villages face problems in the supply of both biomass and kerosene for cooking, most often due to nonavailability. In transitional and modern villages only about 20 percent of households face frequent disruptions in the supply of cooking fuels, often related to biomass supply.

Poorer households tend to face greater problems in the fulfillment of their cooking energy needs. About 28 percent of households with a per capita income of under Rp 5,000 per month face difficulties, especially with regard to biomass supply, mainly due to scarcities or high costs. Only about 20 percent of households in the higher income groups (Rp 5,000–Rp 14,999) experience those problems, with regard to both biomass and kerosene. In dealing with them, these households tend to look for substi-

TABLE 18.
Problems of Energy Supply and Availability: Energy for Cooking

Categories of Households	Total Sample of Households	Households Facing Problems (percentage of total households)	Of Households Facing Problems (percentage)							
			Type of Energy		Type of Problem			Solution		
			Biomass	Kerosene	Non-Availability	Expensive/scarcity	Other[a]	Substitute	Reduce Use	Other[b]
By District										
Bandung	206	31.5	47.6	52.4	57.1	42.9	–	63.5	14.3	22.2
Ciamis	101	13.9	78.6	21.4	35.7	28.6	35.7	35.7	–	64.3
Cirebon	74	13.5	70.0	30.0	–	60.0	40.0	30.0	40.0	30.0
Garut	92	28.3	76.9	23.1	11.5	88.5	–	42.3	3.8	53.8
Serang	60	13.3	87.5	12.5	12.5	87.5	–	12.5	12.5	75.0
By Village Type										
Traditional	46	50.0	56.5	43.5	65.2	21.7	13.0	65.2	8.7	26.1
Transitional	312	18.3	66.7	33.3	21.1	68.4	10.5	54.4	19.3	26.3
Modern	175	23.4	58.5	41.5	43.9	56.1	–	34.1	2.4	63.5
By Income (Rp/cap./month)										
Under 5,000	201	28.4	73.2	26.8	26.3	64.3	9.4	43.9	12.5	43.6
5,000 – 7,999	153	19.6	50.0	50.0	46.7	34.5	18.8	50.0	10.3	39.7
8,000 – 14,999	124	21.0	57.7	42.3	65.4	30.8	3.8	53.8	19.2	27.0
15,000 and above	55	12.7	57.1	42.9	57.1	42.9	–	28.6	–	71.4
All Households	533	22.7	62.0	38.0	37.2	55.4	7.4	49.6	12.4	38.0

Notes: [a] Usually because biomass was too wet due to rain.
[b] Look for supply elsewhere or continue to buy at higher prices.

tutes. In contrast, those problems are felt by only about 13 percent of households in the highest income group (Rp 15,000 and above). When facing problems, most of these households look for supplies elsewhere or continue to buy at higher prices.

In terms of fuels for lighting, of the 533 households surveyed, about 21 percent experience frequent supply disruptions. Of those households, 64 percent face problems of kerosene supply and 36 percent in relation to supply of electricity. The kinds of problems faced by households using kerosene for lighting are nonavailability and increase in prices. For those using electricity, the problem is almost exclusively related to blackouts.

When facing difficulties in the supply of energy for lighting, about 40 percent of households reduce their consumption. This is in contrast to their behavior when experiencing problems in the supply of energy for cooking, where only 12 percent of households reduce their consumption.

Table 19 exhibits the differences in magnitude and nature of the problem of energy for lighting as faced by different categories of households. In Garut, about 37 percent of households experience problems in the supply of energy for lighting, primarily kerosene. The proportions of households facing the problem are about 23 to 24 percent in Bandung and Serang, and only about 8 to 9 percent in Ciamis and Cirebon. In the latter two districts, increase in the price of kerosene constitutes the main source of the problem. While in Ciamis about 78 percent of those households continue to buy at higher prices, in Cirebon about 67 percent and in both Bandung and Garut more than 40 percent of those households tend to reduce their consumption of energy for lighting.

Households in traditional villages face greater problems in fulfilling their energy needs for lighting (48 percent of households) than those in transitional and modern villages (19 percent). Traditional villages usually are more dependent upon kerosene, and nonavailability of kerosene seems to constitute a major source of the problem. When facing disruptions, most of these households (68 percent) reduce their kerosene consumption for lighting. In modern villages, of households experiencing problems of energy supply for lighting about 64 percent use electricity; blackouts are the main problem. These households tend to return to the use of kerosene.

Lower income households tend to face greater problems than higher income households in the fulfillment of their lighting needs. As can be expected, kerosene supplies are the main problem for lower income households, whereas blackouts in the supply of electricity are a problem for higher income households.

When facing problems, higher income households substitute kerosene

TABLE 19.
Problems of Energy Supply and Availability: Energy for Lighting

Categories of Households	Total Sample of Households	Households Facing Problems (percentage of total households)	Of Households Facing Problems (percentage)							
			Type of Energy		Type of Problem			Solution		
			Kero-sene	Electri-city	Non-Avail-ability	Black-outs[a]	Expen-sive	Substi-tute	Reduce Use	Other[b]
By District										
Bandung	206	24.3	50.0	50.0	48.0	50.0	2.0	50.0	44.0	6.0
Ciamis	101	8.9	100.0	-	22.2	-	77.8	11.1	11.1	77.8
Cirebon	74	8.1	100.0	-	16.7	-	83.3	-	66.7	33.3
Garut	92	37.0	70.6	29.4	47.1	23.5	29.4	32.4	41.2	26.5
Serang	60	23.3	57.1	42.9	7.1	42.9	50.0	42.9	21.4	35.7
By Village Type										
Traditional	46	47.8	90.9	9.1	77.3	9.1	13.6	9.1	68.2	22.7
Transitional	312	18.6	69.0	31.0	41.4	31.0	27.6	37.9	39.7	22.4
Modern	175	18.9	36.4	63.6	9.1	57.6	33.3	57.6	18.2	24.2
By Income (Rp/cap./month)										
Under 5,000	201	24.4	75.5	24.5	44.9	22.4	32.7	24.5	49.0	26.5
5,000 - 7,999	153	18.3	75.0	25.0	39.3	25.0	35.7	35.7	35.7	28.6
8,000 - 14,999	124	21.8	51.9	48.1	37.0	44.4	18.6	48.1	37.0	14.9
15,000 and above	55	16.4	22.2	77.8	-	77.8	22.2	77.8	-	22.2
All Households	533	21.2	63.7	36.3	38.9	34.5	26.5	37.2	39.8	23.0

Notes: [a] For those using electricity
[b] Look elsewhere for supply.

for electricity, but lower income households tend to reduce their consumption of energy for lighting.

Use of Charcoal

Of the 533 households surveyed, about 40 percent use charcoal, exclusively for ironing. None of them use charcoal for cooking. The average per capita consumption of households using charcoal is only about 0.05 kg per day. Consumption tends to increase with income from 0.03 kg per day for the lowest income group to about 0.07 kg per day for the highest income group. Per capita charcoal consumption declines with household size. It is highest for households in traditional villages (0.08 kg per day) and in the firewood-producing district, Ciamis (0.11 kg per day).

Caloric and Moisture Content of Tree Species

During the field survey, 929 samples of biomass were collected from households to examine their caloric and moisture contents on the day they were actually used. The samples, 10 cm in size, include 101 species. Of the 12 most used species (see Table 20), five belong to the group of fruit-bearing trees often found in home yards or gardens in rural areas. These species are frequently used as fuel in Serang and Cirebon, where brushwood and forestlands are scarce. Forestry resources, such as *Tectona grandis*, are used as fuel in the two firewood-producing districts, Garut and Ciamis.

Coconut trees are widely used as a source of fuel supply (Table 20). The caloric values of different parts of coconut trees vary inversely according to moisture content. Most of the fruit-bearing tree species have a caloric content between 3,200 and 3,500 kcal/kg at a moisture content from 26 to 38 percent, a relatively narrow caloric variation with no apparent pattern relative to moisture content.

Average caloric contents for fuelwood calculated in earlier studies in Indonesia (3,500 to 4,700 kcal/kg) appear to be too high compared with the average found in actual use of biomass by the sample households (3,100 kcal/kg). Households in the middle-level income groups (about 60 percent of all households in the sample), use biomass with an average caloric value of about 3,200 kcal/kg (see Table 12). This figure can reasonably be applied in further studies.

Efficiency in the Use of Energy

During the survey, no field experiments were undertaken to examine the relative efficiencies of different conversion devices and of the

TABLE 20.
Tree Species and Fuel Characteristics

Species	Number of Samples*	Average Caloric Content (kcal/kg)	Average Moisture Content (percent)	District in Which Used As Fuel
1. Cocos Nucifera	196			
Trunk	22	3,200	33.1	Serang, Cirebon
Stalk	81	2,850	36.8	Ciamis, Serang
Coconut Shell	17	4,080	30.0	Ciamis
Coconut Fiber	55	2,730	39.1	Ciamis, Serang
Leaves	21	3,280	34.2	Ciamis, Serang
2. Bamboo+	100	3,430	30.1	Bandung, Garut
3. Albizia+	78	3,160	27.7	Garut, Ciamis
4. Thea sinensis	68	3,410	34.4	Bandung
5. Mangifera indica#	26	3,430	30.9	Cirebon, Bandung
6. Pterospermum javanicum	17	3,300	34.9	Ciamis
7. Parkia sp	15	3,230	22.6	Cirebon
8. Eugenia aqucae#	13	3,290	30.2	Bandung, Ciamis
9. Nephelium lapaceum#	13	3,380	26.0	Serang
10. Tectona grandis+	12	3,450	37.8	Ciamis
11. Artocarpus integra#	11	3,480	33.1	Ciamis
12. Samaea sp#	11	3,720	28.7	Cirebon

* Number of samples taken during the survey reflects the frequency of use of the respective species.

+ These trees are primarily found on forest lands.

Fruit-bearing tree species found in homeyards or gardens.

different fuels for specific tasks. The survey instrument included items such as amount of food cooked (such as rice), method of cooking, as well as the types of cooking stoves and appliances. However, since the energy used in preparing rice cannot be separated from the energy to cook the whole meal, and also because types of cooking stoves and appliances were not standardized, statistics based on those items failed to provide any useful data on relative efficiencies.

TABLE 21.
Heat Equivalence in Use and Efficiency Factors of Biomass and Kerosene

Source/study	Useful Energy for Cooking		Equivalence kg biomass 1 kerosene
	Kerosene (MJ/l)	Biomass (MJ/kg)	
1. CSIS Study	(1.9e)·37.2	e*·13.4	5.3
2. A. Martono (1974)	17.6	3.3	5.3
3. ITB Study/Filino Harahap (1978)	(3.2e)·36.3	e·19.7	5.8
4. USAID Study/Paul Weatherly (1980)	(1.9e)·37.2	e·14.6	4.8

* e stands for efficiency of biomass as a fuel for cooking.

At a more aggregate level, however, the study included energy input for cooking in households using only biomass and households using only kerosene. If it is assumed that the effective energy output for cooking is equal for average households in the two groups, the relative efficiency between biomass and kerosene as a cooking fuel is about 1 to 1.9 as derived from the respective heat inputs (24.0/12.7, see Table 12). If the heat content of biomass is 13.4 MJ/kg and that of kerosene is about 37.2 MJ/liter, the above relative efficiency would produce a "joule equivalence in use" of 5.3 kg of biomass for one liter of kerosene.

That figure is close to an estimate from a 1974 experiment by Andrini Martono (see Table 21). According to results of laboratory experiments in an Institute of Technology Bandung (ITB) study by Filino Harahap (1978), the relative efficiency between biomass and kerosene as a fuel for cooking is about 1 to 3.2, and joule equivalence is about 5.8. According to a U.S. Agency for International Development (USAID) study by Paul Weatherly (1980), involving several field experiments to measure fuel efficiency, one liter of kerosene equals 4.8 kg of biomass. Given the assumed heat contents of biomass and kerosene of 14.6 MJ/kg and 37.2 MJ/liter, respectively, the relative efficiency of the two fuels was also 1 to 1.9.

REFLECTIONS ON METHODOLOGY

This survey had both design and implementation deficiencies. Comparison with large-scale surveys sponsored by USAID and the

Department of Mining and Energy (DME) would highlight methodological problems encountered in rural energy surveys in general. (See also John Arnold 1980, Weatherly 1980, Ketenagaan 1982, Romir Chatterjee 1981, and Freerk Wiersum 1979 for related discussions.)

Deficiencies in Sample Design

The CSIS survey was carried out primarily to generate information on the pattern of rural household energy consumption. The study hypothesized that this pattern was influenced by household characteristics such as income level, wealth, size of household, and economic activity. It was also hypothesized that households exhibiting the same major characteristics ("twin" households) might have different energy consumption patterns if they were located in villages with different characteristics; that is, that households' fuel selection was influenced by the overall village environment including the supply and availability of energy sources. Moreover, the supply of energy to the village was considered a function of the broader ecological environment. For practical purposes, the district in which the village was located was taken to represent this broader environment.

From this perspective, the sample should have been designed with more systematic regard to ecological types and to households as the sampling unit. Certain deficiencies are evident:

1. The district is essentially an administrative unit and may not represent a uniform ecological environment. Variation in the ecological characteristics of villages in one district may be greater than between some villages in that district and some villages in another district.
2. Therefore, instead of predetermining the number of districts to be included, this should have been determined after the sampling procedure of villages, based on village geographic and ecological characteristics, such as
    a. coastal village, close to forestland;
    b. coastal village, close to agricultural sites;
    c. inland (low plain) village, close to forestland;
    d. inland (low plain) village, close to agricultural sites;
    e. inland (high plain) village, close to forestland; and
    f. inland (high plain) village, close to agricultural sites.
3. For each of those six environments, an equal number of villages of different socioeconomic characteristics (traditional, transitional, modern) could have been represented in the sample. The

selection of different village typologies proportional to their numbers for West Java as a whole has no solid basis.
4. The three village categories are based on the sum of rankings assigned to a number of characteristics. Hence, villages with the same total ranking could have different characteristics. Selection of villages should have depended more on a few major characteristics such as physical infrastructure or dominant economic activity, as revealed by detailed rankings in the Village Statistics.
5. Households should have been selected more systematically. The 1-percent sampling of households in each village has no solid basis. The sample should consist of a reasonable number of households in each village to provide for a reasonable range of incomes and occupations. The difficulty is with the sample frame. No household income data, however crude, are readily available before going to the village. Here, an equal number of households in each village should have been used in the sample.

Sample design ultimately depends both on the purpose of the study and on financial or time constraints (see Table 22). The USAID-sponsored study was conducted to determine the energy use in households in three of the ten rural electrification project areas. The three areas — Klaten in Central Java, East Lombok, and Luwu in South Sulawesi — were selected on the basis of widely varying availability of local resources.[3] The survey attempted to relate the patterns of household energy consumption to the demand on local natural resources. Information was collected to examine factors influencing fuel choice for particular end uses and significant variables relating to the level of fuel consumption. This served as the background for understanding changes likely to result from electrification and for predicting impacts of electrification which affect energy use. Site selection, therefore, did not pose a major methodological problem. Within each site, three villages — poor, average, and prosperous villages — were selected, based on the project supervisors' own observations and interviews with regional political leaders. Three already electrified villages in the Klaten area were added to this sample. Within each of the 12 villages, 15 households were selected to provide a range of occupations and incomes. The study resulted in an environmental assessment for rural electrification projects (Weatherly 1980).

The rural household energy survey sponsored by the DME was carried out as part of the first phase of a USAID-supported project entitled Energy Planning for Development in Indonesia. The goal was to construct a profile of household energy use for different income groups. The information was meant to be used in examining the possible effects

TABLE 22.
Sample Designs of Three Rural Energy Surveys in Indonesia

	USAID (late 1979)	CSIS (early 1980)	DME (late 1980)
Households	179	533	793
Villages	12	40	153
Districts	3	5	13
Provinces	3	1	5
Households per Village	15 (equal)	4-69 (range)	5 (design: equal) 4-18 (actual)

of income changes, fuel substitutions, or changes in technology for policy-making. The data were collected for analysis based on a modified macroscale Reference Energy System (RES) framework. Developed at Brookhaven National Laboratory, this framework is a standardized vectorial analysis of energy flows from sources, through conversion and transfer subsystems, to utilizing devices and end uses. Estimates of future energy supply-demand balances at the national level were projected using this framework.

The survey covered eight regions, three in West Java, two in Central Java, and one each in the East Java, North Sumatra, and South Sulawesi provinces. The regions happen to be those with high concentration of population. Site selection was not a major issue because the survey was assigned to be carried out by one university in each region. Households were selected randomly from the list of rural census blocks defined in the National Socioeconomic Survey (SUSENAS) carried out by the Indonesian Central Bureau of Statistics (BPS) in January 1980. This subsample was representative for the regions insofar as the SUSENAS sample itself was representative. One hundred households were randomly selected from one census block contiguous to each of the eight universities. This sampling procedure eliminated the expensive task of assembling a sample frame and allowed for future comparison with income and other relevant household data collected by SUSENAS. Since fieldwork had to be done within one month, regions were selected for accessibility. A cross section of income groups among households of each district was not intended, so disaggregation into three income classes of equal size was done retroactively based on the rank distribution of sample households' income.

Compared with the two other surveys, the CSIS survey was definitely large enough for its purpose. Its relatively large size may compensate to

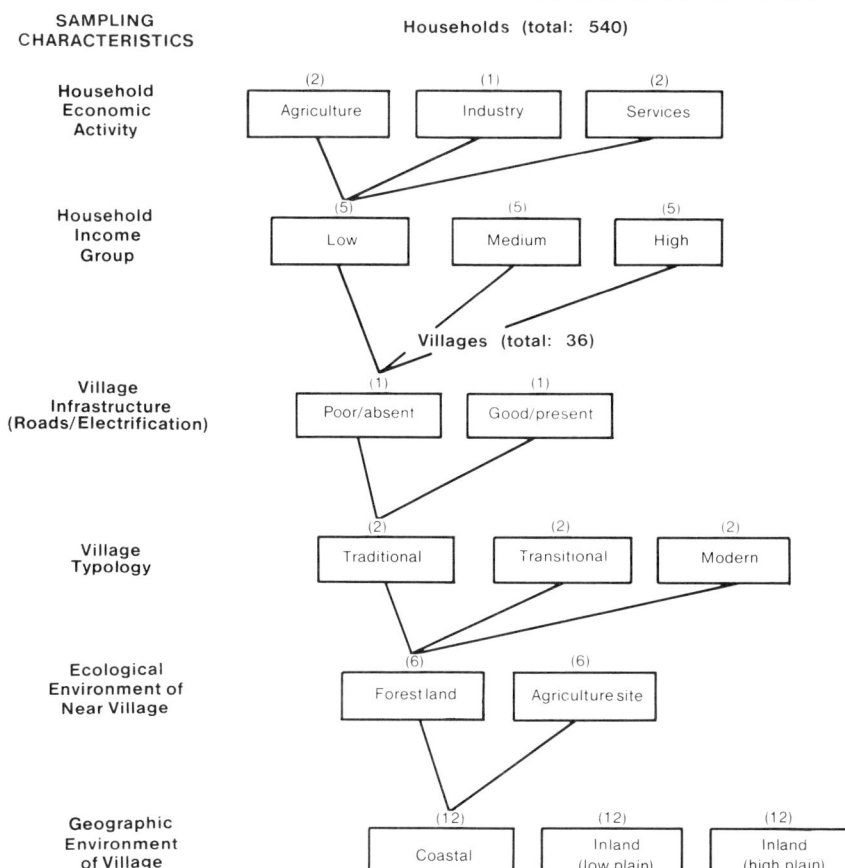

Figure 3. Illustrative Sampling Design by Ecological, Village, and Household Characteristics

some extent for deficiencies in sample design. However, if households had been selected more systematically, a smaller sample size might have produced the same results, or the same sample size might have captured more variations in household and environmental characteristics. A design such as Figure 3, adapted by financial and time constraints, would be a better alternative.

Deficiencies in the Survey Instrument

The CSIS test questionnaire was based on the one used in the USAID-sponsored rural energy survey but was found to be too long and too extensive. A shift in questions from the household to village interview was proposed to keep the schedule concise (see Arnold 1980). The questionnaire finally developed was much shorter and limited to five sets of data (Table 6). Even so, several questions could have been deleted.

Household Characteristics. Accurate household income data obviously were difficult to collect during short interviews. Data to cross check reported income, including land owned, area farmed, cattle owned, crops produced, and household expenditures were recorded for three consecutive interview days. The information was not as valuable as foreseen. Income for agricultural households fluctuated greatly. Household expenses depended on the interview days. For instance, most shopping was done either once a week, on "market" day, or once a month.

Fuel-Use Habits. Data on cooking, lighting, and other activities requiring energy were also less useful than anticipated. Data on daily menu, cooking stoves, and appliances were not used in the analysis because of the difficulties of standardizing items, quantities, and observations. Information on preparation and handling of fuels, especially firewood (whether cut in small or large pieces), did give some insight into fuel uses, although the time of visit influenced the accuracy.

Fuel Consumption. Daily fuel consumption for cooking, lighting, and ironing was recorded by field measurements over four consecutive days to produce data on daily consumption for three days. The same procedure had been used in the USAID-sponsored survey. Although it significantly increased survey time, this information was more reliable because (a) respondents' estimates of expenditure for fuels did not accurately account for noncommercial fuels, (b) villagers didn't use standardized methods to weigh their fuels, and (c) they used many diverse fuels. The repetitive measurements allowed by this procedure also revealed day-to-day variations in consumption (found to be ± 30 percent in the USAID-sponsored survey).

Indeed, most surveyors spent most of their time doing measurements and therefore were unable to observe other household activities as well. To simplify the procedure, a proposal has been made to limit fuel measurements to an initial and final record for several days. That proposal would require separation of data by activity and adequate representation of the various typical fuel sources (Arnold 1980).

Using the same procedures, the DME rural energy survey took measurements three times over five days to produce consumption data for four days. If conducted systematically, this procedure is better than the one used in the CSIS survey.

Specific Fuel Data. Information collected on sources of supply, supply problems, and fuel choice was useful. Respondents possessed reasonable knowledge of the biomass species used if they collected them themselves. If they purchased their firewood, they usually could not identify the species. The few questions on electrification were too general to provide useful information.

Fuelwood Data. Information on biomass potential of home yards and gardens was collected by identifying and counting fuel-producing trees and measuring their diameter. Justification for this effort would depend on its relevance to specific planning of fuelwood programs. Otherwise, information on biomass species used as fuel would indicate the relative importance of home yards and gardens as energy sources. Data on biomass potential could be collected at the village level. The CSIS survey included a separate village questionnaire, but it wasn't administered adequately, perhaps because surveyors were preoccupied with the household survey.

All in all, the questionnaire could have been much simpler to lessen surveyors' burden. Each of the 25 CSIS surveyors visited 21 to 22 households four times each in four consecutive days. The DME survey employed about 40 surveyors for 793 households, or about 19 to 20 households per surveyor, visited three times each over five days. The USAID-sponsored survey assigned one surveyor to each village (15 households) who stayed in the village for one month as a "participant observer." Each was to observe and record data on various village or household activities and conditions. Sometimes, however, a surveyor's intense involvement in village life reduced his objectivity and thus introduced a source of bias in the survey methodology (Arnold 1980).

Constraints in Implementation

Selecting households in the field became a major problem during the survey. Leaving that selection task largely to the surveyors, the leaders of the village, or their secretaries, as was done in the CSIS and USAID sponsored surveys, could have biased the results, especially since surveyors met members of various households through the village leader or his assistant. The latter person often stayed during the interview or assisted with measurements, another likely source of bias.

Surveying poor households was difficult. Very poor families might not cook at all at home, but ate in other households in return for services. Measuring fuel used by some poor families was difficult if they gathered their fuel immediately before preparing their food. And some of the poorest households apparently prepared more food than usual when the surveyor visited: another source of inaccuracy.

POLICY IMPLICATIONS AND SUGGESTIONS

The survey results show the variations in energy consumption patterns among household groups, by income, by household size, by village typology, as well as by districts. Thus, policies related to fuel price and availability could affect each household group differently. Even within household groups variations exist.

Variations Within Household Groups

Consumption of biomass and kerosene, the two major rural energy sources, varies within household groups (see Table 23). The degree of variation is the so-called coefficient of variation (the standard deviation as a percentage of the mean value). According to the USAID-sponsored survey, the variation in the daily household energy consumption was about ± 30 percent. If the mean value of biomass consumption is 2 kg per capita per day for one household group, a coefficient of variation of 30 percent means that about two-thirds of the households in the group consume between 1.4 and 2.6 kg per capita per day. Thus, a coefficient of variation of 60 percent suggests that the distribution is quite dispersed.

In most cases, the variations of biomass consumption are about 80 percent in terms of megajoule inputs (Table 23). The much larger variations in terms of kilogram inputs suggest the great variations in the quality of biomass used. Consumption of kerosene for lighting also varies greatly within household groups. Most households use only very small amounts of kerosene for very minimal lighting and only a few households use larger amounts of kerosene to generate a reasonable amount of lighting. This was observed especially in the Garut district. Consumption of kerosene for cooking varies less within household groups, except in Serang. Unlike the need for lighting, the need for kerosene for cooking appears to be more or less standard.

Kerosene Consumption

Development in the rural areas tends to raise the consumption of kerosene, both for cooking and lighting, according to the survey results. A study by Strout based on national data estimated the income elasticity of kerosene consumption in the rural areas of Indonesia at about 0.83, higher than that for the urban areas (0.56) (Strout 1976). A larger percentage of higher income households already use only kerosene for

TABLE 23.
Variations of Consumption Within Household Groups
(Coefficient of variation, in percent)

	Biomass a		Kerosene b	
Categories of Households			Cooking	Lighting
	(kg input)	(MJ input)	(liter input)	
By Income				
Less than 5,000	83.0	87.2	44.4	81.8
5,000 - 7,999	75.0	82.5	65.6	81.3
8,000 - 14,900	56.1	53.6	47.4	76.5
15,000 and above	163.9	78.4	62.3	71.4
By Size				
Less than 5	114.2	71.7	54.2	85.0
5 - 7	79.2	83.5	56.7	76.9
8 and above	72.7	79.7	45.8	58.3
By Village Typology				
Traditional	39.2	60.5	55.9	64.7
Transitional	111.2	83.3	69.7	93.3
Modern	89.5	86.0	56.8	75.0
By District				
Bandung	82.4	83.9	51.4	70.6
Ciamis	134.1	74.2	42.3	83.3
Cirebon	97.0	88.8	66.7	70.0
Garut	71.8	81.6	50.0	120.0
Serang	52.4	55.0	100.0	78.9
All Households	110.2	81.8	62.9	86.7

Note: The table is derived from consumption data of households using only biomass and using only kerosene for cooking, as well as households using only kerosene for lighting.

a For households using only biomass for cooking.

b For households using only kerosene for cooking or for lighting.

cooking. Per capita consumption of kerosene also increases with income. Similarly, as a village becomes more modern, more households shift to kerosene as a cooking fuel. In the Bandung district, where the physical infrastructure (especially roads) has significantly improved over the last decade, more than half of the households have changed to kerosene as a fuel for cooking.

Lighting in rural households is still a scarce but desirable service: it indicates a better quality of life in the households. As income increases, consumption of kerosene for lighting increases at a much faster rate than does consumption of kerosene for cooking. (Income elasticity of kerosene consumption for lighting is higher than that for cooking.) However, this is partly because consumption of kerosene for lighting starts from a very low base.

A sudden abolishment of kerosene subsidies would have rather severe effects upon rural households. An increase in the price of kerosene, used for cooking, would affect higher income households more than lower income households. Lower income households, as well as higher income households, could easily switch back to the use of biomass if sufficient biomass resources are available in the vicinity. An increase in the price of kerosene, used for lighting, would affect all income groups, particularly lower income households for whom changing to electricity might be difficult.

Abolishment of kerosene subsidies does have some national rationale, but in view of possible negative effects upon rural households, it should be undertaken gradually. A better organized kerosene distribution system could mediate that effect by reducing the distribution costs. Observations show that the rural population in the survey areas pay 50 to 100 percent more for kerosene than the official price at the depot. In some rural areas in Ciamis, for example, the price of kerosene was three times higher than the official price while the survey was being carried out.

Rural electrification appears to have greater appeal now because it might help resolve problems created by reducing kerosene subsidies, and might help lower the growth rate of kerosene consumption. As rural areas become more industrialized, demand for lighting increases. Therefore a good use for part of the subsidies on kerosene would be the financing of a more rigorous rural electrification program. Whereas kerosene subsidies cannot favor either lower income or higher income households, a progressive price structure for electricity could be introduced through a cross-subsidy scheme to benefit needy households.

Another part of the subsidies on kerosene could be reallocated to firewood-farming programs. According to the survey, demand for firewood is still large.

Biomass Consumption

Biomass is still a major part of the rural household energy budget. No time series data on the quality of biomass are available. However, observations suggest that the quality of biomass used as household fuel tends to deteriorate over time. Part of the reason, as suggested by the CSIS energy survey of rural industries, is that better quality firewood is being traded commercially to meet the need of rural industries. (See also Soesastro 1980). Even industries in the firewood-producing district, Ciamis, found good quality firewood scarce with large amounts being diverted to other districts, such as Cirebon, where the market price was considerably higher.

Problems of biomass supply therefore need serious attention. Only 13 percent of the households purchase their firewood. Most collect firewood from home yards, gardens, and estates. Programs could be designed to develop biomass supply, including processing biomass into higher quality fuel such as charcoal. Where demand exists, supply can be created, if prices are right. Potential income opportunities could also benefit the rural population.

During the survey, many higher income households were observed purchasing their firewood, even though kerosene was more economical. A higher income household would pay about Rp 30 per person per day for firewood (2.5 kg at the price of Rp 12.50 per kg) compared to Rp 19 per person per day for kerosene (0.53 liter at the price of Rp 35 per liter in early 1980) for cooking. When kerosene prices increased in early 1982 (to about Rp 80 per liter) the daily per capita cost increased to Rp 42, so firewood became more economically desirable.

Efficiency in the use of biomass is another important area for policy. Introduction of more efficient stoves should be reexamined, in spite of past failures. At the present efficiency level, (with one liter of kerosene generating useful energy equivalent to 5.3 kg of biomass) the cost differential of kerosene and biomass at current prices may not be large. If, however, kerosene prices were raised to their present production costs (Rp 132 per liter), the cost of kerosene would be twice that of biomass, or Rp 66 (at Rp 12.50 per kg for 5.3 kg), to produce the same amount of useful energy. Experience and data are inadequate to determine the potential responses of biomass substitution and price to a change of this order in kerosene prices. Nevertheless, it is obviously important to increase the efficiency of biomass as a fuel to reduce waste of resources nationally and to move toward the most effective long-term mix of fuels.

The Study's Use

Since the CSIS study was not done under contract by any government agency, it may not have a direct or immediate impact on policy. A report has been produced as well as a series of papers on specific policy issues using the data and information generated by the survey (Soesastro and Raymond Atje, eds., forthcoming).

Preliminary conclusions of the study, based on the survey, were delivered at the Second National Seminar on Energy organized by the Indonesian National Committee of the World Energy Conference. The papers presented emphasized conclusions as given in this chapter (Raymond Atje 1979; Soesastro 1979). The main recommendation of the seminar was acceleration of rural electrification programs.

NOTES

1. Rupiahs per current US $1.00:

1977	1978	1979	1980	1981	1982
415	625	627	627	634	650

2. In early 1980, when the official price was set at Rp 25 per liter, the mean values of kerosene prices reported in the survey were Rp 38 per liter in Cirebon, Rp. 39 in Serang, Rp 40 in Bandung, Rp 44 in Garut, and Rp 51 in Ciamis.
3. All three are agricultural areas with rice as the main crop. All are well supplied with kerosene. Klaten has a high population density, supported by intensive rice agriculture, with almost no forested land. East Lombok is the most traditional area and has relatively small landholdings. It is not as densely populated as Klaten, has a much less productive agricultural base and home garden cultivation also is not widespread. Luwu is a transmigrant area. It still has forests and has the lowest population density among the three areas. The land ownership is a uniform 2 ha per family for new settlements.

REFERENCES

Abdoellah, O.S.
    1979   *Penelitian Kayu Bakar di Desa Salamungkal, Kecamatan Paseh, Kabupaten Bandung* (Study of Firewood in Salamungkal Village, Paseh County, Bandung District). Internal Report. Bandung: Lembaga Ekologi Universitas Padjadjaran.

Arnold, J.H., Jr.
    1980   A Revised Methodology for Energy Demand Surveys. Prepared for the Workshop on Energy Assessment Methodologies, Board on Science and Technology for International Development of the National Academy of Sciences, Jekyll Island, January.

Atje, R.
    1979   Peranan Kayu Bakar dalam Pemerataan (The Role of Firewood in the Promotion of Greater Equity). In Hadi and others.

Atje, R.
    1980   Konsumsi Energi di Dektor Rumah Tangga Desa (Energy Consumption by Rural Households). *Analisa* IX(2):158–174.

Chatterjee, R.
- 1981 Energy Consumption in Rural Households, Appendix D in *Energy Planning for Development in Indonesia*, vol. 3, prepared by Energy/Development International.

Dick, H.
- 1980 The Oil Price Subsidy, Deforestation and Equity. *Bulletin of Indonesian Economic Studies* 16(3):32–60.

Direktorat Jenderal Ketenagaan, Departemen Pertambangan dan Energi RI.
- 1982 *Hasil Lokakarya Survai Energi Pedesaan*, Jakarta.

Hadi, S., and others.
- 1979 *Penggunaan Kayu Bakar dan Limbah Pertanian di Indonesia* (The Uses of Firewood and Agricultural Residue in Indonesia). Paper presented at the Seminar of the Indonesian National Committee of the World Energy Conference, Jakarta, April.

Haeruman, H., and others.
- 1977 *Studi Konsumsi Sumber Daya Enersi Pedesaan, Terutama Kayu Bakar di Propinsi Jawa Barat* (The Study of Rural Energy Consumption, Especially Firewood, in the Province of West Java), Bogor: Institut Pertanian Bogor.

Harahap, F.
- 1978 Penelitian dan Pengembangan dalam Bidang Pemanfaatan Energy-Surya di Institut Teknologi Bandung (Research and Development on the Utilization of Solar Energy at the Institute of Technology Bandung), Appendix III. Paper presented at the Seminar of the Indonesian National Committee of the World Energy Conference, Jakarta, May 25–26.

Komarudin.
- 1977 *Penelitian Konsumsi Kayu Bakar di Desa Babakan, Kecamatan Ciomas, Kabupaten Bogor* (Study of Firewood Consumption in Babakan Village, Ciomas County, Bogor District). Thesis, Institut Pertanian Bogor, Faculty of Forestry.

Martono, A.
- 1974 Beberapa Pemikiran tentang Kebijaksanaan Penggunaan Energi bagi Keperluan Rumah Tangga (Some Thoughts on Policies on Energy Uses for Household Needs). Paper presented at the Seminar of the Indonesian National Committee of the World Energy Conference, Jakarta, 24–27 July.

Nasendi, B.D.
   1978   *Analisa Konsumsi Sumber Daya Energi Pedesaan Khususnya Kayu Bakar di Daerah Aliran Sungai Citanduy Jawa Barat* (Analysis of the Consumption of Rural Energy Resources Especially Firewood in the Citanduy Watershed in West Java), Institut Pertanian Bogor, Post-Graduate School.

Soesastro, H.
   1979   Mencari Energi Pengganti Minyak Tanah Sebagai Unsur Pemerataan (In Search of Substitutes for Kerosene Towards Greater Equity). In Hadi and others.

Soesastro, H.
   1980   Basic Energy Budgets of Rural Households in Indonesia. *The Indonesian Quarterly* 8(1):21–38.

Soesastro, H.
   1980   *Peranan Energi di Sektor Industri Pedesaan Java Barat* (The Role of Energy in the Rural Industries of West Java). Mimeo. Centre for Strategic and International Studies (CSIS), August.

Soesastro, H., and others.
   1983   *Energi dan Pemerataan* (Energy and Equity). Jakarta: CSIS.

Strout, A.M.
   1978   *The Demand for Kerosene in Indonesia*. Mimeo, July.

Sumarna and Sudiono.
   1973   *Konsumsi Kayu Bakar oleh Rumah Tangga, Industri dan Perusahaan Jawatan Kereta Api di Jawa Timur* (Consumption of Firewood by Households, Industries and Railways in East Java). Bogor: LPHH.

Weatherly, W.P.
   1980   *Environmental Assessment of the Rural Electrification I Project in Indonesia*. Prepared for the USAID, Embassy of the USA, Jakarta, December.

Wiersum, K.F.
   1976   *The Fuelwood Situation in the Upper Bengawan Solo*, Upper Solo Watershed Management and Upland Development Project, Solo.

Wiersum, K.F.
   1979   Methodology of Fuelwood Surveys with Reference to Indonesia Data. Prepared for FAO's FSP 42: *Forestry for Local Community Development*, Rome, May.

4
Philippine Rural Energy Resource and Consumption Survey

Planning Service
Ministry of Energy
The Philippines

INTRODUCTION: THE GENESIS OF THE SURVEY

The Planning Service of the Ministry of Energy (MOE) was just finishing its third annual iteration of the country's 10-year energy plan in 1978, when it embarked on a nationwide survey of 1977 energy consumption of households, classified as rural or urban. It had been observed that the initial formulations of the energy program did not adequately address the issue of rural energy needs and potentials. The base data available up to that time were not sufficient to provide profiles of energy consumption patterns by economic sector. No data were available for the rural household sector, the agricultural industry sector, the inward-looking manufacturing and processing industry sector, the transportation sector, or the various export sectors. The rural household survey, then, would be only one of several modules in an overall energy mosaic for the entire economy.

Motivated to obtain indicative statistics for planning and program formulation purposes as quickly as possible (preferably within six months) and limited by a budget of $37,871[1] (from the USAID grant to The Center for Non-Conventional Energy and Development), the Planning Service staff decided to save on survey machinery and administrative overhead by taking advantage of a nationwide household expenditure survey conducted periodically by the National Census and Statistics Office (NCSO) of the National Economic and Development Authority (NEDA). NCSO had the largest sample size available: 30,000 households nationwide. A questionnaire had to be developed in time for the NCSO annual survey of households. In three weeks, after consultations between MOE staff and NCSO officers, the first draft of an eight-page questionnaire was completed.

OBJECTIVES OF THE SURVEY

The first objective of the survey was to establish an approximation of how much of what form of energy was used in which region for what purpose. The ultimate objective was development of policies to define the long-term role of nonconventional fuel forms and development of the inputs to satisfy rural energy requirements. Both commercial and noncommercial energy forms were covered, accessed by both urban and rural households. However, discussion of survey results here is confined to the rural households.

The survey also sought to determine whether selected socioeconomic variables in the respondent's environment significantly influenced energy consumption patterns, whether biomass fuel usage correlated well with distribution of biomass resources in each region, and whether regional consumption of other fuels was correlated with region-specific resource and infrastructure availabilities. Another objective was provision of baseline information for future policy studies and concerns.

Since this was the first attempt at collecting data on household energy consumption, survey results should be treated with caution. Certain biases in sampling practices and design may have resulted in overestimated rural consumption. This is evident with respect to commercial energy consumption, for which secondary data available with the Ministry provide a counter check. Accordingly, the Ministry is confident of the overall picture of noncommercial energy use among rural households but looks to future surveys to refine specific statistics.

CHARACTERISTICS OF THE STUDY AREA

This island nation's total land area is approximately 300,000 sq km or 30,000,000 ha. Of that, 98 percent is contained within the 11 largest islands. Discussions herein are confined to 12 of the 13 regions since the population of the other one, the National Capital Region, is entirely urban (see Table 1). These regions are subdivided into provinces (in 1977 there were 72); and each province, into municipalities and cities. Several *barangays*, the smallest political unit in the Philippines, make up each municipality or city. In 1977 there were 40,439 barangays in the country.

The climatic conditions in the different parts of the archipelago vary as much as does the topography and location. Four types of climate are identified according to presence or absence of a dry season and the maximum rain period:

Type A. Two pronounced seasons — wet from May to October and dry the rest of the year. Regions I, III, and IV belong to this type. Mountain ranges shield these regions from the northeast monsoon and part of the tradewinds, but they are prone to the southwest monsoon. Part of Regions VI, IX, and X also belong to this type.

Type B. No dry season and very pronounced maximum rain period

TABLE 1.
Regional Profile: Land Area, Population, and Labor Force, 1977

Region	Land Area	POPULATION			Labor Force
		National	Rural	Urban	
I. Ilocos Region	7.2	7.8	9.2	4.6	8.4
II. Cagayan Valley	12.1	4.6	6.0	1.7	4.7
III. Central Luzon	6.1	10.0	10.4	9.2	10.2
IV. Southern Tagalog	15.6	12.4	12.9	11.2	22.8
V. Bicol	5.9	7.6	9.2	4.2	7.5
VI. Western Visayas	6.7	9.9	10.7	8.0	10.7
VII. Central Visayas	5.0	8.1	8.4	7.2	9.5
VIII. Eastern Visayas	7.1	6.2	7.4	3.6	6.3
IX. Western Mindanao	6.2	4.9	6.1	2.3	3.2
X. Northern Mindanao	9.4	5.5	6.5	3.2	5.4
XI. Southern Mindanao	10.6	6.5	7.1	5.1	6.6
XII. Central Mindanao	7.8	4.9	6.1	2.4	4.8
TOTAL	99.8	88.2	100.0	62.6	100.0
TOTAL	300,000	42,071,000	28,765,000	13,306,000	12,224,964
PHILIPPINES	sq km	persons	persons	persons	persons

Note: Figures are given as percentages of total. Balance percentages (National and Urban) pertain to National Capital Region. Labor Force total pertains to the 12 regions only.

from November to January. These regions are located along or very near to the eastern coast and are not sheltered from northeast monsoon, tradewinds, and storms. Region V, part of Region VIII, and part of Regions X and XI belong to this type.

Type C. No pronounced season, relatively wet from May to October, and dry the rest of the year. Maximum rain periods are not very pronounced; dry seasons last from one to three months. Areas are partly sheltered from northeast monsoon and tradewinds but open to the southwest monsoon or at least to frequent storms. Regions II, III, V, VIII, and the rest of the Visayan regions belong to this type.

Type D. Rainfall evenly distributed throughout the year. Part of Region V and the rest of the Mindanao regions belong to this type of climate.

In 1977, the estimated population of the Philippines was 44.4 million (including the National Capital Region). Rural residents made up 68.5 percent and urban residents, 31.5 percent. Population density was 148 persons per sq km. Central Luzon (Region III), with 231 persons per sq km, is the most densely populated region after the National Capital Region. Cagayan Valley (Region II) is the least densely populated region, with 53 persons per sq km.

In 1975, 4,764,000 of the 6,859,000 families in the Philippines lived in rural areas. By 1978, the total number of rural families was expected to be about 5,120,000. Mean rural annual household income was ₱4,745 (US $630) (NEDA 1979). On a regional basis, rural families in Western Mindanao (Region IX) received the highest average annual income and in Northern Mindanao (Region X) the lowest average annual income.

An examination of the consumption pattern of rural households revealed that the utilization of rice hulls, charcoal, coconut, firewood, and biogas, the principal forms of noncommercial energy[2], is related to land utilization, agricultural crops (mainly palay and coconut), livestock, and agricultural infrastructure and logistics in the different regions.

In 1977, of the total land area of the country 12.2 percent was devoted to *palay* (rice) production, 9.1 percent to coconut, and more than 50 percent of the country's total land area was covered by forest. Regional analysis showed that Bicol (Region V) ranked first in utilization of regional land area for palay production (20 percent), while Southern Mindanao (Region XI) led in relative land area devoted to coconut production (14 percent).

Major lowland Regions II, III, and XII ranked highest in palay production, Central Luzon being the biggest palay producer with 16 percent of total palay production. Central Visayas (Region VII) had the least share in palay production, only 2 percent. The Mindanao regions (IX –XII) had the highest coconut production, with 68 percent of the country's total nuts harvested; Region IV was next, with 18 percent.

Total rice milling capacity was 6,173,000 metric tons (T). More than 18 percent was in Central Luzon, with Cagayan Valley next at 14 percent. The country's total sawmilling capacity was placed at 7,156,000 board feet. Sawmills in Cagayan Valley produced 19 percent, and Southern Mindanao, 17 percent.

Agricultural residues quantified in the study are rice hulls and logging wastes. Based on palay production and milling capacities, rice hull production was estimated at 1,031,000 T and logging residues amounted to 6,299,000 m^3. Cagayan Valley had the highest rice hulls produced while Southern Mindanao generated the biggest share in the total production of logging residues (38 percent).

In 1977, Central Visayas (Region VII) raised the most hogs, with 12 percent of the total hog population; Central Mindanao (Region XII) had the least, with 3 percent. That same year, Southern Tagalog led the other regions in cattle production, with an 18 percent share, and in poultry production, with a 21 percent share. Carabao population was highest in Cagayan Valley, with a 16.5 percent share.

In 1977, the Philippines had about 3.6 million ha devoted to palay production, with only 39 percent being served by various irrigation systems. Southern Tagalog had the largest irrigated area (16 percent of the total area irrigated) while Central Visayas had the least, 6 percent.

Electricity was provided to 652,664 rural households by electric cooperatives in 1977. That number represented 21 percent of the potential household connections that year. (Potential household connections refer to the number of households in areas serviced by electrical cooperatives.)

The number of public markets in 1977 was estimated to be 1,344, with an average of 112 per region. This estimate is based on 1982 data of the National Food Authority (NFA), excluding the number of new licenses issued by the Ministry of Human Settlements from 1978 to 1980.

SURVEY DESIGN AND PROCEDURE

Rural households, as defined by NCSO, are those found in areas not having these attributes:

1. Cities and municipalities having a population density of at least 1,000 persons per sq km in their entirety.
2. *Poblaciones* or central districts of municipalities and cities having a population density of at least 500 persons per sq km.

3. Poblaciones or central districts (not included in items 1 or 2) regardless of population size with
   a. a network of streets in either parallel or right angle orientation;
   b. at least six establishments (commercial, manufacturing, recreational, or personal services); and
   c. at least three of these features:
      (i) a town hall, church, or chapel with religious service at least once a month;
      (ii) a public plaza, park, or cemetery;
      (iii) a market place or building where trading activities are carried on at least once a week; and
      (iv) a public building like a school, hospital, puericulture and health center, or library.
   d. Barangays with at least 1,000 people living in areas having the features listed in item 3c whose livelihoods are predominantly nonfarming/fishing.

Sampling Design

The sampling design and the selection of sampling frames were patterned after the ones used in NCSO's 1977 Annual Survey of Establishments and its 1978 Integrated Quarterly Survey of Households. The frame for the 1978 survey was the 1975 Census of Population. Sampling for NCSO surveys is designed to provide data at the provincial level so a two-stage, stratified-sampling scheme was used by province. The primary sampling units of the survey are the barangays. A sample barangay is chosen according to its number of households.

The NCSO methodology groups barangays in a province by six economic activity strata, namely, (1) palay, (2) corn, (3) other agricultural products, (4) fishing, (5) manufacturing, and (6) other economic activities. The general procedure for this first stratification is as follows: manufacturing is considered the major activity stratum if 20 percent of its working people are engaged in manufacturing. Otherwise, the barangay is classified by the stratum with the greatest proportion of the working population. Sample households were first allocated to each of the six strata at the national level. Then using those emerging national levels as controls, they were grouped by province. A predetermined overall sampling ratio was held constant within each stratum at both the national and provincial level.

The final size and allocation of the sample, however, depended on the number of barangays drawn into the sample. The number of sample barangays for a stratum in a given province was obtained from the equation

$$b_h = f_h \frac{(N_h)}{(K_h)}$$

where

f_h = the overall sampling ratio for the hth stratum (same for all provinces),
N_h = the estimated number of households in h^{th} stratum at the time of the survey, and
K_h = the predetermined number of sample of households per barangay, namely 6.

Overall sampling ratios by rural stratum are fishing, 1/100; palay, corn, and other agricultural products, 1/300; and manufacturing and other economic activities, 1/250.

A minimum of two sample barangays was selected for each stratum with three or more listed barangays. In strata with only one or two listed barangays, all the barangays were selected as samples. The specified number of sample barangays per stratum in each province were identified by population size, starting with the most populated.

Then households in the sample barangays were designated as either engaged (substratum 1) or not engaged (substratum 2) in the predominant economic activity stratum of the barangay. Then three sample households per substratum were randomly selected. If less than three households were listed under substratum 2, the samples in substratum 1 were increased to equal six samples per barangay. The main NCSO Survey to which the Energy Survey was a "rider" covered 30,000 households. However, only 15,000 households were interviewed for the MOE Energy Survey out of which only about 85 percent gave valid responses.

Data Gathering and Analysis

Data were collected using a structured questionnaire in personal interviews with housewives and household heads. The energy usage household questionnaire gathered five basic kinds of information:

1. Total energy consumption and expenditures per household for 17 energy items;
2. Electrical consumption for various appliances per household;
3. Consumption of fuels other than electricity;
4. Availability of animal wastes as fuel source; and
5. Energy consumption of transport facilities per household.

Questionnaire Design. The format, design, and content of the questionnaire were revised through (1) discussions with statisticians of the NEDA Inter-Agency Committee on Survey Design, (2) discussions with the field operations supervisors of the NCSO household division, and (3) feedback from mock interviews that provided other details necessary to obtain information more quickly and accurately. Time constraints meant the mock interviews replaced actual field testing. Field workers were trained during that testing process.

The NEDA Inter-Agency Committee on Survey Design was consulted to standardize and simplify the questionnaire for maximum response effectivity. The NCSO consulted with the MOE about the final questionnaire form (such as number of questions, length of interview, appropriate use of technical terms, and various units of measure). The final questionnaire thus was a compromise formulated by energy accounting experts, statistical experts, and field survey experts. Technical precision might have been traded for comprehension and higher effective probability of response, to comply substantially with standards for statistical significance.

Training and Operations. Survey training was done in three sessions, each of two to three days. First, training of trainers and NCSO Central Office field supervisors (three from MOE and six from NCSO) included specifying how the two training sessions would be conducted. Second was the training of regional and provincial census officers in several regions and provinces. In turn, under the supervision of the trainers, the regional and provincial census officers trained 752 municipal census officers and contractual field enumerators.

Discussion of the concepts and procedures of the energy consumption survey followed an operations manual specifically designed for the survey. The manual contained principles and instructions as guidelines for gathering accurate information about the energy usage of every sample household. Training sessions included discussions of survey details, the use of materials, and mock interviews.

The municipal or barangay census officers had primary responsibility for gathering field data. Where an overload of interviews was expected, census field workers were hired. Field assignments included updating maps of the area, picking and interviewing sample households, and verifying data. Their wages varied according to the spread of households within an assigned area and related transportation costs.

Data gathering took about two and one-half months (from February to mid-April 1978). Included were enumeration; editing-coding, handling, and folioing of questionnaires; and verification of edited-coded questionnaires. The provincial census officers and provincial statisticians checked the accomplished questionnaires for completeness. Missing or incorrect data were input and blank questionnaires were considered nonresponses and discarded. More editing and reverification of data and codes were done by regional census officers. Then the questionnaires were bundled off to the NCSO for further editing of data and codes and then were sent to the MOE's Planning Service Division.

Data Processing. A computer firm was hired for data processing, including conversion of data to a common energy unit and relating energy data to sociological factors. A range of possible answers to each item was set using inferred limits from other related data. Answers outside this preset range were excluded from the final estimation. Altogether 15 percent of all respondents were rejected, but additional households were not interviewed to replace them due to budget constraints.

Full data analysis, originally scheduled for completion within 14 weeks, was interrupted when a fire gutted the building housing the computer firm's office. About 50 percent of the contracted work was accomplished including folioing, coding, and keypunching of survey data. The remaining phases contracted for were development of programs, entry and validation of data, generation of tables, and preparation of reports on energy consumption in relation to family size and income. Of the $9,600 contract amount for analysis, 85 percent was spent on data processing.

The information gathered directly from respondents expresses fuel consumption volumes in the units of measure used in the various localities or nearest marketplace. Those units were translated into one common unit of measure for aggregation using conversion formulas and methodologies (Appendix A). The barrel-of-oil (BOE) equivalent was used, primarily because of habit and practical convenience. In general, all non-oil fuels were converted to the volume of an oil product that would have been required to do an equivalent amount of work or generate an equivalent amount of heat. That oil product volume was then converted to the crude oil equivalent volume based on calorific conversions. The reference oil is crude oil with a heating value of 18,900 British Thermal Unit per pound (BTU/lb.).

Some Limitations

The survey's use of the national statistical office might have caused these problems:

1. Additional time commitment for both interviewer and respondent on an already voluminous socioeconomic questionnaire.

2. Possible compromises between interviewer and regular respondent-clientele.
3. Use of a possibly inappropriate NCSO standard sampling procedure for this specialized energy study.
4. Chance of unacceptable field practices by interviewers.
5. Use of only NCSO definitions and household economic stratifications for cross-survey comparisons and correlations.

The study has some overall limitations. First, it did not account for human and animal energy used for rural domestic chores. Moreover, since some energy items do not have standard units of measure, the interviewers had to determine the amount consumed through knowledge of local units of measure. In addition, as with all sample surveys, some room must be allowed for possible sampling errors and memory lapses of the respondents. At times respondents consciously underreport or overreport, for various reasons from wanting to impress to not caring, or just wanting to be finished with a lengthy questionnaire. Interviewers were advised to elicit details related to a specific item to help them evaluate the response.

The study's most significant limitation was its noncompletion as designed. The data processor's inability to complete the analysis meant inferior surrogate variables had to be used for regression analysis. For instance, regional household averages were used instead of individual household data. Finally, the season in which the survey was undertaken may also have biased the aggregated results.

RESEARCH METHODS AND ENUMERATION OF FINDINGS

The survey data were aggregated nationally and regionally to estimate energy source mixes and household applications and to study income effects on energy consumption patterns. Three statistical methods were also employed to relate the data to other regional statistics: regression analysis, Spearman's rank correlation analysis, and Lorenz distribution correlation analysis, borrowing from the concepts used in standard Lorenz curve analysis and the Gini-coefficient. In the process, results were compared where possible with those obtained in other rural surveys, particularly the microregion studies conducted by Fernando Manibog (Verde Island), Bruce Koppel (Bicol River Basin Region), and a nationwide National Electrification Administration (NEA) survey. The findings and methods of analysis are grouped and discussed below under three headings: first-order aggregations, regression results, and correlation results.

First-Order Aggregations

The first-order aggregates simply refer to the extrapolated summation of results to derive volume estimates for energy source mixes, household applications, and income effects. A spaghetti diagram (Fig. 1) tracing national energy flows from source to application for rural households was constructed based on survey data for sources and on extrapolated data from a 1979 energy consumption survey of low income urban households (MOE study 1979) for allocation by end use and inherent loss. Noncommercial fuels not covered by that survey (such as woodwastes and rice hulls) were assumed to be cooking fuels since cooking is their only household use.

The typical rural household consumed an equivalent of 4.1 BOE per year or a per capita annual consumption of 0.6 BOE, using the Planning Service's computed average family size of seven. Total consumption by rural households in 1977 was equal to 21 million BOE. Nationally, 60 percent of rural household energy consumption was noncommercial. Firewood (55.5 percent), charcoal (24.9 percent), and coconut shells (17.9 percent) together accounted for 98 percent of noncommercial energy used. Kerosene (47.9 percent), gasoline (39.5 percent), LPG (liquefied petroleum gas) (4.7 percent), and electricity (4.5 percent) accounted for 97 percent of total commercial energy forms. The per household monthly electricity use of 42 kWh, or 0.07 BOE, was close to the 48 kWh figure obtained by the NEA in a survey of 1976 consumption.

Variations across regions in the sources of energy for household functions are indicated in Tables 2 to 4 as regional aggregates and on a per household basis. Five regions (Southern Tagalog, Northern Mindanao, Ilocos, Western Mindanao, and Central Luzon) accounted for 67 percent of rural household energy. The percentage of commercial to total energy consumed varied widely among those regions, from 18 percent in Western Mindanao to 47 percent in Central Luzon. Only one region, Cagayan Valley, consumed more than 50 percent commercial energy.

Energy Applications. Useful energy accounted for 45 percent (9.5 million BOE) of total energy inputs to rural households in 1977 (derived by applying secondary information as efficiency factors). Energy wasted in transmission and distribution losses and inefficiencies intrinsic to appliances made up 55 percent (11.6 million BOE) of the total energy inputs to rural households.

Cooking accounted for most — 60 percent — of rural household usage. Cooking, ironing, household transport, and lighting collectively accounted for over 98 percent of total rural household energy consump-

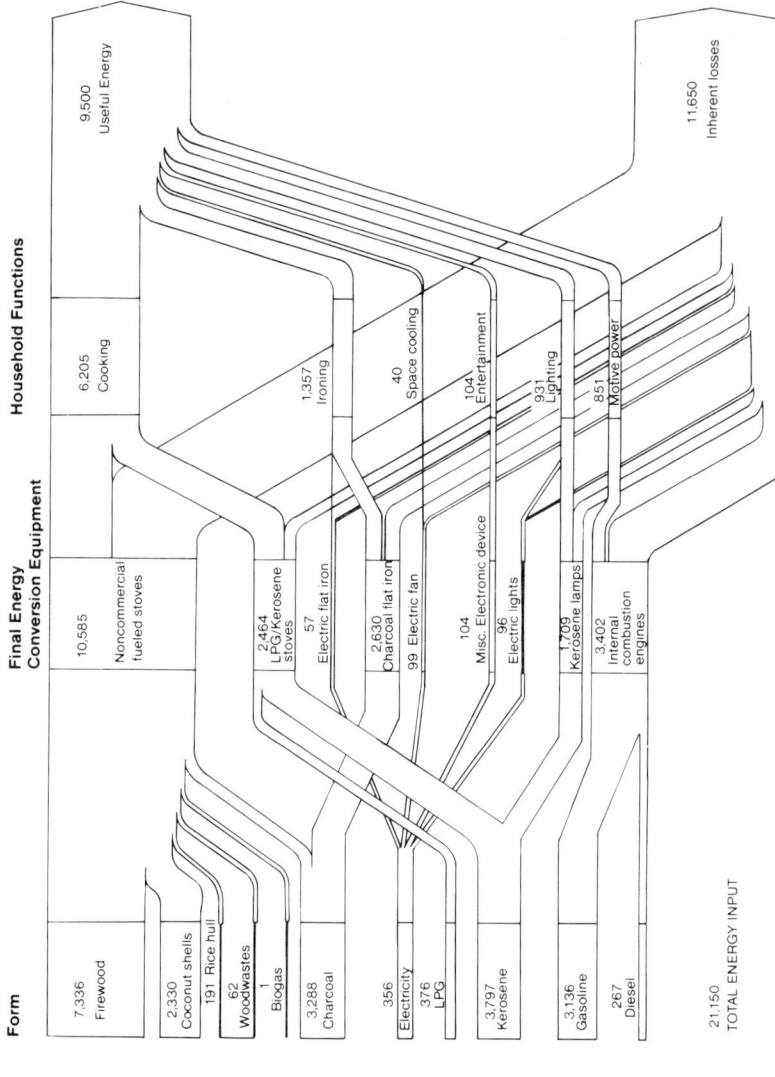

Figure 1. *Estimated Energy Flow and Inherent Losses Among Rural Households (in MBOE)*

TABLE 2.
Total Estimated Energy Consumption of Rural Households, 1977

		Commercial		Noncommercial		Total	
		'000 BOE	Percent	'000 BOE	Percent	'000 BOE	Percent
I.	Ilocos	919	41.6	1,291	58.4	2,210	10.4
II.	Cagayan Valley	291	60.1	193	39.9	484	2.3
III.	Central Luzon	936	46.7	1,070	53.3	2,006	9.5
IV.	Southern Tagalog	1,987	39.9	2,991	60.1	4,978	23.5
V.	Bicol	610	44.8	753	55.2	1,363	6.4
VI.	Western Visayas	508	40.9	735	59.1	1,243	5.9
VII.	Central Visayas	404	30.5	922	69.6	1,325	6.3
VIII.	Eastern Visayas	462	43.3	606	56.8	1,068	4.9
IX.	Western Mindanao	402	17.8	1,854	82.2	2,256	10.7
X.	Northern Mindanao	804	29.4	1,928	70.6	2,732	12.9
XI.	Southern Mindanao	98	28.0	252	72.0	350	1.7
XII.	Central Mindanao	512	45.2	620	54.8	1,132	5.4
	TOTAL	7,932	39.6	13,215	60.4	21,147	100.0

TABLE 3.
Rural Energy Consumption by Source, 1977
(Regional Aggregates)

Energy Source	Highest Consumption		Lowest Consumption	
	Region	Percentage	Region	Percentage
Noncommercial	IV	22.6	II	1.5
Firewood	IV*	14.7	II	1.2
Charcoal	IV	40.2	XI	1.0
Coconut shells	X	65.9	XI	0.1
Ricehulls (only in two regions)	IV	87.9	III	12.1
Woodwaste	III	54.1	X	0.05
Commercial	IV	25.0	XI	1.2
Kerosene	IV+	23.7	–	–
Gasoline	IV	25.4	XI	0.3
LPG	IV‡	41.0	–	–
Electricity	III	33.4	VIII	0.02
Diesel	XII	18.9	I	0.04

*Region I was second, with 13.8 percent.

+The other two top users were Region V, 10.5 percent, and Region III, 7.9 percent.

‡The other two predominant users were Region 1, 23.3 percent and Region III, 23.2 percent.

tion in 1977. Firewood alone supplied 60 percent of cooking needs. (See also Koppel 1980). Except for a small amount of charcoal, used mainly (80 percent) for ironing, noncommercial fuels were used exclusively for cooking.

Reported uses of gasoline and diesel were mainly for household transport. It is surprising that the more expensive product, gasoline, accounted for 91 percent of this amount. The cheaper product, diesel, supplied the 9 percent residual. Perhaps this is because only few rural families can afford the front-end expenditure on the more expensive diesel engine. A significant portion also might have been used for irrigation and agricultural cargo. However, the raw data alone do not indicate how large that portion might be.

Kerosene was used for cooking (2.1 million BOE) and also for lighting on a 55:45 ratio. LPG was used almost entirely for cooking (0.4 million

TABLE 4.
Energy Consumption per Rural Household by Energy Item,
(GJ), 1977

	\multicolumn{13}{c}{R E G I O N}												
	I	II	III	IV	V	VI	VII	VIII	IX	X	XI	XII	Philippines
Noncommercial Energy													
Firewood	12.4	1.6	7.8	9.4	3.1	6.0	9.4	7.1	33.3	3.4	3.5	11.3	8.6
Woodwaste	0.2	-	0.4	-	-	0.1	-	-	-	-	-	0.1	0.1
Charcoal	3.2	2.0	4.8	11.4	3.5	1.3	3.3	2.4	3.5	2.7	0.5	2.8	3.8
Coconut Shells	-	-	0.1	3.0	3.2	0.6	0.2	0.1	0.7	24.3	0.1	0.2	2.8
Rice Hulls	-	-	0.3	1.4	-	-	-	-	-	-	-	-	0.2
Biogas	-	-	-	-	-	-	-	-	-	-	-	-	-
Subtotal	$\overline{15.8}$	$\overline{3.6}$	$\overline{13.4}$	$\overline{25.2}$	$\overline{9.8}$	$\overline{8.0}$	$\overline{12.9}$	$\overline{9.6}$	$\overline{37.5}$	$\overline{30.4}$	$\overline{4.1}$	$\overline{14.4}$	$\overline{15.5}$
Commercial Energy													
Electricity	0.9	-	1.5	0.8	0.1	0.1	-	0.3	1.0	0.3	0.1	0.5	0.4
Petroleum	10.3	5.4	10.3	16.4	8.0	5.4	5.3	7.3	7.8	12.4	1.5	11.4	8.9
Gasoline	(5.4)	(0.2)	(4.4)	(6.9)	(2.5)	(2.6)	(1.0)	(2.3)	(3.2)	(9.0)	(0.1)	(3.2)	(3.7)
Kerosene	(3.8)	(2.9)	(4.4)	(7.8)	(5.5)	(2.8)	(4.1)	(4.9)	(4.2)	(3.1)	(1.4)	(6.8)	(4.4)
Diesel	-	(2.2)	(0.4)	(0.4)	-	-	(0.1)	-	(0.2)	(0.1)	-	(1.2)	(0.3)
L P G	(1.1)	(0.1)	(1.1)	(1.3)	-	(0.1)	(0.1)	(0.1)	(0.2)	(0.2)	-	(0.1)	(0.4)
Subtotal	$\overline{11.2}$	$\overline{5.6}$	$\overline{11.8}$	$\overline{17.2}$	$\overline{8.1}$	$\overline{5.5}$	$\overline{5.3}$	$\overline{7.6}$	$\overline{8.8}$	$\overline{12.7}$	$\overline{1.6}$	$\overline{11.9}$	$\overline{9.3}$
T O T A L	27.0	9.2	25.2	42.4	17.9	13.5	18.2	17.2	46.3	43.1	5.7	26.3	24.8

Note: ( ) = signifies product subtotal.

BOE). The only other commercial energy form was electricity, used for these "internal" household functions: media access, 29 percent; electric fan, 28 percent; lighting, 27 percent; and ironing, 16 percent.

Income Level Effects. In the original presentation of the survey results, regional consumption data were disaggregated into 13 household income levels following NCSO classification. However, since some of the income levels either had minimal (below 0.01 BOE) or no corresponding consumption, the Planning Service regrouped the data into three annual income classes, namely, low (up to ₱2,500), middle (above ₱2,500, below ₱8,000), and high (₱8,000 and up). These three household income classes were used to compare energy consumption to income in the 12 regions.

This comparison shows that in all regions, low-income families accounted for most household energy consumption as well as most noncommercial energy consumption. The high-income households invariably consumed the least, in both respects (Figs. 2 and 3). The main reason may simply be the fact that low-income families outnumbered middle-income and high-income families in the random selection, if not in actual field conditions. Low-income families accounted for 65 percent of total consumption of noncommercial fuels; middle-income families accounted for 32 percent; and the high-income bracket accounted for the 3 percent residual.

The middle-income group consumed more than three times as much fuel (commercial and noncommercial) per household as the lower and higher income brackets. Except for Region XI, commercial energy consumption per household dropped as family incomes changed from medium to high (see Fig. 4 and Appendix B). This outcome can be explained by any combination of the following assumptions: (1) higher income families either use more efficient energy appliances, (2) or have smaller family sizes, (3) or have better ability to contract some household functions outside (such as cooking by dining out).

In all cases except three (Regions VI, VII, and XI), commercial energy consumption per household increased as families moved from low- to middle-income levels (Fig. 4 and Appendix B). Low-income families facing severe budgetary constraints are not normally able to afford more energy-efficient appliances that utilize commercial energy forms, or they use more of the noncommercial forms because of tradition and economics. Manibog's survey of Verde Island suggested that per capita energy use from all sources increased with income.

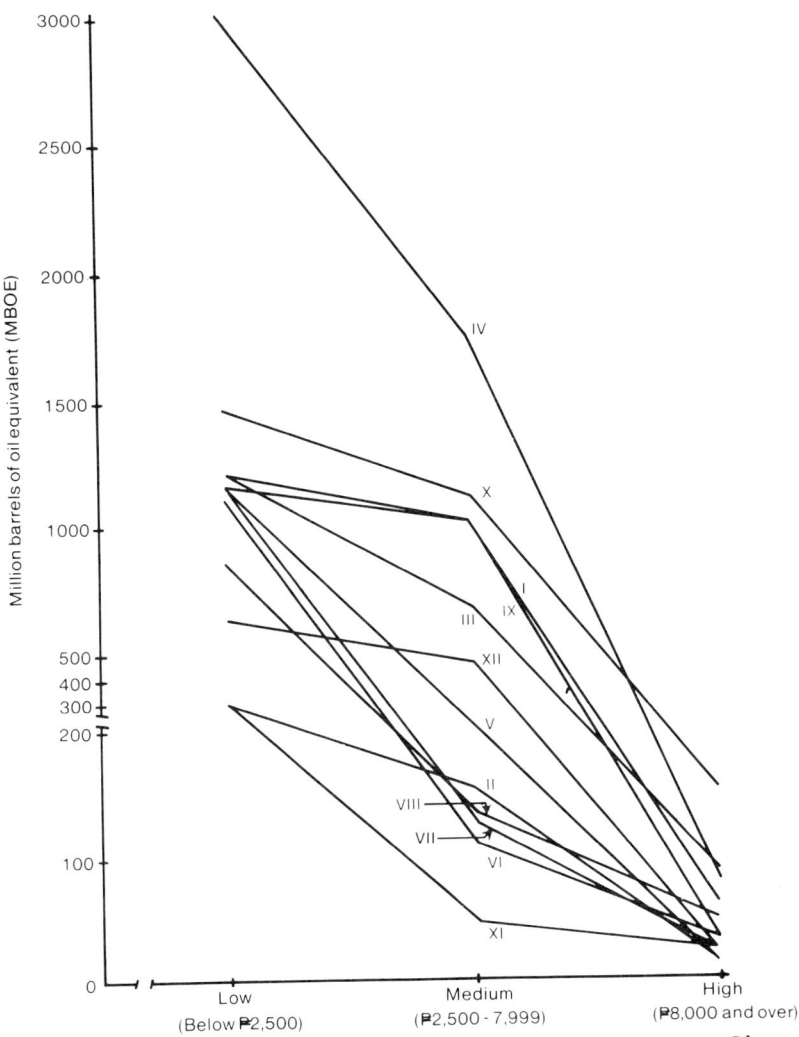

Figure 2. Total Energy Consumption of Rural Households, by Income Class and Region (in MBOE)

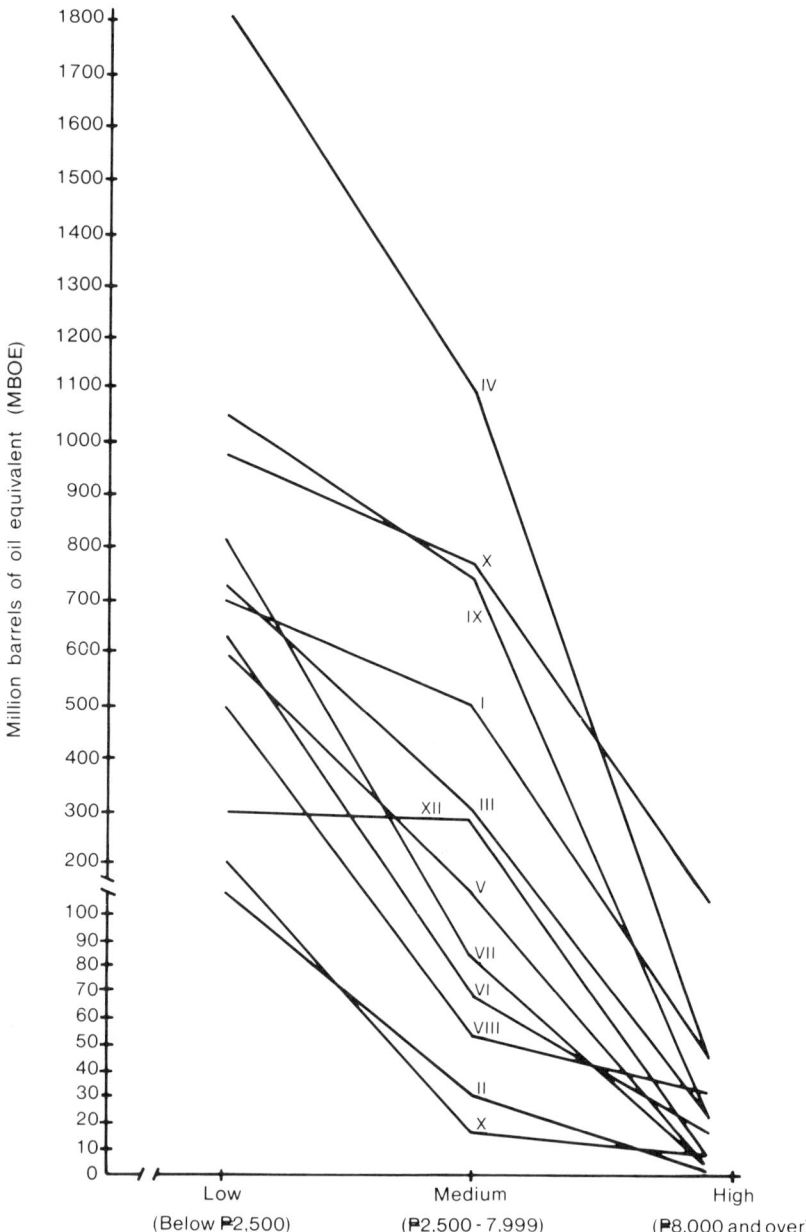

Figure 3. Noncommercial Energy Consumption of Rural Households — by Income Class and Region (in MBOE)

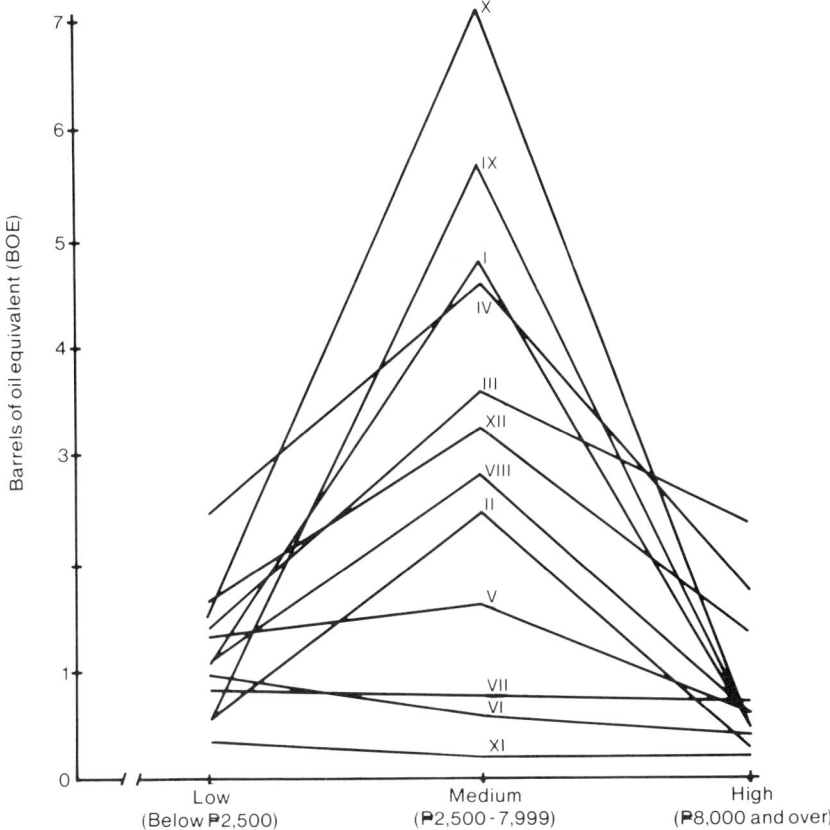

Figure 4. Commercial Energy Consumption per Household — by Income Class and Region (in BOE)

Variables Influencing Rural Energy Consumption

Regression analysis basically attempted to determine if and how significantly aggregate and fuel-specific energy consumption at the regional level were influenced by family size, household incomes, and regional energy prices. The regional consumption of total energy as well as per fuel product were regressed against several socioeconomic variables separately as well as in combination, using secondary and survey data for the 12 regions.

Eleven dependent and four independent variables were developed from the data. The independent variables were (1) middle-range family size per region (estimated by largest family reported in the region divided by two, assuming a normal distribution function (because of the loss of the raw questionnaires); (2) average annual household income per region (from NEDA and MOE statistics); (3) price per energy item or average imputed value, estimated from the regional energy bill by energy item divided by regional energy consumption by energy item in BOE; and (4) number of public markets in the region. Dependent variables were (1) per household total energy consumption; (2) per household noncommercial energy consumption (total and separately for firewood, charcoal, and coconut shell); and (3) per household commercial energy consumption (total and separately for gasoline, kerosene, diesel, electricity, and LPG).

Of the 70 regression exercises, only eight turned up statistically significant coefficients. That was to be expected because of the aggregate nature of the data and the low number of observations relative to the number of variables considered. Interestingly, most of the cases where the coefficients turned out significant involved the commercial energy forms of electricity and gasoline. Selected regression findings follow.

1. Alone, regional mid-range family size proved to be an insignificant determinant of consumption levels for any one energy product. Manibog's survey concluded that overall energy use was a direct function of family size although exhibiting diminishing increments with size beyond a household size of six.

2. Regressing total energy consumption against income did not result in any statistically meaningful coefficients. On the other hand, Manibog discerned a modest association ($r = 0.44$) between income level and energy consumption, although cautioning that the bulk of energy used was not commercial and therefore, nonmonetized.

3. Alone, the level of regional income proved significant only in association with firewood and coconut shell consumption but with contrasting results. The runs yielded a negative coefficient in the case of coconut shell consumption, suggesting that coconut shell could perhaps be the only true "poor-man's" fuel. With firewood, a positive coefficient resulted (0.0013). Koppel's work supports this finding. He observed that firewood was consumed even as the level of living improved. On the other hand, Manibog had concluded that, unlike family size, income level was not a good indicator of wood consumption levels.

4. From survey data, regional kerosene, gasoline, or diesel consumption did not seem to be significantly affected by regional income levels. Manibog reached the same conclusion on household use of gasoline and diesel but suggested the contrary for kerosene, for which he found higher levels of household use accompanying higher household incomes.

5. The consumption effect of price alone was significant only for gasoline and electricity as energy forms. Coefficients were negative and

expectedly low (-0.0190, $t = -2.3$; -0.00115, $t = -2.9$, respectively). The NEA Survey of 1977 tends to support this finding. Respondents confirmed that they would either reduce or completely stop electricity consumption, with a doubling of electricity rates.

6. Together, income and price regression coefficients were significant only in the case of gasoline and electricity. In both cases, the consumption response to price was expectedly negative (-0.02, -0.001, respectively), gasoline sensitivity exceeding that of electricity by a factor of 20 ($t = -2.9$, -2.8, respectively; $F = 6.97$ and 4.42, respectively). Electricity consumption responded positively to income although low and insignificant (0.0001, $t = 0.82$). Gasoline, on the other hand, yielded a negative and statistically significant coefficient (-0.0004, $t = -2.49$).

7. Price and family size coefficients were significant only for gasoline. The price effect was also negative and consistent with the coefficient obtained in item 6 (-0.0011; $t = -2.79$, $f = 4.49$). The effect of family size as expected was negative and insignificant (-0.040; $t = -0.864$).

8. Taken together at a time, regression coefficients were statistically significant only for electricity and gasoline. Electricity consumption correlated negatively with price (-0.00112; $t = -2.68$; $F = 2.89$); positively, though insignificantly, with income (0.0001; $t = 0.59$); and negatively, as well as insignificantly, with family size (-0.031; $t = -0.64$). As a family's income increases, more electricity appliances are afforded, whose use can be shared simultaneously by family members (and even neighbors). In the case of gasoline, the income, price and family size coefficients were all negative (-0.0005, -0.020, -0.05, respectively) and except for the third were all statistically significant ($t = -2.48$, -2.84, -0.62; $f = 4.46$). It is not easy to explain why income level or family size might affect gasoline consumption inversely.

Regional Energy Resource Endowments and Other Factors

Rank correlation and distribution correlation analysis attempted merely to establish statistically whether regional variations in consumption of different fuels were sufficiently explained by differences in regional availability of biomass fuels (as indicated by the hectarage planted to certain crops, the 1977 production of selected crops, and livestock population), or the registered number of infrastructure logistics (electrification, agricultural mills, irrigation pumps, motorized boats, or public markets). Data on regional agriculture and infrastructure were obtained from officially published statistics or MOE's estimates extrapolated from official statistics and working data available from the respective government agencies.

Rank correlations (Spearman's r) and the Gini-coefficient provided the first-order measures of correlation significance. In general, the closer the rank correlation statistic was to "1" the higher the correlation significance. In similar manner, the closer the Gini-coefficient was to "0" the higher the correlation significance.

The first distribution correlation exercise sought to establish whether the selected agricultural resources closely matched the distribution of rural population. While the fit was not perfect, there was sufficient equitable distribution of key agricultural resources among the rural populace on a regional basis as evidenced by the encouraging (below 0.30) Gini-statistics for backyard hog population ($G = 0.20$) and palay production ($G = 0.27$) (Fig. 5 and Appendix C). Thus, if it were economically feasible, those two resources become logical candidates to supply any contemplated rural decentralized energy system. Gini statistics for forestry and coconut hectarage with respect to population are 0.34 and 0.46 respectively.

In rank correlations, regional and per capita firewood consumption were negatively correlated with regional forest area ($r = -0.21$, $t = -0.70$ and $r = -0.14$, $t = -0.45$, respectively). A relatively high Gini-statistic (0.40) also indicated that the regional distribution of firewood consumption did not closely match the regional distribution of forest hectarage. Regions accounting for only 25 percent of forest area accounted for 95 percent of firewood used.

Regional firewood consumption seemed to correlate better with the density of public markets (obtained as the ratio of the number of public markets to land area). Even though the rank correlation is not conclusively high ($r = 0.5$), the t statistic of 1.85 argues it is nevertheless statistically significant. The survey revealed that over 33 percent of all firewood consumed was purchased from the markets. Additionally, the Gini-coefficient correlating the regional distribution of public markets with the regional distribution of firewood consumption was a low 0.24. In the Bicol study, on the other hand, Koppel found that people who reported not using firewood come from market towns where they could have access to other fuels.

Regional charcoal consumption correlated remarkably with the density of public markets ($r = 0.77$, $t = 3.85$). Higher positive coefficients were obtained from correlation of charcoal consumption with fuelwood consumption ($r = 0.61$, $t = 2.43$) than with coconut production ($r = 0.40$, $t = 0.62$). These relationships are closely associated with the high consumption of charcoal in the Southern Tagalog Region. They suggest the role of public markets in charcoal distribution, and the possibility that wood may have been a more popular source of charcoal than coconut shell.

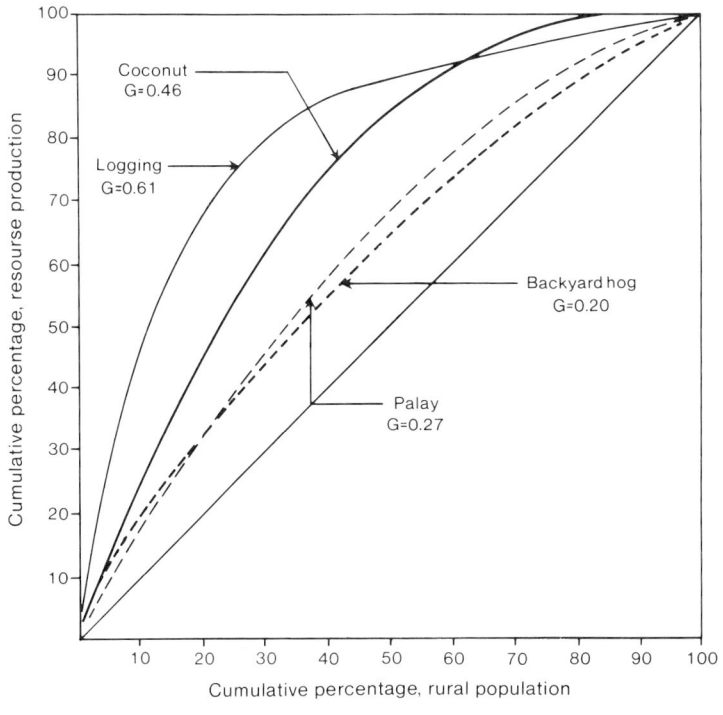

Figure 5. Distribution of Resource Production in Relation to Rural Population

Four regions, accounting for only 25 percent of logging wastes, claimed 95 percent of woodwaste consumption. Per capita woodwaste consumption correlated well with sawmill capacity across regions ($r = 0.67$, $t = 2.39$) but did not seem to vary significantly with the amount of logging residues ($r = 0.38$, $t = 1.30$).

Electricity consumption levels across all regions were negatively but not significantly related to the extent of electrification in a region and the number of electrically connected households ($r = -0.45$, $t = -1.59$). Still, for the four regions in Central and Northern Luzon (Regions I–IV), electricity consumption and household coverage correlated perfectly. In 1977, Regions I, III, and IV represented the three most electrified regions in the country, while Region II was one of the least electrified regions. Therefore, these regions represent the extreme cases, possibly explaining the close correlation. Obviously, factors such as proximity to appliance service outlets and retail power rates may have influenced electricity consumption levels.

The level of electricity consumption by rural households did not appear to be a function of the accessibility or penetration of the electricity system, whether this was measured in pole density, housewire density, or distribution line density. (A survey undertaken by the NEA covering 1976 consumption revealed that only 53 percent of households that were within access of a system were actually connected.) Electric poles per sq km (and other surrogate penetration indexes) correlated rather negatively with rural electricity consumption, per household and per capita.

The studies suggested that the more electricity is consumed in a region, the less its rural households have need for diesel fuel. A significant negative correlation ($r = -0.79$, $t = -3.64$) between electricity use and diesel consumption emerged. The per capita comparisons suggested an even more pronounced substitution relationship ($r = -0.92$, $t = -6.64$). It is possible that where centralized electricity is not easily accessible, rural families rely on petroleum fuels for generating localized electricity to run electrical appliances. A corollary finding, which supports the observed diesel-electric trade-off, is that rural consumption of diesel fuel correlated inversely though not significantly with the household electrification level regardless of whether diesel consumption was measured in regional aggregates, per household, or per capita.

Interestingly, like diesel, rural firewood consumption also correlated negatively with electrification coverage. (A moderately negative relationship also resulted from the Koppel study.) When the number of electrified households is ranked with firewood consumption in total, per household, or per capita terms, a significant inverse relationship was uniformly found.

For many regions, middle-income household gasoline and diesel consumption dwarfed that of higher income households. Normally, gasoline and diesel motive fuels are regarded as "affluence" fuels. Reported gasoline and diesel consumption might have unwittingly included nonhousehold applications as well as transportation uses: ranking of regional rural-road density with regional consumption of these liquid fuels yielded negative correlation coefficients. Total length of feeder roads in the regions does not explain reported regional consumption of motive fuels (gasoline, diesel, gasoline/diesel) either. Rank-matching regional gasoline consumption with the number of motorized *bancas* in each region also gave a low correlation statistic.

Therefore, consumption of those fuels was tested for correlation with other possible applications such as electricity generation and irrigation pumping. Gasoline consumption, either in absolute, per capita, or per household terms yielded high rank correlation coefficients with household electricity generation. This suggests that for those households who generate electricity, gasoline seemed to be the popular choice over diesel. The front-end costs of gasoline generators are generally lower than diesel types.

CONCLUSIONS AND FUTURE STUDIES

The statistical exercise added depth to the aggregate analyses and yielded some instructive findings for purposes of future energy policy.

1. The regression exercises identified energy forms, electricity and gasoline, that exhibited significant consumer responses to price increases. The regression analyses also indicated relationships between firewood and coconut shell consumption and income levels.

2. The distribution correlation tests singled out hog population and palay hectarage as the two agricultural assets whose distributions most closely match the regional distribution of rural population in 1977, making them potential resource candidates for decentralized rural energy programs for the future.

3. The correlation analysis suggested that the institution of public markets may have a significant role in offsetting regional resource imbalances in certain fuels, for example, firewood and charcoal.

4. Correlation analysis also revealed that regional sawmill capacity rather than the level of logging residues provided a meaningful indicator of woodwaste consumption, and that the size of coconut plantation hectarage could be a reliable barometer for estimating regional coconut shell consumption.

5. Electricity consumption surprisingly correlated insignificantly with various measures of regional electrification (household connections, pole and wire densities). The study yielded an inverse relationship between electricity use and diesel consumption but a positive one between electricity use and gasoline consumption. Transportation variables failed to explain gasoline/diesel consumption levels. This suggested that reported regional fuel consumption (particularly gasoline) may actually have included fuel used for electricity generation and other rural activities by the rural households.

The study was rather exhaustive with the rank and distribution correlation exercises. Since regression analysis using household data was not completed, the Planning Service worked primarily on rural energy consumption data by region, by income class, and by family size. Moreover, the true energy flow in the rural areas might not have been depicted by the use of the results of the 1979 energy consumption survey on low-income urban households as basis for extrapolations in this study.

A new Rural Energy Needs Survey will soon be undertaken to update and refine the findings of the 1977 survey and to determine

1. Level of energy consumption of household per fuel type;
2. Energy applications in different household activities (such as lighting, cooking, and ironing);
3. Variations in energy usage as a function of household income, family size, and occupation of household head;
4. Ratio of energy expenditure to household expenditures; and
5. Constraints and inducement to the use of nonpetroleum fuels.

Unlike the 1977 rural energy needs survey instrument, the questionnaire for the new survey traces the flow of energy in the rural setting from energy form to the final end use. Specifically, the survey identifies the household devices and appliances and corresponding amount of fuel used by principal task, such as cooking, lighting, and motive power. The respondents also are asked to identify and rank their most preferred indigenous fuels as well as to record reasons why such fuels are preferred and constraints of using indigenous fuel instead of petroleum fuels or electricity. The respondents also are asked to assess the availability of indigenous fuel supply within their barangay.

Except for these modifications in the survey questionnaire, the methodology used in the survey of 1977 rural energy needs will still be employed since the difficulties experienced were due to mainly the loss of the accomplished questionnaires rather than the survey methodology.

Further surveys will be undertaken by the Planning Service, notably energy consumption of the agricultural crop production sector and the land transport sector. Also, a study involving the electric utilities in the rural areas is currently in progress. For these and future surveys, stricter implementation measures would be adopted. Concrete guidelines have been set up to evaluate the financial and technical capability and monitor the performance of research firms contracted to undertake the survey. Accomplished questionnaires upon encoding would immediately be turned over to the Planning Service to enable the staff to undertake further analysis when needed.

In spite of its limitations, the Rural Energy Needs Survey of 1977 together with microlevel studies done by interested researchers have provided the government with its first insight into the rural energy mosaic. As an initial attempt at generating base data of the rural energy sector in the Philippines, the survey:

1. Acquainted government with the numerical boundaries of the sector, generating the first broad estimates of rural consumption of different energy forms in the country;
2. Provided the government with its first information on the household functions to which these resources were applied; and
3. Confirmed the potential of the noncommercial fuel sector as a major supplier of rural energy needs. Specifically, additional funds and efforts were directed towards utilization and commercialization of biogas and rice hulls after they were revealed as promising decentralized suppliers of rural energy needs.

APPENDIX A
Common Measures and their Conversions to Standard Measures

Type of Fuel	Unit of Field Measurements	Unit Used in the Presentation of Result	Conversion Factor
Premium Gasoline	liter	BOE*	0.006289 BOE/l
Regular Gasoline	liter	BOE	0.006289 BOE/l
Diesel or Crude	liter	BOE	0.006289 BOE/l
Kerosene	liter	BOE	0.006289 BOE/l
Oil for Lighting	liter	BOE	0.006289 BOE/l
Cooking Gas or LPG	kg	BOE	0.007982 BOE/kg
Firewood	kg	BOE	0.001520 BOE/kg+
Woodwastes	kg	BOE	0.001520 BOE/kg
Charcoal	kg	BOE	0.004561 BOE/kg
Coconut Shell/Husks	kg	BOE	0.003041 BOE/kg
Rice Hull	kg	BOE	0.001520 BOE/kg
Electricity	kWh	BOE	600 kWh/BOE‡

Source: Planning Service, Ministry of Energy.
* Barrel-of-Oil Equivalent.
+ Assumes average moisture content at 46 percent.
‡ At thermal efficiency factor of 33 percent based on National Power Corporation actual thermal efficiency of 32-35 percent.

APPENDIX B
Energy Consumption Per Rural Household by Income Class - Below ₱2,500
(In GJ), 1977

	\multicolumn{12}{c|}{REGION}											
	I	II	III	IV	V	VI	VII	VIII	IX	X	XI	XII
Noncommercial Energy												
Firewood	8.6	1.8	7.2	8.5	3.3	6.2	9.7	7.6	22.4	3.2	3.9	6.8
Woodwaste	0.1	-	0.4	-	-	0.1	-	-	-	-	-	0.2
Charcoal	2.9	2.0	5.1	6.3	3.2	1.4	2.5	1.6	2.9	2.8	0.6	2.3
Coconut Shells	-	-	0.1	4.7	2.3	0.5	0.2	0.1	0.9	11.5	-	-
Rice Hulls	-	-	0.2	0.8	-	-	-	-	-	-	-	-
Others (Biogas)	-	-	-	-	-	-	-	-	-	-	-	-
Subtotal	11.6	3.8	13.0	20.3	8.8	8.2	12.4	9.1	26.2	18.0	4.5	9.3
Commercial Energy												
Electricity	0.3	-	1.1	0.4	0.1	0.1	-	-	0.1	0.2	-	0.2
Petroleum	6.4	3.9	7.3	14.2	8.0	5.9	5.3	6.4	4.0	8.7	1.8	9.9
Gasoline	(1.7)	(0.2)	(1.6)	(5.2)	(2.8)	(2.9)	(1.1)	(1.9)	(0.2)	(5.3)	(0.2)	(3.0)
Kerosene	(3.9)	(2.5)	(4.7)	(7.5)	(5.3)	(2.9)	(4.1)	(4.5)	(3.3)	(3.2)	(1.6)	(6.7)
Diesel	+	(1.3)	(1.6)	(0.4)	-	-	(0.1)	-	-	(0.1)	-	(0.1)
L P G	(0.8)	+	(0.7)	(1.1)	-	(0.1)	-	-	-	(0.1)	-	(0.1)
Subtotal	6.7	3.9	8.4	14.6	9.1	6.0	5.3	6.4	4.1	8.9	1.8	10.1
TOTAL	18.3	7.7	21.4	34.9	17.9	14.2	17.7	15.5	30.3	26.9	6.3	19.4

Note: ( ) = signifies product subtotal.

APPENDIX B (Continued)
Energy Consumption Per Rural Household by Income Class - ₱2,500 - ₱7,999
(In GJ), 1977

	\multicolumn{12}{c}{R E G I O N}											
	I	II	III	IV	V	VI	VII	VIII	IX	X	XI	XII
Noncommercial Energy												
Firewood	27.3	1.0	10.6	13.4	2.0	5.2	9.2	5.6	98.6	1.2	1.4	28.6
Woodwaste	0.4	-	0.5	0.2	-	-	-	-	-	-	-	-
Charcoal	3.0	2.8	4.8	31.9	6.9	0.8	1.7	4.0	7.1	2.1	0.2	4.9
Coconut Shells	-	-	+	-	12.4	0.1	-	0.4	0.1	102.3	0.1	0.8
Rice Hulls	-	-	0.5	4.1	-	-	-	-	-	-	-	-
Others (Biogas)	-	-	-	-	-	-	-	-	-	-	-	-
Subtotal	30.7	3.8	11.6	49.6	21.3	6.1	10.9	10.0	105.8	105.6	1.7	34.3
Commercial Energy												
Electricity	3.7	0.1	2.6	2.0	0.1	0.2	-	-	1.5	1.0	-	1.3
Petroleum	26.0	14.2	18.8	26.6	9.2	2.9	5.0	17.0	32.6	41.1	0.9	17.7
Gasoline	(20.1)	(0.1)	(0.06)	(14.3)	-	(0.4)	(0.3)	(6.9)	(18.5)	(38.1)	(0.1)	(4.5)
Kerosene	(4.1)	(6.2)	(0.65)	(9.6)	(9.0)	(2.0)	(4.7)	(9.4)	(12.2)	(2.2)	(0.8)	(7.4)
Diesel	-	(7.7)	(0.41)	(0.7)	(0.2)	(0.1)	-	-	(1.1)	(0.1)	-	(5.6)
L P G	(1.7)	(0.2)	(0.38)	(2.0)	-	(0.4)	-	(0.7)	(0.8)	(0.8)	-	(0.2)
Subtotal	29.7	14.3	21.4	28.6	9.3	3.1	5.0	17.0	34.1	42.1	0.9	19.0
T O T A L	60.4	18.1	33.0	78.2	30.6	9.2	15.9	27.0	139.9	147.7	2.6	53.3

Note: ( ) = signifies product subtotal.

APPENDIX B (Continued)
Energy Consumption Per Rural Household by Income Class - ₱8,000 and Over
(In GJ), 1977

	REGION											
	I	II	III	IV	V	VI	VII	VIII	IX	X	XI	XII
Noncommercial Energy												
Firewood	8.1	-	3.3	5.1	0.8	3.3	0.7	0.1	15.7	4.2	5.1	10.0
Woodwaste	-	-	-	-	-	-	-	-	-	-	-	-
Charcoal	10.3	0.1	2.5	11.2	0.1	0.7	-	30.8	0.4	3.9	0.1	2.3
Coconut Shells	-	-	-	-	-	6.5	-	-	0.2	143.3	-	-
Rice Hulls	-	-	-	-	-	-	-	-	-	-	-	-
Others (Biogas)	-	-	-	-	-	-	-	-	-	-	-	-
Subtotal	18.4	0.1	5.8	16.3	0.9	10.5	0.7	30.9	16.3	151.4	5.2	12.3
Commercial Energy												
Electricity	0.1	-	1.3	1.1	0.1	0.1	1.1	-	-	0.7	-	4.0
Petroleum	2.9	0.2	12.7	9.6	0.3	1.9	3.6	3.6	2.9	1.4	0.9	4.0
Gasoline	(0.1)	-	(8.5)	(2.2)	(0.2)	+	(2.4)	(3.4)	(0.1)	-	-	-
Kerosene	(0.9)	(0.1)	(2.1)	(2.8)	(0.1)	(1.9)	(0.1)	(0.2)	(1.2)	(1.4)	(0.9)	(4.0)
Diesel	-	(0.1)	-	-	-	-	-	-	-	-	-	-
L P G	(1.9)	-	(2.1)	(4.7)	-	-	(1.1)	-	(1.6)	-	-	-
Subtotal	3.0	0.2	14.0	10.7	0.4	2.0	4.7	3.6	2.9	2.1	0.9	8.0
T O T A L	21.4	0.3	19.8	27.0	1.3	12.5	5.4	34.5	19.2	153.5	6.1	20.3

Note: ( ) = signifies product subtotal.

APPENDIX C
Summary Results Regional Distribution Correlation

		Gini Coefficient	Significant (/)
1.	Firewood Consumption	Forest Hectarage	0.40	
2.	Firewood Consumption	Number of Public Markets	0.24	/
3.	Woodwaste Consumption	Logging Waste and Residues Production	0.90	
4.	Woodwaste Consumption	Sawmill Capacity	0.64	
5.	Charcoal Consumption	Number of Public Markets	0.21	/
6.	Coconut Shell Consumption	Coconut Production	0.34	
7.	Coconut Shell Consumption	Coconut Hectarage	0.34	
8.	Ricehull Consumption	Palay Production	0.51	
9.	Ricehull Consumption	Palay Hectarage	0.44	
10.	Forest Hectarage	Rural Population	0.34	
11.	Palay Hectarage	Rural Population	0.23	/
12.	Coconut Hectarage	Rural Population	0.46	
13.	Backyard Hog Population	Rural Population	0.20	/
14.	Logging Production	Rural Population	0.61	
15.	Coconut Production	Rural Population	0.46	
16.	Palay Production	Rural Population	0.27	/
17.	Diesel Consumption	Number of Rural Households Electrified	0.75	
18.	Electricity Consumption (Generated)	Diesel Consumption	0.92	

NOTES

1. In 1977, the foreign exchange conversion was approximately ₱7.50 for every U.S. dollar. Informal consultations reveal that a Bangladesh study covering 23 villages cost $12,000 to complete with 26 technical people doing field survey work for 3 to 4 months. In Indonesia, a study of 40 villages cost $30,000. In Thailand, two nationwide studies of 697 and 200 villages each, cost $60,000 per study for field and analysis work of one and one-half months, featuring a more comprehensive questionnaire.
2. Noncommercial energy refers to the energy forms frequently used in traditional sectors, largely unmonitored. Such fuels are widely bought and sold and include heat-supplying sources such as wood, charcoal, wood wastes, plant and animal residues. Commercial energy is energy obtained from organized energy supply industries, such as electric power and petroleum products.

REFERENCES

Koppel, Bruce.
    1980   A Preliminary Analysis of Fuelwood Consumption in the Bicol River Basin. In Atje and others, *Energy Analysis in Rural Regions: Studies in Indonesia, Nepal and the Philippines.* ERD Program Report I-80-2. Resource Systems Institute, East-West Center, September, Honolulu.

Manibog, Fernando.
    1979   *Patterns of Energy Utilization in a Philippine Village: Sources, End-Uses and Correlation Analyses.* A Draft Report presented to the International Energy Agency, Organization for Economic Cooperation and Development, December, Paris, France, and the Rockefeller Foundation, New York.

Philippines, Ministry of Energy (MOE).
    1978   *Energy Usage Survey of Households.* Unpublished.

Philippines, MOE.
    1979   *Regional Profile 10-Year Energy Program, 1979–1988.* Manila: MOE.

Philippines, MOE.
    1979   *Study on Energy Consumption and Conservation Practices of Urban Households in 1979.* Manila: Unpublished.

Philippines, National Economic and Development Authority (NEDA).
    1978   *Regional Development Information.* Manila: NEDA.

Philippines, NEDA.
  1980 *1980 Philippine Statistical Yearbook.* Manila: National Economic and Development Authority.
Philippines, NEDA, National Census and Statistics Office.
  1978 *Philippine Yearbook 1978.* Manila: National Census and Statistics Office.
Philippines, NEDA, National Census and Statistics Office.
  1980 *Integrated Survey of Households Bulletin.* (Series no. 48) Manila: National Census and Statistics Office.
Philippines, National Electrification Administration (NEA).
  1977 *Nationwide Survey on Socio-Economic Impact of Rural Electrification.* Manila: NEA. February.
Philippines, NEA.
  1978 *Annual Report.* Metro Manila: NEA.

5
Analysis of Rural Energy Development in Thailand

Surapong Chirarattananon

INTRODUCTION

The 1980s mark Thailand's third decade of national development. The Fifth National Economic and Social Development Plan provides the framework for planning and executing activities for the next five years. During this important period of "structural readjustment," Thailand will develop into a semi-industrialized country.

Efforts of the first two development plans, spanning 1962 to 1972, concentrated on the construction of economic infrastructures. Typically, energy resources were imported without constraint, so that over three times as much energy was consumed in 1972 as in 1962. Moreover, the early pattern of petroleum fuel consumption as well as the composition of the refinery output contributed to price differentials distinctly favoring diesel fuel, greatly augmenting its use.

The energy crisis of 1973–74 erupted during the middle of the Third Plan, causing serious repercussions. The gradual adjustments in domestic fuel prices delayed the effect on the overall economy in spite of quantum leaps in imported fuel cost. The government appeared to absorb the deficit to slow inflation and to support the economy. The country's trade balance deficit was offset largely by the export of agricultural products, while agriculture consumed less than 10 percent of the imported oil. By the end of the Third Plan, no significant shift had occurred in the pattern of energy consumption, except for a decrease in the rate of growth. Petroleum imports increased by 50 percent during that time period (an annual average of approximately 9 percent).

In spite of the disturbance caused by the energy crisis, agriculture continued to grow at an average rate of 4.8 percent per year during the Fourth Plan (1977 to 1981) — a slightly slower rate than the 5.2 percent per year during the previous 15 years. That continued growth in agriculture hinged on the 4 percent per year expansion of land under cultivation. By the beginning of the Fourth Plan, available land was rapidly diminishing. Even so, since the rate of expansion of land under cultivation was higher than the rate of population growth, on the average, landholding per capita had been increasing. As a result, farm mechanization, officially endorsed for a decade, expanded, if only slowly, up until the time of the Fourth Plan.

The effects of the energy crisis and the tardy adjustment in fuel prices (reflecting more expensive imports) were gradually felt in the economy during the period of the Fourth Plan. Economic demand and consumption grew excessively, contributing to a high inflation rate, while foreign borrowing and investment loans increased to offset the growing deficit in the balance of trade. The economy continued to grow at even a higher rate than in the previous plans. The industrial sector was growing at an exceptionally high growth rate too.

Concerned economic planning authorities as well as energy authorities drew up the Fifth Plan based on a "structural readjustment toward more uniform economic and social equity." This plan incorporates continuing growth of agriculture by increasing yield. Even so, by the end of this plan's time period, industry probably will play a stronger role in the economy than agriculture. Provincial capitals are envisioned as bases for industrial development centered on small-scale rural industries.

The National Economic and Social Development Board (NESDB) and National Energy Administration (NEA) initiated a series of intensive energy studies and planning activities in 1978, about four years after the beginning of the energy crisis. This tardy official response reflects the uncertain perception of the problem and a gap in the understanding of the situation. In 1978, NEA conducted a major review of Thailand's energy demand and consumption for the previous year. By 1979, NEA had completed a plan for accelerated development of energy sources other than petroleum and it had been approved by the NESDB. While still providing for a slight increase in annual petroleum consumption, this plan emphasizes reducing the dependency on petroleum to 45 percent over 10 years. Lignite deposits, hydropower, geothermal power, and almost all other renewable energy sources were to be explored and developed. The plan also includes an estimated contribution of wind energy, solar energy, agricultural wastes and biogas, and fast-growing wood species from fuelwood forests.

The Energy Master Plan study, commissioned by the NEA, with aid from the Asian Development Bank, is a comprehensive study of problems of energy consumption and energy resource development over the next 20 years. The objective of another study also commissioned by the Ministry of Finance is to develop a comprehensive plan on energy pricing policy and on optional pricing of commercial fuels, mainly petroleum and products from natural gas.

The country has responded to the changing energy constraints with great resiliency and reasonableness even in the face of the state authorities' initial uncertainty. Consumption of petroleum fuels increased substantially in 1980–81 and then stalled in 1981 and into the first half of 1982. Larger industries began converting to lignite, coal, and natural gas. Several other measures were imposed, such as closing all petrol stations for certain hours and days. In addition, discoveries of natural gas reserves and of a small petroleum reserve have stimulated further development and growth. Definite plans for the utilization of the natural gas are pending, but methanol and liquefied petroleum gas (LPG) may be substitutes for diesel and gasoline in rural areas.

Until 1979, rural energy consumption had been estimated by secondary sources. At that time, a pilot survey on rural energy consumption was initiated. National follow-up surveys were completed in 1980. They made available the first reliable estimates on fuelwood consumption and other issues. Unfortunately, the connection between energy utilization and rural development has not been delineated in any official study or development plan.

Commercial energy in rural areas, where over 72 percent of the population lives, accounts for only 13.5 percent of the total commercial energy used, even including the 4.8 percent used in agriculture. Commercial energy consumption is highly concentrated in the developed, heavily concentrated industrial, and heavily populated urban areas. Several policies concerning energy supply, such as rural electrification, mention a connection between energy and rural development, but development policies have not yet aligned themselves to those concerns.

THE RURAL ECONOMY AND NATIONAL DEVELOPMENT

As of January 1980, the Thai population was estimated at 46.15 million, or an average of 90 people per sq km. Traditionally agrarian, the population is concentrated in the three main flat land areas adjacent to rivers and their tributaries: (1) the southern part of the Northeastern Region, (2) the central plain, and (3) the Southern Region. In the last decade, even though some people have migrated to the more economically developed industrial areas near Bangkok Metropolis, the overall distribution pattern has not changed (see Table 1).

The largest administrative unit in Thailand is the province (*Changwad*), made up of districts (*Amphur*). A district comprises communes (*Tambol*). In turn, several villages (*Muban*) form a commune.

Regional landholding patterns are given in Table 2. The Ministry of Agriculture and Cooperatives has classified agricultural land into 19 agroeconomic zones based on soil, average monthly rainfall, average monthly temperature, and matching agricultural activities (see Fig. 1).

TABLE 1.
Population by Region, 1980

Region	Area (sq. km.)	Population (million)	Density (Person per sq. km.)	Rural Population (million)
Bangkok	1,549	5.00	3,228	–
North	156,375	9.59	61	7.37
Northeast	170,227	16.09	95	13.66
Central	35,704	4.93	138	4.50
East	36,395	2.83	78	2.57
West	50,103	3.58	71	3.21
South	70,153	5.82	83	4.58

Source: Figures recompiled from NEA, Report on the 1980 Rural Energy Survey, Part 1, 1982, and the January, 1980, statistics released by NEA.

Legislation on agroeconomy stipulating the zonal division provides the legal basis for concerted agricultural development. In addition, the Ministry of Agriculture and Cooperatives provides an agricultural advisor to every commune to coordinate government projects at the grass roots level. Price movements and other information are disseminated through these advisors.

Rice continues to be the major crop, but production has diversified with more than 10 cash crops each reaching a value of more than one billion bahts[1] (฿) per year. From 1969–1970 to 1977–1978, production of cassava and sugarcane increased over three and four times respectively. However, only rubber, tobacco, and coconut yields increased over that time. Fertilizer used increased from an average of 3.7 kg/rai for all field crops in 1969–1970 to 8.1 kg/rai in 1977–1978. The 4 percent average annual expansion of land under cultivation until the mid-1970s declined to 3.5 percent in the late 1970s. Water resources development and the construction of some 60,000 km of access roads accounted for some growth.

In Thailand, livestock are used primarily for drawing loads. The secondary use is as sources of meat for domestic consumption. Livestock also are important for the dairy industry and export. Although increasing international and domestic demand has stimulated livestock growth, bullocks and buffalo are still commonly used in most agricultural areas

TABLE 2.
Landholding and Ownership Patterns by Region, 1978

Type of Land Use	North	Northeast	Central	South	Total
Percentage of Persons Owning Land	77.2	91.7	65.7	92.3	82.8
		(million rais)			
Total Land Area	106.25	106.39	64.74	43.86	321.25
Total Land Holding	23.99	46.44	30.13	13.25	113.80
Housing Area	.58	.95	.71	.47	2.71
Paddy Land	15.80	33.62	16.92	4.74	71.09
Other Field Crops	6.20	8.80	8.52	.09	23.61
Under Fruit Trees and Tree Crops	.73	.32	2.22	6.90	10.16
Vegetables and Flowers	.08	.06	.06	.02	.23
Idle Land	.24	1.81	.31	.61	2.98
Pasture	.01	.18	.29	.05	2.50
Unclassified*	82.27	59.96	34.61	30.62	207.45

Source: Ministry of Agriculture and Cooperatives, Department of Agricultural Economics, 1978.

*Most of this land is considered forest land.

for draft power. And the number of draft animals stopped declining after the 1973 energy crisis.

Forestry was once a major agricultural activity, but it has rapidly decreased due to exploitation. And attempts at reforestation have not met with substantial success. Large-scale illegal logging has been halted only recently and illegal clearing of forested land for agriculture continues, although at a lower rate than before. The total value of forestry products reached ฿5 billion in 1979 but has declined since then. See Table 3 for estimated forested area by region as of 1980.

Of the 60,296 registered industrial establishments in Thailand in 1979, about 99 percent employed fewer than 100 persons. Clearly, small-scale industries play a significant role in the economic structure. The

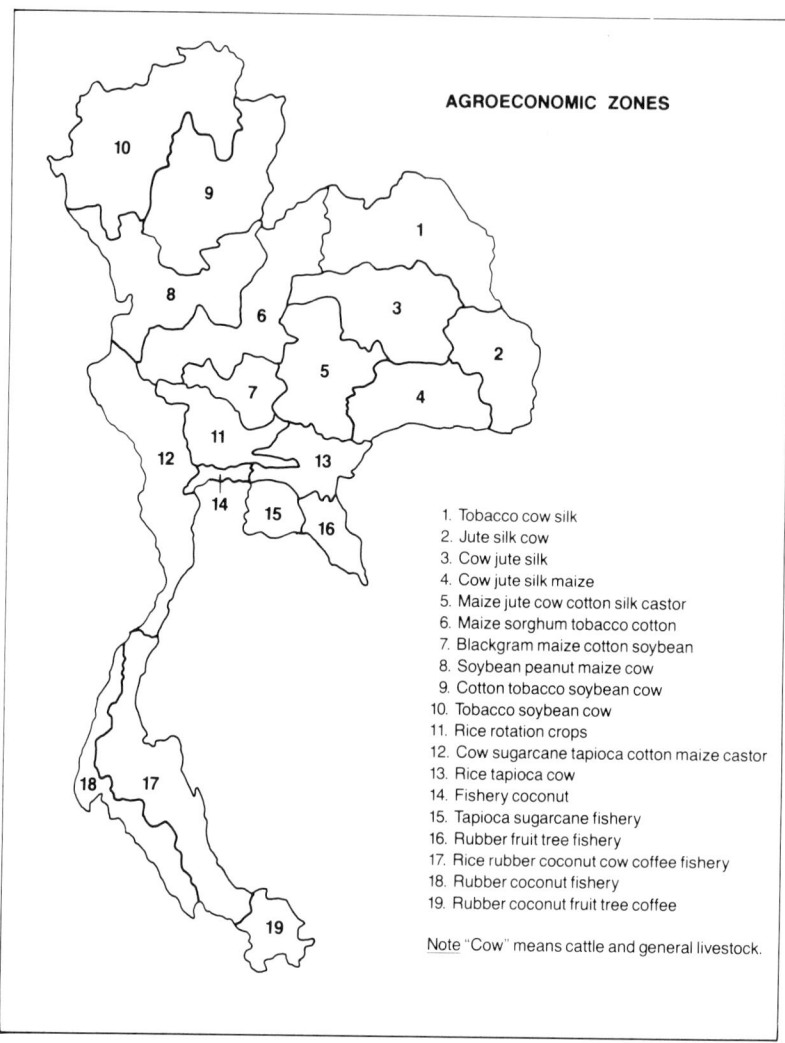

Figure 1. Agroeconomic Zones

principal rural industries are weaving, umbrella making, ceramics, cassava, saw mills, sugar, rice mills, food processing, salt, and ice making. Other such industries concentrate on wood and bamboo handicraft, niello wares and bronze wares, and metal casting—all for household utensils—as well as tobacco and rubber-drying plants. Small oil-extracting plants produce animal and vegetable oils for local consumption, but that industry is dominated by large-scale factories located near Bangkok.

TABLE 3.
Forested Land, by Region, 1980

Region	Forest Area	Natural Annual Regeneration	
	(million rais)	(million rais)	(percent)
North	33.74	.66	1.9
Northeast	13.72	.28	2.0
Central	9.40	.22	2.3
South	8.33	.21	2.5
Total 1980	65.19	1.36	2.1
Total 1970	82.75	1.73	2.0

Source: Compiled from statistic No. 125, Center for Agricultural Statistics. Ministry of Agriculture and Cooperatives.

Note: A 1970 FAO estimate lists forest land area as 145.98 million rais, with marginal forest land (village woodlots) as 4.56 million rais.

Small-scale rural industry answers local needs for goods and provides both permanent and seasonal employment to rural communities, thereby reducing migration to already crowded cities. It can easily adapt alternative production techniques, thereby benefiting technology diffusion. If properly planned, small-scale industry can provide a valuable link to more complex industries.

Rural electrification is regarded as important by development authorities for increasing economic productivity and living standards. As of 1980, 36 percent of Thai villages (or 18,511 of 54,400 villages) had electricity. The Fifth Economic Plan projects 92 percent. Three projects organize rural electrification. The first provides electricity to 18,500 villages by the existing supply grid expansion plan. The second project provides electricity to 10,900 villages under the rural job creation scheme and other special schemes. The last project, the Accelerated Rural Electrification Project (ARE), provides rapid expansion of electrical supply to the rest of the country in five phases from 1977 to 1994. The first phase, covering 5,200 villages, is scheduled for completion by 1982. The second phase, covering 8,000 villages, began in 1981. Requirements for electricity projected for the second phase are shown in Table 4.

TABLE 4.
Requirements for Electricity, Second Phase,
Accelerated Rural Electrification

Classification		Number
Rice Mills		1,675
Electrical Power (Repair Shops, etc.)		6,644
Water Pumping (Agriculture)		2,094
	Subtotal	10,413
Households		753,953
Stores		57,670
Water Pumping (Household)		11.871
	Subtotal	823,474

THE FIFTH ECONOMIC AND SOCIAL DEVELOPMENT PLAN

Over four consecutive plans periods, the GNP increased from ฿60,000 million in 1961 to ฿817,000 million in 1981, and per capita income rose from ฿2,200 to ฿17,200. However, the structural transformation toward a semi-industrialized economy with substantial international trade has been accompanied by serious economic and social problems. The Fifth Plan summarizes certain problems.

1. Economic uncertainty and a worsening financial climate. A high consumption demand, resulting from rapid economic growth, as well as an 11.7 percent average inflation rate over the last five years, and related serious adverse effects have created hardships for people living on fixed incomes. Increasingly more importation because of high consumption demand has caused a serious deficit in the balance of payments, even after efforts to obtain international loans.

2. Degradation of the natural resource base and the environment. Forests have been depleted at an alarming rate. Agricultural yields have diminished due to soil degradation and erosion. Fishery yields have decreased from the inland sources and from the seawaters within exclusive economic zones. And agglomeration of industries has congested the cities, especially Bangkok.

3. Social problems. Society in general has not been able to adjust to

the results of rapid economic transformation. This has created cultural problems, changes in societal value, narcotic problems, and social security problems.

4. Poverty in rural areas. An average economic growth of over 7 percent in the last two decades has benefited mainly urban people and people living in the Central Region. Some of those benefits reach the rural people living in the irrigated areas, but most rural residents have not received proportionate benefits. Significantly, up to 25 percent of the rural residents still live in poverty.

During the last two decades of development, the gross regional product in the Central Region and in Bangkok have increased, at the expense of other regions. The serious disparity in the regional per capita income is aggravated by disparate incomes of people in different production sectors. A 1979 manpower study by NESDB reported that the per capita annual income for the higher income group (just 12.5 percent of the people) was ฿55,670, while the per capita income of the farmers (68.5 percent of the people) was only ฿6,160. NEA also conducted a survey of direct interviews that documented another set of income estimated for the rural population (see Table 5). The general growth in the economy has reduced the incidence of poverty, but poverty remains a significant rural problem and is particularly acute in certain areas, such as in the northernmost part of the North Region and in the Northeast Region. The impoverished population in the Northeast is estimated to be 6 million people.

One problem often linked to poverty is nutrition and health. According to a 1980 FAO report, the per capita per day calorie supply for an average Thai is about 2,200 Cal (see also Table 6).

Children's nutrition is different. Even with a food surplus, an alarmingly high rate of malnutrition in children exists throughout the country. A 1980 survey by the Ministry of Public Health found that nearly 3.6 million children at preschool age (53 percent) suffered from insufficient intake of protein and calories in various degrees. The problem is most severe in the Northeast Region, where 1.7 million children (59.5 percent) in that age group are affected.

The Fifth Plan specifies interrelated economic and financial reform and structural adjustment. Emphasized are (1) increased private and governmental savings; (2) reduced demand for nonessential goods and reduced waste; and (3) increased national productivity, including more efficient use of resources and reduced imports of raw materials and oil. A lower population growth rate is projected as well as higher quality human resources. Reforms at the grass roots level in social service struc-

TABLE 5.
Income Distribution of Rural Households, 1980

	Income Class ('000 ฿)					Average Income	
	0-24	24-48	48-72	72-120	120 up	Household	Per Capita
Region	(percentage)					('000 ฿)	
North	65.6	19.4	7.6	4.2	3.1	23.5	4.4
Northeast	81.5	12.3	3.2	1.7	1.4	15.7	2.5
Central 1	39.	24.0	13.7	13.7	9.6	34.2	5.8
Central 2	53.0	29.5	5.4	6.0	6.0	27.7	4.3
South	66.7	20.0	8.9	3.7	0.7	21.9	3.8
National	68.8	18.4	6.0	3.9	2.9	21.5	3.7

Source: NEA Rural Energy Survey, 1980.

TABLE 6.
Per Capita Food Consumption, Rural Inhabitants, 1980

	Rice				
Region	Glutinous	Non-glutinous	Calories per day	Vegetables	Meat and Fish
	kg/year			kg/year	
North	73.1	95.0	1,740	48.9	12.3
Northeast	138.1	51.3	1,960	33.4	15.0
Central 1	0.6	166.8	1,730	49.7	12.2
Central 2	1.0	141.9	1,470	49.0	11.8
South	0.6	140.4	1,450	49.3	15.4
National	77.4	94.7	1,780	42.6	13.8

Source: NEA Rural Energy Survey, 1980

TABLE 7.
Energy Consumption for Final Year of Each Plan

Unit: equiv. mil. liters of crude oil

Source	Plan 1 1966	%	Plan 2 1971	%	Plan 3 1976	%	Plan 4 1981	%
Petroleum products	2,920	60	6,408	71	9,580	79	12,072	70
Hydroelectric	489	10	669	7	1,229	10	2,084	12
Lignite	130	3	185	2	272	2	1,288	8
Others*	1,296	27	1,791	20	1,024	9	1,707	10
Total	4,835	100	9,053	100	12,105	100	17,151	100

Source: NEA

*Others includes fuelwood, charcoal, paddy husk, bagasse, and other agricultural wastes.

tures emphasize self-help through education and training as well as cooperation among members of the society.

ENERGY ISSUES AND ENERGY POLICY

The growth in energy consumption and shifts in its composition during the first four plan periods are seen in Table 7, although estimates of noncommercial energy use in the table should be viewed with reservation. Energy consumption grew by 13 percent per year during the first two periods (1962 to 1971). The Third Plan supported rural electrification, with total electricity demand increasing by 14 to 15 percent each year. With the energy crisis, energy consumption continued to rise but at a lower rate of about 9 percent a year.

In the Fourth Plan, the pace of hydroelectric development increased and production and consumption of lignite accelerated. Increased exploration activities in petroleum and natural gas deposits resulted in discovery of the natural gas deposits in the gulf of Thailand. By 1981, energy consumption growth reduced to 6 percent per year. Price structure also has influenced petroleum consumption (see Table 8).

The low energy prices contributed to the high growth of demand in transportation, manufacturing, and construction sectors and subse-

TABLE 8.
Energy Prices, 1970-1980

	1970	1973	1975	1978	1979	1980
Average Import Oil Price (฿/bbl)	43	64	232	286	409	650
Electricity (฿/kWh)	.28	.28	.46	.61	.61	.90
Premium Gasoline (฿/l)	2.10	2.30	3.62	4.22	5.60	9.80
High Speed Diesel (฿/l)	.98	1.05	2.33	2.64	4.88	6.54
Fuel Oil (฿/l)	.70	n.a.	1.44	1.61	2.90	3.61

Source: Petroleum Authority of Thailand (PTT).

TABLE 9.
Market Prices of Petroleum Products, 1981-1982

Products	Market Price (Bangkok) (฿/l)	Products	Market Price (Bangkok) (฿/kg)
Premium Gasoline	13.45	LPG, large tank	9.46
Regular Gasoline	11.40	LPG, small tank	9.99
Kerosene	6.12		
High Speed Diesel	7.37		
Low Speed Diesel	7.12		(฿/ton)
Fuel Oil 600	4.70	Asphalt	5,075
Fuel Oil 2,000	4.43		

Source: PTT

Note: One liter of LPG is equivalent to .57 kilogram.

quently to the high average rate of inflation during 1975 to 1980. Perhaps the most serious problem is the 5 to 10 percent of energy from petroleum products used in Bangkok Metropolis alone, for private personal transportation compared with a mere 4.8 percent for agricultural production.

The import bill from petroleum and its products matched the trade deficit in 1981 at 42 percent of the export value. Diesel engines using the cheaper fuel are replacing gasoline engines in industry and transporta-

TABLE 10.
Petroleum Consumption, 1980.

Sectors	Petroleum (million liters)	Percentage
Agriculture	1,208	9.9
Tertiary and Others	3,864	31.6
Transportation	5,068	41.5
Manufacturing	1,968	16.1
Construction	108	.9
Total	12,216	100.0

tion and also in agriculture. LPG is being used in some modified gasoline engines for the same reason (see Table 9). Several studies of price policy have been commissioned recently, but a comprehensive study is needed to investigate the long-term effects, especially as related to development and to poorer sections of the population.

An estimated surplus of LPG and methanol production from natural gas, after 1986, also causes concern about energy pricing. Natural gas has been produced since October 1981 and up to 150 million cubic feet have been used daily to produce electricity. Increased use of LPG as well as methanol may be promoted for rural areas. Petroleum price adjustments in 1980 and 1981 may well be sufficient to compete with the import and to curtail consumption demand. Consumption statistics for 1980 are given in Table 10. The 1981 consumption of petroleum and its products amounts to 12,070 million liters. The drop from 1980 consumption seems to substantiate the curtailment.

The Fifth Plan aims to reduce petroleum imports by 3 percent annually and limit overall energy consumption growth to 4.8 percent each year. The Plan specifies removing subsidies to adjust energy prices and promoting efficient energy use by rewarding energy conservation. In addition, development of appropriate energy sources for rural areas will be accelerated.

AGRICULTURAL AND RURAL DEVELOPMENT POLICY

Problems facing the agricultural sector specifically are summarized in the Fifth Plan as reduced production growth rate, unjust pricing and faulty market structure, and underdeveloped farmers' institutions.

Agricultural Policy and Energy Implications

In the face of these problems, these five policies seem reasonable.

1. Readjust the agricultural infrastructure. Increase productivity by increased yield and optimal land use according to agroeconomic zone structure.
2. Promote just prices for agricultural produce, through venues for produce and a new market structure.
3. Accelerate land reclamation and improve distribution of land ownership.
4. Increase agricultural assistance from financial institutions.
5. More integrated rural institutions (NESDB 1982).

Certain goals for agricultural production have been set.

1. An overall annual growth of 4.5 percent. In particular, this means 4.7 percent for crop production, 4.2 percent for livestock production, 5.5 percent for fishery, and 0.3 percent for forestry.
2. An increase in production efficiency. Paddy yield should increase from 343 kg/rai to 355 kg/rai in the North Region, from 224 to 293 in the Northeast Region, from 302 to 320 in the Central Region, and from 285 to 290 in the South Region. Annual increases of between 1.4 to 8.9 percent are projected for other crops.
3. More efficiently used natural resources. Soil improvement and improvement in the present irrigation system (without significantly increasing irrigation area) are planned, as is a reforestation program of 300,000 rais per year.
4. An 11.7 annual percent increase in agricultural loans (or ฿5,700 million from the ฿41,600 million base of 1982). An estimated 4.23 million households will benefit from average loans of ฿7,200 per year.

Neither fuel nor energy is mentioned explicitly in the agricultural policy. Nevertheless, increasing agricultural production entails some yet undefined increase in energy consumption. Using crop production growth projections (4.7) percent and the elasticity of energy demand based on production growth (1.26), annual energy demand growth can be projected at 5.9 percent. However, microstudies of particular agricultural subregions are needed on energy use patterns and requirements for agricultural production. Perhaps agroeconomic zones are appropriate sizes for sample studies. Without an appropriate prior study, attempts to enforce any alternative energy pricing policy or to convert the use of the present fuel type to LPG, methanol, or even gasohol may not produce the desired result or may even create hardship on the farmers.

Rural Development Policy

Serious disparity exists in the distribution of development benefits as indicated by various economic and social conditions. To alleviate this disparity, the present policy guidelines for rural development emphasize people or human development rather than economic development, which has always received the highest priority in the past. The principle of this policy has a five-point guideline: (1) being area-specific, (2) assisting people to achieve a reasonable livelihood, (3) emphasizing assistance to increase the ability for self-help, (4) identifying and alleviating the problems facing the impoverished, and (5) involving the people in designing solutions to their own problems as much as possible.

Development of Specific Areas and Towns. One section of the Fifth Plan focuses on development of five specific areas (including towns) to distribute more equitably centers of economic activities. These five specific areas and activities to be emphasized are

1. The three provinces in the eastern seaboard: developing basic industries, especially natural gas.
2. The provinces in the West Region: sugarcane and pineapple canning.
3. The lower part of the Northeast Region and the upper part of the North Region: improving soil and water resources, regenerating forest areas, and alleviating poverty.
4. The border provinces in the South Region: social problems and social security.
5. Principal cities, towns, and rural communities: developing economic centers.

Development of Impoverished Areas. Based on a series of studies of indicators associated with agriculture, three levels of rural economy were identified.

1. In the *progressive* agricultural areas, opportunity already exists for increased productivity. These areas include the irrigated areas of the Northern Region and the central plain and those areas with long rainy periods affording plantation of cash crops with higher return than paddy. In those areas the already high overall economic return per plantation area still can be improved.

2. In the *semi-progressive* agricultural areas, agriculture relies more on rain, but income supplements are possible by allocating some areas for cash crops. These areas evidence real income increases by up to 90 percent in the last decade as well as a moderate increase in yield.

184 Rural Energy to Meet Development Needs

3. Where *subsistence* agriculture is practiced, only limited opportunity exists for obtaining monetary income. The affected people cannot benefit from development or technological innovations. For example, they do not benefit from price increases because they do not produce more than they themselves consume, nor can they utilize the modern production techniques.

Conventional development methodology sufficiently deals with the first and second categories. In existing projects, priority is given to encouraging agroindustry, small-scale, and household industry and to distributing industry to rural areas. Improved methods include land reform; land banks and credits, prices, and taxes; and agricultural cooperatives. Private sectors are given greater roles in increasing production of rubber, cotton, maize, livestock, and fisheries as well as in developing technologies such as irrigation and crop varieties and production tools. Independent institutes would govern quality and weight, including variables such as crop insurance, securities and stock distribution, and warehouses.

The first step toward development of impoverished areas is to identify those areas through a methodology applicable, in principle, to all regions. The district is the appropriate level for this purpose because the information is relatively complete but not too aggregated. Targets are identified according to data on land utilization and on the rate of return, size of landholding, occupation, population size, ecological condition, soil, amount and type of water, and amount of forest area. Using those criteria, 216 districts and 30 subdistricts in 37 provinces have been identified comprising approximately 7.5 million rural inhabitants. The regional distribution of impoverished areas is shown in Table 11. The strategy for development is based on these interrelated programs adapted for specific areas: basic health facilities and services, rural job creation, village-level food and nutrition activities designed to increase village productivity, and development research in problem areas to overcome production obstacles. A basic readjustment program specifies reorienting educational philosophy, increasing peoples' involvement in rural development, and employing bottom-up planning initiated by commune councils.

Energy and the Development of the Impoverished Rural Areas. Two identifiable activities in the development plan concern energy: village woodlots and planting of fast-growing trees on public land, and buffalo banks. Tree biomass is recognized as an essential resource for domestic use, especially for cooking. Animal power is recognized as essential for agricultural production. So introducing the concept of buffalo banks is appropriate, but soil variations and preparation difficulties as well as

TABLE 11.
Impoverished Areas, by Region

Region	Province	District	Sub-district
North	16	65	7
Northeast	16	129	18
South	5	22	5
Total	37	216	30

particular labor requirements should be investigated. Introduction of biogas in these areas and a follow-up study also might help elucidate some of the persistent social problems other areas have experienced in the past when biogas was introduced.

Rural development is a complex process involving both explicit, measurable physical components and implicit components, such as social institutions. The physical components and their relationships could be quantified as a first step in understanding the development process. Data is required on economic transactions between identifiable sectors as well as on the form and size of energy and its mode devices. Interrelationships between those sectors would be identified and expressed in the type and size of the intermediate economic transactions, constrained by the implicit production processes. An acceptable methodology is discussed in Surapong Chirarattananon (1982).

RURAL ENERGY: SOURCES AND CONSUMPTION

One important rural energy survey was conducted by the Regulatory Division of NEA in 1980. A second survey, conducted as part of the program for the renewable nonconventional energy project, supported by USAID, has been summarized by John Arnold (1982). A third survey with modest area coverage has been initiated by the Thailand Institute for Scientific and Technological Research (TISTR). Only the results from the NEA rural energy survey will be described here because details of the other two surveys were not available.

The 1980 NEA Rural Energy Survey

The NEA rural energy survey was conducted primarily to establish baseline energy data and to determine energy program priorities. A pilot survey to test the questionnaire and to familiarize personnel to the field

TABLE 12.
Distribution of Households in the Survey

Region	Provinces	Villages	Households
North	6	70	700
Northeast	7	85	850
Central 1	3	15	150
Central 2	4	15	150
South	4	15	150
Total	24	200	2,000

conditions was conducted in June 1979. Another survey using a revised questionnaire and improved methodology was started in April 1980 but then abandoned. A final national survey was conducted in August 1980. In the third, more refined version, respondents' answers could be categorized and recorded on premade forms for direct coding onto computer card. This third survey version is referred to as the 1980 NEA rural energy survey, and the NEA publicizes information from only this third version of the survey. A report on household energy consumption compiled from data from the survey was published in March 1982, in the Thai language. However, some information already obtained from the questionnaire is yet unpublished. This survey will be conducted every three years as part of overall energy planning and monitoring, primarily as a household survey.

A stratified sampling procedure was devised by the National Statistical Offices (NSO) for three levels. First, provinces that satisfy NSO socioeconomic classification criteria in each geographical region were selected based on probability proportional to population size. Then village samples were taken systematically using the size of villages as criterion. At the third, or household level, random sampling was done — 10 households from each village. Table 12 shows the resulting distribution and Figure 2 illustrates geographical coverage.

Ten interviewers and two supervisors were assigned to the Northern Region and to the Northeastern Region. Five interviewers and one supervisor were assigned to each of the other regions. Altogether 35 interviewers and seven supervisors were used. Each interviewer was provided with a tape measure and a balance.

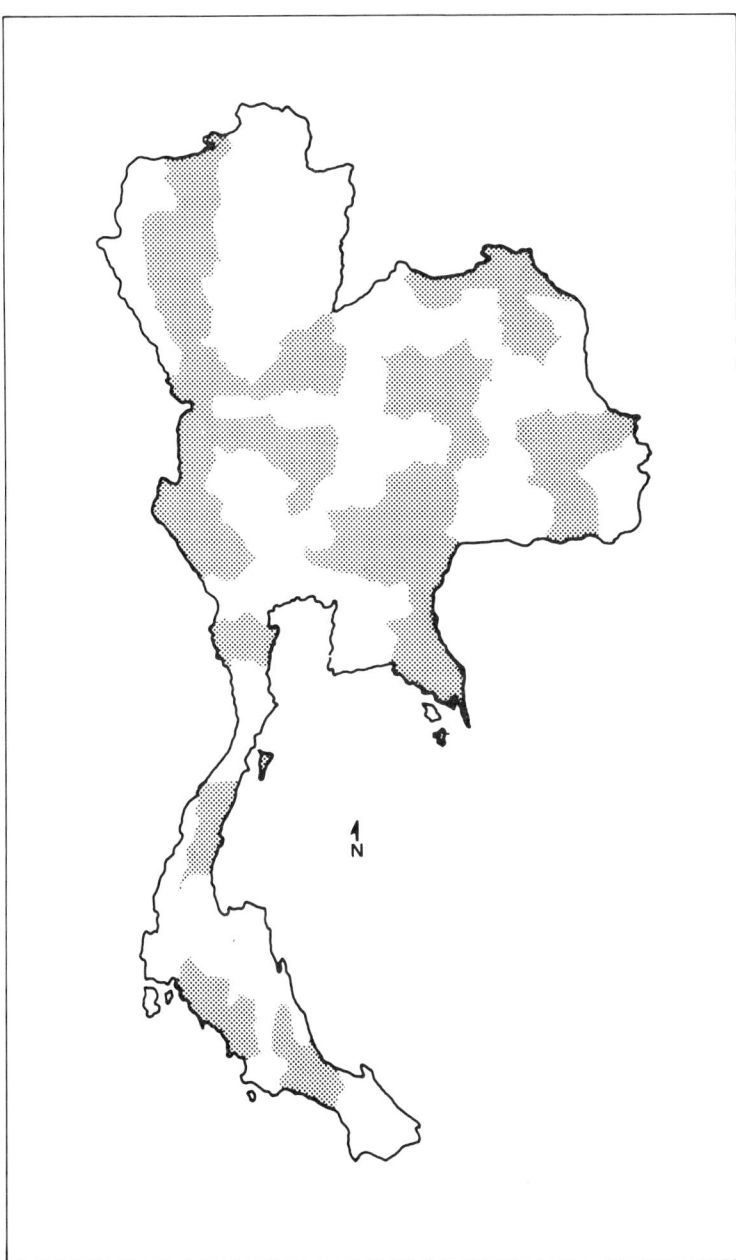

Figure 2. Geographical Coverage, 1980 NEA Rural Energy Survey

Two sets of questionnaires were used. The household questionnaire, covering energy-related activities in individual households, was handled by the interviewer. The village questionnaire, to be used with individual village leaders, was handled by the supervisor, but the information obtained was unreliable because of communication problems.

The village questionnaire covered these items: identification and general information, production resources, social services and public utilities, manpower utilization and occupation, cropping, and energy resources and consumption. The household questionnaire had more interrelated items. Household information included household size; occupation and income; fuel (type and quantity) used in cooking, heating, lighting, and other devices; acquisition of fuel, method, distance, time consumed, and transportation; and cooking utensils, devices, and fuel use. Various transportation questions involved mode of transportation, distance, time spent, and frequency; personnel transportation; and use of use;vehicle and fuel. Information requested on agriculture included landholding and land use, agricultural production and transactions, tractor use and fuel, water pump and fuel, and other agricultural machines and fuel. Information also was requested about human and animal labor in agriculture, labor and fuel requirement for each crop, and fertilizer and chemical products. Seasonal effects and other factors related to time and duration were based on recall by respondents. Supervisors had immediate access to data.

Rural Energy Sources

The major sources of energy in rural communities are biomass, petroleum, and human and animal labor. Quantitative information obtained in the survey on solid fuel use is expressed in the weight unit of kilogram and for liquid fuels in the volume unit of liters. The heat content or calorific content of each type of fuel is given in Appendix 1. Each type of fuel is defined next.

Fuelwood. The branch or trunk of assorted tree biomass, excluding leaves, with varying moisture content. A particular tree species or group of species for fuelwood usually predominates in a particular locality, but no distinction is taken in the report.

Charcoal. The carbonized product from fuelwood, obtained by a partial combustion of fuelwood in controlled atmosphere. Fuelwood or rice husk or dry leaves or the combination are used as starter fuels for the combustion in charcoal kilns.

Paddy Husk. The outer covering of rice grain, obtained when the rice is milled.

Paddy Straw. The trunk and stem of rice paddy. During harvest, the root and about 15 centimeters of stem are left on the field to be burned some time later. The part of stems harvested with paddy is piled on the household ground. Some paddy straw is used to feed cattle.

Coconut Branch. This includes the fiber from the coconut husk and the husk itself, after the flesh is removed, as well as the branches, fresh or dried.

In Thailand, very little use is made of other agricultural residues, even corn cob or jute stem. Cattle dung is usually left untouched. Occasionally dung from where cattle are tied up overnight is collected for compost. Substantial agricultural residue is burned away.

The commercial fuels in use include gasoline (usually the standard grade with octane rating 85), high-speed diesel, kerosene, and LPG. Kerosene is used mainly for lighting, and LPG for cooking. Commercial fuel cost varies according to distance from Bangkok and the accessibility of the location. Up to ฿0.2 per liter is added for provinces several hundred kilometers from Bangkok. Charcoal is a commercial fuel in established communities, including Bangkok. The price differential for charcoal is great: it is cheaper in rural areas. Calcium carbide, which produces acetylene gas when reacting with water, is used in rural communities mainly for lighting. Dry cells are used in radios and electric torches.

Figure 3 shows the pattern of transformation of energy resources to rural end uses on a national basis for 1980. The regional breakdown of rural household energy uses by source is shown in Tables 13 to 18. Published information is limited to domestic use and household industry as well as several tables on energy use in agriculture. Information on energy use in transportation has not been published, so statistics for that use were obtained from a part of the pilot survey on personnel transportation.

The most striking overall feature is that energy consumption for domestic purposes — 67 percent — is nearly the same for every region, and that the significant differences of energy use are found in agriculture. Heat, generated almost totally from biomass sources, is used mainly for cooking except in the Northeast and North Regions where household heating is significant. Mechanical work, used almost exclusively for agriculture, is generated from petroleum fuels and obtained from human and animal labor.

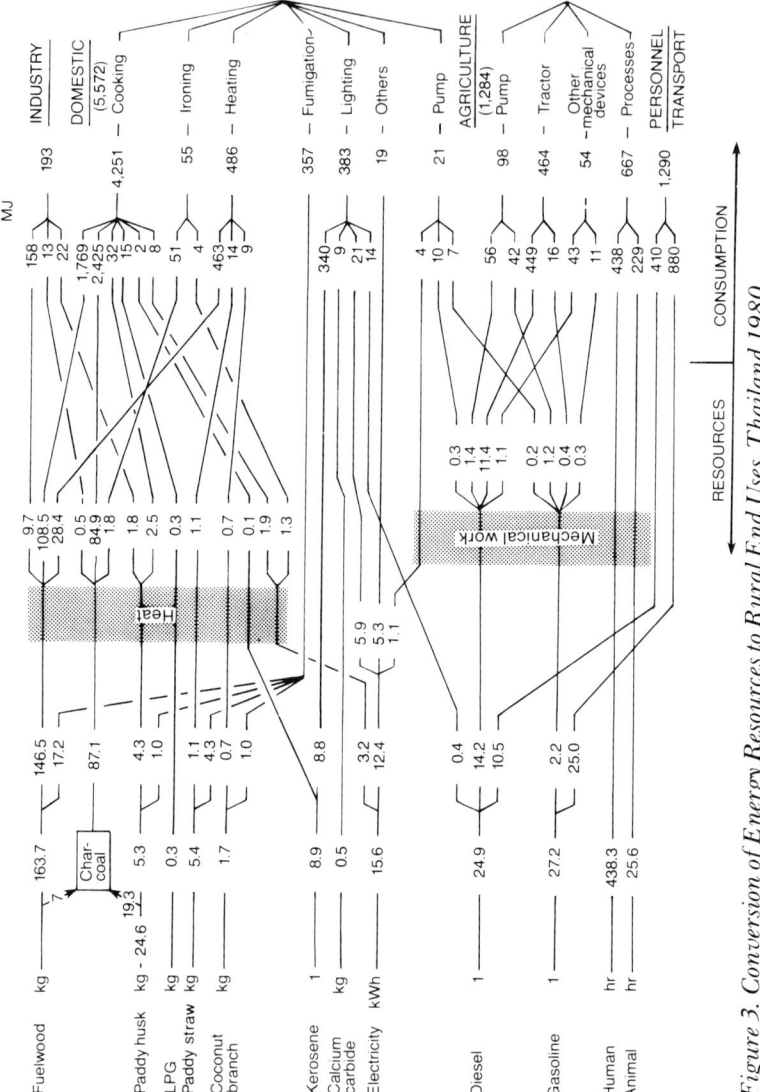

Figure 3. Conversion of Energy Resources to Rural End Uses, Thailand 1980. (Annual per Capita)

TABLE 13.
Rural Household Energy Consumption, <u>NATIONAL</u>, 1980. (MJ/capita)

USE	Fuel-wood	Char-coal	Paddy husk	Paddy straw	Coco-nut branch	LPG	SOURCE Kero-sene	Gaso-line	Die-sel	Elec-tri-city	Dry-cell	Calc. Car-bide	Sub-total	Per-cent-age
Cooking	1,769	2,425	32	-	-	15	2	-	-	7	-	-	4,250	70.8
Lighting	-	-	-	-	-	-	340	-	13	21	-	9	383	6.4
Ironing	-	51	-	-	-	-	-	-	-	5	-	-	55	0.9
Industry	158	13	22	-	-	-	-	-	-	-	-	-	432	7.2
Heating	463	-	-	14	9	-	-	-	-	-	-	-	486	8.1
Fumigation	280	-	12	13	13	-	-	-	-	-	-	-	357	6.0
Appliances	-	-	-	-	-	-	-	-	-	19	-	-	19	0.3
Pump	-	-	-	-	-	-	-	7	10	4	-	-	21	0.3
Subtotal	2,670	2,489	305	67	22	15	342	7	23	56	-	9	6,004	100.0
Percentage	44.5	41.4	5.1	1.1	0.4	0.3	5.7	0.1	0.4	0.9	-	0.2	100	-

TABLE 14.
Rural Household Energy Consumption, NORTHERN REGION, 1980. (MJ/capita)

USE	Fuel-wood	Char-coal	Paddy husk	Paddy straw	Coco-nut branch	LPG	Kero-sene	Gaso-line	Die-sel	Elec-tri-city	Dry-cell	Calc. Car-bide	Sub-total	Per-cent-age
Cooking	1,326	3,049	–	–	–	4	–	–	–	5	–	–	4,384	75.4
Lighting	–	–	–	–	–	–	201	–	11	28	–	12	251	4.3
Ironing	–	37	–	–	–	–	–	–	–	5	–	–	43	0.7
Industry	49	29	–	–	–	–	–	–	–	–	–	–	220	3.8
Heating	340	–	–	21	13	–	–	–	–	–	–	–	374	6.4
Fumigation	399	–	–	99	17	–	–	–	–	–	–	–	515	8.9
Appliances	–	–	–	–	–	–	–	–	–	23	–	–	23	0.4
Pump	–	–	–	–	–	–	–	3	1	1	–	–	5	0.1
Subtotal	2,109	3,115	144	120	30	4	201	3	12	63	–	12	5,811	100.0
Percentage	36.3	53.6	2.5	2.1	0.5	0.1	3.5	0.0	0.2	1.1	–	0.2	100	–

TABLE 15.
Rural Household Energy Consumption, <u>NORTHEASTERN REGION</u>, 1980. (MJ/capita)

USE	Fuel-wood	Char-coal	Paddy husk	Paddy straw	Coco-nut branch	LPG	Kero-sene	Gaso-line	Die-sel	Elec-tri-city	Dry-cell	Calc. Car-bide	Sub-total	Per-cent-age
Cooking	2,485	1,647	-	-	-	1	1	-	-	1	-	-	4,135	63.9
Lighting	-	-	-	-	-	-	421	-	12	9	-	10	452	7.0
Ironing	-	22	-	-	-	-	-	-	-	1	-	-	23	0.4
Industry	179	16	-	-	-	-	-	-	-	-	-	-	583	9.0
Heating	853	-	12	14	-	-	-	-	-	-	-	-	879	13.6
Fumigation	346	-	26	11	9	-	-	-	-	-	-	-	393	6.1
Appliances	-	-	-	-	-	-	-	-	-	4	-	-	4	0.0
Pump	-	-	-	-	-	-	-	1	-	-	-	-	1	0.0
Subtotal	3,862	1,685	415	23	23	1	422	1	12	16	-	10	6,471	100.0
Percentage	59.7	26.0	6.4	0.4	0.4	0.0	6.5	0.0	0.2	0.2	0.0	0.2	100	-

TABLE 16.
Rural Household Energy Consumption, CENTRAL 1 REGION, 1980. (MJ/capita)

USE	Fuel-wood	Char-coal	Paddy husk	Paddy straw	Coco-nut branch	LPG	Kero-sene	Gaso-line	Die-sel	Elec-tri-city	Dry-cell	Calc. Car-bide	Sub-total	Per-cent-age
Cooking	1,766	2,822	273	–	–	23	32	–	–	22	–	–	4,938	76.0
Lighting	–	–	–	–	–	–	149	–	61	39	–	15	264	4.1
Ironing	–	64	–	–	–	–	–	–	–	7	–	–	72	1.1
Industry	148	4	–	–	–	–	–	–	–	–	–	–	365	5.6
Heating	86	–	–	70	1	–	–	–	–	–	–	–	157	2.4
Fumigation	301	–	8	198	5	–	–	–	–	–	–	–	511	7.9
Appliances	–	–	–	–	–	–	–	–	–	44	–	–	44	0.7
Pump	–	–	–	–	–	–	–	1	110	29	–	–	139	2.1
Subtotal	2,307	2,890	494	268	6	23	181	1	171	140	–	15	6,495	100.0
Percentage	35.5	44.5	7.6	4.1	0.1	0.4	0.3	0.0	2.6	2.2	0.0	0.2	100	–

TABLE 17.
Rural Household Energy Consumption, CENTRAL 2 REGION, 1980. (MJ/capita)

USE	Fuel-wood	Char-coal	Paddy husk	Paddy straw	Coco-nut branch	LPG	Kero-sene	Gaso-line	Die-sel	Elec-tri-city	Dry-cell	Calc. Car-bide	Sub-total	Per-cent-age
Cooking	946	3,262	98	–	–	42	–	–	–	21	–	–	4,369	79.3
Lighting	–	–	–	–	–	–	341	–	13	38	–	3	395	7.2
Ironing	–	113	–	–	–	–	–	–	–	13	–	–	126	2.1
Industry	21	–	–	–	–	–	–	–	–	–	–	–	184	3.3
Heating	132	–	–	–	–	–	–	–	–	–	–	–	132	2.4
Fumigation	108	–	1	91	–	–	–	–	–	–	–	–	200	3.6
Appliances	–	–	–	–	–	–	–	–	–	47	–	–	47	0.9
Pump	–	–	–	–	–	–	–	33	16	9	–	–	59	1.1
Subtotal	1,206	3,375	263	91	–	42	341	33	28	128	–	3	5,510	100.0
Percentage	21.9	61.2	4.8	1.6	0.0	0.8	6.2	0.6	0.5	2.3	0.0	0.1	100	–

TABLE 18.
Household Energy Consumption, SOUTHERN REGION, 1980. (MJ/capita)

USE	Fuel-wood	Char-coal	Paddy husk	Paddy straw	Coco-nut branch	LPG	Kero-sene	Gaso-line	Die-sel	Elec-tri-city	Dry-cell	Calc. Car-bide	Sub-total	Per-cent-age
Cooking	1,291	2,635	-	-	-	44	-	-	-	6	-	-	3,975	75.2
Lighting	-	-	-	-	-	-	399	-	4	20	-	7	430	8.1
Ironing	-	79	-	-	-	-	-	-	-	3	-	-	81	1.5
Industry	432	-	-	-	-	-	-	-	-	-	-	-	633	12.0
Heating	31	-	-	-	1	-	-	-	-	-	-	-	32	0.6
Fumigation	79	-	-	-	34	-	-	-	-	-	-	-	113	2.1
Appliances	-	-	-	-	-	-	-	-	-	15	-	-	15	0.3
Pump	-	-	-	-	-	-	-	2	-	6	-	-	8	0.1
Subtotal	1,832	2,713	201	-	34	44	399	2	4	49	-	7	5,286	100.0
Percentage	34.7	51.3	3.8	0.0	0.6	0.8	7.6	0.0	0.1	0.9	0.0	0.1	100	-

Rural Energy Development in Thailand 197

Leucaena forage planting, part of a 10-ha area, Kanchanaburi, Thailand. Wood used for pottery. (R. Van Den Beldt)

Fuelwood and charcoal together account for about 85 percent of the rural household energy supply. Paddy husk, paddy straw, and other agricultural wastes used for heat production contribute only 7 percent of the energy supply. The magnitude of rural household energy use for cooking leads to only two possible types of solutions: interfuel substitution and conservation by more efficient usage.

Comparison of rural household energy use by region shows no major difference in the composition of the energy resources. Specific regional variations among resources and end uses are discussed in the following sections.

Fuelwood and Charcoal. Fuelwood and charcoal are used to produce low-grade heat used for cooking in households or raising temperatures of materials in household and rural industry. As more convenient fuels become available, such as LPG in urban areas, people use them significantly more. Even so, fuelwood and charcoal demand cannot be expected to reduce swiftly.

Fuelwood is available from two sources, national forests and private or existing village woodlots. Scant information is available on the amount of

Table 19.
Household Fuelwood Sources

Sources of Fuelwood	North	North-east	Central 1	Central 2	South	National
	Percentage of Households					
Free Sources						
Within own compound	64.9	35.7	85.2	31.2	5.2	35.4
Other's compound	3.9	21.4	2.3	0.9	2.4	14.4
Nearby forest (less than 5 km)	21.2	26.9	6.9	61.0	44.1	31.7
Distant forest	5.8	14.6	3.9	0.7	45.8	16.3
Sub total	95.6	98.6	98.4	96.6	97.5	97.9
Commercial Sources						
Within the village	0.8	0.2	1.5	1.7	-	0.4
Out of the village	0.2	0.2	0.2	2.7	-	0.4
At own premise, from salesperson	3.3	1.0	-	1.0	2.5	1.4
Sub total	4.4	1.4	1.7	5.5	2.5	2.1

tree mass in each source, but the 1980 NEA survey (Table 19) shows very wide variation among sources of fuelwood in different regions.

The extent of the production of charcoal for household use by households producing their own charcoal is shown in Table 20. The availability of charcoal is directly related to the availability of fuelwood, but significantly more charcoal is obtained through commercial sources. The yield of charcoal from fuelwood by weight varies from approximately 1 to 5 to 1 to 4.

Household Energy Demand

Cooking. Cooking accounts for 50 percent of total rural energy use and 75 percent of household use; charcoal and fuelwood contribute 57 and 42 percent heat respectively. The composition of charcoal and fuelwood used for cooking varies among regions, but the total amount of

Wood for pottery manufacture, Ratchaburi, Thailand. (R. Van Den Beldt)

heat generated remains fairly constant, with a low of 3,975 MJ per capita per year for the South Region to a high of 4,938 MJ per capita per year for the Central Region.

Tables 21 and 22 give percentages of households using each type of fuel and each kind of cooking stove. Some households use more than one type of fuel. For example, fuelwood usage overlaps with charcoal in the Northeast Region. Electricity is used almost exclusively for cooking rice and perhaps for boiling water. This information forms a basis for estimated fuel use probabilities.

Multiple use of different types of stove is less common than of different fuels. The charcoal bucket stove is used most. A fairly fuel-efficient stove, its heat efficiency has been tested at up to 30 percent, with an average of 22 percent. The wood bucket stove, a modified charcoal bucket stove, has a recorded heat efficiency of 15 percent. Individual and regional differences make fuel efficiency difficult to establish during actual cooking. Standards for "cooking effect," or average heat requirement for cooking by locality and for each type of stove, would be valuable. However, for the purpose of estimating the total fuel requirement, the average fuel

TABLE 20.
Household Fuelwood Sources for Production of Charcoal

Sources of Fuelwood for Charcoal	North	Northeast	Central 1	Central 2	South	National
	Percentage of Households					
Free Sources						
Within own compound	43.9	61.2	32.3	47.7	48.9	50.0
Other's compound	2.2	4.8	1.4	5.1	9.9	4.8
Nearby forest (less than 5 km)	6.6	9.6	7.3	7.6	14.9	9.1
Distant forest	9.0	3.2	3.9	8.0	6.8	6.4
Sub total	61.7	78.8	45.0	68.4	80.5	70.3
Commercial Sources						
Within the village	6.3	4.8	7.8	7.6	8.2	6.5
Out of the village	6.6	5.7	20.9	6.2	5.9	7.0
At own premise, from salesperson	25.4	10.7	26.4	17.7	5.5	16.2
Sub total	38.3	21.2	55.0	31.6	19.5	29.8

requirement for each type of stove may be used. Experiments on and observations of fuel requirements have been attempted in an USAID study, simulating normal cooking situations with charcoal bucket and wood bucket stoves (Arnold 1982). Table 23 summarizes some of the study's results. A weight-averaged figure for each region times 365 gives the annual requirement per capita, as shown in Table 24. The statistics in Table 24 imply the amount of consumption of one type, if it completely substitutes for the other.

The fuel requirement data obtained by the USAID program (and shown in Table 23) may not be acceptable standard data for Thailand, nor are the data obtained by NEA, but they are in good agreement. The specific usefulness of the data in Table 24 is their use as substitution factors. For example, the total requirement for charcoal for cooking, if it is used instead of fuelwood in the Northern Region (see Table 14) would

TABLE 21.
Household Cooking Fuels

Region	Fuelwood	Charcoal	Paddy Husk	LPG	Kerosene	Electricity
			Percentage of Households			
North	36.8	66.1	–	0.6	–	4.1
Northeast	75.2	62.9	–	0.1	0.1	1.3
Central 1	45.2	74.7	2.7	4.8	2.1	14.4
Central 2	28.9	78.5	2.0	3.4	–	10.7
South	39.3	71.9	–	3.7	–	5.2
Average	51.7	68.0	0.5	1.5	0.2	4.8

TABLE 22.
Household Cooking Stoves

Region	Wood Bucket	Charcoal Bucket	Husk Stove	Gas	Kerosene	Electric	Three Rocks
			Percentage of Households				
North	38.9	60.8	–	0.3	–	1.0	2.0
Northeast	33.3	61.9	–	0.1	0.1	0.3	24.8
Central 1	28.1	70.6	3.4	4.1	2.1	2.1	2.7
Central 2	21.5	77.2	3.4	4.0	–	2.0	3.4
South	32.6	70.4	–	3.0	–	–	5.2
Average	32.6	65.6	0.7	1.4	0.3	0.8	11.7

be 106.74 + (81.03)(153/165) = 160 kg, based on the consumption data of NEA and the substitution factor from the USAID program. Using the latter figure, the requirement in the same region is given in Table 24 as 153 kg. If this type of calculation is carried out for all the other regions, it will be seen that the data on fuel requirement for cooking obtained by NEA agree quite well with those obtained by the USAID program, although the NEA figures are generally slightly higher.

Another substitute fuel for cooking may be LPG. This is a commercial

TABLE 23.
Fuel Requirements in Normal Cooking, Charcoal Bucket and Wood Bucket Stoves

Province	Region	Median Fuel Requirement			
		Per Capitum*		Per Capita	
		Wood	Charcoal	Wood	Charcoal
			kg/day		
Kampangphet	North	.62	.35	.54	.31
Chiang Mai	North	.49	.62	.43	.55
Lampang	North	.48	.32	.42	.28
Udorn Thani	Northeast	.73	.36	.64	.32
Roi Et	Northeast	.76	.18	.66	.66
Srisaket	Northeast	.65	.29	.57	.66
Korat	Northeast	.55	.32	.48	.28
Chantaburi	Central 1 East	.62	.32	.54	.28
Petchaburi	Central 2 West	.50	.32	.44	.28
Songkla	South	.49	.15	.43	.13
	Average	.70	.34	.61	.30

* Capitum refers to one adult equivalent: children under 12 years old are counted as one-half the adult equivalent.

item though, and its use requires considerable investment over the fuel cost. No reported estimate of the requirement of LPG for cooking or for other types of stoves and fuels similar to the study by USAID has been available.

Paddy husk has been used recently for cooking mostly with a new stove type, a commercial product. This stove is very local in character and is usually produced and marketed in the vicinity of the users. However, the stove and paddy husk fuel have limited cooking application. Kerosene is not commonly used for cooking — it may be used under very special circumstances. Fuelwood in the three-rock type of stove seemed to be used commonly in the Northeast Region. The stove costs nothing, it can be set up anywhere, and the practice of cooking with three-rock stove has

TABLE 24.
Annual Fuel Requirements in Normal Cooking,
Wood and Charcoal Stoves

Region	Wood	Charcoal
	kg/person	
North	165	153
Northeast	209	96
Central 1	197	103
Central 2	159	103
South	156	48
Average	183	104

been handed down. But the use of the three-rock stove obviously is a major waste of energy resources.

Together, the data in Tables 21 and 22 provide some basis for estimating the probable use of each type of stove and the corresponding fuel. The information required for such estimate is the relative frequency of use of each type of stove. An estimate of the cooking effect can be obtained by using the frequency information together with the information on the fuel use and the interfuel substitution factors. Information on fuel requirement and the results in fuel shift due to interfuel substitution obtained this way is much more accurate than the estimate based only on fuel use and the heat content of fuels. However, this type of approach requires microlevel data and field trials.

The process of cooking food consumes more than 140,000 terajoules of energy annually in the form of low-grade heat through burning of fuelwood and charcoal. This amount is almost one-half of the total energy provided by the nonpetroleum sources and is approximately 20 percent of all energy consumed in the country. Price controls cannot slow the pace of consumption; fuel scarcity only creates hardship. Without proper planning and appropriate action, this consumption may rapidly deterio-

TABLE 25.
Use of Lighting Fuels

Region	Kerosene	Diesel	Electricity	Dry Cell	Calcium Carbide
			Percentage of Households		
North	72.8	2.7	33.8	69.5	4.1
Northeast	85.4	4.5	13.5	64.8	37.5
Central 1	51.4	17.8	39.7	74.0	17.8
Central 2	69.8	1.3	30.2	83.9	5.4
South	83.0	0.7	18.5	67.4	14.8
Average	77.4	4.0	23.4	69.7	19.9
			Annual Use Per Person		
Average	8.7 l	0.3 l	5.9 kWh	5.3 units	0.5 kg

rate the environment and create hardship for the impoverished at the same time.

Lighting. Lighting is considered one of the basic utilities. The state provides public lighting as a basic social service, and the state must assist in providing access to the means for private lighting. The energy requirement for lighting is relatively small. The unit of measure for lighting is the lumen. An arbitrary set of standards for household lighting can be established utilizing illumination engineering together with a microlevel study and field measurement. And a macrolevel survey (even by questionnaire) of the number of lighting devices per household will furnish information on the total lighting effect. The NEA survey and their subsequent publication provide the information shown in Tables 25 and 26. Estimates of energy required for lighting are easier to make than those for cooking because with cooking, the actual heat required in cooking and the device efficiency vary according to local practices.

Kerosene is the main fuel used for lighting, particularly in villages with no electricity. The main device used with kerosene is a simple lamp made from a tin can. Diesel oil is also used with tin can as a substitute for kerosene when the latter runs out. Usually households that use a diesel substitute for kerosene already possess a mechanical device using diesel. This is evident by the high percentage of households using diesel oil for

TABLE 26.
Use of Lighting Devices

Region	Kerosene Lantern	Kerosene Tin Can	Kerosene Pressurized	Calcium Carbide	Electric Torch	Fluorescent	Incandescent
			Percentage of Households				
North	8.2	75.5	10.2	4.1	69.5	32.0	26.4
Northeast	6.3	89.9	12.2	37.5	64.8	13.2	5.7
Central 1	13.0	69.2	14.4	17.8	74.0	40.4	24.6
Central 2	8.7	71.1	13.4	5.4	83.9	31.5	19.4
South	3.7	83.7	3.7	14.8	67.4	18.5	8.1
Average	7.1	81.4	10.8	19.9	69.7	23.1	14.0

lighting in the Central 1 Region (Table 25). The discussion regarding Tables 13 to 18 clearly shows that the Central 1 Region consumes the most petroleum per capita, particularly in agriculture, and the predominant fuel type is diesel. The lantern and pressurized kerosene lamp are more expensive devices, with higher percentages of use exhibited by the Central 1 Region.

Electric torch and gas lamps fueled by calcium carbide are used mostly outdoors. Gas lamps are used most often for fishing, catching animals, or collecting rubber latex at night.

Ironing. Ironing is a minor activity for the rural inhabitants. Ironing devices either use electricity or charcoal. Table 27 shows the percentages of households using each type. Although this activity is associated with social status and social values, new no-iron clothing and changing social values probably will mean no substantial increase in ironing requirement in the near future.

Household Industry. Household industry includes charcoal making, food catering, household animal feed-preparing, as well as weaving and dyeing (see Table 28).

In most cases, the use of electricity for household industry cannot be distinguished from its use for other household purposes. This also pertains to lighting fuels. The type of fuel mix indicates that most fuel uses covered in Table 28 are for low-grade heat. In this case, paddy husk use appears prominently both in frequency and quantity.

TABLE 27.
Use of Ironing Devices

Region	North	North-east	Central 1	Central 2	South	Average
Fuel		Percentage of Households				
Charcoal	11.3	10.3	19.9	26.9	28.9	16.3
Electric	12.0	4.3	19.9	22.2	8.2	10.4

TABLE 28.
Fuels Use for Household Industry

Region	North	North-east	Central 1	Central 2	South	Average
Fuel		Percentage of Households				
Fuelwood	3.1	5.0	3.4	1.3	2.2	3.5
Charcoal	1.7	0.6	1.4	–	–	0.7
Paddy Husk	7.2	17.4	12.1	6.7	18.5	13.1

Heating. The need for heating exists in North and Northeast Regions and in the mountain areas in the West Region. The temperature in many areas could fall below 10°C, although usually for short periods. This, together with the inadequate shelter provided by thatched houses or leaf-walled houses makes heating necessary (see Table 29).

Fumigation. Fumigation serves two purposes, getting rid of insects from the household area and cleaning of the ground within the household vicinity. Fuelwood from falling twigs is swept together with leaves into a clearing and lit to create smoke. The extent of the practice of fumigation is indicated by the information in Table 30.

Appliances. Use of electrical appliances not only indicates the extent of rural electrification but also the level of affluence of the household and the level of commercial influence. Table 31 provides information on the types of appliances commonly used and the percentage of households possessing them. Radio has become an essential household device. Most households without electricity have a dry-cell powered radio. Most radios

TABLE 29.
Fuels Used for Heating

Region	North	North-east	Central 1	Central 2	South	Average
Fuel			Percentage of Households			
Fuelwood	32.7	81.0	13.7	16.1	4.4	44.0
Paddy Straw	1.7	0.9	4.8	–	–	1.1
Coconut branch	2.7	24	2.1	1.3	0.7	2.0

TABLE 30.
Fuels Used for Fumigation

Region	North	North-east	Central 1	Central 2	South	Average
Fuel			Percentage of Households			
Fuelwood	8.6	22.6	4.8	2.7	2.2	12.1
Paddy Husk	0.1	1.9	1.4	0.7	–	1.0
Paddy Straw	3.5	1.8	4.8	4.7	–	2.6
Coconut branch	0.9	1.2	0.7	–	2.2	1.1

are portable semiconductor type. Broadcasts are often heard in the agricultural field.

Since East Asia has emerged as one of the main electronic centers and an electrical appliance industry has been established in Thailand, prices of these appliances have gone down. All items in Table 31 are locally produced and are available in district centers or even community centers. Radios and televisions are common — televisions may be almost as common as motorcycles.

Pumps. Water for household consumption is usually either carried by small carts pushed by human labor, or hand carried. Only wealthier households use pumps to lift water. Often pumps are those already used in agriculture for irrigation. However, electrical pumps are used exclu-

TABLE 31.
Use of Electrical Appliances

	Radio		TV	Refrigerator	Fan	Rice Cooker	Pan
	Electric	Dry Cell					
Region	Percentage of Households						
North	6.2	80.8	8.6	3.3	15.0	4.0	1.3
Northeast	1.7	78.3	2.2	1.1	3.8	1.2	0.4
Central 1	9.6	82.9	21.9	7.5	21.9	14.4	1.4
Central 2	6.0	88.6	15.4	8.1	20.8	10.7	1.3
South	4.4	70.4	8.9	3.0	7.4	4.4	0.7
Average	4.3	79.6	8.0	3.3	10.7	4.6	0.9

TABLE 32.
Usage of Pumps

Region	North	Northeast	Central 1	Central 2	South	Average
Fuel	Percentage of Households					
Diesel	0.3	0.1	4.1	2.0	–	0.7
Gasoline	1.0	0.5	0.7	7.4	0.7	1.7
Electricity	2.1	0.1	6.9	6.0	4.4	2.5

sively for household purposes. Table 32 shows the percentage of the households possessing each type of pump.

Income Level and Household Energy Consumption

The NEA rural energy survey report classifies all nonpetroleum fuels, except electricity, as noncommercial energy sources. Table 33 is constructed from that survey report.

Usually data such as those in Table 33 show a monotonic increase of commercial energy consumption with increased income. A similar relationship would be expected for the noncommercial energy consump-

TABLE 33.
Household Consumption of Energy Per Capita, by Income Group and Region

Income Group ('000 ฿)	Region					Average
	North	North-east	Cent-ral 1	Cent-ral 2	South	
Commercial						
0 - 24	293	343	410	348	285	306
25 - 48	281	1,620	569	682	540	816
49 - 72	381	473	440	1,403	1,993	858
73 -120	465	327	791	678	867	666
Over 120	657	339	816	1,089	75	766
Noncommercial						
0 - 24	5,267	5,908	6,850	5,460	4,878	5,619
25 - 48	6,176	6,628	6,155	5,401	3,923	5,761
49 - 72	5,711	5,167	5,623	2,274	5,410	4,832
73 -120	5,535	5,824	4,769	2,688	6,335	4,681
Over 120	5,841	8,667	4,618	4,589	4,170	5,615
Total						
0 - 24	5,560	6,251	7,260	5,808	5,163	5,925
25 - 48	6,457	8,248	6,724	6,083	4,462	6,577
49 - 72	6,092	5,640	6,063	3,677	7,403	5,690
73 -120	6,000	6,151	5,560	3,366	7,202	5,347
Over 120	6,498	9,006	5,434	5,678	4,245	6,381

Source: National Energy Administration, 1980

tion and total energy consumption with increased income. However, the data in Table 33 do not exhibit the expected relationship. The foremost and probably most significant reason is the stratification of income levels. Table 33 shows that about 69 percent of rural households belong to the lowest income class (฿0–24,000). Stratification with more even distribution of the number of households in each class might produce an entirely different result. In the present stratification, too few households are included in second and higher levels to be adequately represented. The most positive relationship envisaged, between income and commercial energy consumption, can be confirmed by the data in Table 33 if households above the first level are grouped together. Then commercial energy consumption increases as income increases.

Most energy-consuming activities in rural households are traditional activities requiring traditional fuels. In some uses and some regions, however, commercial fuels have replaced some of the traditional fuels. The fuel mix and fuel use pattern are affected by a number of determinants, but where commercial fuels are used, monetary income may be the most significant determinant. For example, among households with access to electricity it would be informative to see the composition of cooking fuels stratified according to income class. The focus of interest will be the patterns of use of LPG and electricity as cooking fuel, in competing with charcoal. Knowledge about patterns of electricity use competing with charcoal for ironing and kerosene for lighting would also be of interest. For supply and price policy planning with respect to commercial fuels, the next phase of rural energy studies must explicitly examine these substitution trends and their determinants.

Energy Use in Agriculture

The use of petroleum fuels for agricultural machines is significant (Fig. 3) though the total volume consumed is small. As a national average, the contribution of petroleum fuels to mechanical work in agriculture is nearly 50 percent. This varies to as high as 87 percent in Central 1 Region and as low as 16 percent in the Southern Region. The Central 1 Region, as indicated earlier, has the highest per capita value added in agriculture. Energy is used in agriculture in this region at about four times the national average. Composition of agriculture in the Southern Region is considerably different from other regions.

Except for the South, variation among regions in human and animal labor inputs to agriculture is small. Tasks powered by human and animal work are not easily separable from petroleum-fueled tasks. However, water lifting is one kind of task powered exclusively by engines in Thailand — no human or animal-powered pumps are used. How production and productivity are affected by the use of petroleum-powered machines is not yet clear, but definitely petroleum use is an integrated part of current agricultural production.

The preliminary information from 1979 rural energy surveys made available by NEA on energy and fertilizer use in crop production in each region is analyzed in Tables 34 to 39. The statistics given were obtained by scaling of the original data. The tonnage of crops in the table does not necessarily agree with production information obtained from other conventional means. The most useful information, perhaps, is the per area use of each energy input. The statistics for per area energy use and per area fertilizer use for all crops combined and those for paddy only are considered sufficiently reliable. Other statistics are indicative only due to the limited sample base.

As the tables show, the Central 1 Region has the highest rate of use of fuels and fertilizer; the Southern Region has the lowest use of fuels per area. Productivity (as measured by yield), use of energy in the form of

TABLE 34.
Fuels, Labor and Fertilizers in Production of Crops, Northern Region

Crop	Area Planted '000 rais	Yield kg/rai	Production '000 tons	Gasoline		Diesel		Human Labor		Animal Labor		Compost kg/rai	Fertilizer kg/rai			
				l/rai	total '000 l	l/rai	total '000 l	hr/rai	total '000 hr	hr/rai	total '000 hr		N	P	K	Total
Paddy	19,350	302.8	5,858	0.7	14,126	3.2	62,310	113.0	2,187	5.2	1.0	.35	3.2	3.2	–	6.3
Rubber	–	–	–	–	–	–	–	–	–	–	–	–	–	–	–	–
Maize	3,996	314.9	1,258	0.2	919	3.3	13,028	67.2	269	3.2	12.8	.05	0.2	0.2	0.2	0.5
Kenaf	–	–	–	–	–	–	–	–	–	–	–	–	–	–	–	–
Cassava	1,127	623.6	703	0.8	925	3.2	3,597	115.8	131	0.4	.5	–	2.6	2.6	2.6	7.9
Sugarcane	456	494.2	225	1.2	524	4.4	2,024	226.0	103	0.7	.3	–	2.0	1.6	3.0	6.6
Tobacco	688	71.6	49	0.6	427	0.0	14	25.5	17	0.9	.6	.65	0.3	0.8	1.1	2.2
Coconuts	–	–	–	–	–	–	–	–	–	–	–	–	–	–	–	–
Cotton	455	300.0	1	2.3	1,051	4.2	1,920	178.7	81	8.2	3.7	–	5.9	5.9	5.9	17.6
Nuts, Beans & Sesame	1,615	127.7	206	0.1	1,082	1.3	2,099	202.7	327	0.8	1.3	–	–	–	–	–
Others	1,931	333.0	643	1.3	2,529	3.3	6,275	107.0	207	1.3	2.4	–	–	–	–	–
	29,618				21,583		91,267		3,322		122					

TABLE 35.
Fuels, Labor, and Fertilizers in Production of Crops, Northeastern Region

Crop	Area Planted '000 rais	Yield kg/rai	Production '000 tons	Gasoline l/rai	Gasoline Total '000 l	Diesel l/rai	Diesel Total '000 l	Human Labor hr/rai	Human Labor Total '000 hr	Animal Labor hr/rai	Animal Labor Total '000 hr	Compost kg/rai	Fertilizer N kg/rai	Fertilizer P kg/rai	Fertilizer K kg/rai	Total
Paddy	40,606	164.8	6,690	.1	3,248	2.6	104,357	125.0	5,076	11.8	478	5.6	3.2	3.2	–	6.4
Rubber	–	–	–	–	–	–	–	–	–	–	–	–	–	–	–	–
Maize	817	255.0	208	–	–	5.3	4,306	65.3	53	0.8	.6	0.8	0.1	0.1	0.1	0.2
Kenaf	1,002	151.7	152	–	–	1.0	1,022	203.1	204	6.6	6.6	2.1	2.6	2.6	4.2	9.5
Cassava	4,726	803.4	3,924	0.0	47	4.1	19,329	124.0	586	3.3	15.0	1.7	0.1	0.1	0.1	0.2
Sugarcane	47	713.4	33	–	–	–	–	248.4	12	9.7	.5	8.1	0.5	4.4	1.0	5.9
Tobacco	12	385.7	4.6	–	–	–	–	651.8	8	12.0	.1	–	1.1	3.2	4.2	8.5
Coconuts	–	–	–	–	–	–	–	–	–	–	–	–	–	–	–	–
Cotton	93	212.8	19.8	8.9	828	0.3	28	456.1	42	0.5	.0	–	0.1	0.1	0.1	0.4
Nut, Beans & Sesame	75	83.3	6.3	0.3	25	1.8	138	86.2	65	–	–	–	–	–	–	–
Others	1,357	894.9	1,214	0.0	54	1.7	2,253	163.2	221	4.0	5.4	–	–	–	–	–
	48,735				4,202		131,433		6,208		506					

TABLE 36.
Fuels, Labor, and Fertilizers in Production of Crops, Central 1 Region

Crop	Area Planted '000 rai	Yield kg/rai	Production '000 tons	Gasoline		Diesel		Human Labor		Animal Labor		Compost kg/rai	Fertilizer N kg/rai	Fertilizer P kg/rai	Fertilizer K kg/rai	Total
				l/rai	Total '000 l	l/rai	Total '000 l	hr/rai	Total '000 hr	hr/rai	Total '000 hr					
Paddy	9,424	397.7	3,768	.9	8,105	17.7	166,996	97.2	916	1.1	10	.07	6.0	6.0	–	11.9
Rubber	–	–	–	–	–	–	–	–	–	–	–	–	–	–	–	–
Maize	655	154.8	101	0.1	65	1.8	1,159	44.9	29	0.3	.2	–	1.5	1.5	1.5	4.5
Kenaf	–	–	–	–	–	–	–	–	–	–	–	–	–	–	–	–
Cassava	44	425	18.6	–	–	13.5	587	237.2	10	–	–	–	–	–	–	–
Sugarcane	317	801.1	254	–	–	42.7	13,508	101.9	32	–	–	–	7.4	6.1	11.1	24.6
Tobacco	–	–	–	–	–	–	–	–	–	–	–	–	–	–	–	–
Coconuts	–	–	–	–	–	–	–	–	–	–	–	–	–	–	–	–
Cotton	–	–	–	–	–	–	–	–	–	–	–	–	–	–	–	–
Nuts, Beans & Sesame	480	209.4	101	–	–	26.4	12,670	114.0	55	0.2	.1	–	–	–	–	–
Others	252	384.6	97	0.3	68	7.7	1,942	66.6	16.8	0.2	–	–	–	–	–	–
	11,172				8,238		196,862		1,058.8		10.3					

213

TABLE 37.
Fuels, Labor, and Fertilizer in Production of Crops, Central 2 Region

Area Crop	Planted '000 rai	Yield kg/rai	Production '000 tons	Gasoline l/rai	Gasoline Total '000 l	Diesel l/rai	Diesel Total '000 l	Human Labor hr/rai	Human Labor Total '000 hr	Animal Labor hr/rai	Animal Labor Total '000 hr	Compost kg/rai	Fertilizer N kg/rai	Fertilizer P kg/rai	Fertilizer K	Total
Paddy	16,671	256.0	4,267.0	0.9	14,670	3.2	54,014	113.5	1,892	6.9	114.0	.5	3.1	3.1	-	6.3
Rubber	1,037	41.7	43.3	-	-	-	-	45.3	47	-	-	-	-	-	-	-
Maize	-	-	-	-	-	-	-	-	-	-	-	-	-	-	-	-
Kenaf	-	-	-	-	-	-	-	-	-	-	-	-	-	-	-	-
Cassava	1,589	493.3	784.0	-	-	0.9	1,351	74.4	118	0.5	.8	-	-	-	-	-
Sugarcane	-	-	-	-	-	-	-	-	-	-	-	-	-	-	-	-
Tobacco	-	-	-	-	-	-	-	-	-	-	-	-	-	-	-	-
Coconuts	-	-	-	-	-	-	-	-	-	-	-	-	-	-	-	-
Cotton	-	-	-	-	-	-	-	-	-	-	-	-	-	-	-	-
Nuts, Beans & Sesame	-	-	-	-	-	-	-	-	-	-	-	-	-	-	-	-
Others	1,747	404.6	70.7	7.4	12,994	13.2	22,968	118.6	207	0.7	1.2	-	-	-	-	-
	21,044				27,664		78,333		2,264		116.0					

TABLE 38.
Fuels, Labor, and Fertilizers in Production of Crops, Southern Region

Crop	Area Planted '000 rais	Yield kg/rai	Production '000 tons	Gasoline l/rai	Gasoline Total '000 l	Diesel l/rai	Diesel Total '000 l	Human Labor hr/rai	Human Labor Total '000 hr	Animal Labor hr/rai	Animal Labor Total '000 hr	Compost kg/rai	Fertilizer N kg/rai	Fertilizer P kg/rai	Fertilizer K	Total
Paddy	4,952	293.0	1,451	.3	1,337	1.8	8,814	204.1	1,011	16.3	81	2.0	6.2	6.2	–	12.3
Rubber	4,213	50.2	211	–	–	–	–	99.4	419	–	–	–	–	–	–	–
Maize	–	–	–	–	–	–	–	–	–	–	–	–	–	–	–	–
Kenaf	–	–	–	–	–	–	–	–	–	–	–	–	–	–	–	–
Cassava	–	–	–	–	–	–	–	–	–	–	–	–	–	–	–	–
Sugarcane	–	–	–	–	–	–	–	–	–	–	–	–	–	–	–	–
Tobacco	–	–	–	–	–	–	–	–	–	–	–	–	–	–	–	–
Coconuts	1,869	286.9	536	–	–	–	–	11.6	22	–	–	–	–	–	–	–
Cotton	–	–	–	–	–	–	–	–	–	–	–	–	–	–	–	–
Nuts, Beans & Sesame	–	–	–	–	–	–	–	–	–	–	–	–	–	–	–	–
Others	283	440.1	125	3.7	1,057	–	–	82.1	23	0.9	.3	–	–	–	–	–
	11,317				2,394		8,814		1,475		81.3					

TABLE 39.
Productivity and Energy and Fertilizer Use in Paddy Production.

Region	Productivity (kg/rai)	Total Mechanical Energy (MJ/rai)	Fertilizer (kg/rai)
North	303	311	6.3
Northeast	165	334	6.4
Central 1	398	832	11.9
Central 2	256	333	6.3
South	293	429	12.3
Average	242	386	7.3
Standard Deviation	84	218	3.2

TABLE 40.
Correlation Coefficients, Paddy Production

	Energy	Fertilizer
Productivity	.63	.52
Energy		.58

fuel used by agricultural machines and labor, and use of fertilizer correspond generally. Table 39 converts fuel use and human and animal labor into energy use through standard conversion factors (Appendix 1). The strong, positive correlation coefficients (Table 40) show the correspondence between these parameters.

The NEA survey obtained a direct record of per capita income of the rural population. Using the figures on population size and converting the parameters in Table 39 to per capita basis, a set of parameters is obtained as shown in Table 41. Correlation coefficients are given in Table 42. The figures for income were directly obtained through interview and represent monetary income only. (They do not necessarily coincide with figures obtained by conventional means.)

Overall, diesel fuel is used more extensively in agricultural production than is gasoline. Table 43 provides statistics on fuel used in tractors, pumps, and other mechanical devices. The total commercial energy use for crop production represents only about 3 percent of the national energy consumption. The heavy dependence on diesel oil, as mentioned earlier, is probably attributable to the prevailing price structure.

TABLE 41.
Per Capita Income and Paddy Production, Energy, and Fertilizer Use

Region	Income (฿)	Paddy Production (kg)	Energy (MJ)	Fertilizer (kg)
North	4,350	795	853	17.3
Northeast	2,470	490	1,024	19.2
Central 1	5,840	1,953	4,087	58.6
Central 2	4,280	826	1,074	20.3
South	3,800	317	465	13.3
Average	3,564	674	1,095	20.4
Standard Deviation	1,211	639	1,453	19.0

TABLE 42.
Correlation Coefficients for Paddy Production, Per Capita

	Production	Energy	Fertilizer
Income	.68	.64	.61
Production		.78	.34
Energy			.80

As seen earlier, the favorable growth in crop production during the 1960s and 1970s was due mainly to expansion of land under cultivation. The subsequent slowdown in expansion of cultivated land has reduced the production growth rate. To increase future crop production, yield must increase. It is technically possible to increase yield to two times the present level. The NESDB's estimate of the economic return to fertilizer cost of about two confirms the economic feasibility of yield increase through appropriate use of fertilizer.

Projecting energy demand for crop production is a complex process requiring additional information. A regression analysis on the data in Table 41 gives 2.0 as the figure for the income elasticity of energy demand in crop production. To understand more clearly how widespread use of agricultural machines and energy is, the following information is necessary: the number and type of each machine, the use of machines in their respective tasks, and the amount of fuel used to perform each one's tasks. In order to make an infrastructural provision for agricultural production, reasonable utilization rates for these machines must also be known or be estimable. For example, using a pump to lift water, the amount and depth of water available, the area to be irrigated, and the machine type, number, and performance must be known to be able to estimate the amount of fuel required. In a free market economy, fuel price also heavily influences production mix.

Energy needs for fishing have a distinct character from those of the rest of the agricultural sector. The energy consumed in fishing is mostly through commercial fishing; the energy consumed in subsistence fishing is presumed negligible. Still the NEA estimates that the fishery subsector consumes about 875,000 liters of gasoline and 383 million liters of diesel fuel, or about 4.8 percent of the national consumption.

Appendices 2 to 6 show machinery, fuels, animal labor, and human labor used in agricultural production distributed according to amount of cultivated land by household. In the Northern Region (Appendix 2), for example, 48 households have landholdings of from 4 to 6 rais. Land used for production of all crops in this category amounts to 202 rais, while only 23 households use tractors. The total land used in each category does not coincide with the total landholding. This indicates either the existence of a certain extent of idle land or cultivation on leased land. Total land under cultivation in the Northern and the Central 1 Regions exceeds the size of total landholding.

The use of tractors, tractor fuel, and animal labor for agriculture are plotted in Figures 4 and 5 by region. The graphs indicate that tractors usually replace animal labor as farm size increases, in all except the Northeast Region. The use of animal labor declines quite sharply when farm size exceeds about 12 rais. The use of fuel, as represented by the MJ/HH plot, exhibits a general increase with increase in farm size except in the Southern Region. Use of animal labor in all regions rises to a peak at 10 to 20 rais and then falls off with larger land size.

In the Northeast Region, the prevailing rainfed agriculture probably does not encourage cultivation of larger land plots. An examination of the percentage of households using pumps and the corresponding energy use for pumps confirms this. The frequency of use and extent of use (as indicated by the total amount of fuels used with pumps) appear highest in the Central 1 Region; they appear very low in the Northeast and Southern Regions. However, the different composition of agriculture in the Southern Region means a comparison with other regions is inappropriate.

Unfortunately, the NEA publication does not provide a separate set of data for each major crop like those in Appendices 2 to 6. This information is extremely significant, as it implies the possible substitution of

TABLE 43.
Gasoline and Diesel Use in Agricultural Machines (000 liters)

Region	Gasoline				Diesel			
	Tractor	Pump	Other	Total	Tractor	Pump	Other	Total
North	7,660	10,090	3,910	21,660	79,500	8,200	440	88,140
Northeast	500	2,910	710	4,120	53,770	4,330	30,200	88,300
Central 1	3,620	3,450	1,180	8,250	175,400	18,930	1,980	196,310
Central 2	1,450	21,820	4,540	27,810	55,340	15,480	270	71,090
South	1,310	1,060	10	2,380	7,600	–	1,230	8,830
National	14,540	39,330	11,450	64,220	371,610	46,940	34,120	452,670

Source: NEA, 1980 Rural Energy Survey.

tractors for animal labor in agriculture, liters of fuel per hour of labor, and the likelihood of substitution in terms of land size under cultivation. Tractors are used commonly for other tasks than paddy cultivation, but animal labor is used mainly in paddy cultivation. Without providing detailed data for each crop, an estimation of the substitution potential for the use of tractors in place of animal labor is untenable.

Within each land size category, the data on the use of agricultural machinery and animal labor stratified according to income or according to access to irrigation will provide valuable information on the cultivation pattern and the potentiality of an alternative pattern. The appendices provide valuable information on the probable use and extent of use of agricultural machines, and similarly of animal labor. Further stratification according to common determinants will facilitate an analysis on projected agricultural energy use, independent of a time series analysis.

Rural Industry and Energy Use

Only rough estimates of the production output to fuel consumption ratio are available for certain industries. Sometimes more than one type of production technology is used in one industry, but data on the number of establishments for each type are not available. The total energy consumption arising from rural industry is very low; that is probably why it commands little attention. An estimate of the energy consumption of some of the more energy-intensive industries is shown in Table 44.

John Arnold's rural energy surveys provide illuminating insight into the energy consumption pattern of some of the common rural industries. He attempts to provide an information base for introduction of alternative technologies, and he covers some of the more energy-intensive rural industries. However, the data are inadequate for estimating national patterns.

Charcoal Making. It prevails in every rural area. Charcoal is preferred to fuelwood and is more commonly used in affluent areas. Three types of charcoal kilns are used, each in varying sizes. Paddy husks make starter fuels and provide a cover for wood during carbonization. The energy yield ratio of charcoal in terms of fuelwood input may reach 50 percent when paddy husks use reaches 50 percent by weight of the fuelwood input. Larger kilns (used commercially) are more efficient, giving higher yields. Households also produce charcoal for their own use.

Rice Milling. This is one of the oldest industrial activities. The large number and the wide spread of rice mills reflect the extent of rice cultiva-

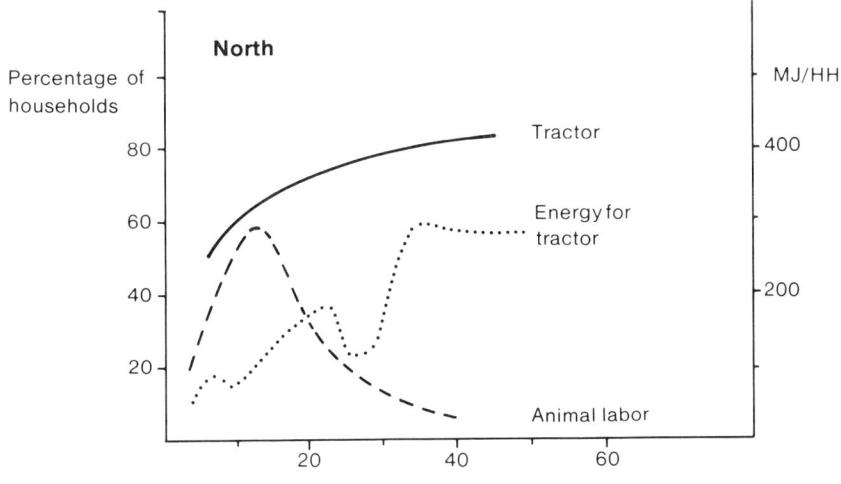

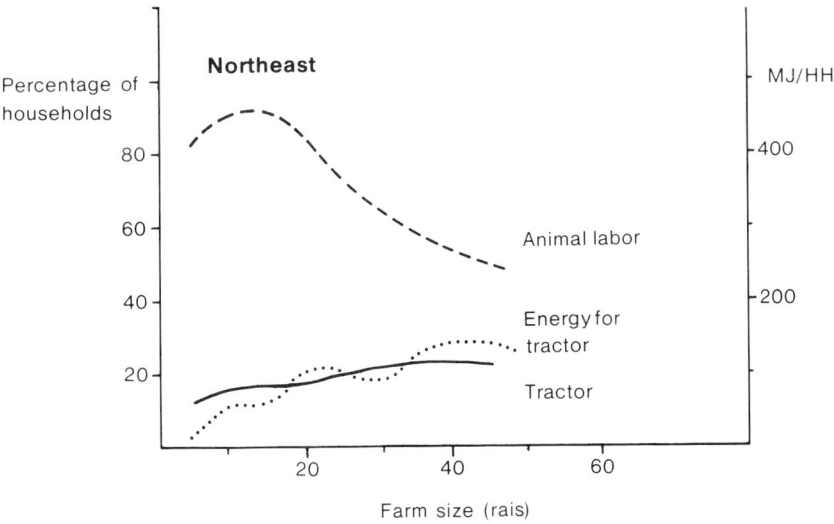

Figure 4. Use of Tractor and Animal Labor: North and Northeast Regions

tion. Rice mills are located both within urban areas and in rural areas. Small mills producing 1 to 2 tons per day are powered by diesel engines and are located in remote rural areas. Medium mills producing 10 to 20 tons per day, also located in rural areas, are also diesel powered. Larger mills are mostly located in the center of the area where cultivation is intensive and are usually easily accessible by paved roads. These larger

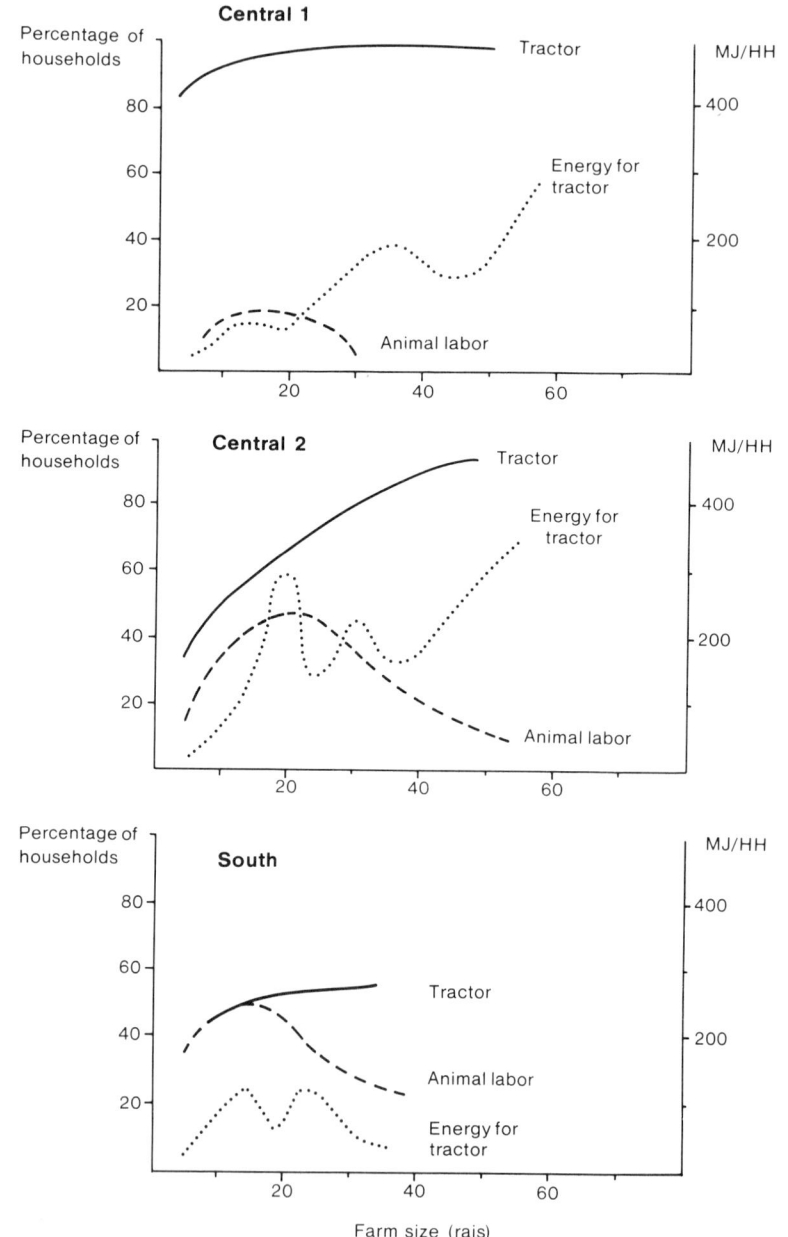

Figure 5. Use of Tractor and Animal Labor: Central 1, Central 2, and Southern Regions

TABLE 44.
Annual Energy Consumption of Some Rural Industries, 1980

Industry	Production	Energy Consumption				
		Diesel ('0001.)	Fuel Oil ('0001.)	Electricity (mil. kWh)	Fuelwood ('000 tons)	Others ('000 tons)
Charcoal Making	3,384	–	–	–	15,140	rice husk (no figure)
Rice Milling	15,758 (paddy)	53	–	–	–	rice husk 2,258
Cassava: Chipping and Pelletizing,	10,580	92,000	–	–	–	–
Flour Milling	557	–	34,810	119	–	–
Ice Making	14,600	–	–	1,292	–	–
Total		92,053	34,810	1,410	15,140	

Source: Compiled from NEA data, and J. Arnold, 1982.

mills use rice-husk-fired boilers, using a design many decades old. The rice millers' association reports a milled-rice-to-paddy ratio of 0.6. Of the 0.22 ton of husk produced from one ton of paddy, 0.159 ton is used in the boiler. The diesel units use about 1 liter of fuel to produce 55 kg of milled rice.

Cassava Chipping and Pelletizing. Up to 95 percent of the cassava produced is exported and chipping and pelletizing are required by the importing countries. The diesel-powered units used for that purpose are locally produced and use about 1 liter of fuel per 715 kg of pellets. Locally cassava is used as flour. The flour mills use 1 liter of fuel oil and 3.4 kWh of electricity to produce 16 kg of flour.

Ice Making. This uses electrical compressors that produce 11 kg of ice per 1 kWh of electricity consumption. Ice can be found everywhere except in the remotest rural areas. The industry uses rice husks extensively as insulating material, but no reliable estimate of the amount of rice husks used is available.

Brick Making. Most local, small-scale brick makers utilize rice-husk-fired kilns, producing 1.6 kg of brick to 1 kg of rice husks. The wood-fired kilns are located mostly in the Southern Region, where rubber wood is used as fuel. That consumption is 1.3 kg of wood per 1 kg of brick.

Pottery. Most of the ceramics produced and used in rural areas are not glazed. Glazed ceramics are produced in modern electrical factories. Ceramic production requires about 1.8 kg of fuelwood for 1 kg of ceramic. Rice-husk-fired kilns require about 1.6 kg of rice husks to 1 kg of ceramic.

Salt Production. No information is available for the amount of electricity used to pump water. Fuelwood is used for boiling away the water, but a recent shortage of fuelwood in the salt-producing areas due to over-cutting of trees has forced a change to other fuels. In one case, sawdust is used producing 1 kg of salt for 2.7 kg of sawdust. In these types of small-scale industries where the heat is required to boil water, a variant of the so-called improved stove is used. The stove can accommodate a number of water pots and has a chimney attached at the end opposite to the fuel input. The stove has proved to be fuel efficient for this use and commercial units are available, although cumbersome to transport.

Cotton and Silk Weaving. Done by hand, this is largely a household activity. Factories with up to 50 manually operated looms can be found in the central area of the silk-producing country. In this industry, dyeing yarn is made in the factory or in the household. The requirement for boiling water is similar to salt production and the same type of stove is used.

Rubber Drying. One kg of rubber crumb requires 0.9 kg of rubber wood for use in the drying tower. Rubber wood is usually available since old rubber trees are continually replaced by new trees.

Energy in Transportation

Although the 1980 NEA rural energy survey questionnaire covers transportation, no report in any form has been published. The information here on energy use in trasnportation is taken from the 1979 Pilot Survey (1980). Table 45 shows the fuel composition used for personal transportation stratified according to income class. Unfortunately, the pilot survey includes only one province for each region, and no detailed information on energy for goods transportation for each region is published.

TABLE 45.
Fuels used in Personnel Transportation, 1979

Income Class (thousand ฿)	Gasoline (l)		Diesel (l)		Energy (MJ)	
	per HH	per head	per HH	per head	per HH	per head
0-24	94.1	15.6	4.1	0.7	3,470	580
24-48	153.9	25.6	52.8	8.8	7,480	1,240
48-72	412.1	68.5	118.3	19.6	19,130	3,180
72-120	351.8	58.4	667.2	110.8	38,550	6,400
120 up	103.2	17.2	-	-	3,630	600
Average	155.3	25.8	63.0	10.5	7,930	1,320

Personal transportation occurs as part of household activities, such as going to community health centers and to movies. These activities relate to general household final demand, not to production. Relating personal transportation to agricultural production would be futile. The number of determinants involved may be innumerable, and the relative attribute of each may not be estimable.

Generally, energy consumption in personal transportation increases as income increases (see Table 45). The heavy use of gasoline compared to diesel corresponds with the national trend toward more personal transportation. Motor bicycles, most private cars, and some pickup trucks use gasoline. The number of samples for the upper income classes may not adequately represent the class. That may be why statistics show small fuel consumption in the uppermost income class.

CONCLUSIONS

Research to date has provided the basis for bringing together, in this chapter, regional profiles of rural energy consumption by principal end uses, of energy sources by principal types of resource, and of intermediate energy forms and quality. Several analytic scenarios relevant to policy purposes can be derived from these data and are treated in a companion study (Chirarattananon 1982).

In the section on firewood and charcoal use in different types of stoves, the chapter has illustrated how data on energy technologies and fuel use obtained through local studies can be combined in hybrid form with nationwide survey data to arrive at preliminary estimates of fuel substitution possibilities.

In similar manner, in order to bring social and economic development objectives to bear on energy technology development, further microscale studies will be required using regional attributes as frameworks for analysis. Three principal foci for such further development research are identified.

1. To meet the Fifth Plan's objectives of reducing poverty and improving income distribution, action-oriented studies of social structure and energy constraints in agroeconomic zones categorized as impoverished are recommended.

2. To achieve improved land use and increased agricultural yields as targeted by the plan, changing energy requirements and use patterns should be ascertained through comparative microstudies in principal agroeconomic zones. If well designed and organized, these studies can contribute directly to farmers' ability (including that of small-scale farmers) to increase production and yields. They also can provide essential data for rural energy supply planning and price policy formation at regional and national scales.

3. Conservation and interfuel substitution were pointed out as the only means of curbing or reducing the large quantities of energy used to meet rural cooking needs. User-oriented design and development studies are now indicated to improve stove efficiencies, to fit cooking devices more closely to the characteristics of available fuels (while recognizing the influence of local cooking practices), and to increase availability of and access to preferred fuels.

Chapters 8 to 10 directly address such cooking technology and fuel needs, while the book's concluding chapter suggests methods for structuring these three recommended areas of development research.

NOTES

1. Bahts per current US $1.00:

1979	*1980*	*1981*	*1982*
20.43	20.63	22.00	23.00

APPENDIX 1.
Conversion Factors

Fuel/Energy Form	Unit	Conversion Factor (MJ/unit)
Commercial		
Crude Oil	liter	38.50
Gasoline	liter	35.16
Diesel	liter	39.24
Kerosene	liter	38.86
LPG	liter	46.70
Fuel Oil	liter	41.40
Jet Fuel	liter	36.05
Natural Gas	1,000 cu. ft.	923.13
Lignite	kilogram	16.60
Electricity	kWh.	3.60
Dry Cell	piece	1.675×10^{-3}
Calcium Carbide	kilogram	20.26
Non-Commercial and Semi-Commercial		
Charcoal	kilogram	28.57
Fuelwood	kilogram	16.31
Paddy Husk	kilogram	12.41
Paddy Straw	kilogram	12.41
Coconut Branch	kilogram	12.79
Bagasse	kilogram	8.00
Human Labor	working hour	1.0
Animal Labor	working hour	9.0

Source: Most of the figures are taken from NEA (1982).

Note: These figures, except for human and animal labor, represent estimates of the calorific heat value of the fuel. For fuelwood, an average humidity content is used. No distinction is made for the species of trees from which charcoal and fuelwood are taken. The actual useful work produced in a working hour by human labor is given as .1799 MJ and by animal labor, 1.346 MJ. The figures for human and animal labor are quoted alongside those for fuels in the text. Since the conversion factors used for fuels are calorific values and not the useful energy at end-use, the factors for human and animal labor have been inflated by about 5.6 times to reflect an assumed efficiency of the machines of 18 percent.

APPENDIX 2
Energy Use in Agriculture, with respect to Land Size, NORTHERN Region

Land Size (rai)	No. of House-hold	Land Hold-ing (rai)	Land Used (rai)	Animal Labor Household Using			Household Using			Tractor Fuel Use				
				No.	%	hr/rai	No.	%	Cu %	Gaso-line*	Die-sel*	MJ/rai	MJ/HH	
2	181	57	25	1	1	2	8	4	4	6	59	100	4	
2-4	39	106	92	8	21	9	20	51	13	36	146	76	10	
4-6	48	228	202	16	33	17	23	48	19	143	1,427	303	86	
6-10	66	496	484	35	53	22	28	42	24	599	415	77	53	
10-15	62	721	979	38	61	12	47	76	32	321	1,547	74	102	
15-20	60	986	939	17	28	7	38	63	36	220	2,905	129	172	
20-25	44	918	1,218	14	32	13	43	98	41	342	2,973	106	182	
25-30	27	718	708	5	19	15	21	78	43	179	1,608	98	98	
30-40	61	2,011	1,889	9	15	10	48	79	47	27	5,710	119	318	
40-50	37	1,602	1,620	1	3	5	32	86	49	973	4,064	120	274	
50-60	27	1,454	1,209	1	4	5	21	78	51	-	3,732	121	207	
60-80	32	2,153	2,488	1	3	4	32	100	53	454	7,100	118	417	
80-120	20	1,880	2,057	-	-	3	20	100	56	614	6,673	138	401	
120	3	1,112	1,224	2	68	2	3	100	56	-	2,265	73	126	
Total	707	14,441	15,158	148			385			3,914	40,624	114	2,449	

	Pump							Household Using			Other Mechanical Devices				Human Labor
Household Using			Fuel Use								Fuel Use				
No.	%	Cu %	Gaso-line*	Die-sel*	MJ/rai	MJ/HH		No.	%	Cu %	Gaso-line*	Die-sel*	MJ/rai	MJ/HH	Hr/rai
4	2	2	130	61	276	9		6	3	3	–	–	–	–	309
4	10	4	393	19	158	21		7	18	6	–	–	–	–	298
9	19	6	119	520	122	35		12	23	9	23	–	4	1	241
15	23	10	524	167	51	35		24	36	15	461	21	39	24	224
20	32	13	1,123	609	65	90		28	45	19	6	–	–	–	316
8	13	15	136	38	7	9		21	35	22	2	24	1	1	112
7	16	15	288	391	21	35		30	68	26	337	52	11	20	183
6	22	15	370	91	23	38		15	56	27	8	9	1	1	87
12	20	16	306	472	15	41		37	61	31	21	12	1	2	95
9	24	17	510	480	23	52		20	54	32	15	–	–	1	77
8	30	17	70	305	12	20		15	56	33	536	16	16	28	108
9	28	18	561	611	18	62		30	94	36	171	81	4	13	104
8	40	18	599	326	17	48		16	80	38	405	10	7	21	80
2	68	19	27	114	4	8		3	100	38	14	–	–	1	44
121			5,156	4,194	23	489		264			1,999	225	5	112	

*In liters.

APPENDIX 3
Energy Use in Agriculture, with respect to Land Size, NORTHEASTERN Region

Land Size (rai)	No. of Household	Land Holding (rai)	Land Used (rai)	Animal Labor Household Using				Household Using			Tractor		Fuel Use		
				No.	%	hr/rai		No.	%	Cu %	Gaso-line*	Die-sel*	MJ/rai	MJ/HH
2	70	26	8	5	7	85		1	1	1	–	170	834	8
2-4	14	36	86	11	79	39		2	14	4	10	122	144	6
4-6	29	139	186	24	83	36		5	17	7	–	231	49	11
6-10	74	570	579	65	88	33		9	12	9	–	1,365	93	64
10-15	100	1,189	1,894	100	100	33		14	14	11	–	960	27	45
15-20	102	1,699	1,986	102	100	28		16	16	12	125	1,616	34	80
20-25	102	2,191	2,370	82	80	21		25	25	15	3	2,356	39	110
25-30	72	1,919	1,631	49	68	26		13	18	15	31	2,100	51	99
30-40	107	5,519	3,153	75	70	20		20	19	16	7	2,961	37	138
40-50	63	2,729	2,180	35	56	16		16	25	17	–	3,231	58	150
50-60	55	2,896	2,503	35	64	19		12	22	17	–	1,695	27	79
60-80	36	2,397	1,754	18	50	15		9	25	17	–	1,067	24	50
80-120	18	1,735	827	7	39	7		2	11	17	–	1,642	77	76
120	5	1,168	987	2	40	3		2	40	17	–	1,837	86	85
Total	844	22,213	19,604	620				146			178	21,353	43	1,000

231

\multicolumn{7}{c	}{Pump}	\multicolumn{8}{c}{Other Mechanical Devices}	Human Labor
Household Using			Fuel Use				Household Using			Fuel Use				
No.	%	Cu %	Gaso-line*	Die-sel*	MJ/rai	MJ/HH	No.	%	Cu %	Gaso-line*	Die-sel*	MJ/rai	MJ/HH	Hr/rai
-	-	-	-	-	-	-	4	6	-	-	-	-	-	3,267
1	7	1	-	-	-	-	-	5	-	-	-	-	-	251
2	7	3	8	-	1	-	2	7	5	-	1	0	0	224
5	7	4	6	26	3	2	8	11	8	138	-	8	6	196
5	5	5	16	62	3	4	7	7	7	31	1,206	43	57	173
6	6	5	281	540	16	37	15	15	9	47	3,225	64	152	176
6	6	5	309	70	6	16	11	11	10	5	496	8	23	111
8	11	6	45	257	7	14	7	10	10	60	579	16	29	133
9	8	6	16	286	5	17	18	17	11	-	1,383	17	64	108
4	6	6	27	-	-	1	4	6	10	-	370	7	17	111
2	4	6	14	69	2	4	9	16	11	-	491	8	23	105
4	11	6	320	366	15	30	6	17	11	-	1,965	44	91	9
3	17	7	12	61	3	3	2	11	11	-	2,277	107	106	174
-	-	7	-	-	-	-	-	-	11	-	-	-	-	45
55			1,154	1,719	6	128	93			281	11,993	25	569	

*In liters.

APPENDIX 4
Energy Use in Agriculture, with respect to Land Size, CENTRAL 1 Region

Land Size (rai)	No. of Household	Land Holding (rai)	Land Used (rai)	Animal Labor Household Using				Household Using			Tractor Fuel Use				
				No.	%	hr/rai	No.	%	Cu %	Gaso-line*	Die-sel*	MJ/rai	MJ/HH
2	22	1	1	–	0	–	–	0	–	–	–	–	–
2-4	4	10	5	–	0	–	2	50	7	3	34	320	10
4-6	3	13	18	–	0	–	3	100	17	–	384	837	103
6-10	8	56	95	1	13	3	8	100	35	101	1,049	470	306
10-15	17	200	187	1	6	2	14	82	50	51	2,622	567	717
15-20	11	186	184	2	18	15	7	64	52	–	2,265	482	608
20-25	13	273	282	2	15	3	10	77	56	–	3,617	503	972
25-30	6	159	240	1	17	1	6	100	60	253	1,713	317	521
30-40	13	421	503	–	0	2	13	100	65	19	7,528	588	2,028
40-50	13	558	344	–	0	–	8	62	65	–	5,941	677	1,597
50-60	18	929	564	–	0	–	11	61	64	–	10,500	750	2,822
60-80	9	580	600	–	0	–	9	100	66	–	11,266	737	3,028
80-120	8	912	1,477	–	0	–	8	100	68	1,201	30,048	827	8,364
120	1	160	545	–	0	–	1	100	69	30	3,376	225	915
Total	146	4,400	5,055	7			113			1,658	80,343	636	21,991

233

	Household Using			Pump Fuel Use				Household Using			Other Mechanical Devices Fuel Use				Human Labor
No.	%	Cu %	Gasoline*	Diesel*	MJ/rai	MJ/HH	No.	%	Cu %	Gasoline*	Diesel*	MJ/rai	MJ/HH	Hr/rai
-	-	-	-	-	-	-	1	5	5	-	-	-	-	244
-	0	-	-	-	-	-	1	25	8	-	-	-	-	12
3	100	10	-	738	1,608	198	-	0	7	-	-	-	-	303
1	125	11	-	5	2	1	5	63	19	-	-	-	-	109
9	53	24	359	2,256	541	693	8	47	28	30	128	33	42	234
4	36	26	425	1,334	365	461	5	45	31	3	4	2	2	77
5	38	28	152	-	19	37	8	62	36	9	129	19	37	128
4	67	31	134	83	33	55	3	50	37	-	-	-	-	229
4	31	31	-	1,978	154	532	10	77	42	99	60	12	4	117
4	31	31	273	501	85	200	4	31	41	43	7	5	12	65
4	22	30	191	214	27	104	6	33	40	29	137	11	44	82
2	22	29	-	433	29	117	6	67	42	202	36	14	58	69
7	88	32	-	1,122	30	302	8	100	45	67	408	13	126	90
1	100	33	44	-	3	11	1	100	45	62	-	4	15	33
48			1,578	8,669	78	2,710	70			544	909	11	375	

*In liters.

APPENDIX 5
Energy Use in Agriculture, with respect to Land Size, CENTRAL 2 Region

Land Size (rai)	No. of House- hold	Land Hold- ing (rai)	Land Used (rai)	Animal Labor Household Using			Household Using			Tractor Fuel Use			
				No.	%	hr/rai	No.	%	Cu %	Gaso- line*	Die- sel*	MJ/ rai	MJ/ HH
2	31	5	2	-	0	-	-	0	-	-	-	-	-
2-4	5	12	6	-	0	-	1	20	-	10	-	56	2
4-6	5	22	43	1	20	16	3	60	10	-	35	32	9
6-10	7	50	51	1	14	6	4	57	17	-	151	116	40
10-15	12	140	187	6	50	12	6	50	23	-	427	90	112
15-20	17	281	250	5	29	14	7	41	27	136	1,373	235	394
20-25	15	322	341	7	47	21	8	53	32	-	546	63	144
25-30	8	211	314	3	38	11	7	88	36	14	854	108	228
30-40	13	424	240	4	31	15	3	23	35	-	649	106	171
40-50	7	293	386	2	29	18	7	100	38	-	986	100	260
50-60	10	512	459	-	0	-	7	70	41	-	1,425	122	375
60-80	5	327	328	2	40	16	2	40	41	-	847	101	223
80-120	8	797	869	2	25	5	7	88	43	113	1,899	91	527
120	6	1,412	487	1	17	4	2	33	43	-	1,216	98	320
Total	149	4,808	3,962	34			64			273	10,408	106	2,805

	Household Using Pump		Pump				Household Using Other Mech		Other Mechanical Devices					Human Labor
				Fuel Use						Fuel Use				
No.	%	Cu %	Gaso-line*	Die-sel*	MJ/rai	MJ/HH	No.	%	Cu %	Gaso-line*	Die-sel*	MJ/rai	MJ/HH	Hr/rai
1	6	3	-	-	-	-	1	6	3	-	-	-	-	7
1	20	6	19	-	107	5	2	0	3	-	-	-	-	110
4	80	15	165	117	242	70	2	40	7	-	1	1	-	239
3	43	19	306	-	211	72	3	43	13	-	-	-	-	161
9	75	30	985	557	302	379	7	58	22	197	-	37	47	134
6	35	31	288	547	126	212	6	35	25	154	2	22	37	86
10	67	37	498	293	85	195	8	53	29	431	-	45	179	235
6	75	40	511	996	182	382	4	50	31	-	-	-	-	114
2	15	37	292	-	43	69	2	15	29	-	47	8	12	87
4	57	38	190	195	37	96	2	29	29	-	-	-	-	103
6	60	40	480	-	37	113	5	50	31	72	-	6	17	68
3	60	41	275	-	30	65	1	20	30	-	-	-	-	53
3	38	41	95	186	12	71	5	63	32	-	-	-	-	107
1	17	40	-	20	2	5	-	0	31	-	-	-	-	88
61			4,104	2,911	65	1,735	46			854	50	8	215	

*In liters.

235

APPENDIX 6
Energy Use in Agriculture, with respect to Land Size, SOUTHERN Region

Land Size (rai)	No. of Household	Land Holding (rai)	Land Used (rai)	Animal Labor Household Using				Household Using			Tractor	Fuel Use			
				No.	%	hr/rai		No.	%	Cu %	Gaso-line*	Die-sel*	MJ/rai	MJ/HH
2	14	4	1	-	-	240		2	14	14	-	3	118	1
2-4	4	10	26	2	50	22		2	50	22	-	21	32	6
4-6	10	45	66	3	30	14		8	80	43	-	89	53	30
6-10	16	120	104	9	56	13		10	63	50	86	166	35	48
10-15	27	323	348	13	48	15		15	56	52	-	415	56	143
15-20	18	301	283	7	4	10		8	44	51	-	246	34	72
20-25	6	126	239	5	83	20		4	67	52	102	279	70	123
25-30	10	262	79	1	10	3		1	10	48	-	12	6	4
30-40	10	319	198	4	40	15		2	20	45	-	140	28	41
40-50	4	177	42	-	-	6		1	25	45	-	15	14	4
50-60	1	57	109	-	-	-		1	100	45	-	5	2	1
60-80	11	764	333	4	36	10		-	-	41	-	-	-	-
80-120	1	105	100	-	-	-		-	-	41	-	-	-	-
120	3	663	150	-	-	-		1	33	41	-	43	11	13
	135	3,215	1,154	48				55			248	1,434	30	481

	Pump							Other Mechanical Devices							Human Labor Hr/rai
	Household Using			Fuel Use				Household Using			Fuel Use				
No.	%	Cu %	Gasoline*	Diesel*	MJ/rai	MJ/HH		No.	%	Cu %	Gasoline*	Diesel*	MJ/rai	MJ/HH	
–	–	–	–	–	–	–		–	–	–	–	–	–	–	3,175
–	–	–	–	–	–	–		–	–	–	–	–	–	–	685
1	10	4	200	–	107	52		2	20	7	–	–	–	–	377
–	–	–	–	–	–	–		5	31	16	–	1	0	1	197
–	–	–	–	–	–	–		5	19	17	1	231	26	67	173
–	–	–	–	–	–	–		1	6	15	–	–	–	–	119
–	–	–	–	–	–	–		6	100	20	–	–	–	–	104
–	–	–	–	–	–	–		–	–	–	–	–	–	–	94
–	–	–	–	–	–	–		–	–	17	–	–	–	–	115
–	–	–	–	–	–	–		1	25	17	–	–	–	–	99
–	–	–	–	–	–	–		–	–	–	–	–	–	–	24
–	–	–	–	–	–	–		–	–	–	–	–	–	–	62
–	–	–	–	–	–	–		–	–	–	–	–	–	–	189
–	–	1	–	–	–	–		–	–	15	–	–	–	–	15
1			200	–	3	52		20			1	232	4	68	

*In liters.

REFERENCES

Agrawal, S.C.
   1981   *Rural Energy System in Two Regions of Uttar Pradesh, First Phase Report*. Honolulu: East-West Center, Resource Systems Institute.

Allentuck, J.
   1981   Energy Models for Developing Countries. A series of papers in *Proceedings of the International Symposium on Non-conventional Energy*, July–August, Nairobi.

Arnold, J.
   1982   *Rural Energy Surveys: The Thailand Experience*. Meta Systems Inc.

Chankong, V.
   1982   A Methodology for Large-Scale System Planning and Analysis. Presented at the Conference on Operations Research, August, Khon Kaen, Thailand.

Charnsangavej, C.
   1980.   *Problems Facing Small-Scale Rural Industry*. Proceedings of the Conference on Appropriate Technology for Rural Development, Chiang Mai University, Chiang Mai, Thailand.

Chirarattananon, S.
   1980   A Preliminary Study on Rural Energy and Energy Implication for Rural Development. Presented at the ESCAP conference on Renewable Energy Technologies, Bangkok.

Chirarattananon, S.
   1982.   A Methodology for Rural Energy Analysis and Planning. Presented at the Conference on Operations Research, August, Khon Kaen, Thailand.

Government of Thailand (GOT). Ministry of Commerce. *Commercial News* various issues.

GOT. Ministry of Commerce. Department of Business Economics.
   1982   *Consumer Price Index and Wholesale Price Index 1982*.

GOT. Office of the Secretary to the Prime Minister.
   1982   *Rural Job Creation Program, Summary Report* and other official documents.

Koppel, B. and R. Morse.
   1980   Influence of Variations Among Rural Energy Users and Agroclimatic Zones on Policy and Technology Alternatives. A conceptual diagram in *Preliminary Plan on Energy and Rural Development*. Presented at the Research Implementation Workshop on Energy and Rural Development, February, Chiang Mai, Thailand.

Krairit, A.
    1980   *Evaluation of Small-Scale Industries.* Proceedings of a Conference on Appropriate Technology for Rural Development, Chiang Mai University, Chiang Mai.

Laparojkij, P.
    1982   Energy Problems and Development of Energy Resources. Presented at Thammasat University, February, Bangkok.

National Economic and Social Development Board (NESDB).
    1982   *Document of the Fifth National Economic and Social Development Plan.*

National Energy Administration (NEA). Regulatory Division.
    1980   *Pilot Study on the Assessment of Rural Energy Delivery Systems in Thailand.* In Thai.

NEA.
    1982a   *Report on the 1980 Rural Energy Survey: Part 1, Household Energy Consumption.*

NEA.
    1982b   Personal communication — a set of tables on fuels, labor, and fertilizers used in agricultural production, by major crops and by planting area.

Palapatapi, S.
    1980   A Concept in Rural Development. (A background paper on the Fifth Economic and Social Development Plan.) Presented at the Conference on Appropriate Technology for Rural Development, Chiang Mai University, Chiang Mai.

Pathak, B.S.
    1980   *Energy Balance and Utilization of Agricultural Waste on a Farm.* Tata Energy Research Institute.

Phillips, D.T., A. Rarindran, and J. Solberg.
    1976   *Operations Research.* New York: John Wiley and Sons, Inc.

Saenanarong, S.
    1980   *Thai Geography*, 4th edition. In Thai.

Satitvithayanont, S.
    1981   *Agricultural Geography.* Bangkok: Prae Vithya Publications. In Thai.

Shaner, W.W., P.F. Philipp, and W.R. Schmehl.
    1982   *Farming Systems Research and Development: Guidelines for Developing Countries.* Boulder, Colorado: Westview Press.

Sirivadhnakul, T. and S. Tadyu.
    1980   Fuelwood and Charcoal. Presented at ESCAP/FAO/UNEP expert group meeting on fuelwood and charcoal, Regulatory Division, NEA.

Smith, K.R., M.T. Santerre, and C.S. Schlegel.
   1980   Criterion Framework and Indicators for Comparing and Evaluating Alternative Energy Technologies. Presented at the Research Implementation Workshop on Energy and Rural Development, Chiang Mai.

Sriplung, S. and K. Khatikarn.
           Role of Operations Research in Agriculture. Presented at the Conference on Operations Research, August, Khon Kaen, Thailand. In Thai.

Sujamnong, P. and others.
   1981   *Integrated Rural Development Strategies for Thailand.* Bangkok: Thai Vattanapanich Publications. In Thai.

Wanprasert, C.
   1979   *Thai Society.* Bangkok: Prae Vithya Publications.

6
Integrated Rural Development Planning and Energy Priorities: Participatory Surveys in India Microregions

T. M. Vinod Kumar

GENESIS OF MICROREGION PLANNING STUDIES

India's rural conditions are far from satisfactory in spite of over 30 years of experience in planned development since independence. More and more people living in rural areas are becoming impoverished (Rajni Kothari 1979, ILO 1979, GOI 1979, ILO 1980, V.M. Dandekar and N. Rath 1979, B.S. Rao and V.N. Deshpande 1982, and Center for Policy Research and Family Planning Foundation 1983). Growing poverty, unemployment, disease, and inadequate access to basic needs and services (T.M. Vinod Kumar 1980) along with growing feelings of inequities and injustice make an explosive mix in rural communities. Energy resources and their use are part of rural development, so global rationing of energy resources only worsens this trend.

In India's rural areas, biological productivity (the rate of appearance of energy and matter as living tissue) given the constraint of current technology is sustained by energy subsidies of chemical fertilizers, diesel fuel, and electricity supplied from outside the area. Scarcities of these nonrenewable energy resources weaken productivity and add to rural tension. Popular rhetoric of local and national politicians as well as emerging policy directives for rural development echo these concerns, but in spite of a *panchayat raj*[1] grass roots political structure and an administrative network reaching to the villages, poverty remains a serious problem. Rural areas must be rapidly transformed and rural consumption patterns strengthened. The newly created markets will facilitate overall national development, including integrated rural and urban development.

Since the start of national planning, several gaps have been evident in knowledge for integrated rural development planning and policy formation. Many studies have been made both within and outside the government. This chapter concerns one such study designed as an intensive planning exercise to guide and inform policymaking and limited local interventions.

The study areas Mahnar block in Vaishali district and Hariharganj block in Palamau district are located in the state of Bihar in eastern India.

Research work was initiated by the Centre for the Study of Developing Societies (CSDS), Delhi. The study was part of a collaborative research program with the Council for Social Development (CSD), New Delhi, which studied Ranaghat Block II in West Bengal, and the Center for Development Research (CDR), Copenhagen, which studied Sirsi block in Karnataka.

Even though it is rich in natural and mineral resources, Bihar is considered one of the most underdeveloped states in India. The total geographical area of the state is 173,870 sq km, with a population of 56,353,000 in 1971 and 69,823,000 in 1981. The river Ganges divides Bihar into two unequal parts: the rich north and the poor south. The north, including Vaishali district, is characterized by fertile land, dense population, and plain terrain. The southern part has less fertile land, less cultivable area, sparse population, and inhospitable, hilly terrain. Palamau district is located there. A larger percentage of socio-economically backward scheduled castes and scheduled tribes[2] live in south Bihar than live in the north.

In order to identify development priorities and strategies in contrasting regions, the CSDS team studied the nature of underdevelopment in one block in each of these parts of Bihar. The blocks selected for study with the concurrence of state and district officials are neither the richest nor the poorest. In their respective regions, they may be considered as average, nontribal blocks. The collaborative study should therefore provide direct insight and data on policy issues of energy and rural development in north and south Bihar. Of course, the two blocks selected cannot be considered representative of India's heterogeneous settlement structures, ecosystems, and varying human and natural resource endowments.

Conceptual, practical, and policy problems in integrated rural development were highlighted. Energy's crucial role in development in these blocks became evident during the field work. Delineating the existing system and energy priorities in these specific geographical areas will therefore benefit our wider investigation of energy and rural development policy issues.

INTEGRATED RURAL DEVELOPMENT

Integrated rural development may be defined as a complex intersectoral and area-specific process for improving the living standards of low-income people and for making their development self-sustaining (Uma Lele 1975, D.M. Nanjundappa 1981). Central to this process are the desired society and alternate life-styles as conceived by the client group, the principal consumers of the developmental inputs. Hardly any worthwhile research is available on this subject.

Energy is all-pervading: it is a productive factor in all employment- and income-generating ventures as well as basic need services. Energy and rural development, then, make up an integral package, not isolated items or services. Energy is one important component in the *nature-society-person* system, the system studied in each community development block described herein.

In the center of this system are *persons*, here defined as adult males and females, aged males and females, children, and destitutes in the client group. *Nature* presents the ecosystem, a dynamic relationship between biotic and abiotic systems, in which rural development should take place. When a person uses nature to meet his needs, nature transforms itself to natural resources. The mutually dependent people living in each development area constitute the *society*. If the rights of man are to be preserved (Pierre Dansereau 1970, UN 1978), the interrelationships of person, nature, and society must be intelligently controlled. Technological interventions occur between society and nature; organizational and psychosocial interventions between person and society; biophysical interventions between person and nature.

This system should address the solving of various vulnerabilities like famine and starvation, thirst, epidemics, lack of energy resources, and social conflicts by securing basic needs for all people. Instead of being highly vulnerable, the system should be self-sufficient and self-reliant; it should have a high degree of autonomy. Autonomy means using local resources for local rural development, countering the widespread application of grants, subsidies, and bureaucratic and political favoritism that often act against the spread of financially viable, economically feasible, and socially meaningful rural development projects. Mass participation of local people responsive to ecosystem development can reverse the trend of dependency on imported fuels and technologies and on proliferating bureaucratic institutions that curb local initiatives and often disregard the local interest.

STUDY APPROACH AND EXPERIENCE

The research started with group meetings at all 12 panchayats of Mahnar block. Only the one urban area in the block had no meetings. The short-term intent of the project was to study, plan, and implement action programs. The long-term interest was to develop inputs for policymaking in rural development. These panchayat meetings were utilized to understand local problems so that the research could concentrate on these issues and research instruments be designed. Panchayat meetings also provided an overall view of the political structure of the block necessary for initiating further studies.

Sixty-one populated and seven depopulated villages are located in Mahnar. The people engage in 50 different occupations; 46 castes are represented. The project was discussed among a group of *Mukhya* and *Sarpanch*[1] and at a larger gathering of important local leaders in each panchayat. Bridges of understanding and friendship were built between researchers and the villagers. Problems posed by the local people encompassed several sectors of rural economy and social services. An alternative energy source for lift irrigation and reclamation of floodable area were prominent among them.

The underprivileged people in the panchayat community, although present in these meetings, were the least vocal group. Therefore, it was necessary to identify problems of specific groups in addition to the collective problems. Study of the structure of cooperation was also necessary as well as the potentials for promoting and strengthening collective action.

Next, in-depth interviews were conducted among officials and non-officials responsible for providing government services. The officials interviewed included the block's development officer, agriculture officer, veterinary officer, educational officer, public health doctors, bank managers, engineers, wholesale merchants, and bus owners. The household census of their own villages became the responsibility of the volunteer workers (educated unemployed).

In order to develop understanding of rural living patterns in these heterogeneous blocks, the research team prepared three questionnaires corresponding to the matrix person-nature-society. Every person was covered in a comprehensive household census. Educated unemployed people as well as local leaders volunteered to help with the household census without any remuneration. About 200 volunteers came from different villages. Depending upon their caste background, they were charged with the hamlets or villages where they lived.

The content of the questionnaire was mostly factual: existing problems, assets, resources, basic needs, demography, production, and income and expenditure. The last two kinds of information were harder to gather. The prestructured household questionnaire consisted of 118 variables and more than five computer cards per household. Included were household identification; family type (nuclear, joint, or extended); type of residence; ownership of land, vehicles, and animals; land leased in and out; land sold; and land irrigated. Also included were use of fertilizer, seed, credit, loans and sources; crops cultivated, produced, and sold; and household and agricultural expenditure. Information was gathered on implements owned, time spent to collect water and fuel, and drinking water source and movements, as well as age, sex, employment, religion, caste, occupation, income, and disease.

Nature was studied through a village questionnaire and by maps. The village census was the responsibility of village-level worker, panchayat workers, and sectoral officers. All of them were trained and supervised by the research staff. The local workers collected information with the help of health, educational, agricultural, veterinary, and revenue staff and supervisors. These local officers uncovered several primary information sources, in addition to secondary data they already maintained. Secondary data included village area and population; service facilities; people's distance from and movement to facilities; roads, linkage, and geographical features; and land utilization and irrigation. Crops and yields were noted as well as marketing, utilization of agricultural inputs and loans, employment and wages, statistics about students, literacy, births and birth attendances, deaths, and vaccinations and immunization.

Society was studied by means of almost daily meetings of officials and survey supervisors. Each panchayat was viewed as a political arena. Patterns of leadership and the nature of political groupings during each panchayat election were studied as well as the nature of caste or class politics. The same topics were studied related to block *samiti* (council) and state assembly elections. Aspects of socioeconomic dependency were studied with special reference to the *jajmani*[3] system, caste, and class. Sources of cooperation and conflict among various sections of society were analyzed.

Casual meetings at the Mahnar market place and at the Jandaha Bazar during the nine-month field study yielded sufficient historical details about elections in the assembly, at the block samiti, and for the panchayats. Analysis of those interviews together with the household census provided sufficient data to study the local power structure, and its factions and conflicts.

Household questionnaires were printed for Mahnar block by December 1979 and the interviewers were trained thoroughly in their use. The project had one jeep and two motor bikes. The bikes were of great help in approaching villages on nonjeepable paths. More than 200 interviewers were contacted almost every day to scrutinize, collect, and often return the questionnaires for clarification. Filling in 10,200 questionnaires took about seven months (one month more than estimated). About 80 percent of the work was done by local people; the rest was completed by the project staff.

246 Rural Energy to Meet Development Needs

No uniform result emerged from the research strategy of household census. In certain panchayats, the level of cooperation was high, but often it was rather low. Student volunteers from one local college attended training with great interest and fanfare, but they were the least effective in covering villages allotted to them. Many volunteers quit: only about one-half of those who attended training completed the work they had undertaken. Some volunteers lost questionnaires or left a village without safeguarding papers. In several panchayats, the Mukhya accompanied interviewers almost daily to supervise their work. Elsewhere, Mukhyas or Sarpanch were not as effective. In addition frequent festivals hampered progress. Even so, the survey was completed more or less on time.

This survey approach was successful because

1. A comparatively high number of educated unemployed youth in Mahnar block volunteered to support this work.
2. The population was dense and the block was characterized by large, compact villages.
3. At least two members of the research team were natives of the study region. One member belonged to the block and was politically active. His past association with the local residents and their leaders helped enormously in getting local cooperation.
4. The daily contact with all the nearly 200 interviewers made possible better relationships with the villagers and helped the supervisors participate in village social activities and define some of their pressing problems. In this process, the study achieved one of its secondary objectives by contributing to the establishment of a few schools and one college in the Hariharganj study area.
5. The questionnaire took only 30 minutes to an hour to complete, was simple in itself and, since the interviewer was usually from the village, interviews were easy to administer.
6. Volunteers easily established good rapport with respondents since they both were residents of the same village and usually were from the same caste group.

The experience of Pilot Research Project in Growth Centres conducted by the Government of India during the Fourth Plan period in 20 blocks widely distributed over the country guided the village survey strategy and the questionnaire's design. One research team member had had experience in that project.

The household questionnaire was coded by trained project staff for computer analysis by block, panchayat, caste, landholding status, and income group. Manual analysis of village and society questionnaires was possible because the units of analysis were fewer in number.

General problems of data collection were as follows:

1. The unconventional survey strategy meant that the content of the questionnaire had to be kept simple and factual, not attitudinal.
2. Reliance on voluntary data collection meant that interviewers were not experienced and their output was not strictly controlled.
3. The progress of survey work was slow.
4. Editing and coding questionnaires were time consuming.
5. The cost of making the survey was reduced considerably by using volunteer interviewers, but more intrablock travel was required because they needed greater supervision by the more experienced researchers.
6. Management and analysis of the more than 160,000 computer cards from the household census were difficult and costly tasks.

The strategy for survey research used in the second block Hariharganj was the same as for Mahnar block, but implementation differed somewhat because of the nature of the study area. Hariharganj is two and one-fourth times larger in area than Mahnar and had three times as many villages. In Hariharganj, the villages were generally small in size, many of them in forested areas. Lack of roads made many of them inaccessible. Two out of the 11 panchayats were cut off during the rainy season.

Low overall educational development in the area made difficult the identification, recruitment, and training of enough educated unemployed people to conduct the household census; however, their dropout rate was very low. Further, inaccessibility of the villages prevented constant supervision. Village and panchayat officials supervised by block development officers spent several days and nights completing their work. Research workers, together with the few available educated volunteers, visited various panchayats and worked together to complete the household census in time. The survey of about 7,900 households and 178 villages started in August 1980 was completed by November 1980.

NATURAL ENVIRONMENTS

The first study area, Mahnar block, is approximately rectangular shaped, 18.5 km north to south and from 5 to 8 km east to west. The total population of the block was 65,390 in 1961 and 79,150 in 1971. Excluding the urban center, Mahnar, the total rural population of Mahnar block was 64,300 according to the 1980 household census conducted by CSDS.

Land and Water Resources

Baya River cutting across Mahnar block on the north and Ganges on the south flood, swell, and spill over into the shallow plains annually. This process removes the rich top soil and waterlogs more than 30 percent of the cultivable land for three to six months to depths of 2 to 5 feet. Eighty-six percent of the rainfall occurs from May through September, and about 6 percent in October. Depletion of tree crops for firewood and lack of forest areas accelerate the process of soil depletion. As of 1980, tree crops and groves covered only 2 percent of the land in Mahnar. Seven percent of Mahnar's 24,800 acres are barren (in Hariharganj, 3 percent is non-cultivable). Even so, rainfall distribution and land characteristics make Mahnar a richer area. It supports more people in larger settlements than Hariharganj.

About 74.3 percent of the Mahnar area is cultivated and another 7.3 percent is cultivable. Close to 47.5 percent of that land is planted more than once. Paddy is the major crop, and wheat and maize follow. Then come tobacco, chillies, gram, sweet potato, onion, potato, and vegetables. Nonirrigated crops, such as sweet potatoes, barley and grams have more area than irrigated crops, such as onions, potatoes, and chillies. More paddy and maize is nonirrigated than irrigated. Irrigation should supplement rainfall for the second and third crops. But even though about 20 percent of the area is fitted with irrigation infrastructure only about one-half of that area is irrigated more than once. Surface irrigation amounts to 306 acres of land: a small percentage compared with the underground potential. Well development is discussed later in a section on energy sources and constraints. The rivers Ganges and Baya irrigate 236 acres, *Chaur* (seasonal pond or lake) irrigates 40 acres, and 7 tanks irrigate 30 acres of land. Power shortages caused by electrical cutbacks and shortages of diesel fuel as well as recurring flooding all contribute to the low utilization figures for the existing irrigation infrastructure.

The second study area, Hariharganj block, consists of 224 sq km. The maximum length of the block east to west is 24 km and breadth north to south is 14.4 km. It is irregular in shape with several protrusions. An entirely rural block, Hariharganj had a population of 51,060 by the 1980 CSDS household census, an increase from 39,360 in 1971. This block is divided into 11 gram panchayats, which vary in geographical area from 1,021 to 7,524 acres. Whereas Mahnar suffers from annual floods, Hariharganj is beset by periodic and seasonal droughts. Still their average rainfall differs by only 2.4 mm.

Rainfall in Hariharganj is totally dependent on local conditions and local winds since the district lies within the retreating range of the southwest monsoon. Droughts, famines, and scarcities have been recorded for Palamau district over the past 120 years (D.M. Singh 1978). In Hariharganj worse droughts are caused by failure of *hathia* (the last rain of the year) than by insufficient seasonal and annual rains (GOI, Planning Commission 1973). Hathia failure occurs about every three years; moderate droughts occur about every two and one-half years.

Hariharganj is essentially a subsistence agricultural economy where mainly paddy, maize, wheat, gram, and arhar (pulses) are grown. Cane sugar and vegetables are also grown. The net area sown is only 32 percent; area sown more than once is 8.9 percent. Nonirrigated cultivation is more prominent. Area irrigated is only 10.7 percent; area irrigated more than once is only 2 percent. Irrigation is mainly used to stabilize the *kharif* (monsoon) crop from hathia failure. Since they require only low amounts of water, gram, maize, and arhar can be grown as rain-fed crops, although yield would be marginally low.

In Hariharganj, 38 percent of the land is forested. Forestry, then, is the most prominent land use. Unlike in Mahnar, supplies of firewood and commercial varieties of timber are abundant, but then the area left available for cultivation is limited, and existing laws do not allow changing land use. About 21 percent of Hariharganj was fallow in 1980; another 10 percent had been fallow for some time. The area's abundant wildlife, such as wild pigs, Indian fox, sambar, and bison, constantly destroy the cultivated land, but laws prohibit trapping or killing wildlife. And factors like these make farmers reluctant to invest on agricultural improvements such as irrigation facilities, fertilizer, and pesticides.

Animal Resources

Table 1 gives the number of livestock available in Mahnar and Hariharganj blocks as of 1980. Hariharganj, with more than twice the area of Mahnar and with fewer people in smaller settlements, has almost twice (1.9) as many animal resources. Bullocks, goats, and cows are the most common animals in both blocks. Goats generally are reared by the landless, marginal, and small-scale farmers and are sold in the markets. Rearing goats for sale is an unacceptable practice for rich farmers.

High summer temperatures burn green fodder, so supplies for cattle are usually inadequate. In addition, nutrients generally are not utilized, so milk yield is very low (1 to 1.5 kg/day) for both local and improved varieties of cows and buffalos. Even with a great demand for milk and milk products, no dairy cooperative or chilling plant has been planned. Programs exist for natural insemination of cattle with improved bulls, but poor upkeep of animals keeps milk yield low. In addition, poor livestock

TABLE 1.
Livestock, 1980

Livestock	Mahnar		Hariharganj	
	Number	Percentage	Number	Percentage
Indian cow (milching)	1,237	5.8		
Improved cow (milching)	98	0.5	5,972	14.9
Cow in dry	652	3.1		
Calves	1,009	4.8	1,989	5.0
She calves	844	4.6	2,884	7.1
Indian buffalos (milching)	932	4.4		
Improved buffalos (milching)	176	0.8	2,852	7.1
Buffalos in dry	938	4.4		
Buffalo calves	570	2.7	956	2.4
She buffalo calves	808	3.8	1,521	3.8
Bullocks	7,147	33.9	9,115	22.8
He buffalos	401	1.9	244	0.6
Goats and sheep	4,776	22.6	8,755	21.9
Hens and cocks	1,373	6.5	4,147	10.4
Pigs	151	0.01	1,509	3.8
Others	0	0	91	0.2
	21,112	100.0	39,985	100.0

Source: CSDS Household Census 1980.

Note: According to the 1977 livestock census, 42.7 percent of the cattle were in milk and 3.1 percent were improved cows. Dry cows made up 54.2 percent of all cattle. In addition, 50.5 percent of all buffalos were in milk and 1.3 percent were improved buffalos. Buffalos in dry made up 48.2 percent of all buffalos.

quality and low yields of cattle introduced by the Integrated Rural Development Program have limited the program's success.

PERSON, SOCIETY, AND RESOURCE DISTRIBUTION

From ancient times an area of religious significance, today, Bihar has a caste-dominated political economy. The 45 enumerated castes have been grouped into nine categories (Table 2) based on relative caste status and occupational similarities. The evident political and economic polarization means that high status Hindu caste groups and rich farmers can effectively take disproportionate shares of government credits, subsidies, and other benefits.

Demographic Characteristics

Population statistics by age groups have a similar ratio in each block (see Table 3a). Neither block is close to its enrollment goal of 90 to 100 percent for primary school and 50 percent for middle school. More female than male students drop out in middle and high school especially in Hariharganj. Indeed, literacy percentages show many more adult female illiterates than adult male illiterates (see Table 3b).

In Mahnar, female deaths during pregnancy and child birth evidenced in sex ratios mean more assistance is needed from subhealth centers. In Hariharganj, the higher ratio of men probably is due to out-migration of working-age males. Child laborers make up 4 to 5 percent of landless and marginal farm families in Mahnar, and 1 percent of small-scale and viable farm families.[4] In Hariharganj their number is negligible. More people in Mahnar (3.4 percent) suffered disease than in Hariharganj (1.9 percent). Fevers were most frequent. Asthma, malaria, and tuberculosis were also frequent. In both blocks, communicable and environmental-based diseases were encountered frequently.

Major primary occupations in Mahnar and Hariharganj were calculated from results of the CSDS household census (see Table 4). The percentage of population with secondary occupations in Hariharganj (12.3) was higher than in Mahnar (5.3) largely because droughts in the southern block affect primary agricultural occupations.

Productive Assets and Technology

The existing distribution of productive assets, technology, and equipment indicates past trends in rural development and can be examined to suggest possible directions of change. If this distribution is compared to

TABLE 2.
Landholding and Social Structure of Study Blocks, 1980

Landholding Groups	Brahmins	Upper Castes (Rajput, Bumihar)	Business Castes	Upward Peasant Proprietary Castes	Service Castes	Low Castes	Scheduled Castes	Scheduled Tribes	Muslims	Not Available	Total
Mahnar											
Landless	0.2	0.6	2.2	3.3	4.8	4.7	20.3	0	2.1	--	38.2
Marginal Farmer	0.9	2.0	0.8	8.2	3.7	2.8	6.3	0	1.4	--	26.1
Small-scale Farmer	0.7	3.1	3.6	4.1	0.6	0.2	0.3	0	0.1	--	12.7
Viable Farmer	<u>1.8</u>	<u>9.5</u>	<u>1.4</u>	<u>8.3</u>	<u>0.5</u>	<u>0.2</u>	<u>0.4</u>	<u>0</u>	<u>0.2</u>	--	<u>22.4</u>
All Households	3.6	15.2	8.0	23.9	9.7	7.9	27.3	0	3.8	--	99.4
Total Land Owned	6.9	44.4	6.7	33.3	3.0	1.3	2.8	0	1.0	--	99.4
Hariharganj											
Landless	0.1	0.3	4.6	1.1	2.7	0.2	18.5	0.1	4.0	--	31.6
Marginal Farmer	0.4	0.5	2.2	4.7	3.9	0	13.8	0	0.8	--	26.3
Small-scale Farmer	0.4	1.3	1.7	5.4	1.9	0.1	5.4	0.1	0.4	--	16.7
Viable Farmer	<u>1.1</u>	<u>8.5</u>	<u>1.9</u>	<u>9.1</u>	<u>1.7</u>	<u>0</u>	<u>3.0</u>	<u>0</u>	<u>0.3</u>	--	<u>25.6</u>
All Households	2.0	10.6	10.4	20.3	10.2	0.3	40.7	0.2	5.5	--	100.2
Total Land Owned	3.3	42.2	6.0	25.5	6.0	0.1	15.3	0.2	1.3	--	99.9

Source: CSDS Household Census, Mahnar and Hariharganj Blocks, 1980.

Note: Figures are in percentages. The landholding unit of marginal farmers is less than one-half hectare irrigated land, of small-scale farmers, one-half to one hectare irrigated land, and of viable farmers, more than one hectare irrigated land. (One unit irrigated land equals two units unirrigated land.)

TABLE 3a.
Selected Demographic Characteristics, 1980

Characteristic	M (%)	H (%)
Age		
0-5 (infants)	15.4	19.7
6-15 (student age)	23.9	24.0
15-59 (working age)	54.9	50.8
60 and over (aged)	5.6	5.4
Females (per 100 males)		
15-45 (reproductive age)	118.4	117.1
46 and over (after reproductive age)	116.7	120.1
Primary Activity		
Minors	26.7	33.6
Students	15.4	22.5
Household work	24.2	22.5
Work outside home	31.1*	30.8*
Retired and unemployed persons	2.6	2.7

TABLE 3b.
Selected Educational Characteristics, 1980

Characteristic	Male		Female		Total	
	M	H	M	H	M	H
School Enrollment+						
Primary	46.3	36.5	41.9	21.2	46.3	30.1
Middle	37.0	20.4	25.1	3.0	32.5	13.6
High	8.0	7.1	4.7	1.1	7.0	4.7
Literacy Rate	40.4	37.7	19.6	11.4	34.1	26.0

Source: CSDS Household Survey, 1980.

Note: M = Mahnar; H = Hariharganj.

*In Mahnar, 5.3 percent of employed persons had secondary jobs; in Hariharganj, 12.3 percent did.

+Figures are number of students per 100 school-age children.

existing rural sociopolitical structures, differential holdings of assets by different groups are revealed (see Tables 5a, b, and c). Such differences are associated with technology attributes such as size, mobility, cost, and labor-intensity. An understanding of these attributes may open new

TABLE 4.
Major Primary Occupations, 1980

Occupation	Landless		Marginal Farmers		Small-Scale Farmers		Viable Farmers	
	M	H	M	H	M	H	M	H
Agricultural Laborers	21.0	17.6	5.9	12.8	–	7.1	–	1.9
Loaders and coolies	4.2	1.8	0.3	1.2	–	–	–	–
Artisans	1.8	1.6	–	1.4	–	–	–	–
Sharecropping	–	–	–	2.7	–	2.9	–	1.7
Petty shopkeepers, services	1.0	2.9	–	0.7	–	0.6	–	–
Cultivators	–	–	11.7	7.7	18.5	14.4	*	22.8
White collar workers	–	–	–	–	3.1	–	*	1.3
Professionals	–	–	–	–	1.3	0.5	*	0.9
Transportation workers	–	–	–	–	0.8	0.6	*	–

Source: CSDS Household, 1980.

Note: Statistics are percentages. M = Mahnar; H = Hariharganj.

*Percentages were calculated together with statistics for small-scale farmers in Mahnar.

paths by which intended beneficiaries of development can gain access to productive assets. Knowledge of such opportunities is important in situations where the political structure in power may tend to protect its group interest, such as caste and size of landholding.

A small percentage of the *landless* people own traditional agricultural implements such as bullocks, plough, harrows, and even persian wheels. These resources are used to cultivate leased land or to assist landowning farmers. Since only a very small number of landless families lease land, use of agricultural inputs is also limited, and only one or two of every 1,000 use them regularly. More of them farm without ever using such inputs. In addition, more landless farmers own bullocks than own agricultural implements: one in 50 in Mahnar and one in 23 in Hariharganj. Those farmers naturally are more employable. About 10 percent of the landless group took out loans to obtain capital assets. In Mahnar, the source was almost always money lenders, but in Hariharganj, money lenders and other institutional credits were used almost equally. In

TABLE 5a.
Ownership of Agricultural Equipment and Bullocks by Landholding Group, 1980

Landholding Group	Ploughs	Harrow	Other Traditional Implements	Persian Wheel/Rope & Bucket Lift	Electric Pump	Diesel	Tractor	Thresher	Improved Plough	Others	Bullocks
Mahnar											
Landless	1.9	1.7	0.8	0.5	0	0	0	0	0	0	1.9
Marginal Farmers	30.3	27.4	7.1	7.6	0.4	0.8	0	0.1	0	0.1	28.0
Small-scale Farmers	74.5	72.3	20.7	32.4	1.5	4.2	0	0.1	0.1	0.1	69.3
Viable & Large-scale Farmers	75.5	76.4	43.4	49.0	10.7	12.7	0.6	2.3	1.4	0.3	87.5
Hariharganj											
Landless	2.9	2.4	0	0.3	0	0	0	0	0	0	4.3
Marginal Farmers	48.1	44.8	1.5	5.5	1.4	0.1	0	0	0	0	50.1
Small-scale Farmers	80.8	78.3	4.2	20.0	7.8	0.1	0	0	0	0	82.4
Viable & Large-Scale Farmers	66.9	81.8	8.4	40.5	29.7	4.1	1.1	1.0	0.1	0.9	94.0

Source: CSDS Household Census, 1980.

Note: Figures are in percentage.

TABLE 5b.
Distribution of Use of Inputs, 1980

Landholding Group	Chemical Fertilizer			Improved Seeds			Pesticides		
	Regular Use	Some-times	Never	Regular Use	Some-times	Never	Regular Use	Some-times	Never
	Percentage								
Mahnar									
Landless	0.2	1.8	17.6	0	1.7	17.7	0.1	1.3	18.0
Marginal Farmers	7.1	40.8	12.4	4.3	40.1	15.1	3.6	30.7	22.9
Small-scale Farmers	7.7	69.4	4.9	5.6	68.6	8.0	4.4	59.0	16.5
Viable and Large-scale Farmers	13.5	79.5	1.8	10.2	80.5	3.7	7.8	74.9	10.6
Hariharganj									
Landless	0	0.3	1.8	0	0.2	1.9	0	0.2	1.9
Marginal Farmers	0.4	17.9	46.2	0.4	13.8	48.6	0.5	7.9	51.4
Small-scale Farmers	1.7	38.6	35.4	1.4	30.9	40.0	0.8	17.7	49.2
Viable and Large-scale Farmers	8.2	60.9	19.6	7.4	51.1	25.7	5.4	28.7	45.0

Source: CSDS Household Census, 1980.

TABLE 5c.
Distribution of Credit by Landholding Size

Landholding Group	Credit: Fixed Capital			Credit: Operating Capital				Credit: Nonproductive			
	Money Lender	Coop-erative	Block	Bank	Money Lender	Coop-erative	Block	Bank	Money Lender	Coop-erative	Block	Bank
	Percentage											
Mahnar												
Landless	10.5	0	0.1	0.1	31.1	0.2	0.2	0	22.2	0	0.1	0
Marginal	14.5	0.4	0.6	0.7	27.2	0.8	0.6	0.1	26.0	0.2	0.1	0.1
Small-scale Farmers	9.9	1.8	0.6	1.6	18.5	1.7	0.9	0.4	17.9	0.2	0.4	0.1
Viable and Large-scale Farmers	9.3	1.4	1.5	2.1	14.5	1.9	1.5	0.6	11.2	0.2	0.2	0.1
Harihargan												
Landless	5.1	0.5	0.2	3.5	3.4	0	0.3	0.3	47.9	0.3	0.2	0.2
Marginal Farmers	6.9	0.6	0.5	2.8	3.6	1.4	2.0	0.5	60.5	0.2	0.1	0.1
Small-scale Farmers	8.4	1.3	1.2	6.6	4.4	2.1	3.7	0.3	58.2	0.3	0.5	0.2
Viable and Large-scale Farmers	9.3	2.4	2.4	16.9	4.4	6.0	7.5	1.2	47.1	0.4	0.3	0.1

Source: CSDS Household Census, 1980.

Mahnar, most landless wanted money to meet recurring expenditures and social (nonproductive) expenditures such as marriage and death ceremonies.

Significant numbers of *marginal* farmers own bullocks and manual- or animal-operated equipment such as ploughs, harrows, persian wheels, and other traditional implements, in higher proportions in Hariharganj than in Mahnar block. A limited number of these farmers, unlike the landless, also own electric and diesel-run equipment (1.4 to 0.1 percent). As with landless households, most marginal farmers get loans from money lenders to meet recurring farm expenditures and nonproductive expenditures, but they also use the loans to obtain capital equipment.

Higher percentages of *small-scale* farmers own agricultural implements and use agricultural inputs and credits. More of these farmers own bullocks and bullock-driven water-lifting devices than do marginal farmers. Small percentages of these farmers own electric pumps and, in Mahnar, diesel pumps. In Hariharganj, no diesel depot is located in the area and supplies are not as available. Diesel-run tractors and threshers also were in limited supply.

Most *viable* and *large-scale* farmers own bullocks, bullock-driven water lifts, ploughs, and harrows. Electric pumps were available among one in nine viable farmers in Mahnar, and among one in three farmers in Hariharganj. Diesel pump sets are owned by one in eight viable farmers in Mahnar and one in 25 in Hariharganj. Money lenders also reach most viable farmers. Less than 20 percent of them use credit for capital and recurring expenditures. In Hariharganj, credit from banks is more common than money lenders — opposite to the pattern in Mahnar.

Overall, fewer Hariharganj farmers use agricultural inputs such as improved seeds, fertilizer and pesticides, for reasons discussed earlier. More farmers there never use such inputs. In Mahnar, more farmers use them sometimes.

ENERGY USE AND CONSTRAINTS

In absolute terms, cooking fuel constituted the largest energy requirement in these study areas. Firewood was the primary fuel of the majority of farm households in both blocks, and cow dung, the secondary fuel. Although Bihar has vast coal resources, its use is not widespread in either block because firewood has been cheaper and more easily available.

Cooking Fuel Sources

For the majority of landless families in Mahnar, leaves and twigs were the primary fuel; firewood was primary for one-third of this group (see

TABLE 6.
Cooking Fuels and Procurement Times, 1980

Landholding Group	PRIMARY FUELS				SECONDARY FUELS				TIME SPENT TO PROCURE FUEL			
	Firewood	Cowdung	Coal	Leaves and Twigs	Firewood	Cowdung	Coal	Leaves and Twigs	Male	Female	Children	Total
	Percentage								Minutes/day			
Mahnar												
Landless	33	8	1	57	6	9	1	15	11	54	16	81
Marginal Farmers	56	13	2	26	12	19	2	12	11	34	14	59
Small-scale Farmers	72	12	3	7	12	36	3	9	9	19	5	43
Viable Farmers	75	14	4	5	14	37	3	7	21	17	5	43
Hariharganj												
Landless	98	*	--	--	*	3	*	*	31	16	--	47
Marginal	99	*	--	--	*	10	*	*	21	16	1	38
Small-scale Farmers	99	*	--	--	*	19	*	*	33	14	--	47
Viable Farmers	99	*	--	--	*	5	*	*	23	5	--	28

Source: CSDS Household Census, 1980.

Note: Other fuels individually accounted for less than 1 percent of primary fuel usage.

*Less than 1 percent.

Table 6). Women and children in the landless group spent considerable time in collecting fuel for cooking. With firewood cost at about Rs.40[5] (US $5.04) per 100 kg, the main buyers of firewood are viable and small-scale farmers.

In Hariharganj, the 38 percent of the total 55,400 acres of land covered by forest is the main source for cooking fuel. Firewood is the primary fuel for nearly all families in the block. Poor farmers collect firewood themselves, but landless people collect and sell it to wealthy farmers. Firewood in Hariharganj was cheaper. It was not generally weighed but sold per head load costing Rs.10 to Rs.15.

In Mahnar, collection of fuel by an average landless farm family took almost double (81 minutes) the time spent by an average viable farm family (43 minutes). In Hariharganj, the average time spent was 47 minutes for landless and marginal farmers, 43 minutes for small-scale farmers and 28 for viable farmers. (There, women and children probably were discouraged by forested terrain and the presence of wild animals.)

Power for Irrigation

The most important barrier for extension of land under irrigation in both blocks is the undependability of diesel and electric power for that use. In Mahnar, farmers use diesel sets more than electric motors, and diesel generators often are used as standbys for electric power. The number of diesel engines in Hariharganj is small because people have to go outside the block for diesel to augment the block's limited supply. (The supply problem was acute as of the CSDS survey.) No petrol or diesel depot is located either in Hariharganj or Mahnar. The most dependable source of energy for irrigation is manual and animal power.

Well types (see Table 7) and depth of water table in the two blocks form a basis for analyzing these sources of power for irrigation. In October, water is available at a mean depth of 2.7 meters in Mahnar and 7.6 meters in Harijarganj. In June, the depths are 7.3 and 10 meters, respectively.

Mahnar has experienced an impressive buildup of wells (see Table 8). Of 14 deep tube wells, one was not in working condition. The unreliable power supply position has meant a bigger increase in diesel engines than in electric pumps. Construction of dug wells has not increased over the last three years.

The command area of an animal-driven irrigation system is 2 to 3 acres; of a manual driven system, 1 to 2 acres. Deep tube wells give a command area of 100 to 125 acres and diesel pump systems give 5 to 6 acres. One well is available for every 13 acres average net area sown. Even so, only 20 percent of the area is irrigated and only 12 percent is irrigated more than once, largely because of a lack of energy. Several indigenous

TABLE 7.
Energy Sources for Irrigation Systems, 1979-1980

Block	Manually Operated	Animal Operated	Diesel	Electric
Mahnar	600	150	379*	---
Hariharganj	350	479	42+	381

Source: CSDS Village Survey, 1979-1980.

*57 in Mahnar town; the rest in 5 panchayats.

+located in 9 of 11 panchayats.

TABLE 8.
Wells in Mahnar, 1977-1980

Type	1977-1978	1978-1979	1979-1980
State deep tube wells	12	13	14
Private wells with diesel pump	361	365	447
Private wells with electric pump	113	113	124
Diesel pump set	321	331	375
Dug wells	1,428	1,428	1,428

Source: CSDS Village Survey, 1979-1980

bucket type lifts are used in Mahnar (see Table 9). Hariharganj has no deep tube wells. Major sources of irrigation and acreage irrigated are given in Table 10. Landholding size and topography as well as energy source limit the command area.

Bullocks, the main draft animal, operate persian wheels or *rahats* (chain lift open buckets). Rahats use two bullocks and irrigate about 1.5 to 2 acres of land. Since available bullocks are underfed and of poor quality, and utilize no improved yoke or harness, they cannot work rahats over a long period, therefore, use of rahats has not become widespread.

Electricity

In 1971, 18 villages in Mahnar had electricity (including seven villages now designated as urban). That number increased to 25 (or 39 percent of all villages) according to the CSDS 1980 survey, compared with 77 of 178 inhabited villages in Hariharganj (43 percent). (Compare those percent-

TABLE 9.
Most Frequently Used Lift Devices, Mahnar

Device	Maximum Lift (meters)	Liters/Hour (000)
Manual		
Scoop swing bucket	0.9 – 1.2	14.0 – 19.0
Circular two-bucket	4.0 – 5.0	12.0 – 14.0
Counter poise lever and bucket	1.2 – 4.0	8.0 – 11.0
Animal-operated		
Rope and bucket	10.0 – 30.0	6.0 – 10.0

Source: A.M. Michael 1978.

TABLE 10.
Sources of Irrigation in Hariharganj, 1980

Source	Number	Acres Irrigated	Average Command Area (Acres)
Wells and tube wells	654	1,306	2
Field tanks	215	1,316	6
Canals	5	300	40
River inlets	2	56	28

Source: CSDS Village Survey, 1980.

ages with the 40-percent goal of the Fifth Plan and the 50-percent goal of the Sixth Plant.) In Mahnar, Palwaia panchayat (located in a floodable area) is the only panchayat with no electric connections. Although there are more domestic than pump set connections, connection to a pump set often is a village's first electric connection (see Table 11).

Erratic and therefore ineffective electrical supply results in a 5-percent collection rate and high arrears for the agricultural sector (see Table 12). No policy exists for compensation by the agency to farmers whose crops fail because power is cut and water is not available at crucial times. Even with several deep tube wells and electrified tube wells, the cropping intensity is only 121 percent and this cropping intensity relies heavily on animal, manual, and diesel irrigation power.

One significant reason for the erratic electric supply is low generation and distribution capacities compared with connected loads. The Barauni thermal generator supplies power to Mahnar block, which has a distribution capacity of 0.5 MW but a connected load of 4.0 MW. Power is supplied in turns so it is not always available at required times. In their study of Bihar, the Central Ground Water Board (GOI 1980) reported fluctuat-

TABLE 11.
Electrical Connections, 1980

	Mahnar	Hariharganj
Domestic	616 (381)	629
Commercial	41 (242)	132
Industrial	13 (94)	29*
Agricultural	803 (967)	381
Total	803 (967)	1,171

Source: CSDS Survey 1980. Figures in parentheses, from Executive Engineer's Office in Hajipur 1979.

Note: Reasons for discrepancies in Mahnar figures are (1) pilferage and technical problems caused 45 percent line losses and unauthorized connections are usually domestic; (2) the EEO's list of commercial and industrial connections includes Mahnar town; (3) some inactive agricultural connections are still in the records.

*15 in Ararua and 14 in Hariharganj.

TABLE 12.
Monthly Electric Accounts, Mahnar, 1979

Load Type	Assessment	Collection (Rupees 000)	Arrears
Domestic	26	10	880
Commercial	50	30	680
Industries	200	150	2,500
Agriculture	90	4	3,000
Total	366	194	7,060

Source: Executive Engineer - Hajipur, 1979.

ing voltage and interrupted power supply. This adversely affects agricultural production. Delay in electrification of state and private tube wells forces farmers to use diesel engines that require more attention and repairs. A current proposal would augment Barauni power generation to 300 MW (a 225-percent increase), but the existing gap is 750 percent. Changes in distribution management must be accompanied by adequate generation expansion. Without reliable electricity, arrears will increase and so will unauthorized connections and pilferage.

Kerosene

Kerosene is used for domestic lighting in Mahnar and Hariharganj blocks. Supply is controlled by the government and the fixed price is

Rs. 1.67 (US $0.21) per liter. It is used even in electrified households and shops because electricity is usually only available in late night or early morning for agricultural purposes. Long-term availability of kerosene depends upon oil import and local production. If supplies of crude import remain uncertain, other renewable lighting fuels will have to be considered for the villages.

Recently opened outlets within panchayats have strengthened the kerosene distribution system, but the profit margin allowed by the government is small. Outlet owners are required to accept delivery from district headquarters without recovering the transport cost in the sale price. Thus owners sometimes sell kerosene on the "black market," so village supplies become scarce even though adequate allotments have been made. Direct delivery to outlets or more economical transportation costs would eliminate this important problem.

Energy for Transportation

Human and animal power are the major transportation power sources in Mahnar. Cycles and cycle-rickshaws enjoy wide use all over the panchayats. Horse carts are available as shared vehicles between the Mahnar bazaar and the railway station. Bullock carts transport most of the goods within the block. The very few auto-rickshaws (three-wheelers) can carry up to eight passengers and their luggage. The buses are usually overcrowded and many passengers ride on top where luggage usually is carried.

Lack of a developed road network in Hariharganj has limited the growth of transportation. Rickshaws are available at major centers and cycles are common everywhere. The rugged topography prevents use of bullocks. Diesel fuels large buses and minibuses throughout the block. (See also Table 13 for vehicle ownership.) Vehicle ownership and ownership of multiple vehicles as indices of prosperity show that Mahnar is more prosperous than Hariharganj. Inaccessibility and rough terrains prevented the spread of vehicular ownership in Hariharganj.

Energy Source by Activity

Noncommercial energy sources are most widely used (see Table 14). A maximum amount of firewood is used, but renewable energy sources such as wind, hydro, and biomass other than wood are not used. (The sun alone is used to dry chillies and grains.)

Collecting cooking fuel, irrigating, and obtaining drinking water generally consume the most time per year. Other activities are generally

TABLE 13.
Ownership of Vehicles, 1980

	Landless		Marginal Farmers		Small Farmers		Viable Farmers	
	M	H	M	H	M	H	M	H
No vehicles	90.1	93.3	68.0	93.2	41.9	89.5	23.9	67.6
Bicycles	9.6	6.5	29.6	5.9	51.7	8.2	63.6	20.8
Bullock carts	0	0	1.0	0.5	1.7	0.9	2.3	4.3
Horse carts	0.1	0	0.1	0	0.0	0	0	0
Motor vehicles	0	0.1	0.2	0	0.2	0	1.1	0.4
Combination of types above	0.2	0.1	1.1	0.4	4.5	1.4	9.1	6.0

Source: CSDS Household Census 1980.

Note: M = Mahnar; H = Hariharganj.

periodic. Human and animal power, although reliable sources, are not used efficiently. They could be utilized more efficiently by using appropriate technology.

REGIONAL PRIORITIES AND CANDIDATE ENERGY TECHNOLOGIES

Regional priorities and prospective energy technologies are identified by investigating the relationships between person and nature (land, water, animals) and between each of those elements and society in each of the study blocks. Since floods in Mahnar and droughts in Hariharganj lead to food scarcity, malnutrition, and ill health for the people of both blocks, an integrated program of flood prevention for Mahnar and one of drought protection for Hariharganj are essential. In addition, irrigation and anti-erosion measures (against sheet erosion in Mahnar and against gully erosion in Hariharganj) have received much attention. Greening these blocks, then, is a priority: a forestry program in Mahnar to protect land from floods and to increase fuelwood resources and one in Hariharganj to counteract erosion.

Development of water resources, especially groundwater resources, is necessary to stabilize Mahnar's second crop and to raise a third crop. It is required to stabilize Hariharganj's first crop and to fight against hathia failures, and then to raise a second crop. Underutilization of the rich, cultivable Mahnar area and the inability to cope with recurring hathia

TABLE 14.
Energy Sources by Activity, 1980

Activity	Human	Animal	Firewood	Coal	Diesel	Petrol	Electricity
Agriculture							
Land preparation	x	x			x		
Cultivation	x						
Weeding	x						
Fertilization	x						
Irrigation	x	x			x		x
Harvesting	x						
Threshing	x				x		x
Winnowing	x						
Milling	x				x		x
Storage	x	x			x		x
Fisheries	x	x					
Cooking	x		x	x			
Lighting					x		x
Water supply (drinking)	x	x					x
Health care and education	x				x	x	x
Construction	x	x	x	x	x		
Industries	x		x	x	x		x
Transportation (goods)							
Field to village	x	x			x		
Village to bazar	x	x			x		
Bazar to markets					x		
Transportation (human)							
Village to bazar	x	x					
Bazar to market					x	x	
Information transfer (radio, cinema)	x				x		x
Total	20	9	3	3	14	2	9

crop failures in Hariharganj are due to the nonavailability of dependable and affordable energy technology for irrigation. The technology utilized should also utilize the available infrastructure like dug wells mostly owned by the marginal and small farmers.

Although sufficient data may not be available from this study to consider social feasibilities, selection of energy technology must take into account purchasing capacity of the client group. Purchasing capacities are indicated for households with income more than or equal to expenditures, identified in the household census. In Mahnar, these households included 45 percent of the landless, 39 percent of marginal farmers, 38 percent of small-scale farmers, and 52 percent of viable farmers. In Hariharganj, these households were 36 percent of the landless, 28 percent of marginal farmers, 54 percent of small-scale farmers, and 80 percent of viable farmers. Except for landless households, many of these households would have the purchasing power to own various recommended technologies. Viable farmers can buy these energy technologies with or without help. Marginal and small-scale farmers would need loans and also could share ownership of these technologies (such as a chilling plant or a cold storage facility) through associations.

Appropriate energy sources for lift irrigation must be identified first, since electric and diesel supplies are not dependable in either block. Field visits and interaction with research institutions in several states of India provided information on frontier technologies available. Field tests and interviews from the neighboring state of Uttar Pradesh suggest that a bullock-powered chain washer pump would be more efficient than the persian wheel for both blocks. With an output power of 0.38 hp, on a lift of 6 meters, such a pump could supply 13,900 liters per hour and would be capable of irrigating 4 to 5 acres per agricultural season. It could be locally manufactured and replacement materials would be locally available. The low-quality bullocks available in these study blocks would supply more water to a larger area more economically with the chain washer pump than the persian wheel.

Feasibility of a small wood-fired steam engine should be investigated as a power source in Hariharganj in view of hathia drought and the block's extensive forest resources.

Windmills also are possible power sources. Two types selected for further assessment in these study areas are (1) a windmill designed for eastern India by the Organization of Rural Poor (ORP) in Ghazipur, and (2) a low-cost water-pumping windmill with a sail type savonius rotor developed by the Indian Institute of Science (IIS), Bangalore (S.P. Govinda Raju and R. Marasimha 1979). A facility is located at Ghazipur for training in the manufacture, installation, and repair of windmills. To reduce cost, the original ORP design incorporating a steel structure can be replaced by steel blades mounted on locally available timber structures. According to a technical feasibility study in which monthly available wind energy was matched to the water requirement for various crop rotations, such a system could irrigate 4 to 5 acres each agricultural season.

These renewable energy technologies could effectively reduce dependency on diesel or electricity. They can be located in dug wells owned by marginal and small-scale farmers. The predominantly illiterate residents can easily learn how to manufacture, install, operate, and repair these technologies. A training program envisaged would reinforce that self-reliance. The small command area with these systems would facilitate water management and resolution of conflicts. The landless agricultural laborer would be the indirect beneficiary of this innovation, since an increase in gross area cultivated would create more job opportunities and increase income.

For lifting water for drinking, the present use of human-operated systems would be sufficient in shallow wells. Technologies involving pedal-driven systems or using wind energy could be effective substitutes for hand pumps and for electricity or diesel-powered pumps in dug wells.

In Hariharganj an ecological balance must be created to even out the benefits of short periods of heavy rainfall. Thus soil and water conservation and forestry should be practiced in an integrated manner, utilizing afforestation, contour bunding, furrowing, and terracing. Water harvesting techniques would conserve moisture in the soil profile, reducing runoff. Water from runoff could be collected in structures minimizing seepage and evaporation, to be utilized during hathia. Together with economic use of uncertain rainfall and more intense cropping, dry land farming is recommended for Hariharganj's drought-prone area. With dry land crops, instead of paddy and sugarcane, less water would be used, less time would be needed for plant growth, and sowing date would be flexible.

Hariharganj's "Food for Work" through the National Rural Employment Scheme is an ideal program for employing people to construct permanent roads and other community projects in the adverse farming seasons. Communal "Dharma Golas" (community warehouses) for the landless and for marginal and small-scale farmers also could be organized to insure supplies of food to drought-affected people. This food security also involves prices of food crops and livestock products. Price stability for milk products (especially for Hariharganj) and crops like potato (especially for Mahnar) depends, in turn, upon the availability of chilling plants and cold storage facilities. Such facilities could improve employment as well as income opportunities. Feasibility studies and design research on solar refrigeration or wind-powered solar pumps are recommended. This drought-prone area has potential for fodder development, which in turn would mean higher milk yields. Organization of milk distribution could improve both production and returns. Livestock varieties should be improved, both dairy animals and sheep, pigs, and poultry. Raising goats, sheep, pigs, and poultry could help augment the domestic income of marginal farmers and the landless households.

Pressure on firewood for cooking in Mahnar could be reduced by development of biodung, biogas, solar pond, and coal briquetting, perhaps on a communal basis. Potential consumers for biodung exist among people whose reported expenditures exceed income, that is, in Mahnar: 55 percent landless, 61 percent marginal farmers, 62 percent small-scale farmers, and 47 percent viable farmers. Biodung can be manufactured by the anaerobic digestion of a variety of water hyacinth locally available in Mahnar, as well as Hariharganj. The abundant supply of biomass during the rainy season could be utilized for manufacturing and storing this fuel for use in winter and summer. The requirements for production and distribution of biodung include polythene bags, technical guidance, training, and mobilization of unemployed landless or marginal farmers who need additional income. Although Bihar is located within the coal belt, coal briquetting is not yet popular, so it is another frontier technology.

These innovations probably would be received more enthusiastically in Mahnar than in the less populated Hariharganj (with its still large forest reserves).

In Mahnar, floods and soil erosion can be controlled through engineering and biological approaches. Engineering options such as annual contour ploughing, permanent terracing and bunding, furrowing, leveling, and construction of drainage structures require huge investments in the Gangetic plain. On the other hand, afforestation is a more gradual process that is self-repairing, self-perpetuating, and solar powered. Species of trees with high transpiration rates recommended for Mahnar by the Forest Research Institute are canes, *Salix temasperma*, and *Eucalyptus tereticornis*.[6] These may be grown on high mounds above water levels. Other species like *Bischofia favanica, Larger stroemia flex-regional*, and *Baningtaria actutangular* are also recommended. With high transpiration rates, the water table is lowered and land can be reclaimed for cultivation. Some new species would in turn generate new small-scale industries such as cane weaving and furniture making.

Afforestation is within the purchasing power of every farmer. Seedlings are available free of cost under social forestry schemes. The only limiting factor in Mahnar is availability of land, since land values have been gradually increasing. Discussion with local people shows that the farmers of chronically flooded areas will generally accept afforestation if it can give good returns. Further road and railway track arboriculture is recommended, as well as on bunds and high land adjoining floodable areas.

For both Hariharganj and Mahnar, new primary and middle schools and also adult education facilities are needed. The health care system should be restructured to emphasize community-level support and vaccination, as provided in the Revised Minimum Needs Program. Those innovations might increase transportation and communication needs. Introduction of walkie-talkies for communicating between health centers should lighten the transportation demand. Another health care need, for vaccine storage, could be met with mini-deep freezers, perhaps powered by photovoltaic cells and a battery system (Central Electronics Ltd. 1979).

Other needs include improvement and extension of housing, which would require effective utilization of local building materials. Some special needs for energy technologies (such as for a solar drier for drying chillies in Mahnar) are of secondary importance.

ASSESSING ENERGY OPPORTUNITIES

Intervention of several agencies in many sectors of rural life to assist with access to basic needs and income and employment generation could potentially bring about a total transformation of the matrix person-nature-society. In Mahnar and Hariharganj, this transformation would be brought about by the introduction of appropriate technological processes (society-nature), and feasible social-psychological processes (person-society) in order to bring about a biophysical process (person-nature) that would achieve the goals of rural development.

In this process, cost and benefits are more probabilistic than deterministic. For example, social mobilization to locate and donate land as well as to participate in construction of subhealth centers in remote villages would be much cheaper than land purchases based on assessed value and contractual building construction. Amount and kind of participation would vary according to each village; government norms and rules alone will not insure effective performance. Villagers' participation would be rewarded by the functions of their health subcenters. Accordingly, the summary of cost and benefits presented next is only indicative.

A windmill manufactured by the ORP, Ghazipur would cost Rs.5,100 per unit (Singh 1978). A chain washer pump manufactured at Naini, Allahabad, would cost Rs.2,200 per unit. Three units each can be transported in a truck for about Rs.2,500, including incidental costs. The training program of 14 person-weeks for eight persons (two supervisors each for one week and six technical workers each for two weeks) would cost Rs.1,400. Since no training is available for manufacture or servicing them, savonius rotor windmills designed by IIS, Bangalore could be installed in a second phase by people experienced in installing ORP windmills. The introduction of the savonius type windmill at a unit cost of Rs.3,000 would reduce the cost by about 50 percent and, after the first installation, the unit cost probably would decrease.

Before introducing a solar-powered chilling plant and cold storage, a detailed technoeconomic feasibility study should be done. The effective demand for cooling units is evident. The one cold storage plant in Mahnar at the time of the survey was not able to service the whole block. Neither block has a chilling plant. Milk production, potato production, and marketing would probably increase if these facilities were available. Mahnar could send supplies of milk to Patna; Hariharganj could send them to Aurangabad and Daltonganj. More production probably would generate more income and employment opportunities. Marginal and small-scale farmers might be able to use a milk cooperative. A chilling plant, in turn, might kindle organization of cattle improvement, fodder production, and veterinary services.

Each seedling for the government's afforestation program would cost less than Rs. 1, but the organizational and demonstrative effort would require considerable planning, as well as participation of local people. One main benefit would be eco-development — prevention of flooding and soil erosion, reduction of dust nuisance and airborne disease, increased agricultural production, and increased income and employment. Costs of the biodung program envisaged probably would be almost nothing except for organization and demonstration, but it would probably generate employment and subsidiary income and also prevent encroachment on forestland.

A coal-briquetting plant would cost about Rs.15,000 with the machinery available from Howrah, West Bengal. This plant would encourage increased income and employment and would reduce dependence on forested areas.

ORGANIZATIONAL STRATEGIES

Implementation strategies for rural development can emphasize the role of bureaucrats, technologists, and politicians, or of the intended beneficiaries. Emphasizing one group's role often means neglecting another's potential. Either inefficient and inflexible approaches not appropriately oriented to goals or unequal or unfair opportunities for utilization of benefits might result. Several factors interact to continue those patterns: caste and class grouping, bureaucracy's nature and the intended beneficiaries' nature, lack of microscale planning, and lack of participation of the poor in planning and implementation.

The social and political power in Mahnar and Hariharganj is concentrated with the rich, high caste Hindus and the bureaucracy is generally sympathetic to the needs of that power structure. Bureaucracy may not always be sensitive enough to goals or achievements. The intended beneficiaries of rural development projects, on the other hand, might be illiterate and less organized. And they might not have participated in problem solving. This project has tried to involve them in the development of solutions by using survey research.

Organizing rural development should be based on a humanistic, participatory, self-reliant, and autonomous process. This approach should be oriented to the community rather than the individual. Even if policies are made for the benefit of individuals in the intended beneficiary group (for example, ownership of pumpset by a marginal farmer), the individuals in that group tend not to take full advantage of these policies due to several socioeconomic and political constraints. A group of marginal farmers can effectively take advantage of such schemes since they can be mutually supportive. Such a group can either coexist harmoniously with other economic groups or can develop toward a conflict over power. In rural Bihar, the harmonious approach — coexistence — seems to be more workable.

Organization for rural development can be designed to depend on outside (governmental) help, or it can be built to be autonomous and self-reliant. A dependent organization tends to decline and degenerate as support is withdrawn. A self-reliant, autonomous organization shows increasing capability with time and maturity. The self-reliant development style, to which the Indian government is committed, is recommended on the basis of this project experience.

We recommend an area-specific, integrated, and intensive organization for rural development as one by its nature of organic evolution utilizing feedback that can determine effective development strategies. We recommend a participatory organizational strategy for Bihar, with beneficiaries involved in analyzing planning issues and problems and their technological solutions. They should be involved in training, implementation, management, and replication of technological modules for development with limited and normal input from outside sources and the government. The resultant organization could take careful advantage of all existing rural credit policies and legislations.

This research project has already laid down the seeds for organizing implementation in the form of good local contacts. Preliminary area development plans have been presented to block residents in panchayat meetings. A larger information base about the local area is available in data tapes. Some generalized project feasibilities have been worked out. Within the plan framework, individual programs need to be designed at a project level through a participatory process. Local organizational details need to be worked out before individual projects are undertaken in a specific area. Neither local problems of implementation nor the long-range perspective can be fully anticipated before embarking on each project.

At least a five-year stay at the block is required. Although implementation is possibly their secondary interest, the research team at CSDS is eager to extend their limited help to participating in the organization of local people for project implementation in association with the Institute for Rural Development, Patna. They were not willing to accept the limited help from the Institute of setting up certain technologies without a longer commitment to stay at the study area. However, the CSDS research team has identified local people, institutions for training the local people, credit availability for implementation, and alternative hardwares. The research group is interested in utilizing existing credit policies and regulations, and they seek an administrative and travel grant from Indian institutions. Negotiations are under way and one possible sponsor is awaiting completion of the plan document. Another sponsor declined to support a research organization's involvement with implementation.

LESSONS FOR PLANNING RESEARCH

A combined survey strategy of village and household census was adopted for integrated rural development planning of Mahnar and Hariharganj blocks with 18,100 households and 246 villages. Six research workers completed this task within 13 months with the voluntary assistance of local people and government staff. The friendship and understanding the research workers developed through this participatory process of survey research was a matter of great satisfaction to the researchers as well as to the villagers. The slow pace of work was appropriate for rural life and was conducive to dialogue that enriched the awareness of the research workers. It also helped to identify key people, places, and groups who could participate in the pilot implementation.

Shortcuts to census would have meant sampling or total dependence on secondary data. Neither was acceptable to the research team. No additional funds were spent for the census, so assessing the value added or benefit-cost between 100-percent sample and any smaller sample would be difficult. Only one additional month was spent compared to the time needed for a sample survey. No payment was made in money or kind to the voluntary investigators.

The village census depended on several secondary data. Often the data had to be verified. In addition, new data had to be collected, such as water depth in a village well during the summer and rainy seasons, or locations of particular services. Livestock census and total enrollment of students were available from secondary sources, but no information existed regarding ownership of livestock or whose children attended schools. So a household survey became necessary. Much of the secondary data compiled at block offices was not available from official records there because all copies had been forwarded to district or state authorities.

Sampling is a very widely accepted survey research methodology. However, it was difficult to accept sampling in this work in view of the perceived (though not precisely understood) complexity of village culture and society. Criteria for operationalizing the definition of complexity, and how it is additive in a block or national culture, are unknown. Administrative boundaries might be ideal for systematic data collection, or for drawing sampling frames, but they might be geographic fiction, as in the case of Kerala villages, or they might not reflect tribal or caste lineage, dialect, or other forces of tradition that have been evolving through several centuries.

The physical inaccessibility and cultural heterogeneity of population groups as well as inaccurate secondary data are the most limiting restrictions in sampling design (Gerald Hursh-Cesar and Prodipto Roy 1976). In effect, there is no known statistical probability for selecting a subculture. Theoretical probability samples do not permit personal judgment in selection. In a probability sample, every unit in population must have a known non-zero chance to be included in the sample. Selection is procedural not judgmental; and principles of randomization must operate at every step of the selection process. This is difficult in a block with several villages with over 40 castes or occupations. Often accurate lists of households are not available for villages. Nonresponse substitutions are the rules of sampling, but no theory exists regarding substitution; random samples cannot be substituted. Probability samples are essentially unbiased. However, in development research, the nonprobability biased or purposive sample is the most prevalent form. Therefore, in the past, whenever researchers were commited to implementation through wide popular participation, they used the strategy of household census, especially if the quality of the implementation structure could not be perceived in advance, such as with the Ranaghat Rural Development Project (CSD 1978). A sampling strategy was more acceptable (CSD 1977) for cases such as the Ghonda Project where a well-organized nongovernmental implementing structure was perceived to be functioning at the grass roots. That strategy also seems more acceptable if the purpose of the survey is to find out income distribution and employment for broad policy research (Kumar 1980).

The type of study area selected is also important. In India, community development and extension blocks make up the project areas used for rural development. So they are the appropriate type of unit for studying integrated rural development planning. Areas appropriate for other study purposes might not correspond to blocks. For energy planning, the spatial spread of individual microenergy systems would influence delineation of planning regions. For example, planning areas related to biomass production and use might be derived from the jajmani system of social division of labor and therefore from interhousehold, intervillage, or interhamlet relationships. On the other hand, study areas for supply planning of commercial sources of energy such as electricity, coal, or kerosene would be defined administratively or commercially. For an energy and rural development program combining several microenergy systems an administrative unit would be an appropriate planning area. The some 225,830 gram panchayats in India give too large a universe for selecting appropriate samples. On the other hand, districts numbering less than 400, although administratively viable, are too big and have too many ecological dissimilarities for such studies. The community development block, though, is an appropriate compact unit of administrative and political structure.

India's 5,004 development blocks form an appropriate universe for a sampling strategy for studying energy and rural development, but they must be divided into a typology of homogeneous regions for sample selection. Homogeneity can be based on agroclimatic regions, bioresource patterns, conditions such as recurring droughts or floods, or levels of regional development. The dynamism of the block as it affects energy flow patterns and rural development may vary considerably depending on typology of blocks. For example, the so-called reserve territory blocks are characterized by weak development of productive forces. Blocks of pioneering economic development are characterized by selective exploitation of natural resources, on a limited scale. Large and powerful foci are blocks situated in close range of large or powerful economic complexes.

If energy is considered a delivery of a service, then blocks can be regionalized by energy facility indicators and energy achievement indicators. The pair of canonical variables giving the largest canonical correlation between these groups are used as measures of facility and achievement. After the scores are computed, the blocks can be categorized.[7] Such regionalization and taxonomy are useful applications of quantitative methods. Therefore, a nation could be sampled on the basis of regional typologies. In-depth studies, including census, could be conducted in sample blocks (or equivalent local areas), to formulate plans and policies on energy and rural development. The planning experience gained from such studies could provide valuable insight for solving national problems of energy and rural development faster and more efficiently.

NOTES

1. Panchayats are the primary political divisions. Each consists of several villages represented by elected officials: *Mukhya*, the village executive and *Sarpanch*, the village judiciary.
2. Scheduled castes and scheduled tribes are communities listed in the Indian Constitution whose interests are constitutionally protected. They receive special reservations in Parliament, in educational institutions, and in various employment markets.
3. Jajmani: traditional labor exchange and compensation system.
4. The landholding unit of marginal farmers is less than 0.5 ha irrigated land, of small farmers, 0.5 to 1 ha irrigated land; and of viable farmers, more than 1 ha irrigated land. (One unit irrigated land equals two units unirrigated land.)
5. Rs. 7.93 = US $1.00.
6. Ghosh, R.C., Director, Forest Research Institute, Dehradun, April 28, 1980 and July 25, 1980, personal communications.
7. A similar intrablock regionalization was conducted for delivery of education in a block (P. Dasgupta, T.M. Vinod Kumar, and P. Ramaswamy 1974).

REFERENCES

Central Electronics Ltd.
    1979   *Solar Photovoltaic Project: Annual Progress Report No. 3, April 1978–March 1979*. New Delhi: Department of Science and Technology.

Centre for Policy Research and Family Planning Foundation.
    1983   *Population, Poverty, and Hope*. New Delhi: Uppal Publishing House.

Council for Social Development (CSD).
    1977   *Indicative Plans for Gainsari and Pachperwa Blocks, Ghonda District U.P.* New Delhi.

CSD.
    1978   *Ranaghat Rural Development Project: Feasibility Studies Vol. 1 to 6*. New Delhi.

Dandekar, V.M., and N. Rath.
    1979   *Poverty in India*. Poona, India: Indian School of Political Economy.

Dansereau, P.
   1970   *The Dimensions of Environmental Quality.* Prepared for the Commonwealth Human Ecology Council Conference. Malta.

Dasgupta, P., T.M. Vinod Kumar, and P. Ramaswamy.
   1974   *Planning for Education in a Micro Region: An India Case Study.* New Delhi: The Ford Foundation.

Ghosh, R.C.
   1977   *Handbook of Afforestation Techniques.* Delhi, Government of India: Controller of Publications.

Government of India (GOI), Planning Commission.
   1973   *Task Force on Integrated Rural Development: Integrated Agricultural Development in Drought Prone Areas.* New Delhi.

GOI, Planning Commission.
   1979   *Yojana Special Issue on Poverty.* New Delhi. (August 15, 1979)

GOI, Central Ground Water Board.
   1980   *Status of Ground Water Survey and Investigation for Further Ground Water Development in the State of Bihar.* New Delhi.

Hursch-Cesar, G., and P. Roy.
   1976   *Third World Surveys: Survey Research in Developing Nations.* New Delhi: The Macmillan Company of India, Ltd.

International Labour Office (ILO).
   1979   *Profiles of Rural Poverty.* Geneva.

ILO.
   1980   Poverty and Landlessness in Rural Asia. Geneva.

Kothari, R.
   1979   *Rural Development Issues and Perspectives.* Special Number on Administration of Rural Development Programmes. *Administrative Change* 6(1–2), Jaipur.

Kumar, T.M. Vinod
   1980   *Basic Needs and Provision of Government Services: An Area Study of Ranaghat Block in West Bengal.* World Employment Research Working Paper No 85. Geneva: ILO.

Lele, U.
   1975   *The Design of Rural Development. Lessons from Africa.* Baltimore: The Johns Hopkins University Press.

Michael, A.M.
   1978   *Irrigation Theory and Practice.* New Delhi: Vikas Publishing House Private Ltd.

Nanjundappa, D.M.
   1981   *Area Planning and Rural Development.* New Delhi: Associated Publisher.

Raju, S.P.G., and R. Narasimha.
    1979   *A Low Cost Water Pumping Windmill Using a Sail Type Savonius Rotor.* Fluid Mechanic's Report 79 FM 2. Bangalore: Indian Institute of Science, Department of Aeronautical Engineering.

Rao, B.S., and V.N. Deshpande.
    1982   Poverty: An Interdisciplinary Approach. Madras: Madras Institute of Development Studies and Somaiya Publication.

Singh, D.M.
    1978   *A Manual On Windmill Technology.* Windmill Project. Allahabad: Allahabad Polytechnic, A.T. Unit.

Singh, T.
    1978   *Drought Prone Areas in India: Aspects of Identification and Development Strategy.* New Delhi: Peoples Publishing House.

United Nations.
    1978   *The International Bill of Human Rights. Universal Declaration of Human Rights. International Covenant on Economic, Social and Cultural Rights. International Covenant on Civil and Political Rights and Optional Protocol.* New York.

7
Organizing For Energy Need Assessment and Innovation: Action Research in Nepal

Deepak Bajracharya

A CONCEPTUAL FRAMEWORK FOR RURAL ENERGY DEVELOPMENT

How to *implement* effective measures in dealing with energy-related problems in rural areas of the Third World is a serious concern shared by many people. Elaborations of rural energy issues have been refined recently. Starting with the recognition in the mid-1970s of the dominant role of "noncommercial" fuels in the rural Third World, subsequent rural energy surveys in different parts of the world (see, for example, Chapters 2 through 6 in this book) have brought forth a wealth of information regarding the complexities of fuel consumption patterns, fuel supply potentials, and their implications for rural futures.[1] Several technological measures also have been prescribed to correct the situation. Three of the more commonly mentioned technologies include improved stoves (see Chapter 8), fuelwood plantation (Chapter 9), and biogas (Chapter 10). The potential contributions of these and other related technologies in confronting energy problems in rural areas are undeniable. The challenge, however, is to ensure their compatibility with the needs and priorities of the people in rural areas and consequently their absorption from within.

This challenge is not unique to energy technologies. It is germane to overall rural development efforts. The problems that develop in implementation generally stem from a process of externalization, that is, activities such as identification of rural development issues, determination of priorities, conceptualization and planning of action programs, all take place outside the purview of the people for whom the programs are intended. Even though some rural development theorists recognize how important people's participation is, the "blueprint approach"[2] continues as the basic dogma in rural development efforts, energy-related activities being no exception. The control and decision mechanisms remain foreign to the local populace. At best, local people's interests and needs are intellectualized by the planner and plans that supposedly embrace people's needs are assumed to be "appropriate." Using these criteria, "appropriate technologies" are conceived in such terms as "capital saving," "labor intensive," "small scale," and "basic needs oriented." Ironically, local people and planners alike still see these well-intentioned schemes as "interventions" and development planners tend to be preoccupied more with the process of meeting "intervention targets" than with the process of transforming interventions into innovations.[3]

This chapter advocates the basic premise that the process of conscious transformation must be the focus of rural development efforts and that community organizations play the key role in determining the absorption of innovation within the rural milieu. Within this framework, then, energy technologies, their importance, and their implications are a part of the overall scheme of rural development. The need for the strengthening of local organizations and their capacity to influence external development agencies is understood in the context of these assumed conditions:

1. People in rural areas are already witnessing energy problems among other problems. In their own ways they are seeking solutions that will help curb the problems;

2. People in rural areas have been unable to tackle those problems using only their own internal resources and they need selective external assistance in the form of information flow, capital flow, and material flow to help develop and mobilize their own internal resources; and

3. Direct interactions are needed between insiders (local people) and outsiders (technology designers and suppliers, government support personnel and agencies, donor consultants, and agencies) to bring forth technological processes that match local needs and priorities and meet requirements related to technological parameters and regulations of donor and government operations.

Those conditions provide an orientation to user-directed technology design and technology diffusion. They provide guidelines for fitting the technology to meet demands induced by people's needs and priorities. This approach differs from the one that promotes technology design and diffusion by designers and rural development agents with local people expected to change their organizational structures to fit the criteria and attributes of the new technologies. In place of this interventionist approach, the alternative seeks to help bring about "innovations" within rural communities and help people fend for themselves against the emerging problems.

Effective new means of communication with external policy agencies are essential, and in many instances external assistance is also required. The central dilemma of development, according to Amulya Kumar Reddy (1980), is that, "the traditional [system] has ceased (or is rapidly ceasing) to be adequate, but the modern [system] is invariably inaccessible except to a few." This dilemma then defines the crucial role of science and technology, namely, to "generate a variety of adequate and accessible options." The process to make these options appropriate to the village context is achieved only by promoting interactions of these three principal parties: the user, the designer, and other support agencies of the government or quasi-government institutions. Such interactions are important steps toward internalization and integration of the process of technology development and implementation. Without these interactions, those three parties might look at the process with three different and perhaps inconsistent worldviews.

The processes of interaction are illustrated in Figures 1, 2, and 3. The first stage of interaction deals with the identification of the new technology to counteract the prevailing problems (Fig. 1). Since all problems cannot be solved at once, the people in the rural community would interact among themselves to determine their own priority and express the need for the new technology. A range of technical options can be assumed to exist based on present knowledge. The important test is to see if need and technical options can be matched, or if not, to see if the user and designer can interact together to modify the technology as needed or to set criteria for long-range research and development plans.

The second step is to assess technical feasibility and to work out implementation strategies through mutual negotiations among all concerned parties (Fig. 2). The focus here is on identifying resource needs and determining which parties are in the best position to contribute the required resources. The use of internally available resources should be maximized through the rural people's cooperation. The technical expertise from outside is assimilated within. When internal resources alone are inadequate, external assistance in the form of a loan, a subsidy, or a grant is sought. Negotiations among all three parties involved are clearly quite crucial before implementation strategies are developed.

As Figure 3 shows, implementation strategies are the direct outcome of commitments made by local people as well as by technology designers and donor agencies. The main focus during implementations is to examine whether socioeconomic objectives as determined by local people are fulfilled and whether the technical performance is operated satisfactorily. Monitoring and impact assessment through interactions of all concerned parties are important at this stage to determine whether the program is to be sustained and expanded as appropriate or whether some alternative strategies should be proposed.

The interactive three-stage process just described can be characterized as action research.[4] The emphasis is on the experiential learning

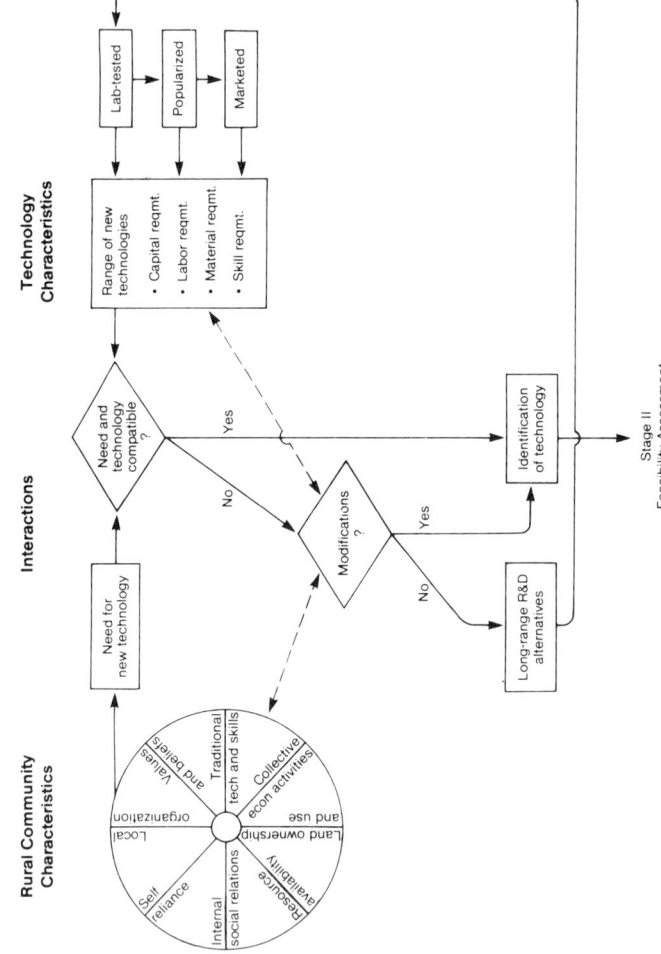

Figure 1. Technology Identification

STAGE 2. FEASIBILITY ASSESSMENT AND NEGOTIATIONS

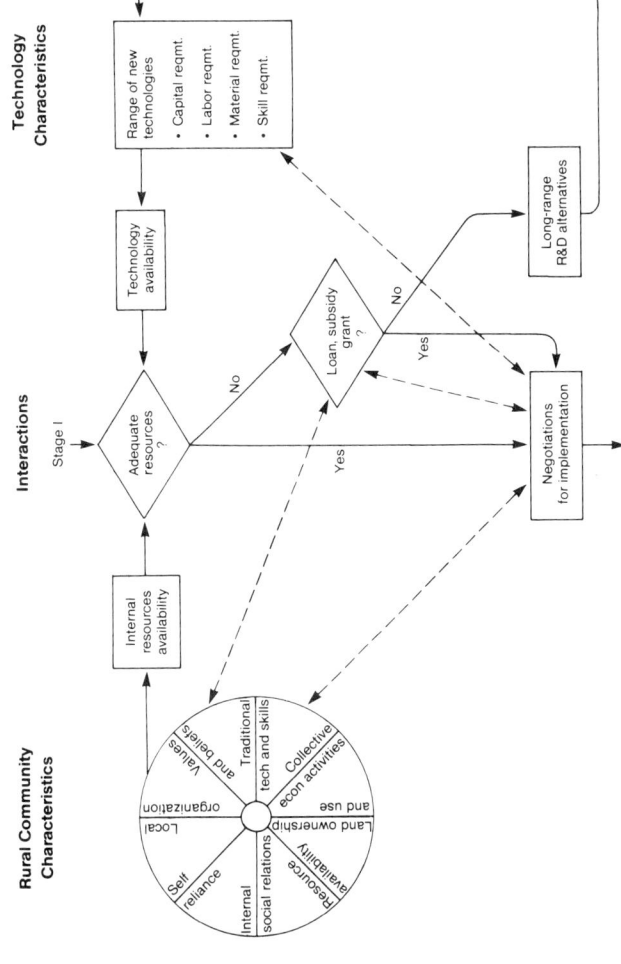

Figure 2. *Feasibility Assessment and Negotiations*

STAGE 3. IMPLEMENTATION STRATEGIES AND MONITORING

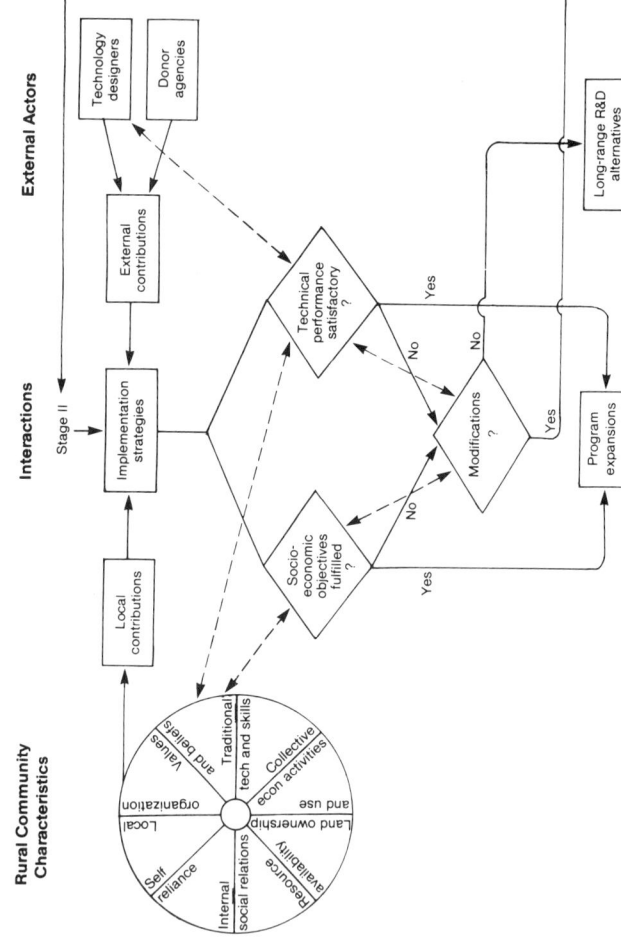

Figure 3. *Implementation Strategies and Monitoring*

approach to feed back results for modifying and improving the operational program as a continuous process of action and reflection influencing each other (Paulo Freire 1973). This principle of operation is based on dialogue among all actors — the local people from rural communities, and the technology designers and the donor agencies from outside the rural communities — to be sensitive to one another's needs and priorities in relation to diffusion of technologies. At the same time, the emphasis is placed on building up local people's capabilities to absorb technologies so that the technologies which are initially interventions are eventually assimilated within as innovations. The end result is then like a marriage between the insiders and the outsiders in a compatible manner for mutual benefit and satisfaction.

The interactive process proposed is not yet an accepted norm; this means that a key individual who acts as an intermediary becomes a crucially important catalyst. He or she facilitates the "marriage." In Asian countries where arranged marriages are still being practiced, the intermediary can be described as a matchmaker (*lami* in Nepali is one such local term). Quite understandably, the lami must understand the perspectives of the insider as well as of the outsider and harbor sufficient trust from both parties before he or she can make the best match.

The crucial question now is how to make the marriage happen within the present rural context and also the operational structure of the government agencies, other nongovernment or quasi-government agencies, and the technology designing institutions. A shift from the legacy of the "blueprint approach" is necessary. Considering the growing recognition that an alternative is required to make rural development activities effective (Bharat Pradhan 1982), the experiential learning approach is proposed here as a worthwhile undertaking. An experience in a Nepalese village is presented below to give credence to this concept. In this case, I acted out the role of a lami (1) in identifying the desired technologies, (2) in negotiating for implementation, and (3) in carrying out implementations by building up local control in operations and decision making.

BACKGROUND FOR ACTION RESEARCH

Chhoprak Village Panchayat in Gorkha district, Western Nepal, provides the setting for the action research program.[5] The panchayat lies from 28°1'N to 28°5'N latitude and 84°34'E to 84°37'E longitude. It is bounded by the Khahre Khola (stream) to the East, the Bhusundi Khola to the West, and the Daraundi River to the South. A mountain ridge marks its northern border. From the Daraundi River basin at the south to the top of the mountain ridge at the north, the altitude varies from 400 to

1,700 meters elevation. The area covered by the panchayat approximates 22 sq km. According to the survey conducted by New ERA[6], Chhoprak Panchayat consists of 925 households indicating a population of about 4,800.[7]

Community Characteristics

The population is made up of several castes and ethnic groups: the Kumhals predominate at the low elevation (500 to 700 m); the Brahmins and Chhetris, at the middle elevation (800 to 1,000 m); and the Gurungs, at the high elevation (1,400 to 1,600 m). Interspersed within are other minority groups including Newars, Barams, Kamis (blacksmiths), Sarkis (cobblers), and Damains (tailors). In terms of socioeconomic status, Brahmins, Chhetris, Newars, and Gurungs would in general be at a relatively high position; Kumhals and Barams at a relatively low position; and Kamis, Sarkis, and Damains, the lowest (i.e., the untouchables).

Subsistence agriculture is the prevalent economic mode of production for all the people, regardless of their castes. The crops produced in the area include rice, maize, millet, wheat, pulses and lentils, oil seeds, and potatoes. As secondary sources of income, (a) about 25 people own small shops for tea and sundries (these are located in mainly two bazaar areas, Chorkate and Maidan); (b) about 25 teachers are employed in one high school, one lower secondary school, and two primary schools; and (c) the Kumhals living at the low elevation fish along the Daraundi River and the Bhusundi Khola.

Fuelwood is the major source of energy for cooking. *Chulos* (semi-enclosed mud stoves) and *agenas* (open hearths) are about evenly distributed. No feeling of immediate fuelwood scarcity is yet noticeable. Kerosene is used sparingly for lighting *kupis* (wick lamps). Homemade mechanical devices are used for rice hulling (*dhiki*), grinding grains (*jato*), and extracting oil and molasses (*kol*). Water supply is available within 15 to 20 minutes walk from the household but is relatively scarce during the dry months. Traditional irrigation canals (*kulo*) are limited and the fields are mostly rainfed.

Access to Chhoprak is not very easy, although it is not as remote as some northern panchayats in Gorkha district. The panchayat headquarters is located at about four hours walking distance from the district headquarters and about three hours walking distance from the nearest motorable road (13 km along the 25 km road from Abu Khaireni to Gorkha).

On the way to Chhoprak Panchayat. (Banoo Bajracharya)

The Energy Project Objectives and Scope

The action research project on energy in Chhoprak was conducted from February to July 1982 as a joint undertaking of the Resource Conservation and Utilization Project (RCUP)[8] and the Resource Systems Institute of the East-West Center. Chhoprak was chosen after a reconnaissance study tour, at the end of December 1981, to five out of six "lead" panchayats in Gorkha district.[9] The factor that influenced the choice was the opportunity to carry out the project in three sites of Chhoprak at different elevations with three different population compositions. The objectives of the project were

1. To test the compatibility of selected energy technologies in village conditions. The technologies included smokeless stove, haybox cooker, ceramic, sawdust and other stoves, solar water heater, solar crop dryer, solar cabinet dryer, solar kiln, hydraulic ram, improved water mill, pedal thresher for paddy, and biogas;

2. To implement technologies in selected communities by determining

their feasibility through local people's participation in all the phases — such as identification of technology, planning for implementation, and decision making for benefit-sharing; and

3. To assess the practicality of extending the methodology demonstrated in Chhoprak to other areas.

Assurance was given to me by RCUP prior to the beginning of the field action program that a maximum of Rs. 67,000 (US $5,100) would be made available for the implementation of energy technologies if and when they are identified as appropriate by the people of Chhoprak.

As initially conceived, three communities were chosen in Chhoprak based on population distribution according to low, middle, and high elevations and the corresponding distribution by castes and ethnic groups. The focus was on (a) the Kumhal group in ward 9 at the low elevation; (b) the Brahmin, Chhetri, Newar, Sarki, and Kami groups in wards 5, 6, and 7 at the middle elevation; and (c) the Gurung group in wards 2 and 3 at the high elevation. Before detailed discussions were started separately with the three groups, a deliberate effort was made to present the rationale of the energy project at the general meeting of the Panchayat Conservation Committee[10] made up of representatives from all the wards of the panchayat together with prominent members representing the farmers' caucus and the women's caucus. Village-level extension workers, such as the soil conservation assistant, junior technician for agriculture extension, junior technician for livestock development and animal health, and the forest ranger were also represented in the Committee. This approach of entry into the village was adopted partly to formalize my presence in the panchayat, partly to familiarize people with the role of energy in the context of resource conservation and utilization, and partly to notify representatives of the wards concerned that I would be coming into their communities for detailed discussions. The endorsement from this general meeting proved to be useful in freeing people from any suspicion regarding the legality of my presence within the panchayat and also in establishing contacts with prominent members of many wards for future meetings.

Subsequent meetings were held in smaller groups of representatives from the selected wards. The discussions emphasized certain specific topics: (a) communication of advantages and disadvantages of all the technologies in terms and concepts understandable by local people; (b) participation in the discussion of as many strata of people as possible, not merely the rural elites from the existing political hierarchy, the traditional social hierarchy, and the opposition factions: (c) promotion of self-reliance (i.e., RCUP assisting locally initiated efforts) instead of fostering the so-called aid mentality (i.e., RCUP solving their problems); and (d) evaluation of traditional internal organizations and viewpoints to facilitate adoption of new technologies.

The discussions were pursued along three types of reasoning. First, each of the technologies can potentially help rural people to improve their existing conditions. Second, all of them are virtually untested and, therefore, in spite of their potential benefits, the people need to decide whether or not they are willing to take the inherent risk associated with their adoption. Third, when people choose one technology and contribute locally available resources, RCUP will contribute in bringing external resources as appropriate.

The responses from each of the three communities after the initial discussions were different. Each of these responses as will be documented below is quite instructive about diffusion and absorption of energy technologies.

RESPONSES OF THE KUMHALS

Need and Technology Identification

Remarkably, after an initial discussion with them about various energy technologies, the Kumhal community identified their own needs for an improved water mill. However, adequate water resources were not available to turn the water wheel. The generation of about 7 hp (5.2 kW) requires, for example, 70 liters/second (1/s) waterflow and 10-meter head (Balaju Yantra Shala, n.d.), but the perennial streams around the community have only about 1 l/s at low head. Even though they recognized the scarceness of water supply they held onto the idea of the mill as the most appropriate energy technology discussed at that meeting. These anecdotes express some of their strong feelings.

> We need the mill to improve the future of our children. We [the women] spend our time every day from the moment the cock crows till eight or nine in the morning to mill the grains. Sometimes, we do not finish the milling in time to cook and feed the children before they go to school and the children go to school with empty stomachs.
>
> Our women are so preoccupied with their housework including the milling of grains that they have no time for their education. We [the men] are so ashamed that they cannot articulate their ideas to you even though you wanted to get their input in the discussions. We hope that by saving on their

labor, we may be able to get them involved in some activities similar to those of the Women's Training Centre in Kundhur [a nearby village].

The whole family including men and women spend the entire day in trying to extract one *mana* out of one *pathi* [8 *manas*] of mustard or sesame seeds whereas we know that in the Katahari mill [at about 45 minutes walking distance] they can get at least twice as much without having to suffer through the physical stress.

The mill owners at Katahari and Taku are making so much money that we think the mill must be a profitable venture.

The mill seems to be the only technology in your list which will benefit all the people from all the five Kumhal villages. If we choose another technology we are afraid that it will benefit one or two groups and leave out others. We don't want conflicts which might split our togetherness.

The idea of having a mill in the community became so strong that everybody was talking about it. Some of them considered operating it with diesel if nothing else could be done, but I warned them about the low profit margin and relative inaccessibility of the fuel. The challenge, therefore, was to find a viable alternative for fuel so that the mill could be run at a profit. For this reason, the process of biogas generation was seriously considered. A complex set of negotiations was, however, necessary for assessing the implications of implementation.

Negotiations During Planning

Three important questions had to be resolved. (1) Is the idea of biogas generation to run the mill (including a rice huller, an oil expeller, and a grinder) feasible from the technical and economic perspective? (2) Who will finance which component and how? (3) How would the community be organized for implementation, decision-making, and control of operations? The role of internal and external actors (including donor agencies and technology suppliers) had to be critically examined and integrated.

Gobar Gas Tatha Krishi Yantra Vikas Private Limited (Gobar Gas Company, for short) was the only institutionalized supplier of biogas plants in the country. Technical consultations were, therefore, carried out with them. From the discussions, several affirmations were possible:

Discussing the terms of Agricultural Development Bank loan. (D. Bajracharya)

1. The company had already tried out successful implementation of running biogas through a dual-fuel engine to operate the rice huller. Addition of the grinder and the oil expeller would technically be no problem.
2. The 7-hp engine can be adapted to a dual-fuel operation to use 80 percent biogas and 20 percent diesel as fuel.
3. Approximately 3 m³ of biogas (100 cubic feet) and 0.2 liters of diesel would be required to run the engine for 1 hour. Hence two biogas plants of Chinese dome design with 20 m³ internal volume would produce sufficient gas for about 5 hours of operation per day.
4. Approximately 300 kg of fresh *gobar* (cow dung) (or an equivalent of 15 tinfuls of gobar; one tin = 20 liter capacity) would be required.
5. The cost of biogas plant construction, the mill machinery, and its installation is expected to be about Rs. 80,000 (US $6,100).

Although, I, as lami, initially made the contacts with the Gobar Gas Company and got them interested in a "possible project," full discussions were carried out by bringing together the company people and the Kumhals. This became possible when the company agreed to send a team consisting of two members of their staff to the village. On the day of their arrival, a meeting was announced in which a large number of Kumhals attended. The technical aspects of the biogas plants were put forth by the company people and very insightful questions were asked by the local people.

The idea of generating gas from gobar to run the mill sounded bizarre to the village people. Moreover, they could not believe that the gobar could be reused as farm yard manure after the gas was extracted. They kept insisting: "If the millet residue after extraction of alcohol (*raksi*) is unpalatable as human food, how can the spent slurry after extraction of gas be useable as good manure?" Since the gobar is greatly valued and was in short supply, their fear of losing 15 tinfuls of gobar per day was understandable. The people from the Gobar Gas Company and I found it difficult to explain the scientific facts — that the nitrogen and phosphorus content in dung (of nutrient value) was unaffected by the generation of methane and carbon dioxide, which together contain only carbon, hydrogen, and oxygen. Eventually, we had to ask them to trust us and promised them compensations if the value of gobar turned out to be useless.

The people also had some very poignant questions about the organizational aspects of operation and maintenance procedures. After a 4-hour meeting lasting late into the night, they apparently were still unconvinced. Yet they did not abandon the whole idea because, as they told me later, they could not believe that we outsiders had come so far to tell them about something that would not work. The consensus at the meeting was to continue to explore and to decide later.

The question of financing was also discussed. The RCUP energy budget alone was not sufficient to cover the entire cost. Sugestions were made to ask the Agriculture Development Bank of Nepal (ADB/N) to help. I was mandated to explore with the officials at ADB/N their program of providing 50 percent subsidy and 50 percent loan to cover the costs of constructing biogas plants. The constraint (I found out later) was that the scheme was being implemented only in the panchayats where Small Farmers' Development Programme (SFDP) was in operation and Chhoprak village panchayat was not one of them.[11] Nevertheless, after long discussions, the concerned ADB/N officials were persuaded to make an exception to the rule. The financial scheme was subsequently developed in this manner: (a) ADB/N would provide about Rs. 45,000 (US

TABLE 1.
Landholding Pattern Among Residents of Chhoprak Ward No. 9

Size of Landholding		Kumhals		Other Castes		Total	
Category	Range (ha)	No of HH	Area (ha)	No of HH	Area (ha)	No of HH	Area (ha)
Marginal	<0.50	21 (20)	7.6 (7)	21 (64)	4.7 (28)	42 (31)	12.3 (10)
Small	0.51–1.00	40 (38)	29.2 (27)	6 (18)	3.6 (22)	46 (34)	32.8 (26)
Medium	1.01–2.00	33 (32)	46.2 (42)	6 (18)	8.2 (50)	39 (28)	54.4 (43)
Large	>2.01	10 (10)	28.0 (25)	--	--	10 (7)	28.0 (22)
TOTAL		104	111.0	33	16.5	137	127.5

Note: (a) Constructed from 1981 cadastral survey records, Department of Survey, His Majesty's Government of Nepal.

(b) Numbers in brackets signify percentages.

$3,400) for the construction of two 20 m³ biogas plants: half as grant and the other half as loan repayable at 6 percent interest over a seven-year period, and (b) RCUP would provide the balance of about Rs. 35,000 (US $2,700) in the form of another grant for the project.

The next big question was related to community organization. As soon as the funding possibilities were confirmed, the Kumhals had to make a decision. Although they were reluctant to seek a loan from the bank (having never done so before), they did not want to let go the opportunity of accepting government assistance (through RCUP) for their own "development." If they did, they felt it might be a long time before the government would turn to them again. At the same time they had ambivalent thoughts about "development." They remembered one "development" project (a new irrigation channel along the Daraundi River basin) that became the center of many land conflicts.

Compounded with their uncertainty about biogas technology and reuse of slurry as manure was their fear that if they provided their land as collateral for a loan, it could be lost. This could mean a substantial threat to their subsistence because they depend heavily upon the products of the land for their livelihood. As shown in Table 1, landholding among the Kumhals on the whole is quite small. Only 10 percent of the Kumhals (10 households) own more than 2 ha. Those who own less than 1 ha constitute 58 percent (61 households) of total population. When the entire ward is considered, the outlook is even worse.

The Kumhals also mistrusted government officials and were relatively isolated from many development activities. In terms of social hierarchy, the Kumhals were considered as belonging to a "backward" group (similar to the *adivasis* in India) and the Kumhals themselves were self-conscious about this. Consequently most of them were visibly restrained in front of people from outside about voicing their opinions about their needs and priorities. They themselves described how, until recently, they tended to keep away from any outside visitors. Their inhibitions were very much linked to the history of suppression, manipulation, and exploitation of the Kumhals by the Kumain Brahmins of the vicinity and others from outside. From that perspective, their reservations about loans were understandable. They felt that they had always been tricked. They did not want to fall into any trap. They asked many questions about the loan: "Why should we take the loan? Isn't the American Government giving money to our government as grant and not loan? Why does not the government give us money in the same way?" Only several discussions convinced them that (a) budget constraints with respect to the energy project meant settling for a smaller and less expensive project if they did not take the loan, and (b) I was not a part of any conspiracy to confiscate their land.

After overcoming those initial reservations about a loan, we had to establish the basic economic soundness of the project. A detailed assessment was worked out in the presence of the people. The people furnished the data for realistic amounts of grains that are likely to be milled by the Kumhals alone. Milling rates were derived from the people's own experiences at the mill in Katahari, which is located about 45 minutes walking distance. Conservatively estimated, revenues from milling those grains, as shown in Table 2, amounted to about US $1,830. Annual costs also were calculated conservatively and this amounted to about US $1,740. The net cash flow in a year amounted to about US$100. This indicated a no-loss situation to the people because they were certain that more grains from outside their immediate community would be milled. If the mill works at full capacity, say 8 hours per day instead of 5 hours as is the case with the amount of grains indicated, the total revenues increase up to US $2,900 and the costs to only US$2,000. In both cases, the benefits do not include such items as saved labor or greater amounts of oil extraction. With these figures in hand, a sufficiently large number of people took interest in the project. Despite some dissenting voices, the general feeling was "We must not give up the project at any cost."

While the process of negotiation continued, some questions lingered in my own mind: Were the Kumhals in a position to cooperate together as a group? Did they have any precedence for community involvement in

TABLE 2.
Annual Revenues and Costs of the Biogas/Mill in Chorkate

	Annual Revenues	(US$)
(a)	Dehusking 73 metric tons of rice (@$1.00 per 180 kg)	405
(b)	Grinding 48 metric tons of grains (@$1.00 per 50 kg)	960
(c)	Expelling oil from 7 metric tons of oilseeds (@$1.00 per 15 kg)	467
	Total revenues	1,832
	Annual Costs	
(a)	Loan repayment (@6%; 7 years)	305
(b)	Labor costs (24 person-months @ $30/mo and 24 person-months @ $15/mo)	1,080
(c)	Diesel costs (300 liters @ $0.50/liter)	150
(d)	Repair, maintenance, replacement	200
	Total costs	1,735

adopting innovations? They were answered by learning about two instances of very important cooperative undertakings in the community. One was a recent purchase of a sugarcane crusher. Before its purchase, sugarcane crushing was done in the same wooden press used for extracting oil from oilseeds, an admittedly very tedious and inefficient system. Somehow, they heard about a machine in the market that did the same job but much more efficiently. What I found remarkable was that the group decided on voluntary contributions from any interested person to accumulate Rs. 2,800 (US $210) for the purchase. The contributions ranged from Rs. 5 or Rs. 10 rupees by the poor households to Rs. 100 or Rs. 200 by the better-off households and up to Rs. 1,000 by one rich household. Despite the variation, each contributor was entitled to do his own sugarcane crushing at no additional cost. The person who contributed Rs. 1,000 was given the privilege of keeping the machine at his house and anyone who wanted the machine had to make necessary arrangements with him. As far as I could gather, the system was working out very well. I also saw the crusher used by noncontributors for a rental fee apparently redistributed to contributors according to the number of shares they hold.

Another interesting community involvement involved the establishment of a community primary school about 20 years ago (it has since then been expanded to a lower secondary school). The idea was first conceived by Jit Bahadur Kumhal while he worked in the Indian army. He was greatly embittered at the time because he could not advance in his position along with his peers because of illiteracy on his part. He decided, "My fellow villagers must suffer no more as I did on account of educational disability." He came back to the village and made a personal commitment to build the school. In spite of great resistance from his own people, Jit Bahadur persisted, and finally succeeded in setting up the school with people's cooperation and contribution. Jit Bahadur today is an uncontested leader of the community and he takes special pride in the success of the school. The lessons learned by Jit Bahadur as he told me in his own words are as follows:

> Getting the people to work together in a project, no matter how good it is, is a difficult task. It is possible, however. The important thing is that those who take the primary responsibility must not expect instant recognition. They must be prepared to receive abuses.

In this context, Jit Bahadur's enthusiasm about the biogas/mill project was exciting. He also had the support of two other strong leaders this time, Bhot Bahadur Kumhal and Rajendra Bahadur Kumhal. I felt that, with the apparent commitment of the trio, the necessary community organization would be forthcoming.

Promises for Local Contributions

Still more specific promises from the villagers were necessary:

1. Who would provide the land for collateral to procure ADB/N loan?
2. Who would provide the dung for the biogas plant operation?
3. Where would the biogas plants and the mill be located and who would provide the land for the project site?
4. How would the contribution of labor be arranged for carrying out collection of locally available materials, transportation of machines and materials brought in from outside, and construction of biogas plants and installation of mill machinery?

Additional public meetings were held to elicit people's commitments. Regarding collateral, it was first decided that contributions from *all* the

TABLE 3.
Distribution of Collateral Land Contributors and Other Supporters

Size of Landholding		Total No of HH	Land Contributors		Other Supporters		Total of Involved Households	
Category	Range		No of HH	%	No of HH	%	No of HH	%
Marginal	<0.50	21	0	0	3	14	3	14
Small	0.51-1.00	40	1	3	23	58	24	60
Medium	1.01-2.00	33	11	33	16	48	27	82
Large	>2.01	10	5	50	3	30	8	80
TOTAL		104	17	16	48	46	65	63

members of the community were neither necessary nor desirable. The decision instead was to have representatives selected from each of the five villages in the community. Written agreement was deemed vital, so altogether 17 people from the five villages placed their thumbprints[12] on a paper committing them to providing land for collateral. Interestingly, all but one of the 17 people were either medium or large landowners by village standards (Table 3). That one small landholder was the teacher, Rajendra Bahadur Kumhal, one of the strongest supporters of the project. Significantly, this community decision freed the marginal and small farmers (58 percent of the total number of households in the community) from the risk of losing their subsistence. Also, since each contributing household provided less than 0.1 ha, or, on the average, about 5 percent of a household's total landholding, no immediate threat to the contributor's subsistence was posed. It was an impressively just distribution of responsibilities decided by the people at their own initiative within the community.

Another kind of risk was assumed by 48 other supporters of the project, who joined in agreeing that if, for any reason, the mill ran at a loss so that repayment to ADB/N using only mill income was impossible, all signatories totaling 65 people would be jointly responsible. Even if, at worst, the mill had to be liquidated, they all would share the burden of repayment to ADB/N so that the collateral land would not be auctioned off by the bank. This idea was confirmed in a public meeting, and everybody agreed to follow the accepted tradition of signing a *kabuliyat tamasuk* (contractual agreement). The mechanism is clearly an effective way of cushioning the risks associated with an innovation. The large-scale and medium-scale landholders were still the principal risk takers (Table 3), although a significant number of small farmers were supportive as well. This is interesting because the benefits from the mill would be shared by not only those contributors but all other residents in and outside the community.

The constraints involving the supply of gobar (300 kg) had less to do with the amount than with logistics. If all the households in the community contributed, their bringing the gobar and taking back the slurry daily would be difficult, mostly because the villages are spaced far apart. Several suggestions were discussed in the meeting. One such suggestion was to locate the biogas/mill near one village whose residents would take the entire responsibility of supplying the gobar on a continuous basis. Several complications were, however, anticipated. Later in the course of that discussion, Santa Bahadur Kumhal, one of the richest men in the community, spoke out, "I will provide all the 15 tins, whether or not others put in their share.... Now that I have spoken openly in front of everybody, I will stick to my word even if I have to purchase more livestock."

The gobar problem was thus solved.[13] Measurements made later indicated that 12 tins of gobar could be collected daily and possibly more during the wet season. For security reasons, however, three other farmers within the immediate vicinity of Santa Bahadur's house were persuaded to provide supplementary amounts of gobar if he alone was unable to supply the requisite amount. Primarily because of Santa Bahadur's willingness to provide the gobar, the assumption was made to locate the biogas/mill near his house. Accordingly he agreed to donate a piece of land (0.05 ha). In return for all his favors, the community agreed to provide him enough gas for 2 hours of lighting every night.

Labor contribution was the next agenda. The first concern was to collect locally available materials for the construction of biogas plants. Twenty cubic meters of stone had to be dug out, 110 bags of gravel (5.5 metric tons) had to be prepared by crushing bigger pieces of stone, and 400 bags of sand (20 metric tons) had to be collected from the river bed. Finally, all of these had to be transported on human backs to the project site. Also, two cylindrical pits had to be dug, each with an outer circle 5 meters in diameter and 1 meter deep with an inner circle 3 meters in diameter and 1 meter deep. Because the construction had to be complete before the monsoon, the schedule for this part of the work allowed only 10 days. Considering the tight schedule and the large volume of labor, the people decided that the work should be contracted out to individuals from the community. The question was how to finance it because a substantial amount of Rs. 3,000 (US $230) was involved. One option was to use the donation of Rs. 7,800 (US $600) to be received by the community from ADB/N when the loan was sanctioned.[14] That option was considered reasonable because Rs. 4,800 would still remain for other expenses. The people particularly wanted to purchase corrugated iron

Preparing stone and gravel, Chorkate. (D. Bajracharya)

sheets to use for roofing the house for the mill machinery (at a cost of Rs. 3,300). The other Rs. 1,500 was saved for any other expenses that might be incurred in the course of time.

The Rs. 1,500 left over was certainly not enough to pay for the requirements for labor. The decision therefore was to have the local people contribute labor on a voluntary basis. These activities included (a) transportation of externally acquired materials, from the unloading point to the project site (2 hours walking distance); (b) unskilled labor requirements for the construction of biogas plants; and (c) semiskilled labor requirements for the construction of the mill house including roofing. The local people did initially agree to commit themselves to these responsibilities.

Complications developed, however, during the construction phase, principally because construction was not complete before the monsoon started. Because the people would not divert their attention from rice transplantation activities to biogas construction, and understandably so, either labor would have to be paid or construction would have to be postponed for four months until voluntary labor contribution could be

Concrete work on digester dome. (D. Bajracharya)

available during the next slack season. The second option was, however, considered not desirable and a decision was made to continue construction while community spirit held strong. The decision necessitated the use of hired labor by spending the Rs. 1,500 that still remained. It also introduced a tricky problem. If workers were to be paid for the remaining phase of construction, those who worked previously had to be compensated in some ways. A sensible formula was worked out. Since the opportunity cost for labor was very high at the time, the workers were paid at the regular rate of Rs. 12 per man-day and Rs. 7 per woman-day. In consultation with the people from Gobar Gas Company, Rs. 500 was considered sufficient to pay wages at this rate and finish the remaining job. Rs. 1,000 was then allocated for compensating those who had contributed labor until that time. Of some 500 person-days to be compensated, it was agreed that up to 5 person-days per household would be considered as voluntary contribution. Any household that contributed more would be compensated. This meant payment for the remaining 200 person-days at the low rate of Rs. 5 per day, regarded not as daily wage but as *khajakharcha* (meal expenses). That ingenious formula probably would not have been possible by any externally imposed means.

Removing earth from inside the digester. (D. Bajracharya)

Satisfying Regulations of External Agencies

In spite of the close involvement of local people in planning and then conviction to continue with the project, problems developed in formalization procedures with external agencies. They were caused mostly by some regulations of external agencies incompatible with realities of rural conditions.

The first problem concerned some ADB/N regulations. ADB/N wanted for their safekeeping the proof of land ownership from all 17 people to ensure that no transaction on the collateral land would take place before the repayment was complete. The local people had one of two choices: (1) acquire copies of the registered proofs of ownership from the Land Reform Office for submission to the ADB/N, or, (2) submit the temporary title of ownership in their possession as provided by the Cadastral Survey Office subsequent to the land survey conducted the year before. People wanted to avoid the first alternative that would have meant a minimum payment of Rs.. 9 per person, plus any under-the-table bonus considered necessary, as well as much harassment from the staff at the Land Reform Office.

The second alternative would have been acceptable if the Cadastral Survey Office had distributed the permanent certificate of ownership (*lalpurja*) for each piece (*kitta*) of land. But the temporary title listed all kittas owned by a farmer on one piece of paper. In order to submit the proof of ownership of the collateral land, the owner had to submit the proof for the rest of his land as well. Considering the mistrust of the people for the officials and particularly the fear of manipulation, parting with the entire title of ownership was unacceptable. When the ADB/N manager from the Gorkha sub-branch came with three other colleagues to the project area to complete all the paper work in one day, the people appreciated the special favor rendered by ADB/N staff. However, more than half of the 17 people refused to submit the temporary land title.

Subsequently, arrangements had to be made for the Chorkate Kumhals to visit the ADB/N offices in Gorkha and see their operations; Kumhals from other areas who had ADB/N loans and submitted certificates of ownership were introduced. Further, assurances were given that if, for any purpose of reference, anyone needed the paper, it would be in their hand at no cost "within five minutes" of their arrival in the ADB/N Office, provided that it was returned again afterwards. Also, they were told that upon receipt of the lalpurja from the Cadastral Survey Office, the temporary title of ownership would be returned, and only the lalpurja for the particular collateral land would have to be kept. They eventually agreed to that option, but grudgingly so.

Another problem developed concerning the land to be used for biogas plants and the mill. Again, an ADB/N regulation required this land be placed as collateral. That required land amounted to no more than one *ropani* (0.05 ha), but it was part of a big parcel (kitta) of about 40 ropanis (2 ha). The ADB/N regulation required that the whole kitta had to be provided, not simply the portion where the site was to be located. The owner as well as other people in the community expressed understandable resistance. Even the bank manager saw the unfairness of taking 40 ropanis from one farmer while the rest of the group together provided only about 25 ropanis. However, he was in no position to disregard the rule and written permissions had to be acquired from the Central Office of ADB/N. Under ordinary circumstances, the correspondence could have delayed the program significantly. However, the bank manager himself traveled to Kathmandu with me and obtained permission to accept only the one ropani land as collateral.

A slightly different problem concerned the construction of the mill house. Although the local people had agreed to contribute fully, it became apparent in due course that this was not feasible. Budgetary allocations from ADB/N and RCUP had already excluded the costs for the mill house construction. An alternative was sought to avoid a complicated budget revision. One loophole was found: in the contribution of ADB/N loan and subsidy, twelve biogas lamps were budgeted but since only four were needed, the remaining eight could be returned to Gobar Gas Company. The Rs. 2,000 thus obtained could be used to construct the mill house. Additional funds were obtained, this time from RCUP (who were providing the expenses for installing the mill machinery according to a quotation from Gobar Gas Company) by including Rs. 500 worth of locally collected materials. Altogether Rs. 2,500 was eventually collected for mill house construction (with full unofficial knowledge by concerned officials of ADB/N and RCUP). Clearly, the two organizations supported the scheme, since they themselves showed us how to bypass their own rules.

Management by Local Control

Local control was emphasized throughout the process, and decision making in all cases rested with the people. Management through local control was not easy, however; it had to be developed. For example, somebody from the community had to be trained for the operation and maintenance of biogas plants and of the mill machinery. RCUP agreed to pay the monthly salary to one local person and send him to the Butwal branch of Gobar Gas Company for training at RCUP's expense. His salary was to be continued until the mill was generating regular income. The people, insisted, however, that two operators were necessary to make the operation more dependable and suggested designating those operators as apprentices who for the interim period would share one full operator's salary. This suggestion became a precondition for the operators' employment. The operators also were required to provide some land as collateral for the ADB/N loan to give them a stake in the mill. If they did not have any registered land in their own name, their family members could do so on their behalf.

The people chose two young men who were subsequently sent to Butwal for a two-week training in the operation and maintenance of biogas plants. The training was scheduled prior to the construction of biogas plants in Chorkate. Mill operation training was separate. One mechanic from the nearby panchayat with experience in the operation and maintenance of mill machinery was brought into Chorkate when the machinery was installed. He stayed in the village for a month at the expense of RCUP and ran the mill while guiding the two local operators on the job.

I also carried out informal training to acquaint the Kumhals with

Final adjustment of dual fuel engine. (D. Messerschmidt)

business procedures outside of Chorkate. For example, to bring materials for biogas construction, I took one operator and one other representative from the community to the Pokhara sub-branch of the Gobar Gas Company to show them how to reach people if any operating problems in the biogas plant should develop. The company's guarantee provided free services for seven years. Similarly, two other people from the community went with me to Bhairahwa to purchase the mill machinery so I could acquaint them with the machinery shops in case they needed spare parts or parts for a new mill. These trips were more valuable than mere words would have been and gave the people experiences they would not have gotten under normal circumstances.

Sharing Benefits and Responsibilities

Responsibilities were shared in decision making through public meetings where the issues were discussed at length and consensus reached. A minimum of one person represented each of the five Kumhal communities. Sometimes, to satisfy this quorum, meetings were delayed for up to four hours, because of a lack of a sense of time, not a lack of interest on the part of the Kumhals. These meetings were extremely important, particularly since they were obviously participatory. The sharing of the responsibility in decision making from the beginning of the idea paid off when the program was implemented.

A more formal organizational structure, the Community Mill Committee, was also established for decision making. All 65 signatories were members. The 17 members who provided the land for collateral made up the executive body; the rest were known as associate members. Both member categories enjoyed a 10-percent discount on their milling activity. Jit Bahadur Kumhal was chairman (also the chairman of ward 9 in the village panchayat and, as mentioned previously, the most respected leader among the Kumhals). He was assisted by Bhot Bahadur Kumhal as the vice-chairman, and Rajendra Bahadur Kumhal as secretary/treasurer. These three people carried out the special responsibility of being principal organizers among the Kumhals.

Three subcommittees also were formed to address three concerns: (1) to ensure provision of all the inputs to the biogas plants as needed, proper maintenance of mill service, and prompt repairs; (2) to check the accounts regularly regarding the inflow and outflow of cash; and (3) to administer the net profit from the mill operation.

Regarding the third point, the net profit from the mill was to be accumulated into the community provident fund. This fund was viewed as a way to share the benefits among all community residents. The fund was to be used to provide social services such as repair of trails or construction of drinking water facilities or irrigation schemes or provision of interest-free loans to the needy or potential investors from the community. The third subcommittee's responsibility was to assess the investment criteria and plan future activities emphasizing the community's principal priorities. That subcommittee also was expected to facilitate use of the fund as seed money and to stipulate reasonable and timely repayment schedules. Each subcommittee consisted of five people representing the five Kumhal villages, with one member from the executive body as chairman.

The Outcome and Some Reflections

After five months of planning, organization, and negotiations, the mill was ready for inauguration on July 11, 1982. The newly elected district panchayat president participated in a formal ceremony attended by dignitaries, RCUP staff, district officers from the various line agencies, class organizations, members of the Chhoprak Village Panchayat, and the Kumhals from Chorkate. The Kumhals' achievements were appropriately recognized as "an example set for the entire district," "a model for RCUP," and "people's participation bearing fruits when cynicism is prevalent." A note of caution also was communicated in the meantime: would the marriage continue to be happy after the honeymoon was over?

Undoubtedly it would be to the people's advantage to make the marriage work. The lami would be gone, and they would have to resolve any problems by themselves. One expediter from the RCUP office in Gorkha would be available once in a week to assist with some technical problems, but his role would be significantly smaller than the lami's. Key decisions and proper initiatives would rest with the people themselves even more.

One month after the inauguration, reportedly 106 *muris* of grains were processed and a revenue of Rs. 1,087 was generated. This was a very positive indication of the mill's usage. But several technical problems had developed: (1) gas leakage at the stopcock, (2) diesel leakage in the engine, (3) the necessity of a grease cup that was not provided earlier on, (4) loose v-belt, and (5) shortage of gobar to fill the tanks. Worst of all, a broken belt had caused serious injury to Rajendra Bahadur Kumhal, the secretary/treasurer (Donald Messerschmidt, pers. comm., 1982).

Thanks to the continuing interest of the RCUP staff, several of their knowledgeable people assessed the problems and suggested appropriate resolutions. Gobar Gas Company corrected some of the technical problems. In the meantime, the Community Mill Committee management remained very effective. When RCUP's energy specialist, Gyani Shakya, visited the site in October 1982, he noted,

> Both the digesters had been filled with dung. There was a big line of customers waiting in line for the mill service. On my last day in Chorkate, there were at least 25 people waiting in line. Man Bahadur, our operator, and Kal Bahadur were both very busy with work. Actually it has become a very impressive site. I have been highly impressed by the spirit of work and enthusiasm of Jit Bahadur Kumhal, hard working operators, and voluntary devotion of Rajendra Bahadur Kumhal. (Gyani Shakya, pers. comm.)

One year after the inauguration, the mill is very much operational. Minor technical problems need continued attention no doubt. Because of long hour usage, the gas production is not sufficient to sustain 80 percent composition in the fuel input. The milling rates have recently been increased to cover the greater usage of diesel (Nirmal Chitrakar, pers. comm., May 1983; RCUP Staff, pers. comm., July 1983). The most encouraging part remains the continued interest of all concerned people and the continued flow of mill customers. The little verse sung by some school children in Chorkate (Donald Messerschmidt, pers. comm., 1982) remains a fitting tribute to the hard work endured by all concerned people:

> Rice is hulled at the Chorkate Mill;
> Kutmiro leaves give cattle their fill;
> Gobar gas fuels the Chorkate Mill.[15]

RESPONSES FROM THE MID-ELEVATION COMMUNITIES

Technology Identification

Unlike the Kumhal community, the mid-elevation communities including wards 5, 6, and 7 are very heterogeneous, as different castes and ethnic groups are represented, such as Brahmins (37 percent), occupational castes including Kamis, Sarkis, and Damains (29 percent), Newars (11 percent), Chhetris (9 percent), and so on (New ERA, 1982). Political factions also were distinctly noticeable. Even people with similar political views were not as cohesive a group as the Kumhals. Partly because the three dominant ethnic groups (Brahmins, Chhetris, and Newars) were politically and socially articulate, people tended to express more in terms of their individualistic needs and priorities. Such factors greatly influenced decision making regarding technology identification.

Notably, an attempt was made by the Pradhan Pancha (elected chairman of Chhoprak Village Panchayat) to make a unilateral decision about technology choice. He wanted a watermill that would be located in ward 6 (which happened to be his political constituency within the panchayat). He was also insistent that no public discussion was necessary because all the people would go along with his decision. I told him, however, of RCUP's regulation that no project was forthcoming unless it was decided at a public meeting. To this he conceded. What was interesting was that at a subsequent public meeting the Pradhan Pancha's suggestion was disregarded on grounds that the mill location would be too far and removed to be useful to wards 5 and 7.

Instead two other technologies were prominent in people's minds: (1) smokeless cooking stove and (2) pedal thresher for paddy. The interest of one group in the smokeless stove had been swayed by the installation of five stoves earlier on as part of another RCUP energy demonstration program.[16] The interest in the thresher by another group was spurred by the curiosity to try out a new technology related to agriculture. After much discussion, the two groups could not decide on the choice of just one

Training in stove making. (D. Bajracharya)

technology as was originally intended. Also, it was apparent during the discussion that a rift between the two groups would be created if they were forced into the choice of one technology alone. This kind of confrontation had to be avoided. In the meantime, I was making a quick calculation in my own mind to see if both technologies could be provided and still stay within the budget available to me. It was conceivable that 150 stoves and six threshers were within the budget range. A suggestion to that effect was made at the meeting. A compromise solution was then reached. Each of the three wards would receive about 50 smokeless stoves and two paddy threshers.

Planning for Implementation[17]

Concerning the stoves, three primary questions had to be resolved first: (1) What should be done about the chimneys? (2) Who would make the stoves? and (3) What would be the local contributions?

Chimneys have been one of the main constraints in diffusion of stove technology in rural areas of Nepal. Some previous attempts relied on the use of old kerosene tins (B.K. Sharma 1981) or on the transportation of baked-clay pipes from Kathmandu to the village site (RECAST, pers. comm., 1982). Both approaches were problematic. Each stove required four tins, the availability of which in the rural area was limiting. Further-

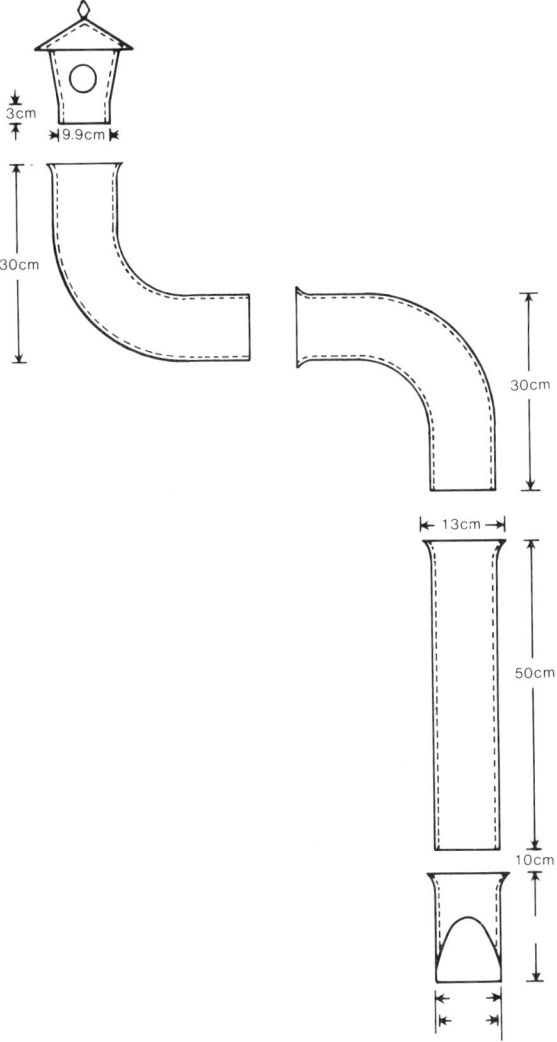

Figure 4. Parts of the Chimney Set

more, flattening and bending them into pipes were cumbersome and time consuming. Baked-clay pipes were quite heavy and bulky to transport over long distances. Thus an alternative had to be found. Fortunately, a group of potters from Kathmandu Valley regularly spent the winter (from January to April) at the lower elevations (Chorkate) of Chhoprak Panchayat to make clay pots for sale. This group expressed definite interest in producing the chimney set according to the designs I got for them from the Research Centre for Applied Science and Technology (RECAST) in Kathmandu. Each set consisted of seven pieces, three of the 50 cm long pipe and one each of the other parts shown in Figure 4. The potters did a remarkable copying job and made about 140 such sets at Rs. 40 (US $3) per set.

At separate meetings in the three wards, 15 persons were suggested for training, five from each ward. A stove expert, financed by RCUP, was brought to the area to train them. During the one-week long training period, however, only four people took a sustained interest. They became the propagators of the new stove design. The training consisted of actual stove making by each trainee. Within a week they felt sufficiently confident to pursue stove making on their own. The trainees were one woman and three men. After the stove program had been underway for a few weeks, some of the low-caste groups (Kamis, Sarkis, and Damains) began to feel deprived of the opportunity because the stove makers did not want to come into their house. All the stove makers happened to be from the high-caste group. Finally, I decided to train one person from that group to make stoves.

The strategy for implementation was programmed this way: somebody from each interested household would fetch the chimney pipes from the potters' site. (This meant a two-hour walk each way.) The stove maker would be informed and on a mutually convenient day, members of the household would collect stones (or bricks) and fetch water for mixing mud paste while the trainee made stoves.

One more person was hired to help monitor and evaluate the stoves being built. The monitor recorded fuelwood consumption by weighing firewood used for cooking (a) in the agena or chulo before the smokeless stove was made, (b) in the first week after the smokeless stove was made, and (c) after several weeks' use of the smokeless stove. The monitor also conducted interviews in each household.

Results and Problem Feedback

In the interviews, household members were asked the reasons for their interest in the stove installation. All types of answers were recorded (see Table 4). As expected, the aspects dealing with smokelessness and fuelwood savings were mentioned most. Other aspects also were interesting and noteworthy. For instance, as shown in Table 4, 38 percent of the respondents admitted their willingness to try out the stove because "some of our neighbors have already tried them out"; 20 percent of the respondents anticipated the ease and comfort in the cooking process; and about 13 percent wanted to risk its adoption because it was new and it might have potential benefits.

TABLE 4.
Reasons for Willingness to Adopt the Smokeless Stove (Prior to its Installation)

	No. of Responses	Percentage
We'll save ourselves from the smoke	54	71
We'll save on fuelwood	47	62
Some of our neighbors have already tried them out	29	38
Cooking will be easier and more comfortable	15	20
We want to try out the new idea	10	13
The government is giving it for free	7	9
It may keep us cool while cooking	5	7
The deteriorating forests may be preserved	4	5
Our son/daughter/brother wanted to have it	4	5

Note: Total number of respondents = 76.

Impressions after installation were also gathered (Table 5). Smokelessness and actual fuelwood saving were noted. Several other benefits also were noted. The complaints, however, are of particular interest. About 50 percent of the respondents complained that cooking on the new stove was slower than on the traditional stove or the open hearth. About 25 percent of them said they were afraid the hot air and the sparks coming out of the chimney head would cause a fire in the thatched roof.

An investigation yielded interesting observations. I found that the cooking time depended largely on habits related to traditional cooking style. A traditional two-mouth chulo, for example, is designed to supply equally intense flames to both mouths (Fig. 5). Also the positioning of three small mounds for holding the pots at the top provided ample air circulation for good combustion. Flames came all the way up to where the pots were located. The flame was controlled by adjusting the position of the firewood. If firewood was placed directly under the pot, it got the most heat. Conversely, if firewood was placed away from the pot, it got less heat.

On the other hand, the smokeless stove was built on a different principle. Firewood was burned at one end of the stove and so only one mouth received direct flame; heat for the others was transmitted by hot gases moving along the flue toward the chimney (Fig. 6). Accordingly, the

TABLE 5.
Ex Post Impressions about the Smokeless Stove.

	No. of Responses	Percentage
Perceived Benefits		
No smoke	54	71
Fuelwood _definitely_ saved	41	54
Fuelweed _maybe_ saved	28	37
Easy and convenient	19	25
Cooks fast	16	21
Easy to wash pots and pans	6	8
Complaints		
Cooks slow	38	50
Afraid that thatched roof might catch fire	20	26
Fire doesn't burn well	11	14
Fuelwood _not_ saved	7	9

Note: Total number of respondents = 76.

second mouth received less heat and so cooking on it was slower. However, if the pots were interchanged (say, after the first pot reached the boiling point) this problem would be easily overcome. When the new method was explained, those people who had complained earlier changed their way. The important point is that apparently simple adaptations require deliberate efforts in order to realize change.

Another related problem was that some chulos had been designed with the opening for firewood placement on the side (Fig. 6b) rather than in front (Fig. 6a). Influenced by their cooking habit, some women pushed the firewood far down the channel close to the second pot hole with the expectation that cooking on this hole could be made faster. Unfortunately, "the fire did not burn well" because it did not get enough air, and the stove did not perform well. Telling people not to push firewood all the way down solved the problem in some cases. In the later constructions, we decided that the design with the opening on the side should be avoided. Instead the oven with the opening in the front should be promoted.

The concern over the fire hazard was another real problem. Its

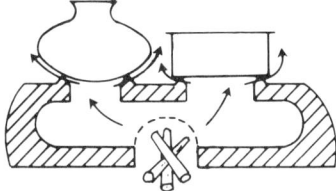

Figure 5. Traditional Chulo (front view)

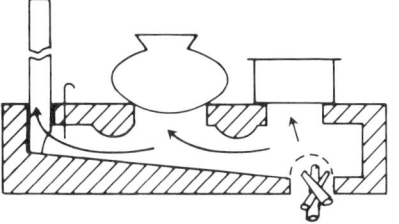

Figure 6a. Improved Chulo — Opening in the Front

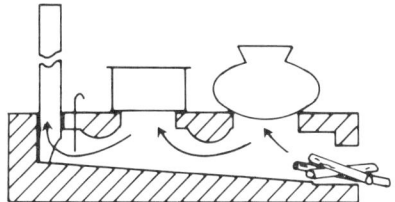

Figure 6b. Improved Chulo — Opening at the Side

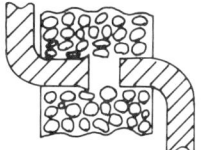

Figure 7. Chimney Parts Incorrectly Assembled

seriousness was realized only after a technical fault almost resulted in a fire in one of the households. The stove maker had left the chimney joints unconnected because the stone wall was quite thick (Fig. 7). More importantly, the joints at one side rested against a wooden beam. Within two weeks of the stove installation, the hot exhaust flame gradually had eaten away the beam. One afternoon the owner noticed smoke was coming out of the wall after the fire had been extinguished. When the chimney was dismantled, he found that the beam had caught on fire, but luckily it hadn't been seriously affected. The news of the fire spread very fast in the entire panchayat.

An immediate step had to be taken to convince people that the fire was an isolated example; corrective measures could be taken; and the roof had not been threatened by the sparks. The first point was proved satisfactorily by dismantling chimney setups in several other households and finding no indications of fire. The corrective measure stipulated was to ensure that the joints were connected in each and every case.

An experiment was undertaken to explore the concern about fire in the roof. A quantity of straw was hung about two to three feet above the chimney head and then, after two or three weeks, was examined for any signs of burning. When no signs were noticeable, many people were convinced. Some, however, insisted on having a tin plate (such as a piece from an old kerosene container) placed over the chimney head so that the sparks would hit the tin plate before anything else. A compromise was reached that we would install the plates if people could get the tin plate on their own and not at RCUP expense. We found out later that some had installed tin plates on their own initiative.

Reflections

The experience with the stove program is clearly different from that of the biogas/mill program. Notably, the difference in the socioeconomic setup of the area led to the identification of different technologies even though initial discussions with each group began at the same plane. The choice of different technologies meant different strategies for organization among people and approaches to implementation. Consequently, the type of external assistance also had to be different.

One commonality is that many practical problems were faced during implementation. If unattended, they would have resulted in abandonment of the program. This commonality reinforces the importance of the lami's role in assessing the situation through direct interactions with the people, in responding on the spot to their needs, priorities and problems, and in channeling external assistance as appropriate. The specificity with respect to locations cannot be underestimated under the circumstances.

Guidelines for future research and development can be developed from the experience. Complaints voiced by the people suggest the emphasis needed for measures toward educating people about a new system of technology and also for developing technical improvements. For example, the questions that emerge from the stove experience are: Can some technical design (besides the fire breakers) be suggested to improve the heat exchange in the second and third pot hole thereby answering the people's concern about longer cooking time on slow burning fire? Can the design of the chimney head be improved in some way to quiet people's concern and fear of fire hazard? Are there some practical schemes of making chimney pipes with local expertise and local materials without having to depend on chance presence of migrant potters? Even though the stove was well received, clearly many problems have not been answered satisfactorily yet. The experiential learning approach clearly is valuable for improving the situation.

RESPONSES OF THE GURUNGS

Technology Identification

The experience with the Gurung community demonstrates how a technology's limitations can lead to disenchantment and then abandonment of an entire project. The process began the same way as in the Kumhal community and in the mid-elevation communities. The technology identification phase, in fact, proceeded quite well. The pattern of the people's behavior was very similar to the one in the Kumhal community; the group was quite cohesive. During the initial discussions on technology choice, they indicated a great deal of enthusiasm in their preference for the water mill:

> A mill is useful because we won't have to get up very early in the morning for dehusking and grinding. With the mills in operation, we can devote more time to such activities as plowing the fields, gathering fuelwood, weeding the corn and millet fields and so forth.
>
> Once the mill is in operation, people will gather around there. We will therefore benefit from the discussions we will have among ourselves. We will be talking about the new thing that benefited the villages and we will be open to more new things.
>
> The mill will make it easy for us to get the brides for our boys in the village. The girls from adjoining villages will be anxious to come to the village that has a mill.
>
> Now that we have the opportunity, we need a big project. We have already been managing somehow to get the fire burning in our stoves. Firewood is not such a big problem. We need the mill so that everybody from the three surrounding villages can benefit.

Despite the community's enthusiasm, an adequate water supply was not assured. A water resources engineer was brought in from the engineering division of RCUP to measure the water flow and the head. An investigation of the various sources around the community soon showed that the stream flow of 10 l/s during the dry months (winter) was significantly less than the needed volume. One exception was noted at one site which recorded about 30 l/s during the dry period. However, two problems were associated with the site, one related to head, and one related to location. A long channel would have to be dug to get the head and a lot of civil work would be needed along an area of massive rocks. The second and more serious problem was its distance from the settlement area. As mentioned by one person, at one-half hour's walk each way, "the time saved by processing grains in the mill will be used up in the coming and going." After a long discussion, that idea was dropped.

At that point the question was whether the Gurungs would be willing to choose another technology. Their disappointment about the water mill was so strong that many people in the community simply lost interest. One young community leader from one village (Chitre) did become interested, however, and he suggested a biogas plant for his community for domestic usage (cooking and lighting). Leaders from the other two villages (Bhange and Pam) indicated their interest in trying out the pedal threshers for paddy. But they were not very enthusiastic: they were interested mainly because the distribution of technology was free.

Planning for Implementation

Paddy threshers required the fewest negotiations because the local people did not have to spend anything out of their pocket. They had to organize among themselves a schedule for everyone's use, but only after the machine arrived.

Biogas installation needed a lot more organization. Eight households would have to work together to set up the biogas plant and share the responsibilities and the benefits of the operation. The basic understanding during the initial discussion was that an arrangement would be worked out with the ADB/N to take advantage of their 50-percent subsidy and 50-percent loan program to cover the costs of biogas installation. Since we were considering the establishment of two 20 m^3 plants, the cost was estimated at about Rs. 48,000 (US $3,700). Subtracting the 50-percent subsidy by ADB/N, and the Rs. 12,000 to be contributed by RCUP, the community together would have a loan of Rs. 12,000, or of Rs. 1,500 per household. Each household would also have to provide the dung and the labor for construction.

The people were hesitant about taking the loan for reasons quite similar to those of the Kumhals. Some suggested that each household could contribute Rs. 1,500 in cash since that would roughly correspond to the kerosene expenses for the 7-year period. So the benefits due to savings on fuelwood usage would be additional. However, not everybody in the community could produce Rs. 1,500 on demand. The people also were uncertain about the supply and collection of dung and proportionate distribution of slurry. Even though one leader assured the others that daily records could be kept, several people were still reluctant to believe that his plan would work.

The third and most serious problem was the concern regarding the collection of materials. Some materials, such as cement, would have to be brought from outside. Sand would have to be transported from the river bed located at the bottom of the mountain (at about 450 meter elevation) to the village (at 1,500 meter elevation) — a particularly awesome task. The Kumhals, living near the river bed, would have to transport the sand on their behalf. Some people pointed out that the Kumhals already had been engaged in similar business since they transported sand over the same distance to the district headquarters. Clearly the Gurungs would have to consult with the Kumhals before they could undertake the responsibility of building biogas plants. So the first meeting ended in uncertainty. They needed to weigh carefully various constraints.

By the second meeting, several days later, four of the eight households had decided not to participate in the program. The other four households were still not sure whether or not to go ahead. The cash contribution was less of a problem for them since they were economically better off than the others. But the transportation problems remained. They did not like the idea of depending on others for transportation. Nobody was prepared to take a full-time, voluntary managerial role. With more discussion, another reservation, concerning the adequacy of dung supplies, became apparent. They were not sure if the dung collected from the 20 cattle in their possession could provide the daily input requirements (150 kg) for one 20 m^3 biogas plant. All these reservations together led to a consensus before the meeting's end not to go ahead with the program.

Still, an important lesson was learned about the inappropriateness of biogas technology presently delivered by the Gobar Gas Company. Both the Indian floating gas tank design and the Chinese dome design that they supplied, required the transportation of huge volumes of foreign materials (the gas tank or sand and cement) for construction. For villages such as Chitre, located away from the road, transporting these materials is too big a burden to bear for a sustained interest in the technology. For the Kumhal community, it was a smaller problem because the plant site was located at low elevation and sand was available at a closeby river. Conditions were clearly different for the Gurung community. The present experience shows that technology designers need to look into the possibility of building biogas plants with locally available materials.

THE CHHOPRAK EXPERIENCES IN RETROSPECT

Lessons Learned

Several interesting points emerge from the Chhoprak experiences. One very striking observation is the three quite varied experiences in three different communities even though they were approached initially in similar ways. In a small administrative and political unit such as a village panchayat, the choice of technology from a limited list of energy technologies was influenced by the composition of the groups and the interaction among people within the groups. Group cohesiveness in one group, the Kumhals, led to the choice of biogas because the scale of that technology meant the group could enjoy its benefits in a communal way. A more fragmented group on the other hand such as the mid-elevation communities preferred a divisible technology — a stove required less concerted group effort. The important aspect in the choice was the group assessment regarding potential benefits, according to the group's own evaluation. When an overall positive return was questionable, such as with Chitre, the program was abandoned altogether.

Each technology carried its own set of requirements that determined the degree of adjustments in local organizations, the type and extent of external assistance, and the forms of internal mobilization of skills and resources. The nature of the chosen technology, therefore, altered each set of necessary steps before implementation. The biogas/mill program, for example, required that agreements be reached (a) among village residents about collateral, dung input, labor contributions, and material requirements; (b) with ADB/N and RCUP regarding loans, grants, and subsidies; and (c) with Gobar Gas Company concerning construction materials (not available locally) and the schedule of construction. With the stove program, yet another set of negotiations and agreements had to be pursued. One important aspect of such negotiations is that often the rules and regulations of outside agencies are incompatible with the local circumstances. This problem is compounded when communication between local people and those from outside agencies suffers because of mutual prejudices, status conflicts, and mutual mistrust.

Problems do not cease to occur even when agreements are reached during planning and negotiations, and implementation is already underway. They arise, for example, due to unexpected technical malfunctions; unanticipated mismatch of local conditions with expectations; misunderstanding by people about operation and maintenance; and the waning interest in the program caused by delays and complications. In the face of these problems, quick measures need to be taken before serious set-backs are felt.

Reflecting on the series of events that took place in Chhoprak, the most important lesson for me was the realization that no specific formula could be derived regarding the appropriateness of technology without a good understanding of the context where the technology is to be implemented. The heterogeneity of the various contexts in this regard is a concern that has to be overcome in order to bring about successful implementation. The unfortunate part is that the extension system in Nepal as it is practiced today does not put nearly as much emphasis on the understanding of the local context as it does on the meeting of physical targets related to various technologies. The important question is whether the role of the lami, as I was playing in Chhoprak, can be instrumental in reducing the shortcomings. Based on my experiences, I have come to realize quite strongly that the role, if extended on an institutional basis, can greatly enhance the effectiveness in introducing innovations within the villages of Nepal. I believe that my presence in the village and therefore my understanding of the people and the village setting, the respect local people felt toward me, the credibility I possessed with people from outside agencies, my knowledge of the technology system, and my contacts with people dealing in the technology — all worked favorably in surmounting the problems to a considerable extent.

Before a discussion of how the process can be extended, it is important, however, to analyze some of the weaknesses of the Chhoprak experience. The principal one stems from the fact that the experience is very much an isolated case as it presently stands. No direct linkage exists of the experience with either the research and development system or the extension system. Any semblance of a linkage exists only to the effect that RCUP and ADB/N have invested money in the area and they have therefore the vested interest in its success. Whether or not the approach taken in Chhoprak will be extended to other areas as a deliberate policy is an open question. No direct mechanism is available for continuous monitoring and evaluation of the technical questions with the intention of improving the operation or for generating new and more appropriate technologies. No direct provision has been made for continued evaluation of needs, for assessment of alternatives, or for strategic planning of action programs in Chhoprak panchayat or other adjoining panchayats.

For the sake of continuity, the following questions should be considered and addressed:

1. What can be done when problems develop in the biogas/mill operation or in stove operation? Can technical problems be solved by an involvement of scientists and technologists? Can social cohesion and excitement over innovation be sustained or, if some problems develop,

can they be mitigated before they grow beyond control?

2. Considering the nonadoption of biogas technology in the Gurung community at the high elevation, can an alternative be found by involving scientists and technologists to overcome constraints such as the transportation of materials over long distance?[18]

3. Considering the high motivation of local people toward innovations, can alternatives be promoted to solve other problems in the area with greater participation, usage of local resources, and reasonable assistance from outside and, above all, will those alternatives promote self-reliance?

4. Can the experiences of Chhoprak be extended to adjoining panchayats? What kind of networking can be developed to involve government line agencies and scientific and research institutes, and to encourage local exercises in innovations?

Until these various aspects are deliberated and pursued, the Chhoprak experience alone is too limited to have any significant impact on the development of the panchayat and other adjoining areas. A systematic approach is needed on an institutional basis if the role of the lami is to be effective. I will return to this subject again. First, we should review some efforts in other circumstances toward building a new set of paradigms for generation of appropriate technologies and their dissemination for rural development.

Research, Development, and Diffusion for Rural Areas

Many concerned people are becoming convinced that the research and development system and the extension system as they exist today have not succeeded in inducing development in rural areas of the developing countries. Uma Lele (1975) summarizes the feelings about extension quite succinctly:

> Extension agents are few and far between, ill-paid, ill-trained, ill-equipped with a technical package, and consequently very poor in quality. That the farmer often knows more, at least about what is wrong with the innovations, and that extension agents often do not follow their own advice have become parts of a folklore of extension in developing countries.

Freire (1973) sees an even more fundamental problem with the very concept of "extension."

> ... In accordance with the concept of extension, [extension agents] transform their specialized knowledge and methods into something

static and materialized and extend them mechanically to the peasants — invading the peasant culture and view of the world — they deny that men and women are beings who make decisions.

What Freire suggests is "dialogue" as the mechanism of "communication" between technical experts and peasants to reflect critically on the problems at hand and "transform" old realities to new ones by merging advanced technology with the empirical methods of the peasants.

Regarding the research and development (R and D) systems in developing countries, Amilcar Herrera (1981) points out quite rightly that they "have evolved with the modern sector of the economy and are closely connected to the R and D systems of the advanced countries. Their paradigmatic determinants are very similar to those of the developed societies." Furthermore he remarks that "the traditional sector exerts very little explicit demand on the R and D system of the underdeveloped countries." With specific reference to agriculture technology, for example, Yujiro Hayami and Vernon Ruttan (1979) emphasize that

> Inadequate recognition of the location specific character of agricultural technology was a major reason for the lack of effectiveness of much of the technical assistance effort of national and international agencies during the 1950s and 1960s. Major emphasis was placed on extension projects designed primarily to transfer materials and practices from the developed to the less developed countries and on the implementation of multipurpose, and frequently superficial, "community development" efforts that met with only limited success.

Most people accept the view that inappropriate techniques have been the result in developing countries. Appropriate alternative techniques are promoted only insofar as "the benefits are obvious and the attack on existing interests is insignificant" and there are definite "limits on the willingness and ability to pursue different policies on a sufficient scale to secure a significant change at the macro level" (Frances Stewart 1978). Based on a review of 696 appropriate technology organizations listed by the United Nations Environment Programme (UNEP 1978), Herrera (1981) concludes that most of the formal organizations "tend to follow the conventional approach to research, with little or no participation of the local population, and very scarce input from the social sciences." The informal organizations, on the other hand, despite their effective linkage with the local population and their successful handling of local problems with local resources, "are unable to generate technologies, or to adapt to local conditions technologies with some degree of sophistication, due to their lack of high level scientific and technological support."

Given these problems, the attempts to shift the directions of the research and development and the extension systems is a major challenge. At various levels, however, many interesting efforts are directed toward (a) assessing problems and priorities in rural areas with direct participation of the local people; (b) identifying alternative ways of problem-solving; (c) generating new technologies of adapting existing technologies to fit local conditions; and (d) building up local capability to promote local control in the implementation of appropriate plans.

The Centre for the Application of Science and Technology in Rural Areas (ASTRA), for example, is based within the prestigious Indian Institute of Science with the principal objective of generating "alternative technologies that facilitate low capital investment, employment generation in rural areas, dispersal of mini-production units to the villages and production of inexpensive goods and services of the mass consumption variety." In order to safeguard against the transfer of inappropriate technologies, ASTRA has adopted the strategies of (a) "consulting the local people, prior, during, and after the innovation process"; (b) "scrutinizing the technologies with theoretical criteria of appropriateness"; and (c) "demonstrating these technologies on the Extension Centre" (A.K.N. Reddy, K.K. Prasad, and K.S. Jagadish 1977).

Research groups from Ethiopia, Iran, Mexico, and the Philippines have worked together in a project, "Research and Development Systems in Rural Settings," with the objective of developing a methodology for

1. The generation of technologies for use by the rural poor through a process that involves their interaction with research groups.
2. The utilization of the capabilities and knowledge of the traditional societies linking these to the R and D systems to optimize the benefits for the rural poor.
3. The assessment of the strategy utilized by the participating research groups to develop technologies for the rural areas, and to undertake a comparative analysis of these strategies in different socioeconomic situations. (Herrera 1978)

Interim reports suggest that "all the teams have succeeded in establishing an active interaction with the peasant communities, and in starting to develop adequate mechanisms to make that participation most effective." At the same time all the teams are receiving strong support from the science and technology organizations of their countries and also from within their own institutions. This is interesting from the standpoint not only of the geographical and cultural spread of the research groups but also their institutional affiliations. For example, two of them are from the educational institutes, one from the national council of science and technology and the last one from an economic development foundation.

Another very interesting attempt has been the project on "Cooperatives for Small Farmers" coordinated by the Royal Tropical Institute, Amsterdam, together with the Cooperative League of Thailand (CLT) and the National Cooperative Council (NCC) of Sri Lanka (Konraad Verhagen 1982). Their principal objectives have been to examine the various cooperative models that can effectively provide small farmers with the "security of income" and "improvement of material living standards." What is particularly interesting about these efforts is their emphases on (a) the full *participation* of small farmers, cooperative staff, board members, and officials in analyzing the problematic situations; (b) the *diagnosis* of remedial measures to overcome problems; and (c) the *action-orientation* to affect the desired changes.

These three examples and several others of a similar nature (see, for example, David Brokensha, D. M. Warren, and O. Werner 1980; David Korten 1980; Alan Jedlicka 1977; and Lele 1975) constitute a new perspective where local needs become a primary focus of problem solving. Local participation and local knowledge systems constitute the basis for problem analyses. Development of the science and technology system is encouraged under these premises. Though external assistance is channeled into the area as required, the principal goal of the effort is maintaining self-reliance for endogenous development. The signals of involvement of various groups of people with different institutional affiliations under different circumstances are quite encouraging particularly from the perspective of building up new paradigms of research and development. Nevertheless, the sobering fact is that most of these efforts tend to be sporadic and the interinstitutional linkages are erratic and very much on an ad hoc basis.

Perhaps in only one area is the system of extension meshed systematically into the research and development system on a nationwide basis. A prototype of such an alternative system is the process of technique transformation in the People's Republic of China. A particular reference to agriculture is described eloquently by John Hawkins (1978). According to him,

> Research and development activities to transform techniques are carried out at virtually every level of society in China, from the central, governmental institute and academy level, to provincial institutes, through *hsien* (county) governmental bureaus, and at

each of the three commune levels (commune, production brigade, production team).

Hawkins concludes that as a consequence of the approach, and the ideological commitment of the government of China (a) high morale and motivation have been sustained among people in the rural areas, (b) decentralization and mass participation have increased, (c) communication channels have been effective and widespread, and (d) the potential for increased and more sophisticated research and development is good.

In very much the same spirit but without yet a strong government commitment as in China, ASTRA also has made an encouraging beginning in this direction. As a matter of principle, ASTRA, which is located in the Karnataka State in India and has an extension in Ungra village, is committed to four mechanisms of technology dissemination:

1. Microdiffusion to the cluster of villages around the Ungra extension center.
2. Mesodiffusion through Karnataka State Council for Science and Technology to the rural areas of Karnataka.
3. Macrodiffusion through rural development organizations and agencies throughout the country and in other developing countries.
4. Long-term diffusion through the establishment of an institute course that produces a new breed of rural development technologists. (Reddy, Prasad, and Jagadish 1977)

The involvement of other kinds of groups in many countries of the Third World indicate positively that the seeds of new paradigms in research, development, and diffusion are sown. In the next section, I will try to look at the mechanisms of how this direction could be forwarded in Nepal.

Suggestions with Particular Reference to Nepal

Many highly placed officers in the government of Nepal do realize the existing inefficiencies of the extension mechanism and the absence of its linkages with the research and development system. Presently, extension work is a vertically directed exercise. Policies, plans, and targets are made in Kathmandu. District officers are instructed to implement the targets and their performance is measured by the extent of target fulfillments, not by the quality of implementation nor by the local peoples' receptivity to the program. The targets are implemented through extension workers, posted to distant villages, who are directed to perform a set of tasks with the local people involved. The extension worker is an underpaid government servant with little authority, his technical competency is barely adequate, and he commands little respect from the local people. The extension system allows little opportunity to the extension worker to feed back the villagers' perspectives and he pays more attention to carrying out orders from his supervisor than to the village people's welfare.

Most of the scarce research and development today ignores the linkage with needs and priorities of local people in rural areas. At best, questionnaire surveys are conducted to rationalize the inputs of local people but rarely do the people get opportunity to participate in planning or determining the priority of research. The prevalent feeling in the village is "Many groups of people from the city come for a few days, ask us all kinds of questions, promise to do a lot things for the village, and then we never see them again."

On the positive side, however, the Sixth Plan (1980–85) prepared by the National Planning Commission (1980) is quite explicit about the objective of "satisfying basic needs" through adoption of such important policies as "people's participation policy," "integrated rural development policy," "environment and land use policy," and "water resource and energy policy." This plan also stresses important aspects of using local resources as extensively as possible and of building up local capabilities as much as possible. It is in line with these aspects that the concept of lami as explained in the Chhoprak case has true promise. This becomes even more so in light of the inefficiencies of the existing extension system. On the question of practicality, three concerns are of specific interest: (1) Is enough manpower available in the country? (2) What kind of support mechanisms are required for the lami to operate successfully? (3) What is the incentive for qualified people to go and live in rural areas?

Regarding the first question, at least three groups of people deserve close examination. The first group is the master's level students who participated in the National Development Service (NDS). In the course of 10 months of stay in the village as a part of their degree requirements, many of them have heightened awareness of problems and realities in the villages. They constitute an excellent source of manpower that can be tapped to undertake the responsibility of the lami. The second group is the educated city dwellers who were originally from the villages. Several of them are disillusioned by "the good life" in the city and are questioning whether they would not be better-off going back to the village. Given the institutional support as a lami, they have the potential of making significant contributions. The third group of people are those who are already a part of the vast network of the extension system in the fields such as health, agriculture (including livestock development), and forestry. Their effectiveness can be greatly increased by upgrading their positions and providing some training and orientation in the lami concept. Based on the availability of these various groups, finding people to become lamis is a less serious problem.

The greater problems are related to the second and third points mentioned above. For the lamis to be effective at the village level, one of two conditions has to be fulfilled. Either the lamis have to be self-motivated, resourceful, and respected (e.g., by villagers and district officers alike); or, institutional support mechanisms have to be created and made accessible to the lamis if and when outside inputs are required. The style of my operation in Chhoprak probably fulfills the first condition. I recognize (and several people have commented) that this is a special case and people with qualifications as mine cannot be expected to work as extension workers as a general rule. In lieu of this, steps have to be taken toward the development of the support system. Three interrelated activities can be envisioned in this respect: (1) training of people to prepare them to act as lamis in the villages; (2) administrative and financial support from government line agencies with respect to village-level planning and implementation; and (3) functional and effective linkages with the R and D system so that (a) the lami's experience is utilized by research institutions in developing criteria for further research, and (b) the research results are tried out at the village level through the lamis.

The concept of the institutionalized support system is already an integral part of Nepal's Decentralization Act, 2039 which was approved by the Rashtriya Panchayat (the national legislative body) in September 1982 (The Rising Nepal 1982). The Act stresses the importance of "Village Development Plans" as the basic components of the District Development Plan and specifies that in formulating the plan, each village panchayat should "consider its own labour, resources and means, and economical (sic!), physical, and technical assistance to be received from His Majesty's Government and the District Panchayat, and treat the projects demanded by each ward as main basis." Furthermore, the Act has the provision for the establishment of "Service Centres" within each of the nine divisions in the District Panchayat "to provide necessary physical and technical assistance in . . . formulating, implementing, and evaluating the plan for the development of the Village Panchayat." The Act specifies quite succinctly that among other things, "the function and duty of the Service Centre shall be . . . to assist in the establishment and proper operation of consumer's committee in the Village Panchayat area, . . . to make available necessary technical service to the Village Panchayat for conducting their development programme, . . . to forward problems and necessities of the

Village Panchayat to the upper levels of administration and to cause them to be included in the plan. . . ." Quite clearly the lami's activities could fit in well within the structure envisioned for the service center.

Another encouraging development is the Appropriate Technology Information Clearinghouse and Village Outreach Project of RECAST. This project which is supported through UNDP funds, is establishing five outreach Centers (one in each development region). The project provides the opportunity to disseminate technical information and to test the applicability of available technologies under village conditions. Pilot projects are being designed to assess village needs and priorities and adapt technologies accordingly. The R and D setup as envisioned here no doubt provides opportunities for linkages between the outreach centers and the lamis.

Given the willingness of the political and administrative structure as demonstrated by the Decentralization Act, 2039 and that of the R and D structure as demonstrated by the RECAST project, it is conceivable that the lami concept can be operationalized effectively at a larger scale than was the case in Chhoprak. The principal constraint that will have to be overcome then is the question of incentives. In that respect, considerations must be given to upgrading the status of the extension worker. At the same time training and workshop from time to time would be necessary partly to familiarize the lami with effective operational methods and partly to give opportunities for several lamis at a time to share their experiences. Furthermore, there has to be a sense of commitment and continuity in terms of moral and technical support from officers of government line agencies and scientists and technologists from research institutions.

As shown in Figure 8, the process of technology development and diffusion in harmony with local conditions requires a concerted effort on the part of different groups of people in different capacities. The lami is essentially a facilitator in this process and one who provides the necessary perspectives to orchestrate the different notes played by different groups into pleasant choruses. Another important aspect of Figure 8 is that the process described is a continuous and evolving one that involves experiential learning for all concerned parties.

A mechanism that provides the opportunity to bring about the learning experience is the process of *gaun sallah*, "village dialogue," as proposed within RCUP. Gaun sallah is a planning strategy, or process, designed to integrate villagers and development officers in a common goal to plan and implement projects (Messerschmidt and others 1983; Messerschmidt 1982). It has similarities with the *sondeo* method of field reconnaissance for planning (Peter Hildebrand 1979). In gaun sallah, this

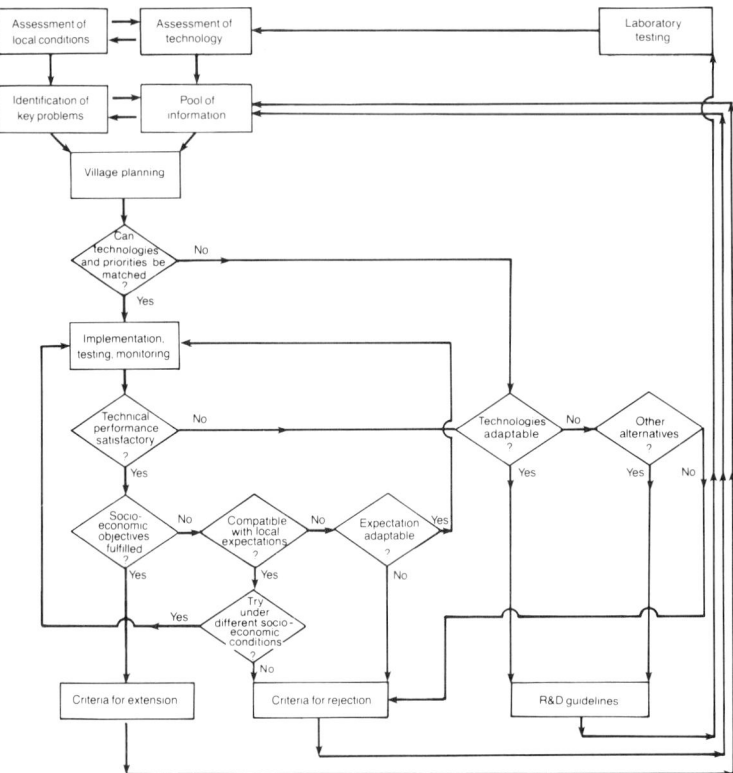

Figure 8. Rural-Based Technology and Development Diffusion System

reconnaissance for planning has been designed to fit the Nepalese context and to elicit the greatest amount of people's participation (including that of villagers and development officers) as possible. The objective is "to elicit panchayat participation, to acquaint planning officers with local conditions for resource development, and to lay the ground for preparation of a draft Resource Development Plan (RDP)." This is done by bringing the development officers to the village panchayat and by facilitating their interactions with local people from as many political and social hierarchies as possible. It is anticipated that a few days' interaction, although short, will at the very least provide a common basis of understanding and hopefully encourage critical and realistic assessment of the village development plan. The assessment includes the appropriateness of the plan in meeting needs and priorities of the villagers, the extent of the villagers' willingness to make necessary contributions for the plan, and the commitment of government offices to allocate funds in the annual budget item.

The importance of the lami lies in providing continuity through intense involvement before and after the gaun sallah. In this respect, it is clear that the gaun sallah has to operate simultaneously with lamis in order to render the lami effective and vice versa. This kind of complementarity ensures the two-way communication that has to be pursued from the conception stage to planning and implementation of rural development projects. A word of caution is in order at this point. The lami is assumed to be dedicated, compassionate, and highly motivated. These qualities have to be cultivated and nourished. To this effect, it is advisable to start modestly and move forward with careful deliberations. Spread-over effects of good results both in terms of village-scale planning and recruitment of lamis could very well form the foundation for future expansions of the program. The task ahead is no doubt mammoth. The challenge to improve conditions in rural areas for human development is however a worthy cause that should inspire new and innovative ideas.

ACKNOWLEDGEMENTS

I am grateful to the Resource Conservation and Utilization Project and the East-West Center Resource Systems Institute for financial assistance that made it possible for me to undergo a very valuable learning experience in Chhoprak. To my wife, Banoo, and daughters, Sepideh and Sharareh (aged 5 and 1 at the time of the research period), who accompanied me to Chhoprak, I can never thank enough for their constant support and courage under harsh conditions. For continued encouragement and timely advice, my thanks are due to Richard Morse and Donald Messerschmidt. I owe special gratitude to Donald Messerschmidt and Nirmal Chitrakar for keeping me abreast of the progress in Chhoprak, long after I had left the place. Last but not least, I thank the people of Chhoprak Village Panchayat who taught me many things.

NOTES

1. For elaboration of rural energy issues, see Eckholm 1975; National Academy of Sciences 1976, 1980; Makhijani 1977, 1979; Reddy and Prasad 1977; Cecelski, Dunkerley, and Ramsay 1979; Brown and Smith 1980; Wood and others 1980.
2. Korten (1980) describes the processes involved in the "blueprint approach" as follows: "Researchers are supposed to provide data from pilot projects and other studies which will allow the planners to choose the most cost-effective project design for achieving a given development outcome and to reduce it to a blueprint for implementa-

tion. Administrators of the implementing organizations are supposed to execute the project plan faithfully, much as a contractor would follow construction blueprints, specifications, and schedules. An evaluation researcher is supposed to measure actual changes in the target population and report actual versus planned changes to the planners at the end of the project cycle so that the blueprints can be revised."

3. Barnett (1953) defines innovation as "any thought, behavior, or thing that is new because it is qualitatively different from existing forms." He emphasizes, however, that the new evolves out of "pre-existing or antecedent knowledge" (e.g., local knowledge, resources, needs) because of the "stimulus" created by events such as a serious problem faced internally or by inducement of new needs, and with the help of the "facilitator" who provides the linkage between what is known and what is sought. Freire (1973) calls this the "transformation of reality" through "critical reflection" on the problems as spurred by the process of "dialogue" between technical experts and peasants. Innovation is, therefore, a "non-mechanical concept [in which] the new is born from the old through the creative transformation emerging from advanced technology combined with the empirical methods of the peasants." This contrasts with intervention that is similar to "the superposition of the new on the old" with total disregard of the cultural background and the technical-empirical methods of the peasants and therefore resembles "cultural invasion."

4. Havelock (1969) defines action research as "the collaboration of researcher and practitioner in the diagnosis and evaluation of problems existing in the practice setting. The action research technique provides the researcher with an accessible practice setting from which he may retrieve data, usually for publication. It provides the cooperating practitioner system with scientific data about its own operation which may be used for self-evaluation."

5. A village panchayat is the smallest political constituency and administrative unit in Nepal. Each one has nine wards. Each ward is represented by one ward chairman and four other members who are all elected by the eligible voters of the ward. The village council is headed by the *Pradhan Pancha* and his assistant, the *Upa Pradhan Pancha*, who are elected by all voters of the village panchayat. Normally a village panchayat consists of a population between 3,000 and 7,000 in an area between 10 and 30 sq km. There is a total of 4,000 village panchayats distributed over 75 districts in the country. Each district is represented by a district panchayat which has nine members elected by respective voters from nine subdivisions within the district.

The district panchayat chairman and the vice-chairman are elected by district-wide voters. Five class organizations (women, farmers, ex-servicemen, youth, adult) have village and district-wide representations also. A district is administered by the Chief District Officer (CDO) with the assistance of many government officers representing many departments.

6. At the time when I was working in Chhoprak, New ERA, a consulting research firm based in Kathmandu, was also involved there on contract from the Resource Conservation and Utilization Project. The scope of their work can be described as "a combined action research strategy including the sociocultural and ecological data, the planning of a variety of villager participation actions and the implementation of select and innovative strategies for income generation and resource management at the village." Although my style of operation was different from New ERA's, we benefited mutually from the sharing of experiences, information, and thinking. For a detailed report on their interesting work, see New ERA, 1982.

7. The 1981 census data on the population of Chhoprak Village Panchayat is no longer applicable because of the change in the boundary. The change was instituted in 1982 prior to the village and district elections. Parts of what used to be Chhoprak have been shifted to another panchayat and parts of yet another are now included in Chhoprak. In general, the size of the panchayat today is smaller in terms of area and population than what it used to be.

8. RCUP is an integrated resource development project that is presently administered in the village panchayats of two river catchment areas (Daraundi and upper Kali Gandaki) covering parts of three Western Nepal districts, Gorkha, Myagdi, and Mustang. The project is coordinated through the Department of Soil Conservation and Watershed Management (DSCWM) of the Ministry of Forest and Soil Conservation with involvement of many other resource-related government line agencies. It is funded by USAID, initially over a five-year period 1980–85.

9. In the initial period of its field involvement, RCUP plans to make an intensive effort in a few village panchayats only. These "lead" panchayats were selected by the district-level Catchment Conservation Committee (CCC) that is established by RCUP to function as the board of governors for planning and implementation of RCUP activities in the district. In the case of Gorkha district, the CCC had chosen six "lead" panchayats.

10. The Panchayat Conservation Committee is seen by RCUP as a board

of directors for decision making in planning and implementation of its activities at the village panchayat level.
11. SFDP is a program of ADB/N directed toward the assistance of the rural poor by providing favorable loans to the needy for adoption of technical options. A small staff is provided at the village panchayat level to identify potentially interested people and prospective projects, and to facilitate loan transactions as soon as possible. The SFDP is ADB/N's intensified effort toward integrated rural development through small farmers' own initiatives or through a cooperative group.
12. Thumb prints are the accepted substitutes for signatures. Most people in this area, as in other parts of Nepal, are illiterate (the national literacy rate being about 20 percent) and are forced to use their thumb prints when signatures are required.
13. The manner in which the gobar problem was solved in Chorkate was quite unique. It was the consequence of circumstances, interpersonal relationships, and attitudes that were very specific to time and location. To a large extent, this particular opportunity provided a basis for organization which was relatively simple. If, for instance, this opportunity had not been available, the other alternative would have been to site the project close to one of the five Kumhal villages and organize with the residents for dung collection. The organizational dynamics in this instance would no doubt be very different. If experiences of the community biogas plant in Fatehsingh ka Purwa, India, are at all indicative (see, for example, Ghate 1980; Bhatia and Niamir 1979), the organization required could have been very complex. In this Indian village, availability of dung had reportedly declined by more than 50 percent within a few months and contribution of cow dung was the most serious constraint. Whether the case in Chorkate could have been different is difficult to speculate. It is important to stress, however, that the delivery of dung can certainly not be taken lightly and a very good organizational basis has to be established during the planning period. This type of reasoning emphasizes even more the importance of close interaction between technologists and village residents in order to determine the commitments on everybody's part and the benefits each will get out of such arrangements.
14. The general procedure set by ADB/N in consultation with Gobar Gas Company was to provide a cash amount of Rs. 3,900 per 20 m^3 plant for use by farmers to collect locally available materials and to supply local labor.
15. I am grateful to Romola Morse for poetic rendition to the English

translation from the original Nepali verse. The verse in Nepali is as follows:

> Chorkāte mā dhān kutna mīla pathāyo.
> Gāibastu aghāunchha kutmiro ghās le.
> Chorkāte mā mīla banyo gobar gyāsa le.

Kutmiro is a local tree specie, the leaves of which are valued for fodder.
16. According to the original RCUP "target" set for the energy subproject in the fiscal year 1981–82, 15 stoves were to be disseminated in the six "lead" panchayats of Gorkha district. The five stoves installed in Chhoprak were a part of this target. The activity that I was pursuing was an addition to the original "target."
17. The following description concerns the planning for stove implementation only. With regard to paddy threshers, committees were formed in each of the three wards to determine how the operation should take place. Unfortunately, the threshers arrived at Terha Kilo, the nearest motorable place from Chhoprak, on the day when I was leaving the place. I was, therefore, unable to watch the progress on how the threshers were received by the people. For this observation I would have had to wait till the harvest time which is around November. This was not possible because of my other commitments.
18. I pursued this idea in an informal way with a friend, Ramesh Manandhar, who was in Nepal to do his fieldwork in partial fulfillment of the Ph.D. degree at the University of Melbourne. I suggested to him, an architect and human settlement planner, who was interested in mud brick domes, that he should look into the possibility of using these domes for biogas construction. It turned out that he pursued the idea with RECAST and Gobar Gas Company, the outcome of which was an experimental plant with mud brick domes that seems to work well (see Manandhar 1983). Clearly this design will have to be tested thoroughly before wide-scale implementation. Early indications are that the costs are reduced drastically. The exciting part for me was the practicality of using the criteria derived from local experience for furthering research and development.

REFERENCES

Agarwal, B.
    1980   *The Woodfuel Problem and the Diffusion of Rural Innovations.*
             London: Tropical Products Institute.

Balaju Yantra Shala (BYS).
   n.d.   *BYS Water Turbines.* Kathmandu.
Barnett, A.G.
   1953   *Innovation: The Basis of Cultural Change.* New York: McGraw Hill.
Bhatia, R., and M. Niamir.
   1979   Renewable Energy Sources: The Community Bio-gas Plant. Presented at the Seminar in the Department of Applied Sciences, Harvard University, Cambridge, Massachussetts.
Brokensha, D.W., D.M. Warren, and O. Werner.
   1980   *Indigenous Knowledge Systems and Development.* Lanham, Maryland: University Press of America.
Brown, H., and K.R. Smith.
   1980   Energy for the People of Asia and the Pacific. *Annual Review of Energy* 5:173–240.
Cecelski E., J. Dunkerly, and W. Ramsay.
   1979   *Household Energy and the Poor in the Third World.* Washington, D.C.: Resources for the Future.
Decentralization Legislation of 1982.
   1982   The *Rising Nepal,* September.
Eckholm, E.P.
   1975   *The Other Energy Crisis: Firewood.* Worldwatch Paper 1. Washington, D.C.: Worldwatch Institute.
Freire, P.
   1973   *Education for Critical Consciousness.* New York: Seabury Press.
Ghate, P.B.
   1980   Action Research in the Community Biogas Programme. Presented at the Energy and Rural Development Research Implementation Workshop, Chiang Mai, Thailand.
Havelock, R.G.
   1969   *Planning for Innovation Through Dissemination and Utilization of Knowledge.* Seventh Printing, 1979. Ann Arbor, Michigan: Center for Research on Utilization of Scientific Knowledge.
Hawkins, J.N.
   1978   Rural Education and Technique Transformation in the People's Republic of China. *Technological Forecasting and Social Change* 11:315–333.
Hayami, Y., and V.W. Ruttan.
   1979   *Agricultural Development: An International Perspective.* Baltimore, Maryland: The Johns Hopkins University Press.

Herrera, A.O.
    1978   *Research and Development Systems in Rural Settings: Background of the Project.* Tokyo: United Nations University.

Herrera, A.O.
    1981   The Generation of Technologies in Rural Areas. *World Development* 9:21–35.

Hildebrand, P.E.
    1979   Summary of the SONDEO Methodology Used by ICTA. Presented at the Conference on Rapid Rural Appraisal, Institute of Development Studies, University of Sussex, Brighton, England.

Jedlicka, A.D.
    1977   *Organization for Rural Development: Risk Taking and Appropriate Technology.* New York: Praeger Publishers.

Korten, D.C.
    1980   Community Organization and Rural Development: A Learning Process Approach. *Public Administration Review.* September/October: 480–511.

Lele, U.
    1975   *Design of Rural Development: Lessons from Africa.* Baltimore, Maryland: John Hopkins University Press.

Makhijani, A.
    1977   Energy Policy for Rural India. *Economic and Political Weekly* XII (33/34): 1451–1460.

Makhijani, A.
    1979   Economics and Sociology of Alternative Energy Sources. Presented at the Regional Seminar on Alternative Patterns of Development and Life Styles in Asia and the Pacific, Economic and Social Commission for Asia and the Pacific, Bangkok.

Manandhar, R.
    1983   A report submitted to RECAST of Tribhuvan University on the Construction of Mud Brick Domed Bio-gas Plant, the First of Its Kind in Nepal, Kathmandu.

Messerschmidt, D.A.
    1982   *Gaun Sallah*: The "Village Discussion" Method of Panchayat Level Planning. RCUP Social Science Working Paper no. 2. Kathmandu.

Messerschmidt, D.A., U. Gurung, B. Deukota, and B. Katwal.
    1983   Gaun Sallah: The "Village Dialogue" Method for Local Planning in Nepal. A discussion paper. Kathmandu: South East Consortium for International Development and the Resource Conservation and Utilization Project.

National Academy of Sciences.
　1976　*Energy for Rural Development: Renewable Resources and Alternative Technology for Developing Countries.* Washington, D.C.
National Academy of Sciences.
　1980　*Proceedings International Workshop on Energy Survey Methodologies for Developing Countries.* Washington, D.C.
National Planning Commission.
　1980　*Draft of the Sixth Plan, Part I (Summary).* Kathmandu, (in Nepali).
New ERA.
　1982　Draft final report, social action/adaptive research program in Gorkha, Myagdi, and Mustang districts, Nepal. Kathmandu.
Pradhan, B.B.
　1982　*Rural Development in Nepal: Problems and Prospects.* Kathmandu.
Reddy, A.K.N.
　1980　An Indian Village Agricultural Ecosystem — Case Study of Ungra Village Part II: Discussion. Bangalore: Cell for Application of Science and Technology in Rural Areas.
Reddy, A.K.N., and K.K. Prasad.
　1977　Technological Alternatives and the Indian Energy Crisis. *Economic and Political Weekly* XII (33/34):1465–1502.
Reddy, A.K.N., K.K. Prasad, and K.S. Jagadish.
　1977　Problems in the Generation and Diffusion of Appropriate Technologies. In *Science and Technology for Integrated Rural Development*, S. Radhakrishna (ed.). Bangalore: Committee on Science and Technology in Developing Countries.
Sharma, B.K.
　1981　*Improved Smokeless Chulhos.* Kathmandu: World Neighbors.
Stewart, F.
　1977　*Technology and Underdevelopment.* London: Macmillan Press.
United Nations Environment Programme (UNEP).
　1978　*Institutions and Individuals Active in Environmentally Sound and Appropriate Technologies.* Nairobi.
Verhagen, K.
　1982　Cooperative Research and Planning with Small Farmers. Presented at International Technical Meeting, Colombo. Amsterdam: Royal Tropical Institute.
Wood, D.H., D. Brokensha, A.P. Castro, M.S. Gamser, B.B. Jackson, B.W. Riley, and D.M. Schraft.
　1980　*The Socio-Economic Context of Fuelwood Use in Small Rural Communities.* Washington, D.C.: U.S. Agency for International Development.

8
Field-Based Assessment and Development of Improved Stoves

M. Nurul Islam

OBJECTIVES AND PARTICIPANTS

Most energy consumed in rural areas in developing countries is for domestic cooking, and the major fuel available is firewood and leaves, agricultural residue, and animal residue, either separate or mixed. Systematic evaluation of biomass-burning stoves used in rural and urban areas is a first step for each country in designing a program on improved stoves. Notably, though, a successful improved stove program alone will not solve the fuelwood problem: increasing fuelwood supply and altering its relative distribution among various socioeconomic groups are also necessary.

Various interrelated groups of people and their considerations are involved in an evaluation of existing stoves and in developing a program and design considerations for improved stoves (see Fig. 1). These people include stove users as well as researchers in the laboratory and in the field in developing and industrialized countries. Policy planners and decision makers in various agencies are also involved in any improved stove program. I hope this chapter will encourage interaction among these groups. My own personal observations about users' needs are included, in the context of available published information, for the many laboratory researchers who have no direct contact with users. Policy-makers will find presented the scope and limitations of improved stove programs as they relate to institutions.

It has been widely believed that traditional methods of cooking with firewood do not utilize 90 to 97 percent of total heat produced. Advocates of improved wood stoves claim that such a stove can achieve an overall efficiency between 20 and 30 percent and that consequently these stoves potentially could significantly reduce fuelwood requirements (NAS 1980). Most improved stoves have been designed primarily to reduce fuel consumption and to reduce smoke, but use of most types has spread very slowly, suggesting that users' needs may or may not be limited to those assumed by designers. Bina Agarwal (1983) has provided a framework of analysis for identifying the factors affecting the adoption of improved wood-burning stoves and the possible effects of such adoption on demand for fuelwood.

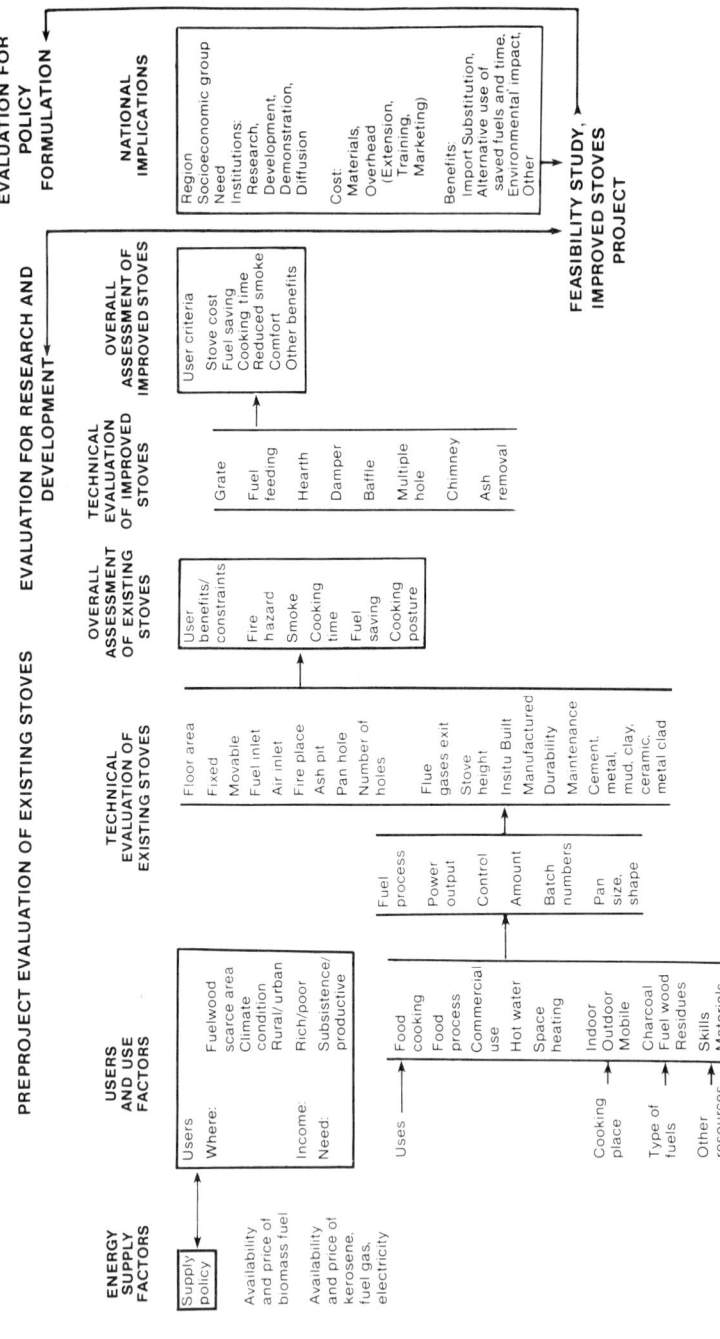

Figure 1. *Evaluation Factors for Development of Program on Improved Wood Burning Stoves*

Existing designs for improved wood stoves are suitable mainly for fuels used by higher income and some middle-income groups in rural areas. Many of these groups have adequate supplies of fuelwood, so they may not appreciate the need for fuel savings. Low-income users, who need the improved stoves most, often use residue as cooking fuel. The new designs would have to be able to handle fuel available to them.

Secondary benefits attributed to diffusion of improved stoves are savings of fuelwood collection time and reduction of environmental degradation (deforestation, soil erosion, nutrient loss, flooding, drought). In order to realize the benefits of stove programs, a compatible match must be found between primary benefits to stove users and secondary benefits that may involve the interests of groups other than users.

Performance of various kinds of stoves in use varies. The chapter undertakes to develop methodologies for evaluating stove types and factors important in designing programs on improved stoves for present and changing fuel situations. Applications of methods are illustrated principally for Bangladesh cooking situations. As a starting point in assessing factors important for stove improvement, the situations of stove users should be considered as well as the characteristics of their cooking activities, fuels, and presently used stoves.

THE COOKING PROCESS

Managing a cooking stove is part of a total cooking process that takes place in a miniature food-processing plant (kitchen). The time and effort involved in preparing the fire and food and in cooking and cleaning up are all important factors in estimating food-processing efficiency. That efficiency can be increased by various activities from designing better cooking pots to finding better techniques for air-drying fuelwood in addition to designing new stoves. This wider perspective suggests that many factors, in addition to cooking efficiency, determine the acceptability of a cooking stove to a user: cost, availability of materials of construction, size and type of fuel available, space available for cooking activities, family size, cooking practices, and types of food to be prepared. Other uses of fire also are involved, such as for light and heat and as a setting for conversation. All these factors vary greatly from region to region. Consequently, acceptability and efficiency of any single stove design may be limited (NAS 1980).

Stove Users and Cooking Stages

Cooking food involves several tasks in three stages: (1) *precooking*—preparation of food (cutting, cleaning, washing) and of the stove for

*Cooking for the wayfarer on a self-designed stove, Nepal roadside shop.
(D. Bajracharya)*

cooking (maintenance of stove, kindling); (2) *cooking* food, including control of the particular dish cooked and of the food-cooking process; and (3) *post-cooking*, including clearing and cleaning the stove area. Overall stove performance depends on all three cooking stages. Of those stages, the actual cooking is considered the primary task, requiring expertise to satisfy family members. The first and third stages include work often considered inferior. One woman may control all three stages or different family members may perform specific tasks, according to the socioeconomic condition of a particular household. Attitudes towards fuel savings are related to women's status in the family and their potential gain from fuel savings.

More female family members are available to participate in cooking in traditional rural households than in urban households. In low- and middle-income households in rural areas, younger, less experienced female members do the unpleasant and hard work, and older and more experienced women do the actual cooking. In some high-income households, hired women helpers contribute in the first and third stages of cooking. With all three groups, even though the owner and user may be different persons, they probably are part of the same household so they would be expected to be equally interested in fuel savings and in comfort.

In high-income urban households, the owner of the stove usually is a different person from the operator. Although family women may supervise, an employed cook does the actual cooking, and helpers perform the other two stages. Sometimes the employed cook does all three cooking stages. Thus, helpers and cooks benefit directly from added comfort but not from greater fuel efficiency. The more convenient kerosene stoves might be used in those households.

In middle-income urban households, the family's women perform the cooking stage while employed women helpers from low-income groups perform the other two stages. Thus, greater fuel efficiency directly benefits the household women and added comfort (such as by reduced smoke) benefits them and the helpers.

In low-income urban households of Bangladesh, the family's women perform all three cooking stages, so they benefit directly from greater fuel efficiency and added comfort. That group would probably be most motivated to use improved stoves designed for their particular needs.

Food Characteristics

Cooking certain foods requires specific stove characteristics. In parts of South Asia, for instance, bread is baked in two stages: first on a hot plate kept on the stove and then by direct heat from radiant flame or hot wall. Since an improved stove utilizing a completely enclosed fire chamber for maximum heat utilization would not permit that second baking stage, it would not be appropriate for or accepted by those users.

Cooking Pans

Pans are made from ceramic, burned clay, cast iron, copper, brass, stainless steel, and aluminum. Aluminum is used most in many areas. Different cooks use pans of particular materials, sizes, and shapes to prepare various dishes. And heat utilization efficiency for a particular stove varies according to those variables. For example, in rural Bangladesh, pans with spherical bottoms are used for cooking on wood stoves. In urban areas, pans with flat bottoms and spherical bottoms are used for cooking on electric, gas, and kerosene burners. On wood stoves, pans with spherical bottoms used heat noticeably more efficiently than did pans with flat bottoms (Nurul Islam 1980).

Cooking Methods and Fuel Consumption

Particular cooking methods also affect stove type used and fuel consumption patterns. Practices of cooking rice describe variables important in field-level evaluation of stove performance. In Bangladesh, parboiling of paddy consumes about 15 to 20 percent of cooking fuel used in rural areas (see Chapter 2). One kg of parboiled paddy yields 0.7 kg parboiled rice and 0.3 kg husk and bran residue. (Unparboiled paddy, dehusked, is called unparboiled rice.) Fuel consumption per unit weight varies according to local methods of paddy parboiling (Islam 1980, A.K.M.A Quader and K.I. Omar 1982).

Field-level experiments in a few rural households determined that mainly two-mouth stoves were used for parboiling and that the heat required for parboiling varied from 4,710 to 6,430 kJ per kg of parboiled rice. According to observation, parboiling in some small-scale paddy processing industries required from 3,500 to 4,500 kJ per kg of parboiled rice (Islam 1980), and they used paddy husk as fuel. At small-scale rice mills, paddy is soaked overnight and a comparatively large quantity of paddy is parboiled in a batch to save fuel, so less fuel is required.

In Sri Lanka, about 17,200 kJ per kg of parboiled rice were required using wood fuel (Jan Bialy 1979), but that might have been because a different type of stove was used. Notably, in one location of northern Bangladesh, where three-stone open fires were used, about 9,900 kJ per kg of parboiled rice were needed (Quader, Omar 1982).

A series of fuel consumption experiments compared parboiled rice with unparboiled rice, by cooking rice according to local practices using excess water that is later drained off. For each type tested, enough rice (1.36 kg) for one average family's meal was cooked in a kerosene cooker. From 2,470 to 3,520 kJ per kg of parboiled rice were required, depending on degree of parboiling and rice's grain size, while only 1,760 to 2,200 kJ per kg of unparboiled rice were needed, depending on grain size (Islam 1980).

Cooking parboiled rice uses more fuel at the household level since fuel is used first for parboiling and then for cooking the rice. Besides more fuel for cooking, parboiling uses more heat to dry parboiled paddy (from moisture content 40 to 45 percent to 14 to 16 percent moisture content) and more energy for polishing. In addition, unmonitored parboiling may result in rice with unpleasant smell, texture, flavor, and color. Even so, users perceive these advantages to parboiled rice: increased head yield; greater ease in husking; more protein, vitamins, and minerals than the same variety of unparboiled rice; greater resistance to insect infection during storage; less breakdown (into gruel) during cooking; not pasty

Two- and three-brick stoves, Sri Lanka. (Kirk R. Smith)

even if overcooked; and more oils in bran. In Bangladesh, only two districts (Sylhet and Chittagong) do not practice parboiling.

Cooking Time and Fuel Consumption

Previous experiments have established a relationship between a stove's fuel economy and the fuel's burning rate. As the burning rate increases, fuel economy decreases (K. Krishna Prasad 1981b). Sometimes, however, cooking time can be unrealistically long with a low-burning fuel (Prasad 1981a). For instance, in a Sri Lankan village, I observed a woman feeding her three-stone stove from three directions (increasing the burning rate) to save cooking time even though more fuel would be consumed. In urban areas a standby kerosene cooker or an electric coil heater might be used instead of a fuelwood stove to save time in the morning. Obviously, sometimes cooking time is more precious than fuel.

STOVES IN USE

Traditional (biomass burning) stoves are considered here to be presently used stoves capable of burning any biomass fuel. Improved (biomass burning) stoves are identified as stoves with better performance levels than traditional stoves for the same cooking requirements and operating conditions. They may or may not be in actual use. Wood-burning stoves are usually classified as *insitu* built and mass fabricated. Each has its own set of merits, so planning their diffusion requires careful consideration.

In-Situ Built Wood-Burning Stoves

In Bangladesh, most wood stoves are built insitu (i.e., they are fixed stoves). Clay and mud are most commonly used to make fixed stoves. Other materials are used such as clay-sand, clay-cement, and brick-cement-sand without significantly affecting heat utilization. Design considerations are stability, strength, and durability. Insitu stoves are usually owner built, although in some villages, experienced, self-taught women-artisans are available for house-to-house construction. These stoves are not built according to engineering drawings, but each component is made according to specific dimensions measured by finger width, hand span, and sizes of cooking pots. After each one is constructed, it is cured, or dried, for a certain period before it can be used. Human labor for construction may be the only cost; material may be free.

Mass Fabricated Wood-Burning Stoves

The most common construction material for this kind of stove is ceramic (burnt clay). Two other materials used to manufacture mass fabricated stoves are cast iron and sheet metal. This stove is designed for short-duration cooking. It does not permanently occupy floor area of the cooking place. In some places, potters manufacture ceramic stoves according to a local design, but a standard model can be fabricated. Users do not have to wait for any curing time. On the other hand, this type of ceramic stove is fragile, unstable, and it must be manufactured outside the household and purchased. Another type of stove component becoming popular is mass fabricated inserts for insitu stoves. Price of mass fabricated stoves depends on costs of materials, labor, and fuel as well as costs for transportation and distribution.

Stoves and Housing Conditions

When the improved stove (e.g., multiple pot stove) requires larger floor area than the existing stove (e.g., three-stone stove) it may not be acceptable where there is a space constraint. In rural areas of Bangladesh during wet months stoves are constructed inside the house. Poor households living and cooking in a single-room accommodation may have a problem finding space for a large stove. The 1973 housing census in Bangladesh recorded 44 percent of rural households living in one-room accommodation and 31 percent in two rooms (GOB 1979). In dry months, stoves are made outside in kitchen yards and availability of space is not critical. A village study in South Africa reports outside summer cooking and inside winter cooking (Marc Best 1979). In Botswana cooking is generally done in open areas; during cold, rainy times of the year, cooking is done indoors (Howard Geller and others 1983).

As of the 1973 housing census, 54 percent of urban households in Bangladesh lived in one room and 23 percent in two rooms (GOB 1979). The number of rooms usually corresponds directly with the economic status. The poorest households, with only one very small room, do not have enough space even for an insitu cooking stove. Most utilize movable stoves or temporary three-brick stoves. Where space permits, during the dry months, the stove is moved outside or a permanent one is constructed outside. Space constraints, however, usually preclude outside cooking in urban areas even during the dry months. Therefore, size of home is an important consideration in designing an appropriate type of improved stove.

Low-income urban dwellers with limited money and space purchase only small quantities of firewood at a time, either daily or weekly. They usually have to burn wet fuel bought from profit-minded retailers who save on labor and storage costs involved in drying wood. Thus low-income urban dwellers must use less efficient fuel that creates more smoke and discomfort, especially during the wet season when they have no way to dry their fuel. Inadequate ventilation also significantly increases smoke problems. On the other hand, higher income urban dwellers with more money and more storage space can purchase greater quantities and the more expensive dry fuelwood. They suffer less discomfort from smoke and probably use less fuel for a given cooking task.

Mobile people choose their stoves and fuel according to various priorities. For example, boatmen, fishermen, and people who live in boats use movable wood stoves. They prefer firewood for cooking because of limited storage space. But if they do not carry a stove with them, they use the three-stone method of cooking on the nearby shore. People with movable stoves who reside in permanent houses in Bangladesh prefer quick-starting kindling fuel, such as leaves, straw, and jute stick, that are more easily and safely used than on a boat.

Design Features of Commercial Stoves

Cooking stoves used for commercial purposes have different operational criteria. For example, tea shops located along roadsides in village marketplaces and in urban areas operate their stoves for longer periods of time each day. Their stoves may be continuously warming two or three kettles. These traditional heavy mass stoves are similar to insitu built improved (heavy mass) stoves developed for domestic cooking. Since they are located in well-ventilated places, they do not have chimneys.

Domestic stoves used commercially (such as for frying rice or pulses to be sold in the market) are traditional stoves modified to accommodate multiple pots over longer time periods and to utilize heat more efficiently. Multiple pot wood-burning stoves for food cooking were observed 80 years ago around temples near Puri, India (J.C. Roy 1904). In addition, I observed similar traditional multipot stoves in a village near Lucknow, India. Multiple pot stoves also are used in some areas of Bangladesh in making ghur (crude sugar) from sugarcane juice.

The immediate cash savings from a more fuel-efficient wood stove would appeal more to commercial users than to domestic users in rural areas. The hundreds of mobile food-selling shops in the region's urban areas prefer lightweight, movable stoves. Domestic users might find that type more attractive, too, because it occupies less space. The Thai bucket stove, for example, is equally acceptable to owners of mobile shops and to urban dwellers.

Variation in Stove Design Characteristics

Table 1 lists stoves used in rural households by feature, season, and place used, according to two rural energy surveys in Bangladesh (Islam 1980, Quader, Omar 1982). One-mouth stoves (those with one pot setting) were most common in both areas for daily cooking. *Tafal*, used for parboiling of paddy — usually in large amounts — and ghur-making, was the other type used in both areas. Two-mouth stoves were used in one location for cooking or for cooking and parboiling of paddy (Islam 1980). At the other location (Quader, Omar 1982), three-stone open fires were used for parboiling of paddy.

It may be noted from Table 1 that in a particular season four different types of biomass-burning stoves (one-mouth, two-mouth, tafal, movable) were in use in some villages. Total number of stoves per household varied from two to five, the number depending on changes in cooking place and specific cooking needs of the household, which may vary year to year.

I observed a family in Comilla district of Bangladesh who had four

TABLE 1.
Seasonal Use Pattern of Stoves, Rural Bangladesh

Number of Stoves by Type

Number	Households	Inside Kitchen (Wet Season)					Kitchen Yard (Dry Season)					Biomass burning stoves per household+
		Biomass burning stoves				Kerosene	Biomass burning stoves				Kerosene	
		One-Mouth	Two-Mouth	Tafal	Movable		One-Mouth	Two-Mouth	Tafal	Movable		
N 1	293	336	80	9	16	8	262	287	12	2	0	3.4
N 2	145	154	52	2	4	1	135	131	6	1	0	3.3
N 3	349	380	88	13	16	2	295	283	18	0	0	3.1
N 4	151	169	63	1	38	7	147	150	26	1	2	3.9
N 5	44	51	7	1	4	0	19	62	5	1	0	3.4
N 6	43	51	21	2	4	2	44	51	2	0	0	4.1
N 7	61	76	31	5	15	0	84	74	13	3	0	4.9
N 8	83	100	35	1	11	4	89	85	12	0	0	4.0
N 9	94	122	23	4	34	9	132	100	5	0	0	4.5
N10	70	79	52	1	13	1	79	80	1	0	0	4.4
N11	94	129	0	0	8	1	80	80	0	0	0	3.2
N12	106	116	41	1	4	1	102	98	5	0	0	3.5
N13	94	129	37	4	3	1	79	50	8	0	0	3.2
N14	63	79	4	5	10	0	25	40	17	3	0	2.9
N15	8	8	0	0	0	0	5	5	0	0	0	2.3
N16	62	72	13	0	9	1	50	46	1	1	0	3.1
N17	172	191	30	0	9	1	149	143	1	0	0	3.0
N18	117	141	30	2	14	1	88	95	9	0	0	3.2
N19	20	21	3	1	0	0	12	22	4	0	0	3.2
N20	93	93	1	0	7	1	93	88	4	33	0	3.4
N21	67	80	14	0	7	0	66	93	0	0	0	3.9
N22	146	183	14	3	28	2	126	110	14	0	0	3.3
N23	445	541	68	0	49	3	425	419	0	1	0	3.4
K 1	455	498	6*	2	8	—	288	52*	18	3	—	1.9
K 2	250	322	2*	—	3	—	145	28*	14	5	—	2.1
K 3	155	150	4*	—	—	—	103	113*	13	—	—	2.5
K 4	94	115	16*	—	—	—	41	198*	11	1	—	4.1

Sources: Village Numbers N1-N23 are from Islam 1980 and Village Numbers K1-K4 are from Quader and Omar 1982.

Note: Three-stone type temporary stoves are marked with *. Villages K1-K4 have no two-mouth stoves. Temporary stoves are used mainly for parboiling paddy.

+Include biomass burning stoves used in wet season.

stoves with the following combination: one each of one- and two-mouth stoves (Figs. 1 and 2, Chapter 2) were made inside the kitchen for use in the wet season. One each of two- stove and three-mouth stoves (Figs. 2 and 3, Chapter 2) were made in the open kitchen yard for use in the dry season. One-mouth stove was used for daily cooking. Two- and three-mouth stoves were used for cooking or for cooking and parboiling of paddy or for parboiling of paddy only. In this type of situation the designer of improved stoves has the following choices: to replace each type of stove with an improved version of the same type, to develop a single improved stove capable of meeting the need of all the traditional stoves in use, or to develop improved stoves to meet each season's needs.

Domestic cooking stoves reported on in an 80-year old review (Roy 1904) have a cylindrical hearth constructed below ground for minimal heat loss. For better mixing of air and more stable structure, the inside of the one-mouth version is shaped like a pitcher. Since they are made of a mixture of dung and mud, ash from combustion acts as a heat insulator. Careful operation of the stove saves on fuel. If a grate is added combustion is better.

Using a two-mouth stove (Fig. 2, Chapter 2) and two pots might reduce cooking time but increase the amount of fuel used. Less fuel is used if the fuelwood inlet port is moved from the center to one side, the pot settings are arranged one behind the other, and flue gases are only allowed to escape through the second mouth. That would heat the second pot more slowly. So the design choice is between more fuelwood or more time spent for cooking (Roy 1904).

My rural energy survey (Chapter 2) found that in 42 percent of the 5,800 one-mouth stoves, the fuel inlet hole was from 15 to 18 cm in diameter. In 40 percent, the cylindrical combustion chamber was from 18 to 20 cm in diameter. The depth (distance from the base of combustion chamber to the bottom of pan) of 55 percent of the stoves was from 38 to 51 cm (see also Islam 1980). According to another survey, in 92 percent of one-mouth stoves, the diameter of the fuel inlet hole was about 15 cm. In 71 percent, the diameter of the combustion chambers ranged from 20 to 25 cm. The depth of each stove surveyed was 38 cm (Quader, Omar 1982).

The diameter of fuel inlet hole is related to the number of fuelwood pieces that can be fed into the stove at once. Generally two to three pieces of fuelwood, each 4 cm wide, are used at one time with adequate clearance for air. The diameter of the cylindrical combustion chamber is related with the size of cooking pan which, in turn, varies according to the family size. The depth of the combustion chamber provides combustion volume, controls combustion air by chimney action, and acts both as an ash collection pit and as an air shield for the combustion zone. Depth has been found to be a crucial factor in heat utilization efficiency, amount of cooking time, and amount of smoke (Islam 1980). Thus, dimension as a design factor affects the overall performance of a stove.

A series of experiments varying the depth of one-mouth stoves revealed that thermal efficiency increases as depth decreases, to an optimum performance depth of 43 cm (i.e., performance with high thermal efficiency, minimal cooking time, and minimum smoke). When the depth was less than 43 cm, thermal efficiency increased slowly, cooking time increased considerably, and thick smoke filled the cooking environment (Islam 1980). Those observations correlate with the cited survey findings that in one area 55 percent of one-mouth stoves were from 38 to 51 cm deep (Islam 1980) and in another area all stoves surveyed were 38 cm deep (Quader, Omar 1982).

In Nabagram, in 50 percent of the 3,200 two-mouth stoves surveyed, the diameter of the fuel inlet hole was from 15 to 18 cm; 49 percent of those stoves were from 53 to 61 cm long and 27 cm wide. The depth of 45 percent stoves was from 51 to 64 cm (Islam 1980). A survey of dimensions of 14 tafals was carried out in Kulaghat. In 72 percent, the diameter of the fuel inlet was 18 cm. In 80 percent, the depth was 28 cm; all were about 64 cm long, and 60 percent were from 38 to 51 cm wide.

Altogether, over 100 designs of stoves are contained in a stove compendium by Guido de Lepeleire and others (1981). This publication recorded for the first time the many designs of traditional stoves. Without diagramatic presentation, previous publications on stoves gave an impression that all traditional stoves were designed as three stones with an open fire inside. Yet that compendium still does not cover all existing designs.

FUELS

Kindling Fuels

Starting a fire without kindling is difficult, yet laboratory researchers who use artificial methods such as using a pilot gas burner or soaking the end of fuelwood in kerosene are not familiar with this key step in wood stove operation. The amount of kindling needed depends on fuelwood used and stove design. Kindling procedures and the quality of kindling fuel itself would vary the amounts of fuel required for lighting the stove. Time and conditions of ignition are also affected. For instance, in Bangladesh fuelwood that has been cut and split into small pieces by hand and dried in the sun is used as kindling with normal size fuelwood pieces. Easily ignitable fuels such as paraffin wax, kerosene, waste paper, small

Coconut fronds hung for drying before burning. Fishing village, Sri Lanka. (Kirk R. Smith)

pieces of dry fuelwood, and dry leaves are used for kindling, but rural people in developing countries generally use locally available natural sources such as leaves, straw, jute sticks, and brushwood. A fibrous byproduct from oil palm production called jha is used for kindling in Western Nigeria (P. Ay 1978). One study reported on the use of brushwood to start dung fires in a South African village. People there light their movable wood stoves outside and then move them back into the kitchen (Best 1979). People in Bangladesh use movable stoves in a similar way. In some rural areas, the same fuels are used for kindling and for main cooking because they are in abundant supply during specific seasons of the year. Generally, wood-burning stoves create heavy smoke at first from incomplete combustion, although after the fuelwood ignites, the smoke lessens considerably. Fuel consumption estimates also vary according to methods used to account for heat used in kindling.

In Bangladesh, usually the stove is lit before the cooking pot is put on the stove, to keep the food from smelling smoky and to make lighting easier. The cooking pot is removed from the stove if the fire needs relighting during cooking. Only the poor put the pot on first to utilize the heat immediately. Stoves with grates light more easily and reduce smoke so they would appeal to people who are able to improve their quality of life.

Biomass Cooking Fuels

Biomass fuels used for cooking are obtained from three different sources: trees, agricultural crops, and animals. Of these, fuelwood is considered superior because its burning quality is better and its storage presents fewer problems (less fire hazard, less space). In cities of developing countries, fuelwood is the most common commercially available biomass fuel for cooking, although charcoal also is used in some urban areas. Except in some special situations, most cooking fuels are still noncommercial in rural areas of developing countries. Usage varies by region and locality, and by season. Substantial use of agricultural and animal residues for cooking, as in rural Bangladesh, signifies fuelwood scarcity. The composition of available biomass fuels may also vary with landholding size and income group (see Chapters 2, 4, and 6).

Processed Fuels

Accommodating different fuels is a difficult design problem. Compacting locally available fuels is the main way to overcome that problem (Prasad 1981a). The procedures and hardware involved in compacting wood, crop, and animal residues have been discussed elsewhere (J. Janczak 1980). Within Southern Asia, research on compaction of residue fuel is continuing at the Thailand Institute for Scientific and Technological Research (TISTR) (N. Pitakarnnop 1981); Institute of Technology Bandung (ITB), Indonesia (Filino Harahap 1981); Jyoti Solar Energy Institute (JSEI), India; and University of the Philippines at Los Baños (UPLB) (E.P. Lozada 1981). In rural areas of Bangladesh people burning dung are found to support it on a stick to control its burning like a piece of firewood.

Processing trees into fuelwood is difficult with only hand tools, so fuelwood gatherers prefer trees with smaller diameters. Furthermore, sizing and splitting dried trees is harder, so green trees are processed often. In urban Bangladesh, fuelwood pieces are from 0.3 to 0.4 meter; in rural areas fuelwood is from 0.4 to 0.75 meter. The different lengths accommodate a particular combustion chamber's depth. Stoves in urban areas are made above ground level; in rural areas the combustion chamber is made below ground level. In rural households where mixed fuels are used, more space is necessary for the greater volume of ash. In urban areas where fuelwood is used, less volume is required for ash collection. The cut length of fuelwood is also related to transportation and storage convenience.

Smoke and Flue Gases

Problems related to smoke and flue gases vary according to house design in addition to the seasonal variations mentioned earlier. These problems are discussed next in the context of one stove study in Ungra village near Bangalore, India (Geller 1981), focusing on roof structure as one among many affecting ventilation.

Approximately 80 percent of the houses in Ungra have tile roofs, 10 percent have thatched roofs, and 10 percent have *malige* roofs over the kitchen. Malige roofs consist of a 30-cm thick mud covering over horizontal wooden rafters. Smoke seeps through thatch and gaps in tile roofs. In homes with tile roofs, a space is often left open between the top of the wall and the roof above to let out smoke. Houses with malige roofs are air tight. In those houses, ceiling vents and, in some, kitchen windows release smoke, but the smoke still reaches uncomfortable and apparently hazardous levels.

In one rural area in Bangladesh, 91 percent of the kitchens and 96 percent of the roofs are made of leaves and straw (Islam 1980). Those materials have good ventilation for the smoke. Throughout Bangladesh, most have thatched roofs.

Fuel Efficiency

Overall stove efficiency equals combustion efficiency times heat utilization efficiency, or:

$$\frac{\text{heat actually produced by normal combustion}}{\text{heat producible by complete combustion}} \times \frac{\text{heat received for cooking}}{\text{heat actually produced by normal combustion}}$$

Provisional standards and methods for analysis of stove efficiency are presented in a recent publication (VITA 1982). Recognizing the need for qualifications related to indexing problems (Chapter 1), overall comparisons between the efficiencies of different fuels used for cooking are cited next.

Electricity is usually considered a superior form of energy to fuelwood, but its use for cooking is nearly as inefficient as fuelwood is in wood stoves. About 75 percent of primary energy is lost during generation and transmission. Given an average cooking efficiency of 70 percent for an electric stove (C.G. Segeler 1966), the overall efficiency of primary energy use becomes only 17.5 percent ($0.25 \times 70 = 17.5$).

Among wood fuels, charcoal is considered to be the best fuel. Estimates of charcoal output per unit of the original wood range from 3 to 12 percent of dry wood by the earth stack method to 12 to 15 percent by pit method (E. Uhart 1975), or as high as 25 percent in some earth stack situations. Given an average efficiency of 50 percent for a charcoal stove, assuming a mid-estimate charcoal/wood ratio, and assuming the heating value of charcoal is 35 MJ/kg and of fuelwood 20 MJ/kg (Bialy 1979), the overall efficiency of primary energy use is only 13 percent:

$$\frac{0.15 \text{ kg charcoal}}{1 \text{ kg wood}} \times \frac{\text{heating value of charcoal}}{\text{heating value of wood}} \times 50\%$$

$$= \frac{0.15 \text{ kg.} \times 35 \text{ MJ/kg}}{1 \text{ kg} \times 20 \text{ MJ/kg}} \times 50\% = 13\%$$

For an analysis of experience in introduction of improved charcoal stoves, see Matthew Gamser and C. Harwood (1983). Efficiencies for gas stoves and liquid fuel stoves are higher, not accounting for transport or other energy requirements outside the stove boundary.

For some users of traditional stoves, an improved wood stove might potentially increase overall efficiency by 200 to 300 percent, but user response would be slow because of their economic conditions and availability of resources. On the other hand, users who are already suffering from a cooking fuel crisis might be operating their stoves at a higher efficiency level than that first user group, because they adjust cooking habits (cooking time, operator's attention) and cooking methods. For them, the potential of increased fuel efficiency might not be more than 50 to 100 percent higher than their existing level. In adopting an improved stove with higher efficiency, some users might modify their cooking practices or shift fuel source from residues to fuelwood. Family nutrition and health as well as the environment might benefit, but effects on fuelwood demand and supply would vary according to local circumstances. Both sets of factors require local assessments.

CHANGING DESIGN PERSPECTIVES

A stove has four functional components: (1) a fuel and or air inlet port, (2) a fireplace (combustion chamber), (3) a pot seat, and (4) an outlet port for combustion products (flue gases). A traditional stove has these shortcomings: incomplete combustion of fuels (less heat and more smoke), and inefficient heat utilization. An improved stove should achieve these objectives: better burning (and therefore more heat and less smoke), reduced heat loss and better heat utilization, and removal of smoke.

Three central features of recently advocated stove designs are (1) a closed combustion chamber; (2) horizontal ducts through which hot combustion gases are passed to maximize heat recovery in a series of pot holes; (3) a chimney to let air in and let smoke out from the stove environment (de Lepeleire and others 1981, T. Acott and others 1980). These design features were developed earlier in India (Nideshak 1958, S.P. Raju 1961) and Indonesia (Hans Singer 1961).

Two recently recorded stove designs have attractive features. The basic technical design of the Thai bucket stove (NEA 1981) is the same as the charcoal-burning ceramic stove developed by the Ceylon Institute for Scientific and Industrial Research (CISIR) in Sri Lanka (Dhammika de Silva 1981). Because it is clad with a metal sheet, the Thai bucket stove is less fragile and more durable. Also this type is suited to either charcoal (used mainly) or to wood fuel. The metal sheet adds to the stove's cost, so a different grade of the same basic design might be more appropriate for low-income consumers. On the other hand, features such as a metal

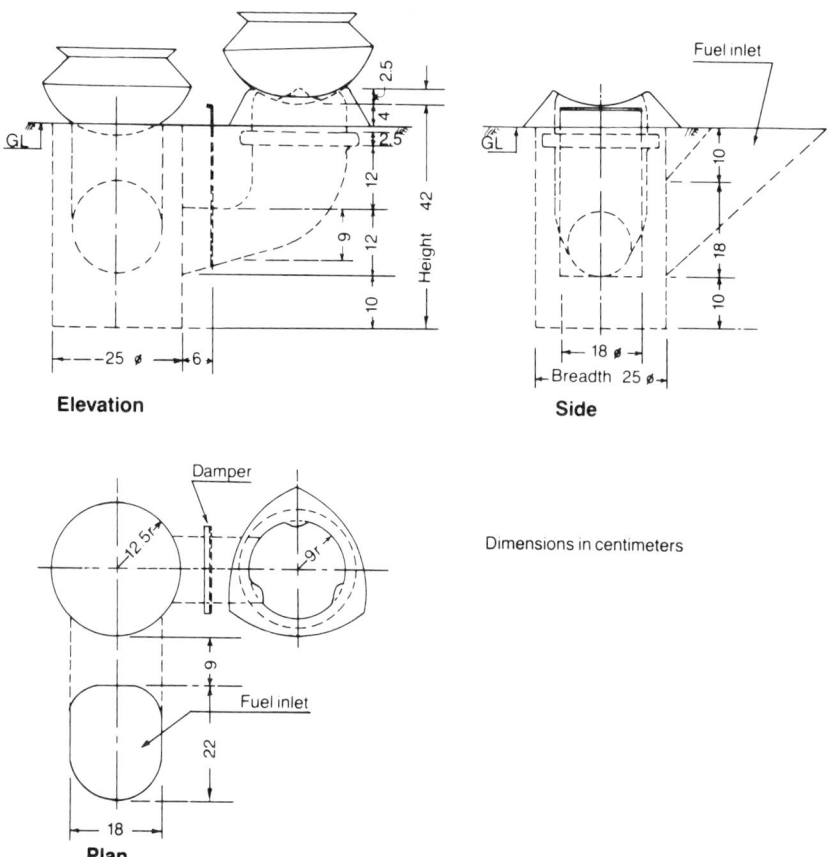

Figure 2. Improved (Nabagram) Stove

framework and tile top (Lozada 1981) would make the stove more attractive to higher income groups, though such external improvements increase the stove's cost by nine times.

A recent design (Fig. 2) developed by Islam (1980) is like the traditional stove described by Roy (1904), except the pots are arranged side by side instead of one behind the other. In a Sri Lankan village near Kandy, I observed a traditional stove similar to this recent design.

These examples are cited to stress that some of the designs of improved stoves claimed by researchers are similar to existing stoves used elsewhere or to traditional stoves developed by rural people through their own experience. This design history might motivate research to discover the conditions (a) that have fostered acceptance of the latter stoves where they are in use, and (b) that have discouraged their wider diffusion.

356 Rural Energy to Meet Development Needs

Metering smoke exposure, bread making, Gujarat. Note chimney-fitted stove. (Kirk R. Smith)

Various components of improved stoves generally cited in the literature are

1. Grate, to distribute air better to achieve better burning.
2. Chimney, to remove flue gases from the combustion chamber thereby indirectly drawing in air for combustion.
3. Damper, to control the flow of air for optimum combustion.
4. Hearth, a type of enclosed combustion chamber that reduces heat loss.
5. Secondary combustion chamber to increase combustion efficiency.
6. Baffles, to divert hot flue gases to cooking pots to achieve better heat utilization.
7. Pot holes designed to fit pots exactly, thereby reducing the heat loss around pots.
8. Multiple pot holes, to improve heat utilization.

A particular design of improved stove, of course, may not have to incorporate all the components.

Stove researchers have considered the chimney a technical solution to the smoke problem. A chimney actually has two functions; one as a duct to carry out smoke and flue gases, thereby creating a draft that permits air to flow into the combustion chamber. A single pot stove fitted with a chimney increases fuel consumption. In a multiple pot stove flue gases pass through horizontal ducts to maximize heat utilization. But even then, to avoid higher fuel consumption the chimney draft must be controlled by using a damper.

A chimney might not be an appropriate innovation in some cases. Sometimes escaping smoke acts as protection for roofing materials against insect attack; sometimes heat from the fuel gases dries crops or heats a space. In hot climates, users would like to remove hot food vapors from the cooking area. A thatched roof with enough clearance at the top of the wall, or a corrugated iron sheet or concrete roof incorporating a vent system, could remove flue gases and vapors from the whole kitchen. A chimney could not fulfill that need.

Individual situations influence successful operation of a stove with a chimney. A laboratory researcher might advocate a certain chimney height that insures a desired air intake, without considering kitchen height. Increasing the chimney height to carry smoke beyond the roof might increase fuel consumption. Examination of particular housing conditions and ventilation should include a quantitative analysis of smoke and gases generated in different kitchens and their impact on health of stove users in order to provide a valid basis for identifying well-designed kitchen systems (see Kirk Smith, A.L. Aggarwal, and R.M. Dave 1983).

Other design choices must take into account particular circumstances too. Raised stoves may be desirable because children are less likely to be burned, but sometimes, as in rural areas of Bangladesh, stoves at ground level may reduce fire hazard in kitchens with combustible biomass roofing or walls. That fire risk is less if roofing is made of corrugated iron sheet, tiles, or concrete. Cooking practices and tools are adapted to particular stove designs and are not easily changed. When the cooking position and matching kitchen tools and appliances must be changed if a new stove design is to be used, the new design might not be accepted.

An improved stove (Sheba Chula) fitted with a grate of wire mesh was developed in Bangladesh to cook with firewood. Rural households using agricultural residues and dry leaves find it unsuitable because grate openings get clogged with unburnt fuel. On the other hand, wider gap of grate might allow the fuel to fall below the grate in unburnt condition.

In a multiple pot-setting stove heat utilization efficiency would be affected if all the pot holes are not in use at a time.

TESTING STOVE PERFORMANCE

Generally stove performance has been rated according to overall thermal efficiency. The overall thermal efficiency is inversely proportional to the amount of fuel consumed. Therefore, to carry out a certain cooking operation, a stove with higher overall thermal efficiency would require less fuel than a stove with lower overall thermal efficiency. Recorded attempts to compare the performance of various stoves using estimates of their overall thermal efficiency often have failed because testing has not been standardized. Estimates of overall thermal efficiency and fuel consumption depend on many interrelated factors. For valid results, these factors must be identical in nature or their corresponding influence must be considered on a similar basis.

Selecting Stoves for Testing

Criteria used for testing stoves have been very weak in the past. In domestic energy surveys, various statistical sampling procedures are used to select a household survey group. A sampling procedure to select representative test stoves, however, is rarely followed. Often no data are available or collected about the stove types used in the survey area and, in most cases, either one kind or only very few existing stoves are selected arbitrarily.

Since domestic cook stoves are built using indigenous technical knowledge and materials, their critical dimensions, mode of operation, and fuel vary greatly. For example, size and shape of one particular three-stone stove depend on the orientation of the individual stones, their height, and the gap between them. Those differences may in turn affect the operation and procedures and then the overall thermal efficiency of the stoves. Virtually every individual stove, then, can be considered a separate design and as a group, three-stone stoves cannot be compared with any particular improved design.

For those reasons, the number of stoves selected for technical evaluation must be representative of the total number in use. Then physical and operational differences of those stoves should be correlated to overall thermal efficiency, additional uses of the stove, comfort, and other user variables. If thermal efficiency of stoves in use varies widely, factors inhibiting emergence of a dominant version should be identified through in-depth discussions with representative users. Then that information could be used in future designs for improved stoves. The indiscriminate transfer of experience from laboratory to various users or from one country to another should be avoided.

Heating Water Tests

Overall thermal efficiency of a stove can be determined experimentally by burning a known amount of firewood and heating a known amount of water in a cooking pot on that stove. The amount of firewood burned is used to estimate the heat input, and the amount of heat received by the water is used to estimate heat output. Water is universally used for this experiment because of its known physical properties, easy availability, and wide applicability in cooking.

Experimental Procedures. These procedures have been followed in various studies:

1. The weight of water used for heating is kept fixed. A particular experiment is considered complete (a) when the amount of water reaches its boiling point; (b) after the water boils, evaporation (simmering at constant boiling point) continues for a predetermined time period; or (c) after the water boils, heating continues (simmering at the boiling point) until all the water is evaporated. The amount of fuel burned in each of the three cases is noted.

2. The time of the experimental run is kept constant. A certain amount of water is heated to boiling or boiling and evaporation. The amount of fuel burned is noted.

3. The weight of fuel burned is kept constant, and a particular run is continued until a certain amount of fuel is burned. Then either (a) the number of times a fixed quantity of water is brought to boiling is recorded, or (b) a given amount of water is heated to boiling and evaporation is continued until the fixed amount of fuel is used.[1]

Factors Affecting Results. Five kinds of factors can affect the estimates of thermal efficiency: variations in fuel, differences in pots used, operator dissimilarities, difference in physical environment, and the sequence of experiments. The variations in the fuel (heating or calorific value, moisture content, species of tree, tree part, and size of fuelwood) are discussed next.

The two types of *heating value* of firewood are gross (higher) heating value and net (lower) heating value. Gross heating value is the amount of heat released by the complete combustion of unit weight of fuel in standard condition, assuming all the water produced during combustion has returned to its liquid state. Net heating value is that amount assuming the water produced remains vaporous. Therefore use of the gross heating value calculations means a lower value of overall thermal efficiency; use of net heating value results in a higher value.

Both gross and net heating values are affected by the presence of *moisture* in the firewood; a correction in the heating value is necessary to allow for the amount of moisture present. In laboratory conditions, either dried firewood or firewood with a known moisture content are used in experiments, but in the field, moisture content of firewood is difficult to determine precisely. A procedure for approximating the moisture content of wood from the relative humidity of air is provided in VITA (1982). The moisture content in fuelwood requires a portion of the generated heat for its evaporation but may or may not positively affect the combustion process directly.

Although the heat value by weight (kJ/kg) of various hardwood and softwood species is approximately the same for oven-dry samples using a bomb calorimeter the variations in density and in physical and chemical composition of different *firewood types* might alter burning characteristics. So heat would be released at different rates. Bark, twigs, and other tree parts also have varying burning characteristics.

Thermal efficiency estimates are also affected by differences in cooking pots. Construction, materials, shape, and size of the pot used affect overall thermal efficiency. So do a particular pot lid, soot deposits on pots, and dissolved solids deposits on the inside pot surface.

A given weight of firewood can be fed into a stove intermittently, in batches, or continuously in the form of long sticks. Burning characteristics of the wood vary according to such feeding rates and overall efficiency is affected. Generally one specific laboratory staff member is responsible for carrying out experiments estimating efficiency. The accuracy of each experiment is tested by the same individual's repeating the tests for reproducibility. That kind of reproducibility is most attainable when the *operator* has very little direct influence on experimental variables. Tests on lighting the stove, feeding firewood, and maintaining the fire depend directly on operator expertise. Therefore, if operators are different, reproducibility test results will probably vary. Variations between scientist operators and experienced rural women operators would be worth studying in the future.

Variations in the *physical environment* (ambient temperature, humidity, wind speed and direction) also affect overall thermal efficiency. Whether or not the experiments are conducted inside a closed room or in an open space also greatly affects the interpretation of the results.

Test Limitations. In some cultures, much cooking involves baking and frying. Those operations are not reflected in simulated water heating tests. The heat input rate is decreased somewhat for simmering in actual cooking, depending on the type of food, the cook's experience, and the cooking time available. The oil used in cooking also varies heat transfer. In the simulated water heating test, the heat input is constant, and no cooking oil is used.

In actual cooking, the objective is proper cooking of food; how much water is evaporated during cooking is not important. Water heating tests estimate overall thermal efficiency by how much heat is required to evaporate the water. That measurement is only a partial indicator of stove performance. Thus, final evaluation of a stove should be based on performance of the desired cooking operation.

Cooking Food Tests

Overall thermal efficiency is not measured directly when evaluating a stove by food cooking experiments. Instead, the amount of fuel consumed for cooking a given quantity of a certain dish is used as a basis for evaluation. That amount of fuel depends on stove performance as well as on the particular dish's characteristics. These characteristics include quality and quantity of uncooked food and the process of cooking.

One series of experiments focused on variations in amount of heat required to cook rice (Islam 1980). Stove type (heating system), cooking pan types, and rice variety, as well as amount of rice cooked in one batch and method of rice cooking were the variables for the study.

Tests A: Different Heating Systems. In these experiments (with one exception), enough rice for an average family's meal was cooked using varying heating systems. (In the experiments using a pressure cooker the amount cooked was limited to 0.9 kg by the vessel's capacity.) Rice was cooked with excess water (drained off after cooking) — the typical method of cooking rice in Bangladesh. Four types of cooking pans were used with each heating system.

Consistency was maintained as much as possible by using the same initial amount of kerosene for each reading when starting a kerosene burner. Also, kerosene flow rate, natural gas flow rate, and electricity input were kept constant. For cooking on a wood stove, firewood of the same size was used each time. In normal cooking, heat input is slowed for simmering. During the experiments, however, the heat input was kept constant to avoid another operator variable. Thus, the estimated heat requirement during the experiments is slightly more than would be normally used.

Since rice is usually the first course, each stove was tested from the cold condition of stove. (With a wood stove more heat is required to cook a particular food item if the stove is cold; the difference is only marginal

362 Rural Energy to Meet Development Needs

Simulated brick stove, East-West Center air pollution lab. (Kirk R. Smith)

with a light electric heater, gas burner, or kerosene cooker.) Usually a few grains of rice are pressed between fingertips to test for doneness. So in these experiments, cooking time depended on each cook's subjective judgment. Readings were checked and repeated if the physical quality of cooked rice or the ratio of the weight of cooked rice to dry rice were markedly different. (See Table 2 for summary of results.)

These observations are possible:

1. With each type of heating system, the amount of heat required

TABLE 2.
Fuel Required for Cooking Rice by Using Different Heating Systems

Heating System	Description	Type of Pan			
		FA	SA	SC	PC
Electric	Time (min)	65	67	94	43
	kJ/kg of rice	3435	2584	3628	2489
	kJ/kJ[++]	0.23	0.18	0.25	0.17
	Cooked rice/rice	3.87	3.27	3.59	3.49
Natural gas top surface burner	Time (min)	42	50	61	25
	kJ/kg of rice	3996	4131	4746	2916
	kJ/kJ	0.27	0.28	0.32	0.20
	Cooked rice/rice	3.30	3.53	3.57	3.47
Kerosene: wick control burner	Time (min)	56	56	63	32
	kJ/kg of rice	4179	3618	3889	3094
	kJ/kJ	0.28	0.24	0.26	0.21
	Cooked rice/rice	3.62	3.54	3.53	3.41
Firewood: One-mouth stove (0.43 m depth of hearth)*	Time (min)	46	44	44	24
	kJ/kg of rice	35890	25110	25110	22510
	kJ/kJ	2.42	1.7	1.7	1.52
	Cooked rice/rice	3.56	3.62	3.55	3.44
Firewood: improved stove[+]	Time (min)	47	45	55	
	kJ/kg of rice	19340	15500	15560	
	kJ/kJ	1.31	1.05	1.05	
	Cooked rice/rice	3.63	3.5	3.59	

Source: Data are from Islam 1980.

Note: FA: Flat bottom aluminum pan--1.14 kg rice cooked with 4.88 kg water; SA: Spherical bottom aluminum pan--1.36 kg rice cooked with 5.9 kg water; SC: Spherical bottom clay pan--1.36 kg rice cooked with 5.9 kg water; PC: Pressure cooker--0.91 kg rice cooked with 2.27 kg water. Type of rice used was *aman* rice, moisture content 16 percent. Caloric value of rice = 14,800 kJ/kg.

* See Figure 1, Chapter 2,
+ See Figure 2.

++ kJ/kJ = Heat input to cook unit weight of rice/Caloric value of unit weight of rice.

Net heating value of fuels (GOB 1976): Natural Gas = 993 kJ/standard cubic foot, Kerosene = 46,055 kJ/kg, Firewood = 15,119 kJ/kg.

to cook a unit weight of rice varied according to the type of pan used.

2. Cooking with a pressure cooker required the minimum amount of heat (fuel) to cook a unit weight of rice, but the expense of

TABLE 3.
Heat Utilization Efficiency of Cooking Devices, by Water Boiling Test

Pan Type and Heat Heating Device (Heat Input Rate)	Utilization Efficiency (percent)			
	FA	PC	SA	SC
Electric coil heater 876 Watt (3154 kJ/hr)	77	63	68	70
Natural gas: top surface burner. Input 6300 kJ/hr (average)	56	57	54	50
Kerosene: wick control domestic burner (10 No. wicks). Input 5000 kJ/hr (average)	54	53	53	50
Firewood (Sundari variety). One-mouth traditional stove* with 0.43 m depth. Input 46,000-48,000 kJ/hr (average).	8	12	13	12.5

Source: Data are from Islam 1980.

Note: Cold start experiment, 5.45 kg water load for pan types FA, SA, SC, and 3.3 kg water load for pan PC, one hour duration. FA: Flat bottom aluminum pan; PC: Pressure cooker; SA: Spherical bottom aluminum pan; SC: Spherical bottom clay pan. Water capacities of different pans: FA = 7.3 kg, PC = 4.4 kg, SA = 7.4 kg, SC = 8.8 kg, .

*See Figure 1, Chapter 2.

the pan might discourage its use, especially since its capacity was limited.

3. The quantity of heat required to cook rice using different heating systems varied inversely with the heat utilization efficiency estimated by water boiling test (Table 3).

According to two rice cooking experiments reported by Roger Revelle (1976), the energy required to bring the cooking water to boiling and then to boil away requisite quantity of water was 2,510 kJ/kg of rice or 17.5 percent of the food energy content of rice. The estimated amount of heat required recorded by Revelle (1976) is valid for a particular type of cooking device, type of heating system, and cooking pan. Results also are limited by rice variety used, by amount of rice cooked in one batch, and by cooking method used.

The heat required to cook rice using an electric hot plate was reported as 2,770 kJ/kg of rice (Mansur N. Hoda 1979). That finding is comparable within the range of tabulated data for an electric heater in Table 2.

Tests B: Different Rice Varieties. In Bangladesh, available rice varieties are classified as either parboiled rice (*Siddha*) or unparboiled rice (*Atap*). Four varieties of parboiled rice and two varieties of unparboiled rice were selected for experiments carried out using one top surface, natural gas type domestic burner, and one domestic burner with kerosene wick control. Rice was cooked using excess water, in flat bottom aluminum pans (see Table 4). The amount of water used each time was based on the experience of the woman researcher. Results of another rice-cooking study using no excess water (all water boiled off at the end of cooking) are summarized in Table 5.

Those two studies provide these observations:

1. The parboiled rice variety required more heat per unit weight than the unparboiled variety. Heat requirements for each parboiled variety depended on the degree of parboiling and on grain size of rice. Heat requirements also varied for each variety of unparboiled rice, depending on grain size.
2. Using one fuel and one variety of rice, the amount of heat required per unit weight of rice decreased as the amount of rice cooked in a batch increased. This scale factor explains why per capita fuel consumption decreases as number of people per household increase (Chapter 2).
3. Less heat was required per unit weight of rice with a gas burner than with a kerosene cooker, probably because heat was utilized differently in each system.
4. The dry method of rice cooking (with no excess water) required less heat per unit quantity of rice compared with the same amount cooked with excess water.

Using these methodologies, sets of readings on rice cooking could be taken for wood stoves. Then the experimental data could be compared with heat utilization data obtained at users' level. Similar experiments could estimate fuel consumed in cooking common dishes, e.g., baking of chapati, cooking rice, meat curry, fish curry, vegetable curry, and pulses.

Tests C: Typical Daily Meals. Performance of a stove also can be evaluated by cooking the typical daily meals of an average family. If the test is carried out for cooking one typical meal, the fire is extinguished immediately after each period of active cooking. Unburned firewood

TABLE 4.
Fuel Required to Cook Different Rice Varieties, Using Excess Water

Type of Rice	Amount of Rice Cooked per Batch (kg)	Type of Fuel							
		Kerosene				Natural Gas			
		WW/WR	kJ/kg Rice	kJ/kJ	WCR/WR	WW/WR	kJ/kg Rice	kJ/kJ	WW/WR
P-1	0.45	4.6	10140	0.80	3.5	4.6	7518	0.51	3.1
	0.91	3.0	4875	0.33	3.0	3.0	4174	0.28	2.8
	1.36	3.0	3519	0.24	2.9	-	-	-	-
P-2	0.45	4.6	6898	0.47	2.9	4.6	5605	0.38	2.8
	0.91	3.0	3554	0.24	2.7	3.0	3079	0.21	3.0
	1.36	2.5	2470	0.17	3.0	-	-	-	-
P-3	0.45	4.0	6492	0.44	2.7	4.0	5193	0.35	3.0
	0.91	3.0	3402	0.23	3.2	3.0	2873	0.20	2.9
	1.36	2.5	2538	0.17	2.8	-	-	-	-
P-4	0.45	4.0	5579	0.38	2.9	4.0	5195	0.35	2.8
	0.91	3.0	3245	0.22	3.3	3.0	2942	0.20	3.0
	1.36	3.0	2504	0.17	3.1	-	-	-	-
UP-1	0.45	3.5	4768	0.32	3.3	3.5	3282	0.22	3.4
	0.91	3.0	2336	0.16	3.2	3.0	2192	0.15	3.1
	1.36	2.5	1760	0.12	2.8	-	-	-	-
UP-2	0.45	3.5	4870	0.33	3.5	3.5	3692	0.25	3.7
	0.91	3.4	2590	0.18	3.5	3.0	2395	0.16	3.1
	1.36	3.0	2199	0.15	3.1	-	-	-	-

Source: Data are from Islam 1980.

Notes: Caloric value of rice = 14,800 kJ/kg. P is parboiled rice; UP is unparboiled rice. Arabic numbers after P and UP indicate a particular variety of rice grain size or degree of parboiling. Moisture content of different varieties were determined as follows: P-1 (12.4%), P-2 (17%), P-3 (16%), P-4 (17%), UP-1 (16%), UP-2 (12%).

WW: Weight of water; WR: Weight of rice; WCR: Weight of cooked rice. kJ/kJ = Heat input for cooking unit weight of rice/Caloric value of unit weight of rice.

Net heating value of fuels (GOB 1976): Kerosene = 46,055 kj/kg, Natural gas = 993 kJ/standard cubic foot.

Water capacities of flat bottom aluminum pans used for rice cooking are: 3.2 kg (for cooking 0.45 kg rice), 5.9 kg (for cooking 0.91 kg rice), 7.3 kg (for cooking 1.36 kg rice).

TABLE 5.
Fuel (Heat) Required to Cook Different Rice Varieties,
Without Using Excess Water

		Type of Fuel							
		Kerosene				Natural Gas			
Type of Rice	Amount of Rice Cooked per Batch (kg)	WW/WR	kJ/kg Rice	kJ/kJ	WCR/WR	WW/WR	kJ/kg Rice	kJ/kJ	WCR/WR
P-1	0.45	4.0	10850	0.73	3.3	2.7	6151	0.40	2.8
	0.91	2.6	4672	0.32	3.0	2.5	3900	0.26	3.1
	1.36	2.4	3452	0.23	3.1	-	-	-	-
P-2	0.45	2.8	6087	0.41	3.3	2.4	4921	0.33	2.9
	0.91	2.0	3402	0.23	2.8	2.3	2943	0.20	3.0
	1.36	2.0	2198	0.15	2.8	-	-	-	-
P-3	0.45	2.5	6289	0.43	2.8	2.4	4784	0.32	3.0
	0.91	2.5	3351	0.23	3.0	2.3	2737	0.19	3.0
	1.36	2.0	2504	0.17	2.9	-	-	-	-
P-4	0.45	2.6	5073	0.34	3.1	2.6	4511	0.30	3.0
	0.91	2.5	3097	0.21	3.1	2.4	2737	0.19	3.1
	1.36	2.3	2436	0.17	2.9	-	-	-	-
UP-1	0.45	2.5	3348	0.23	3.2	2.6	3412	0.23	3.2
	0.91	2.3	2234	0.15	3.1	2.5	2875	0.19	3.2
	1.36	2.3	1995	0.14	3.2	-	-	-	-
UP-2	0.45	3.2	4463	0.3	3.6	3.0	3828	0.26	3.7
	0.91	2.6	2539	0.17	3.3	2.2	2258	0.15	2.9
	1.36	2.3	2131	0.14	3.1	-	-	-	-

Source: Data are from Islam 1980.

Notes: Caloric value of rice = 14,800 kJ/kg. P indicates parboiled rice; UP indicates unparboiled rice. Arabic numbers after P and UP indicate a particular variety of rice grain size or degree of parboiling. Moisture content of different varieties were determined as follows:
P-1 (12.4%), P-2(17%), P-3(16%), P-4(17%), UP-1(16%), UP-2(12%).

WW - Weight of water, WR - Weight of rice, WCR - Weight of cooked rice. kJ/kJ = Heat input for cooking unit weight of rice/Caloric value of unit weight of rice.

Net heating value of fuels (GOB 1976): Kerosene = 46,005 kJ/kg, Natural Gas = 993 kJ/standard cubic feet.

Water capacities of flat bottom aluminum pans used for rice cooking are: 3.2 kg (for cooking 0.45 kg rice), 5.9 kg (for cooking 0.91 kg rice), 7.3 kg (for cooking 1.36 kg rice).

and charcoal are weighed to estimate the amount of fuel consumed for active cooking.

In actual practice, the unburned charcoal is allowed to burn to ashes after cooking. The generated heat of charcoal and the stored heat in the mass of the stove are utilized partially to warm leftover food after meals, to limit fermentation, to help light the fires at the start of the next meal, and to minimize the amount of fresh fuel needed to warm up the stove for subsequent meal cooking. This burning of charcoal is like a thermal power plant's banking operation during its off-peak run. It also provides space heat in some areas and seasons.

Typical laboratory meal-cooking tests should continue over a 24-hour cycle and follow the sequence of normal household cooking. Thus, assuming a typical "cooking cycle," (e.g., breakfast: chapati, pulse; lunch: rice, fish curry; dinner: rice, pulse, vegetable) the results can be arranged to estimate fuel consumption for different meals consumed in a day. The number of meals selected should reflect typical meals of each socioeconomic group. Also, the fuel available at the time of the experiment should reflect the seasonal behavior of fuel consumption. The amount of fuel consumed by a family larger or smaller than an average family could be estimated by cooking a different quantity of food.

Steps to Standardize Stove Performance Tests

Stove researchers have failed to follow any standard methodology in laboratory food-cooking tests in the past. If the amount and type of foods cooked by users are different, laboratory evaluations would have only limited usefulness. Recently, an attempt has been made to standardize the testing of cookstoves' efficiency through development of Provisional International Standards (PIS) for Water Boiling Tests, Controlled Cooking Tests, and Kitchen Performance Tests (VITA 1982). The document also outlines methodologies for data gathering, analysis, and statistical evaluation of test results.

Some observations follow about the PIS in comparison to the test procedures just described. The Provisional International Standard has not suggested any procedure for selection of representative samples of traditional stoves for performance tests. The PIS concept of "high power" and "low power" water boiling tests is well thought out to reflect two contrasting but representative water heating situations. However, water boiling tests should be carried out by simulating all relevant local practices observed during cooking. For example, if the local cooking practice allows unburnt charcoal to burn to ashes without recovering its heat, heat content of charcoal should be considered as consumed.

Usha Rao tending fire, East-West Center air pollution study.
(*Honolulu* Advertiser)

Controlled cooking tests with one or two meals as suggested by PIS may not be able to account for the multiple variables involved in actual cooking. Considering this, in the present text controlled cooking test has been subdivided into cooking of different dishes, cooking of typical daily meals (instead of a particular meal of a day), and estimation of fuel consumption for typical daily cooking cycles. Water boiling tests and controlled cooking tests should be carried out at local stove testing centers to prepare the promoter for possible results to be expected under users' conditions.

Kitchen Performance Test (KPT) at users' level should be continued at three-month intervals for one year to include all possible seasonal variation. It is a difficult task to outline exact procedures for carrying out KPT. Some possible difficulties are cited as examples. People who are really in need of an improved stove may not have sufficient space to accommodate two stoves (existing and improved) for test purposes. This is particularly the case for insitu built stoves. On the other hand, a particular user may not agree to replace the existing stove with a new one without knowing its performance. Those who have sufficient space may not have the felt need for an improved stove. Too much interference with household activities for testing purposes may result in discontinuation of the testing program.

FIELD TESTING: A PIVOTAL PROGRAM ELEMENT

Effective diffusion of improved stoves has been limited in these ways:

1. Research, development, demonstration, and diffusion of improved stoves not considered seriously by decision makers. The gap between technical papers or popular journal articles and implementation has concerned conscientious researchers for many years. With limited resources and capability in the past, researchers attempted to promote improved stoves by a top-down approach. Mechanisms to evaluate users' reactions were inadequate or absent and demonstration processes were ill organized. As a result, the research and development effort could not reach the stage of effective diffusion (see Elizabeth Cecelski 1983).

2. Lack of communication about local innovations. The performance of particular traditional stoves often has been improved at a specific location, but the design has not spread because of lack of appropriate evaluation and communication. Women's personal contacts as principal stove users are the traditional method for communication about stove diffusion. Limited mobility of women beyond their own place of living is one reason for slow diffusion of indigenous improved stoves through this mechanism.

3. Failure to extend designs to relevant users. Several designs for domestic cooking, with multiple pot settings and a chimney, are appropriate for commercial activities in which similar stoves have been used elsewhere for years (Roy 1904).

Benefits from stove improvement efforts may also be limited or diverted by user factors that are poorly understood. For example, the user of an improved stove may use the same amount of firewood she used before because she is using the additional available heat to meet a heretofore suppressed need. On the other hand, if firewood use is reduced the environment might be preserved from degradation specifically related to this factor, but land use might shift to other potentially detrimental uses such as expanded agriculture or timber extraction.

To overcome the limitations of previous diffusion efforts, stove design and fuel consumption experiments should be done at user level. Stove testing centers in local areas could simulate users' fuel consumption patterns according to fuel and food type more effectively than those in national laboratories. Care would have to be taken to select appropriate users who would be willing to cooperate over sufficient periods of time (probably at least one year). Also, care would have to be taken to insure that users' normal cooking patterns were not influenced inadvertently by the researchers. Specific research choices such as the number of families and the number of indigenous stoves in the experiment and the seasonal and daily frequency of observations would have to be carefully planned.

These stages define a procedure for field testing stoves as an integral element in stove improvement programs:

Stage 1. An area known to have been affected by a fuelwood crisis is selected for initial survey and design testing. The criterion of fuelwood scarcity must not be secondary to availability of infrastructure facilities. Therefore, the test center should not be located at an existing institution for convenience's sake. The testing place must be similar to local cooking places, including open air cooking if this is done at certain times of the year.

Stage 2. Information about stove use is gathered through a survey that considers socioeconomic status, cooking place and relation to dwelling house, stove characteristics, cooking pots, types of food and fuel, seasonal variations, and cultural practices. Observation of users' cooking practices in fuel-scarce situations and dialogues about their felt needs are important methods for obtaining design guidance at this stage (Jamuna Ramakrishna and Kirk Smith 1982, Smith and Carol Colfer 1983).

Stage 3. Representative indigenous stoves and stove designs matching local needs, developed locally or elsewhere, are selected for testing at the stove testing center. Performance is tested by a water boiling test simulated to local cooking methods, a food cooking test, a typical meal cooking test, and daily meal cooking test. The actual cooking test is conducted either by a woman researcher or by a village woman with close guidance and monitoring by the researcher.

Stage 4. One of these outcomes is possible, based on Stage 3:

a. Identification of one or more indigenous stoves with good performance.
b. Identification of one or more stoves obtained from outside sources with good performance.
c. Identification of one or more new stoves developed by modifying indigenous and outside stoves.

Stage 5. Limited introduction of stove(s) is undertaken in selected households. At users' level, their performance is evaluated at three-month intervals for at least a year to observe the effect of seasonal variation.

Stage 6. If necessary, based on users' reaction, further modification and testing is undertaken.

Stage 7. Each stove design adopted by users is identified for wider use within the survey area. In addition, based on the background survey described in Stage 2, a program to extend identified stoves into other areas of the country can be established.

Extension agents and locally trained persons would be needed to disseminate insitu stoves effectively. Dissemination of mass fabricated stoves would require training of potters or metal craftsmen, but these stoves could be marketed through existing channels relatively quickly.

The research methodology described here is based on my personal experience (Islam 1980) and on the methodologies outlined by Stephen Joseph and Yvonne Shanahan (1980), W. Stewart (1981), and Shanahan (1982). Based on scientific knowledge, high performance stoves could be developed (de Lepeleire and others 1981, Prasad 1981a). To insure acceptability, such stoves would have to be evaluated locally beginning at Stage 3.

This testing procedure requires a multidisciplinary team. People with field experience in assessment and diffusion of other rural technologies should be included. Effective research liaison should be established with programs for assessing air pollution from biomass-burning stoves (Smith and others 1983). A country in which no previous stove research has been conducted would need three to four years' lead time to organize and prepare research personnel for large-scale development, demonstration, and diffusion of improved stoves. During this period, initial research and development work should be organized at field test centers. Funding requirements for local and national components of the program would have to be determined in the country's institutional context; a multi-year commitment is obviously essential.

INSTITUTIONAL INVOLVEMENT AND RELATED POLICY ISSUES

The initiatives of Non Governmental Organizations (NGOs) working in rural areas have been mainly responsible for the status of stove research in developing countries. In most cases, stove projects were developed in close collaboration with overseas NGOs (Acott and others 1980, Joseph and Shanahan 1980 and 1981, Joseph and J.C. Loose 1981, M. Sarin 1981, Shanahan 1982, A. Soedjaro 1981, Stewart 1981, TERI 1981, and VITA 1981). Among the overseas NGOs, the contributions to stove programs of Aprovecho, the Intermediate Technology Development Group (ITDG) and Volunteers in Technical Assistance (VITA) are noteworthy.

Research in developing countries (as elsewhere) tends to respond to the individual researcher's initiative. Only recently, because of innovative work by researchers at Eindhoven University of Technology (de Lepeleire and others 1981, Prasad 1980 and 1981a, b) has research on woodburning stoves gained some prestige. Most work on stoves has been done through informal institutions and with ad hoc arrangements, so continuity is often related to the original researcher's commitment. Departure of initiators or reduction in funding can mean an untimely end to the project.

During the time that the more thermal efficient multipot, chimney-fitted stove was receiving publicity, an early stove researcher prematurely concluded that local traditional stoves were inefficient. Without further assessment, the multipot design was recommended to rural people as an "improved" stove. The subsequent limited diffusion of that stove caused researchers to seek alternative designs. At the same time, no available study yet assesses the success of indigenous stoves used in any country. That assessment area, then, should have priority in stove research.

Various institutions have different specific roles to play in the many facets of stove research, development, demonstration, and diffusion. A major task for each country is to identify groups of institutions that have or can develop the necessary specific capabilities. Experienced NGOs can act as trainers in the field design, testing, and diffusion process. All participating agencies must cooperate closely.

The current "fuelwood crisis" has been publicized worldwide. As a means to solving fuelwood problems, the design of improved stoves has attracted the attention of aid agencies, policy planners, and researchers of industrialized and developing countries. Over the coming years, efforts to develop and diffuse improved stoves probably will accelerate.

Systematic country-specific evaluation of indigenous biomass-burning stoves should be the basis of a stove improvement program. Development of the program should be considered as part of a national energy conservation program, planned in the context of national energy policy. An improved stove program also could be considered part of a national forestry policy, even while its direct impact on deforestation remains to be established. Results of local biomass fuel surveys as well as assessments of cooking stoves used with fuels such as kerosene, fuel gas, and electricity could aid energy conservation efforts. This chapter has presented methodological guidelines for such assessments.

In developing countries, the ministries of energy and forestry generally do not have any grass roots extension agency of their own. Therefore, diffusion of improved stoves might have to depend on other ministries such as the ministry of agriculture or rural development. Nongovernmental organizations can contribute to testing and diffusion in their operational area. Existing market mechanisms also can be an important route for disseminating specific types of improved stoves. In addition, training and research capabilities in different regions of the country should be made an integral part of a stove improvement program, since the program should address the needs of both rural and urban areas.

ACKNOWLEDGEMENTS

Background materials for this paper were prepared while I was at the University of Sussex under an IDRC sponsored fellowship. Mr. J.A. Barnett and Dr. M. Howes deserve grateful acknowledgement for their valuable comments.

NOTE

1. Experimental controls for water heating tests involve the following considerations. For experiment types (1) and (2), the fire is extinguished after each reading to estimate the amount of unburned fuel (firewood, charcoal). The maximum time required in the first experiment is for case (c), water boiled and completely evaporated; minimum time is for (a), water only boiled; and some amount of time between those two is required for (b), boiling and simmering for fixed times. The experiment is less efficient for the shorter duration experiment because some of the fuel's initial heat is used to heat the stove. When the water is just heated to boiling, it must be weighed at the end of the experiment, since some of it evaporates during boiling.

    Weighing unused fuel is not necessary with the third type of experiment. In that case, determining the end point is difficult. It might be the end of the active burning period with no unburned wood left in the stove, but with some charcoal still burning. It might be when all the charcoal is ashes. If so, the extended charcoal-burning period may cause a cooling effect. If the experiment is limited to the end of the active burning period, the amount of unburned charcoal must be weighed, and then the procedures followed are like those in the first two groups of experiment types.

Notably, if a fixed amount of firewood is burned to bring to a boil several batches of a fixed quantity of water, the overall thermal efficiency of the first batch would be lower than the others because some heat would be used to heat the cold stove. Efficiency would seem to be higher with subsequent batches done with an already hot stove.

REFERENCES

Acott, T., and others.
   1980   *Helping People in Poor Countries Develop Fuel Saving Cook Stoves.* Eschborn: German Appropriate Technology Exchange.

Agarwal, B.
   1983   Diffusion of Rural Innovations: Some Analytical Issues and the Case of Wood-Burning Stoves. *World Development* 11(4):359–376.

Ay, P.
   1978   Fuelwood and Charcoal in West African Forests: Field Research in Western Nigeria. *Rural Energy Systems in the Humid Tropics.* Proceedings of the First Workshop of the United Nations University Rural Energy System Project, Ife, Nigeria.

Best, M.
   1979   The Scarcity of Domestic Energy: A Study in Three Villages. Saldru Working Paper no. 27. Capetown, South Africa: Southern Africa Labour and Development Research Unit.

Bialy, J.
   1979   Measurement of the Energy Released in the Combustion of Fuels. ATO 22. Edinburgh: University of Edinburgh, School of Engineering Science.

Cecelski, E.
   1983   Energy Needs, Tasks and Resources in the Sahel: Relevance to Woodstoves Programmes. *GeoJournal* 7(1):15–23.

de Lepeleire, G., K.K. Prasad, P. Verhaart, and P. Visser.
   1981   *A Wood Stove Compendium.* Prepared for the United Nations Conference on New and Renewable Sources of Energy, Nairobi. The Netherlands: Technical University of Eindhoven, Department of Physics and Mechanical Engineering, The Wood-Burning Stove Group.

de Silva, D.
   1981   Personal communication, Wood and Cellulose Technology Section, CISIR, Colombo, Sri Lanka.

Gamser, M.S., and C. Harwood.
   1983   The Implementation of New Energy Technologies in Developing Nations: Problems and Policies in the Introduction of Charcoal in Papua New Guinea. *GeoJournal* 7(1):35–40.

Geller, H.S.
   1981   *Cooking in the Ungra Area: Fuel Efficiency, Energy Losses, and Opportunities for Reducing Firewood Consumption.* Bangalore: Indian Institute of Science, Centre for the Application of Science and Technology to Rural Areas.

Geller, H.S., B. Leteemane, T.A.M. Powers, and J. Sentle.
   1983   *Prototype Metal and Mud Wood-Burning Cook Stoves for Botswana.* Associates in Rural Development, Inc., Burlington, Vermont.

Government of Bangladesh (GOB), Ministry of Planning.
   1979   *Statistical Year Book of Bangladesh 1979.* Dhaka: BBS Ministry of Planning.

Harahap, F.
   1981   Personal communication, Mechanical Engineering Department, Institute of Technology, Bandung, Indonesia.

Hoda, M.M.
   1979   *Solar Cookers.* Research Report Series no. 1. Lucknow, India: Appropriate Technology Development Association.

Islam, M.N.
   1980   *Village Resources Survey for the Assessment of Alternative Energy Technology.* Prepared for International Development Research Centre, Canada. Dhaka: Bangladesh University of Engineering and Technology (BUET), Department of Chemical Engineering.

Janczak, J.
   1980   *Compendium on Simple Technologies for Agglomerating and/or Densifying Wood, Crop and Animal Residues.* Prepared for the Technical Panel on Fuelwood and Charcoal, UNERG. Rome: FAO.

Joseph, S., and J.C. Loose.
   1981   *Testing of a Two-hole Indonesian Mud Stove.* Report no. 3.3. London: ITDG.

Joseph, S., and Y.J. Shanahan.
   1980   *Designing a Test Procedure for Domestic Wood-Burning Stoves.* Report no. 3.1. London: ITDG.

Joseph, S., and Y.J. Shanahan.
   1981   *Laboratory and Field Testing of Monolithic Mud Stoves.* Report no. 3.2. London: ITDG.

Lozada, E.P.
   1981   Personal communication, Institute of Agriculture Engineering and Technology, University of the Philippines at Los Baños, Laguna, The Philippines.

National Academy of Science (NAS).
   1980   *Firewood Crops, Shrub, and Tree Species for Energy Production.* Washington, D.C.: National Academy Press.

National Energy Administration (NEA).
   1981   Personal communication, Technical Division, NEA, Bangkok.

Nideshak.
   1958   *Report on the Experiments with Various Designs of Smokeless Chulah.* PRAI publication no. 146. Lucknow, India: Planning Research and Action Institute.

Pitakarnnop, N.
   1981   Personal communication, Industrial Fuel Division, Thailand Institute of Science and Technology Research, Bangkok.

Prasad, K.K.
   1980   *Some Performance Tests on Open Fires and the Family Cooker.* The Netherlands: Technical University of Eindhoven, Department of Physics and Mechanical Engineering, The Wood-Burning Stove Group, and TNO, Division of Technology for Society, Apeldoorn.

Prasad, K.K.
   1981a   Wood Stoves: Some Thoughts on the Technology and Its Development. Presented at the NGO Forum on New and Renewable Sources of Energy, August 8–17, Nairobi.

Prasad, K.K.
   1981b   *A Study on the Performance of Two Metal Stoves.* The Netherlands: Technical University of Eindhoven, Department of Applied Physics and Mechanical Engineering, The Wood-Burning Stove Group and TNO, Division of Technology for Society, Apeldoorn.

Quader, A.K.M.A., and K.I. Omar.
   1982   *Resources and Energy Potentials in Rural Bangladesh*, Dhaka: Bangladesh University of Engineering and Technology (BUET), Chemical Engineering Department.

Raju, S.P.
   1961   *Smokeless Kitchens for the Millions.* Revised edition. Madras: Christian Literature Society.

Ramakrishna, J., and K.R. Smith.
   1982   Smoke from Cooking Fires: A Case for Participation of Rural Women in Development Planning. Working Paper no. WP 82–20. Honolulu: Resource Systems Institute, East-West Center.

Roy, J.C.
   1904   *Country Cooking Stove, Shahittya* 15(5):312–324 (published in Bengali in Bengali Year 1311).

Revelle, R.
   1976   Energy Use in Rural India. In *Science* 192:969–976.

Sarin, M.
   1981   *Chulha Album*. New Delhi: The Ford Foundation.

Segeler, C.G.
   1966   *Gas Engineers Handbook. Fuel Gas Engineering Practices.* New York: The Industrial Press.

Shanahan, Y.J.
   1982   *Report on the Workshop on Stove Projects, April 6–11, Sri Lanka.* London: ITDG.

Singer, H.
   1961   *Improvement of Fuelwood Cooking Stoves and Economy in Fuelwood Consumption: Report to the Government of Indonesia.* Expanded Technical Assistance Program no. 1315. Rome: FAO.

Smith, K.R., A.L. Aggarwal, and R.M. Dave.
   1983   Air Pollution and Rural Biomass Fuels in Developing Countries: A Pilot Study in India and Implications for Research and Policy. *Atmospheric Environment* 17(11):2343-2362.

Smith, K.R., and C. Colfer.
   1983   Cooks on the World Stage: The Forgotten Actresses/Actors. Working Paper no. WP–83–5. Honolulu: Resource Systems Institute, East-West Center.

Soedjaro, A.
   1981   Personal communication, Stove project, Yayasin Dian Desa, Jurugsari, J1. Kaliurang, Yogyakarta, Indonesia.

Stewart, W.
   1981   Personal communication, report on Sarvodaya stove project, Sri Lanka.

Tata Energy Research Institute (TERI).
   1981   *Experimental Studies on Firewood Chulahs.* Pondicherry, India: TERI, Field Research Unit (Care Sri Aurobindo Ashram).

Uhart, E.
　　1975　*Charcoal Development in Kenya.* United Nations Economic Commission for Africa, ECA/FAO Forest Industries Advisory Group for Africa (Ind–91/Mr.29, July).
Volunteers in Technical Assistance (VITA).
　　1981　An Evaluation of Thai Cooking Fuels and Stoves. Presented to United Nations Conference on New and Renewable Sources of Energy, Nairobi.
VITA.
　　1982　*Testing the Efficiency of Wood-Burning Cookstoves.* Provisional International Standards. VITA, 1815 North Lynn Street, Suite 200, Arlington, Virginia.

9
Supplying Firewood for Household Energy

Rick J. Van Den Beldt

INTRODUCTION

The role of wood fuel in rural households of the tropics, the difficulty with which it is sometimes obtained, and the often negative environmental impacts of wood fuel gathering for rural energy have been heavily scrutinized research areas recently. T.M. Pasca (1981), citing the FAO report of the Technical Panel on Fuelwood and Charcoal (1981), states that 100 million persons worldwide cannot obtain sufficient firewood to provide even minimal energy needs. Another 1,000 million dwell in areas with lesser shortages, a figure that is expected to double in 20 years. By the year 2000, the FAO (1982) estimates that there will be as many as three billion persons affected by firewood shortages. To meet this growing need, FAO estimates that some 50 million ha of firewood plantations will have to be planted by the turn of the century.

Firewood gathering has been cited as a major cause of deforestation in many parts of the world, directly responsible for environmental problems that forest clearing brings. Notable examples of such claims include widespread soil erosion in Nepal and desertification in the more arid parts of Africa (NAS 1980).

While acknowledging the seriousness of the "other energy crisis" and the danger it poses to the environment in certain locations, this chapter presents a review of the issues, citing local studies in the tropics in order to draw a clearer picture of the character of this crisis. Secondly, the chapter will suggest strategies to alleviate the problem.

FIREWOOD CONSUMPTION ISSUES

Deeper analyses of local fuelwood problems increasingly point to inconsistencies and deficiencies in the "crisis" hypothesis just outlined. First, sources of firewood besides forests, and their capacity to meet annual fuel needs, have not been investigated sufficiently. Second, other detrimental land uses are ignored even where they contribute substantially to environmental problems. Third, subsistence firewood use is often not distinguished from industrial firewood use. Finally, the "crisis" point of view does not take into account use of vegetative fuels in households and the households' willingness to switch to various combinations of available vegetative fuels when firewood is scarce. This latter point is discussed for cases in Bangladesh (Chapter 2), the Philippines (Chapter 4), and India (Chapters 6 and 11) and will not be dealt with extensively here.

Rural Firewood Sources

Since forested areas are popularly viewed as the major fuelwood source, firewood supply projections generally are based on deforestation rates. Elizabeth Cecelski and others (1980), for example, cite a total growing stock of forests in developing countries of about 200 billion m^3, with an annual increment of 4 billion m^3. At present rates of exploitation (of all types), they project that harvest will exceed growth by a factor of two in 20 years. This, they conclude, will have serious impact on the availability of firewood in local situations. The FAO, also taking the global perspective, has estimated that firewood shortages will total approximately 1 billion m^3 by 2000.

Even at local levels, similar exercises are used to estimate reserves of firewood. Although such exercises may be applicable where gathering from forests and savannahs is the major source of subsistence firewood (Howard Dick 1980), they do not adequately estimate firewood resources actually available to rural families. Numerous other local sources may provide most firewood used by rural people for subsistence, such as homelot trees, scattered stands of residual forests, riparian growth, and the like. In parts of Southeast Asia, twigs, dead branches, and culled trees from home gardens ("the artificial forest") are gathered as home fuels (Dick 1980). Failure to account for these sources can lead to misleading conclusions in planning future firewood supplies (J.E.M. Arnold and Jules Jongma 1978).

Although annual production figures for sources around homelots are rarely reported, these sources can be significant, as in the cases of three villages of the Upper Solo Watershed in Java where residents were found to use firewood on a subsistence basis (see Table 1). Forests accounted for less than 3 percent of the firewood used. The largest sources were homelots and fields. Rarely were trees cut down for firewood. Herman Haeruman (1979), in a survey of West Java, found that 35 percent of wood gathered as subsistence fuel came from dead trees and fallen branches. Pruned branches and twigs made up 65 percent.

Homelot plantings are by no means a universal phenomenon in the tropics. Where they do exist, they often provide an economic buffer in hard times. James Douglas (1982) points out that rural people in Bangladesh plant slower growing, spreading trees such as *Samanea saman* in

TABLE 1.
Annual Production of Firewood per Rural Household in
Villages of the Upper Solo Watershed, Java, Indonesia (tons)

	Samin	Wiroko	B. Solo	Average
Home compounds	0.63	0.75	0.80	0.73
Dry Fields	1.00	0.82	0.38	0.73
Other (wet rice fields, roadsides and forests)	0.09	0.01	0.05	0.05

Adapted from Dick (1980), on Wiersum.

homelots so they can prune branches for sale. In hard times, the rights to prune trees are sold to wealthier neighbors before more valuable commodities like livestock or land are liquidated (Nurul Islam, pers. comm., 1982).

Causes of Deforestation

With such apparent low demand for subsistence use of forest-based firewood, what then are the causes of deforestation? Obviously, the question does not have a simple answer. However, were one to examine closely the land uses in areas of severe firewood scarcity such as Nepal, Tanzania, the Sahel, and Java, one could probably identify numerous local practices, exacerbated in some cases by harsh environments, that are more destructive to forest cover than firewood harvesting. Grazing with burning of savannah and grasslands bordering forests, shifting agriculture, and the so-called "selective" logging practice are a few notable examples. In Nepal, for example, two dry tons of animal fodder per head are fed yearly to livestock (Krishna Kumar Panday 1981). An estimated 7 million tons of this fodder are from trees alone (James Brewbaker 1983).

Industrial Use versus Subsistence Use

Forests in certain areas are indeed being mined for firewood. A Mitre Corporation report (W. Park and others 1982) describes a highly organized marketing system — clearly not subsistence use — that supplies the city of Managua, Nicaragua, with firewood from the nearby Las Maderas forest. Arnold and Jongma (1978) state that demand in urban and heavily populated areas can severely reduce forest cover, as in the Gangetic Plain and Central Thailand. In Java, on the other hand, kerosene is available to residents of cities and towns for cooking (Paul Weatherly 1980).

Most felling of forest firewood on Java occurs to supply rural industries, such as limestone reduction, pottery and tile manufacture, and palm sugar processing. As Dick (1980) points out, "it is the existence of market demand, either for sale of firewood, or for use in the manufacture of commodities . . . that creates for villagers the opportunity to exploit forests to generate income." According to Dick, the poorest villages (both economically and in terms of forest cover) are the first to exploit the forest for firewood because of the economic incentive.

Firewood is viewed often as an expensive commodity by rural residents in Java. Marketable pieces of wood often are sold for supplemental income instead of being burned as home fuel. As a result, a substantial portion of fuel burned in households is not wood at all—it is twigs, leaves, and agricultural residue. In this paradoxical situation, moderate increases of firewood supply from planting programs probably would not result in proportionately higher wood fuel use in the home. Instead, the wood would probably be sold on the market or used at home to prepare marketable commodities, such as palm sugar. This is indeed the case in Sikka where residents grow firewood for sale in the market and use the income to buy vegetables that would have been grown on firewood production plots (Viator Parera 1983).

Clearly, one must distinguish between subsistence use and industrial use of firewood when predicting where shortages might occur and when defining objectives of firewood production programs. Otherwise, choices of persons to involve might be inappropriate; so might land to utilize, species to plant, and choice of management techniques.

Other Uses of Wood and Trees

Fuel is by no means the sole use of wood in households of the rural tropics. There is a large demand for wood and other forest products for housing, construction of farm implements and household utensils, and cottage industries such as carving. Although there is little written on this household use of wood—its annual consumption, its source, or the ability of sources to supply yearly increments—it undoubtedly is a significant component of household wood use in the tropics. Indonesian home gardens, a major source of firewood, are also heavily utilized for timber and bamboo production (Anonymous 1973). In Bangladesh, also, such plantings are carefully designed by rural folk for long-term production of fruits and many lesser products as well as firewood (Douglas 1982). These customs might suggest that a fuelwood program would better meet the needs of the people if it also included assurances of a steady supply of raw materials for other household uses.

HOUSEHOLD FUELWOOD PROGRAMS

Simplistically, the most obvious solution to firewood shortages is to plant more trees. Most often, statements conveying the urgency of this task are rather ambitious. Edouard Saouma (1981) states that the amount of money and land area delegated for firewood production must increase fivefold by 2000 if firewood demands are to be met. Cecelski and others (1980), citing Keith Openshaw, state that 1.3 billion ha of land must be planted to produce 13 billion m^3 of firewood.

Even locally, solutions to insufficient supplies of firewood often are viewed merely in terms of the number of hectares to be planted and do not adequately consider methods. As a result, many projects have not reached their goals (Dennis Wood and others 1980). Perhaps the greatest consideration is the nature of planting to be implemented: ubiquitous small homelots, roadside or bund plantings, intermediate-scale scattered village woodlots, or intensive plantation establishments on land designated exclusively for firewood production.

Another consideration is the appropriate management mix to utilize in marketing or in use of firewood products. Such a mix could include sale or use of firewood by individual households, the establishment of village cooperatives as marketing structures, or control of firewood sales for profit by industry or government (with residents assuming the role of firewood farmers or laborers).

A third and no less important consideration is the selection of species and silvicultural methods, perhaps the most overlooked constraint in recent literature. Despite the common claims that "techniques of establishing woodlots are well known" (Raymond Noronha 1981) and "simple" (Matthew Gamser 1980), in truth the status of the literature and experience in the field is deficient, particularly in the area of species selection. This sad situation is evident when one visits firewood plantings and observes gross errors in both species selection and in silvicultural practices.

Careful consideration of these points can result in the identification of initial testable and locally verifiable goals before implementation. Of course, the people involved would have to agree on goals or contribute to their modification.

Identifying Demand, Source, and Production Strategy

A first step in developing a program for assuring firewood supplies to households is to determine market dynamics and firewood sources. In the past, demand has usually been expressed in terms of per capita consumption for a given geographical region. Deficiencies in this approach have been noted by Deanna Donovan (1981) and Hadi Soesastro (1980), who provide lists of reported estimates of firewood consumption in Nepal and Java, respectively. As shown in Chapter 3 (Table 2), estimates of Javanese consumption vary widely because of particular village socioeconomic conditions, availability of fuelwood markets and forests, and experimental techniques. Therefore, per capita firewood consumption figures may not be useful, particularly if they cover an extensive area or a diversity of lifestyles. With few exceptions, most studies of fuelwood in rural households are isolated projects that do not include information about reforestation or fuelwood production strategies.

More refined techniques exist and are discussed elsewhere in this book. Among these is the study in Chapter 3 on 40 villages (1 percent of all villages) in West Java. Another is Islam's study of 23 villages in Nabagram Union, Bangladesh (see Chapter 2), which assesses fuel shortages and suggests programs to ameliorate these shortages (see also Chapter 11). A substantial compendium of fuelwood research techniques has recently been published by FAO (1983); see especially chapters by Russell deLucia and W.B. Morgan and Annexes I (deLucia), II (Anon), and III (Openshaw).

The extent and depth of socioeconomic data collection is limited by financial, personnel, and time constraints. Thus a detailed analysis of regional rural firewood market dynamics may not be financially feasible. However, several approaches have been illustrated in the literature that do produce valid information. First is the structured sample approach used by Soesastro (Chapter 3). Similarly, John Raintree (1980) derived reasonable estimates of price structures, of employment of rural people as firewood harvesters, and sources of wood for sale from interviews of vendors along a 95-km stretch of national highway in West Java.

The second approach is to conduct an intensive research program in one area, using findings to predict trends in similar areas. These often complex programs address a range of socioeconomic questions, and results generally can be adapted for use elsewhere, even though the test area may not prove strictly representative. Survey questions from such intensive surveys often provide the basis for types of questions asked in broader surveys.

The third option is to send into the field a team of experienced people competent in areas such as forestry, agriculture, economics, and social science for a "bird's eye view." Park and others (1982) made a two-week survey of the supply of firewood to Managua, Nicaragua from the Las Maderas forest area using a team of experienced foresters and economists. Such research usually saves much time, but results depend directly on the overall competence of the team.

However the program collection of socioeconomic data is made, the following information is considered important:

1. *The amount of firewood used locally, how it is mixed with other fuels, and its end uses.* Wood is usually only one of many vegetative fuels used by households. Information about how a household allocates firewood resources (household use versus sale in the marketplace) should give a basis for predicting how an increased supply of firewood from plantings would be utilized. As noted, in Java, increased firewood supply sometimes is viewed as a source of income and not for home use.

2. *Land allocation practices in the region.* Identification of landlord-tenant customs regarding division of harvest and land use and other traditional relationships that may affect profitability of the enterprise from the farmer's point of view are useful. It is often worthwhile to examine what is meant by "communal lands." Often such land will be designated for a particular use like grazing and the act of tree planting will cause local resentment. As Gamser (1980) notes, communal lands are often under the control of wealthier and more powerful individuals who enjoy a disproportionate share of benefits from the land. In such cases, choice of communal land for firewood farming would not be appropriate.

3. *Market opportunities for wood produced on plantations.* Initial success of firewood projects may depend on marketability. As necessary, markets can be strengthened by an appropriate demand-side strategy such as development of rural industries and establishing improved transport links to centers of demand.

4. *Labor opportunities and costs of firewood production.* Women have a primary role in firewood production (see Jacqueline Ki-Zerbo 1981, Marilyn Hoskins 1979). Rural people must be convinced that establishing and maintaining plantings are a wiser use of time than gathering firewood from distant sources, or success is not likely. Moreover, this interest must be maintained until the first harvest, sometimes as long as two to four years.

Mixing Government and Community Goals

Participation by the people affected by development programs is generally accepted as essential during planning and implementation stages of rural programs. Many projects in the past have failed without consideration of those human aspects. In Tanzania, for example, the complex land tenure system was not understood; neither were the varying and often contradictory conceptions of the usefulness of forest plantations (Noronha 1981). A World Bank project in Niger was doomed because village woodlots were set up on land the people used primarily for grazing.

Thus even though governments may view dwindling firewood supplies with alarm, rural folk may be more concerned with declining agricultural production, inequitable land tenure, and other unique local problems. A project that is insensitive to local human needs may exacerbate local problems instead of alleviating them. The project is then viewed as a threat rather than a solution.

On the other hand, planning from a central perspective, including a management plan that takes into account the scope and diversity of the problem, is an essential element to program success. Increased people's participation, although important, is not the primary goal of a household firewood project — increased firewood supply is. Rural people probably don't completely understand the benefits to agriculture from reforestation; in many areas they are probably not aware of the economic benefits possible from firewood farming. Similarly, they probably lack many necessary technical skills such as harvesting, nursery methods, and spacing, layout and maintenance of plantations. All too often, failure of "communal" forestry plantings is due more to lack of understanding and commitment on the part of the rural folk to maintain the plantings than to the failure of planners to consider the people's best interest in the face of the nation's as a whole. The success of Korea's program for example was as much due to passage of new laws and strengthening of enforcement capabilities in the early phase of the program as to improved extension and increased "democratic" processes of village participation (Arnold 1982).

A good example of a successful mix of people's participation and "top down" planning is Indonesia's Perum Perhutani. As much as possible, Perhutani's extension efforts support their programs and are made to involve the people in production on their own lands. Managers of a beekeeping enterprise near Sukabumi in West Java using the tree species *Calliandra calothyrsus* as bee forage involve local residents by showing them how to keep their own hives while protecting a watershed reforestation area. Residents make up to Rp 3,000 (US $5) per month in additional income by selling honey to Perhutani (per capita income for Indonesia in 1981 was US $520 per year). In other areas, local people are employed as forest guards and are paid in kind for their services by being allowed to plant crops between rows of trees for the first year. Seeds and fertilizers are provided on low credit terms. So enthusiastic was the response in Purwakarta that Perhutani employees were forced to allow only 0.25 ha per farmer and an additional 0.25 ha for each able-bodied male in his household (M. Soehantro, pers. comm., 1982).

The job of the planner is first to discover ways of making his goals acceptable to the community, region, or nation. If this task proves impossible, the planner must be able to reassess his goals in terms of the stated needs of rural folk. Project managers, even after implementation starts, must be flexible enough to account for and incorporate stated community goals in different regions, especially if the project covers a large area.

Furthermore, the planner must be ready to respond promptly to innovations and initiatives of rural people. In Thailand, for example, one sees 0.1 to 0.2 ha plantings of clonally propagated *Casuarina junghuhliana* throughout the central plain. These mushrooming plantings are spontaneous efforts by small landowners to capitalize on the high value of firewood to charcoal and pottery manufacturers. For the most part they were installed without assistance from government agencies. As the price of fuelwood rises in the tropical world, instances of these "spontaneous" fuelwood plantings will undoubtedly increase.

The people of Toyomerto, a village in East Java, also planted firewood trees with little outside assistance. Noting that crop yields were declining over time, the farmers approached forestry officials for advice. Perum Perhutani gave farmers seeds of *Calliandra calothyrsus*, a fast-growing, low-stature leguminous tree. By 1978, over 100 ha of unmixed calliandra plantings were established by local people in plots ranging from 0.25 to 4.25 ha in size. Fuelwood from the plantings supplies households and local firewood-based industries. Unquestionably, the high demand for firewood and the attractive price the product brings played a major role in the villagers' efforts (H.P. Hariyatno and B.M. Purnama 1982).

SPECIES SELECTION, SILVICULTURAL TECHNIQUES, AND EXPERIMENTATION

Although much is being learned about the socioeconomic parameters of fuelwood use and distribution in the rural tropics, the technology of fuelwood production is deficient in many respects. Traditional foresters often have selected species and silvicultural techniques commonly used in reforestation and industrial plantation programs. This practice has inhibited maximization of firewood production. A few of the more important concepts of firewood planting developed through research at the University of Hawaii and verified in collaborative trials in tropical countries of Asia are presented next.

Environmental Assessment

Collection of adequate environmental data is crucial to the success of firewood plantations, as they are often the only valid basis for selection of species and silvicultural methods. Many species, especially the more publicized fuelwood species, have rather strict environmental limitations. Too often, species are preselected and forced into environments where they do not thrive. This can either lead to outright failure or significant expense of agronomic inputs like fertilizer, lime, and water to salvage the operation.

Often, much of the types of data listed below are not available, or are limited at best. Nevertheless, it will pay the manager to make an effort to search for data. Surprisingly, environmental data exist for rural areas in many countries, some of it going back 50 years or more. In Bangladesh, for example, many weather stations taking rainfall, temperature, and other data have records from the preindependence era.

It is important that data collected be useful; actual field measurements of pH with inexpensive indicator kits are far more useful than misleading labels like "acid" or "lateritic" soils. Some of the more desirable environmental parameters are listed below.

Rainfall and Rainfall Seasonality. At best, these should be taken from the nearest rain gauge and should represent several years. Realistically, however, rain gauges are located at experimental stations and data from them should be used only with careful consideration. More practically, it is probably sufficient in most cases to utilize isohyet rainfall maps prepared by central weather bureaus. Similarly, separation of wet and dry seasons can be obtained from these sources. One should not discount information from local farmers who probably know rainfall patterns and amounts in their areas better than anyone else.

Temperature. Average annual high and low temperatures and an idea of extreme temperatures are useful information. Preliminary data from Hawaiian planting experiments have shown that temperature is highly correlated with tree growth. Placing a tropical species in a temperate environment can be disastrous. The reverse, planting temperate species in the tropics also has been disappointing in areas where tried. Despite this, it is surprising how often one comes across failed planting of *Alnus* in hot lowlands or *Leucaena* in cold uplands.

Elevation and Slope. This information has important bearing on erosion hazard and the thermal suitability of species. Slope is simply mapped on site with simple, inexpensive instruments. Steep slopes should best be planted on the contour to promote soil conservation. Topographic maps, widely available even for remote areas should be consulted for elevation. One must remember that average temperatures decrease as elevation increases.

Soil Data. Minimally, field measurements of pH and descriptions from soil survey publications are indicated. Most countries have government soil bureaus that are often good places to start. Project staff should be equipped with their own portable testing kits, a variety of which are sold at reasonable prices or prepared locally. Many universities and soil bureaus maintain soil testing services at reasonable prices. Useful soil data taken on site should ideally include pH, phosphorus, calcium, nitrogen (if nonleguminous species are to be used) and potassium, with other nutrients included if possible. Information on soil salinity and water logging are vital in many areas.

Indigenous Vegetation. This can be simply mapped on site. Of primary importance here are perennial grasses that are sources of weed problems and fire. One can often use indigenous vegetation (including crops) as a guide for species selection.

Existing Land Use. More properly, these data should be included in the socioeconomic survey. They are included here to re-emphasize the fact that firewood programs that ignore local land uses will be less likely to succeed. Also, by mapping land uses the manager can get a feel for areas available for tree planting and focus environmental data collection efforts on these areas.

Species Selection

The book *Firewood Crops* (NAS 1980) is a valuable reference of numerous species for several ecosystems of the tropics as well as a good source of further references on included species and on researchers and organizations in related areas around the world. Other sources are valuable and much can be gleaned from the literature, as demonstrated by Table 2, compiled by Brewbaker and others (1982) that examines the worth of several nitrogen-fixing trees as fuel according to production rate, fuel properties, adaptability, and co-uses.

Still, reported yields, caloric values, wood densities, and "multiple uses" of firewood species often are not representative of a suitable cross section of environments, good experimental technique, or intended

TABLE 2.
Characteristics of Some Nitrogen Fixing Trees

	Acacia auriculiformis	Acacia mangium	Acacia mearnsii	Albizia falcataria	Calliandra calothyrsus	Casuarina equisetifolia	Gliricidia sepium	Leucaena leucocephala	Prosopis pallida	Sesbania grandiflora
UTILIZATION										
Forage	3*	3	2	3	2	3	1	1	1	1
Fuelwood	1	2	1	3	1	1	1	1	1	1
Pulpwood	1	1	1	1	3	2	3	1	3	2
Green Manure	3	3	2	2	1	3	1	1	2	1
Food	3	3	3	3	3	3	2	1	2	1
ADAPTABILITY										
Acid Soils	1	1	-	-	2	1	-	3	3	3
Cold Soil	2	2	1	2	2	2	3	3	3	3
Drought	3	3	2	3	2	1	2	1	1	2
Min. Rain (mm)	1200	1500	1000	1500	1000	300	1500	600	250	1000
Coppicing Ability	2	-	1	1	1	2	1	1	2	1
PROPERTIES										
Specific gravity	.68	.65e+	.65e	.33	.65	1.00	.75e	.54	.80	.42
Wood yield (ton/ha/yr)	12	20	13e	13	33	15	6e	25	6e	9
Average caloric value (cal/g)	4850	-	-	-	4600	5050	4900	4600	-	-
Average annual height growth (m)	2.6	2.5	4e	5.0	6.0	2.1	2.5e	4.5	2.5e	3.3
Height at maturity (m)	20	30	25	45	10	35	10	20	20	10
DBH at maturity (cm)	40	25	50e	200	20	50	20e	35	60	30

Source: Brewbaker, Van Den Beldt and MacDicken 1981.

*1 = Good. 2 = Average. 3 = Poor.

+"e" refers to estimated values.

product use. Caloric values, for example, are widely reported for many species, but methodologies used to derive them vary widely (A.P. Harker and others 1982). Yields are often reported at a single population or at a series of wide spacings suitable for forestry enterprises in general but not necessarily useful for fuelwood plantings. Reports are rare of a species response to the higher yielding, close spacings highly suited for firewood production.

In selecting species for fuelwood production, three factors must be considered. First is the intended use or uses of the product and the suitability of various species. Second, the proposed species must be appropriate for the particular environment. Finally, the species should be acceptable to the people who are to grow them.

Intended Use. When firewood is the primary use, burning qualities should be considered. These qualities include caloric value (usually expressed as cal/gram) and density (dry wt/green vol), which provide information about the amount of heat in dry wood; and moisture content, which governs adjustment of heating value for the water in the fuel. Other particular information such as ash content should be considered if charcoal making is desired. These data should be matched with organoleptic and preference tests with local residents, if possible. Such tests may include smokiness, taste imparted to food, general burning qualities, and other local standards that define "good" firewood (such as characteristics of splitting, drying, cutting, and bundling). Such standards often vary among localities and therefore deserve special attention.

Further, if a particular species has several potential uses (such as fodder, green manure, gum, food, bee pasture, tannin and other chemical products) perhaps firewood would be best viewed as a secondary product if the other products are valued locally. This approach was used for the village with the beekeeping industry of Java's Perum Perhutani. There, the fast-growing, nitrogen-fixing tree *Calliandra calothyrsus* is grown as a source of bee pasture.

Site Adaptability and Influences. Species designated for firewood plantings are often aggressive, pioneer types that adapt well to suitable sites. On the other hand, their performance can be spectacularly disappointing when they are planted in unsuitable areas. Thus a manager may first choose species tolerant to a comparatively wide range of conditions and focus later, as experience is gained, on species adapted to his particular site.

A particular species can be selected to improve site quality. Nitrogen-fixing trees add nutrients, primarily nitrogen, to the soil through litter-fall. These include the Papilionate and Mimosaceous families of the legumes and other unrelated genera, such as *Casuarina* and *Alnus*. *Leucaena leucocephala* was shown in Hawaii to add 8.5 tons litter/ha/yr under dense plantings (Rick Van Den Beldt 1982). This mass served to introduce 100 kg/ha of nitrogen to the soil and recycle other nutrients like calcium (200 kg/ha), phosphorus (7 kg/ha), potassium (16 kg/ha), and sulfur (12 kg/ha). Other species, notably *Tamarix* sp. are able to store large quantities of salt in their tissues, presumably reducing concentrations in the soil (NAS 1980).

TABLE 3.
List of Most Often Named Fuelwood Species,
in Order of Preference, Bangladesh Survey

1.	Sesbania grandiflora*
2.	Samanea saman*
3.	Trewia polycarpa
4.	Anthocephalus sp.
5.	Trema orientalis
6.	Delonix regia*
7.	Lannea coromandelica
8.	Erythrina sp.*
9.	Pithecelobium dulce*
10.	Albizia procera*

Source: CPR (1982).

*Nitrogen-fixing or leguminous species

A species' possible negative influences on the site must also be considered. Some tree covers may be detrimental to certain soils. Many pines tend to lower soil pH. Some eucalypts can alter certain volcanic soils chemically, making them water repellent. Perhaps more importantly, an introduced species can "take over" an area where it is established, becoming a curse rather than a cure.

Social Acceptance. Generally, the more uses a given species has locally, the better the chance of its acceptance as a fuelwood tree. Usually, species known to residents are more easily accepted by them. As part of a fuelwood-planting program, the Center for Policy Research (CPR) in Bangladesh researched the species used by rural residents for firewood. Of the 10 most often named fuelwood species (Table 3), three were selected by that study as initial species (CPR 1982).

The two most desired species have uses other than firewood; both fix nitrogen and, since both are grown throughout Bangladesh, seed supply is no problem. The flowers of *S. grandiflora*, the species most preferred, are used as food and in religious ceremonies by the Hindu minority. The tree also is a favored ornamental in yard plantings. The species next preferred, *S. saman*, is a desirable shade tree and much valued as a timber species. It produces edible pods with high carbohydrate and sugar content.

Other concerns influence choice of species. For household firewood uses, even species with poor form or undesirable traits such as thorns may be utilized. Where a market demand exists for firewood, either species with desirable form or silvicultural techniques that improve form should be selected. Species should give a fair amount of genetically stable seed, or else be easily propagated vegetatively so as to insure wide planting by residents.

Silvicultural Techniques

Household firewood production has perhaps a rare quality in forestry: a wide variety and quality of sites are available for use, from prime agricultural lands to degraded, eroded sites. Thus, not only is matching of species to site important, but matching management methods to site must be considered. Basically, three options are available.

1. *Block Plantings.* Contiguous plantations of variable size and more or less compact dimensions.
2. *Hedgerow Plantings.* Windbreaks, fencebreaks, or boundary plantings established in long rows.
3. *Integrated Agroforestry Plantings.* Integration of forestry and agriculture on the same land in a coproductive arrangement.

Block Plantings. Block plantings are usually made in areas where availability of land is not a limiting factor. Still, exceptions are found where land is limited. Farmers in Faridpur, Bangladesh, with mostly tenanted landholdings of less than 1 ha were willing to devote small patches of agricultural land (0.1 to 0.2 ha) exclusively to firewood growing (Ken MacDicken, pers. comm., 1982). However, R. Petheram and others (1983) reported that farmers in West Java willing to plant trees in blocks reserved their poorest fields for this purpose in most cases.

Use of blocks of agricultural land for wood or fuelwood production is a growing practice in the tropics as the value of these products increases. I have seen such plantings in many areas of Thailand and Taiwan. Certainly for household production, there is no better or more direct distribution system than plantings on personal landholdings or on tenanted plots if tenanted landlord custom permits.

When considering block plantings, the two most critical factors to management (excluding site, nutrient, and water considerations) are *per hectare* population and rotation age. Fuelwood plantings are most productive with relatively close spacing (i.e., 10,000 to 20,000 stems/ha), and with short rotations taking maximum advantage of the fast-growing

TABLE 4.
Effect of Population on Wood Yields of
Leucaena leucocephala (dry ton/ha)

Population/ha	Age (years)			
	1	2	3	4
5,000	16	35	39	84
10,000	31	74	109	148
20,000	36	84	116	155
40,000	42	69	78	112

Source: Van Den Beldt 1983.

fuelwood species. The effect of spacing on yield was researched at the University of Hawaii, where *Leucaena leucocephala* was grown at 12 spacings ranging from 2,500 to 80,000 stems/ha (Table 4).

Note that, for the first year, the closer spacings yielded the most volume at all four sites in Hawaii. More widely spaced treatments produced low wood volume at first, but rapidly picked up production. In this case the highest wood yields at four years were at 10,000 and 20,000 stems per ha. Yields of 5,000 per ha population are expected to exceed those of higher populations at six to eight years.

Rotation age is not only a function of growth but also of stand spacing. Thus managers can use plant populations to exploit early, rapid growth to obtain high yields over short rotations.

Plant populations can affect physiological parameters of firewood stands as well. Both diameter and height growth decrease with increasing population. As *leucaena* ages and competition pressures increase, the effect is more pronounced. This can become an important factor when fuelwood growers sell wood to rural industries, since splitting costs (in time or in money) can be very high for large diameter trunks. Wood-burning parameters (density and moisture content) also vary as a function of season, age, and position on the tree. Although perhaps not important to household producers, these concerns may be worthwhile where wood is to be marketed as fuel.

Block plantings also are capable of yielding coproducts such as fodder, fruit, and green manure. Often per ha yields of these products are influenced by population. In Hawaii, year-old firewood planting of *Leucaena leucocephala* yielded 3.6, 4.3 and 5.5 dry tons per ha of forage for 10,000, 20,000 and 40,000 stems per ha respectively (Van Den Beldt 1982). If coproducts are considered valuable by the community, the manager could consider experimenting with spacing and cutting cycles to maximize production of the coproduct while keeping yields of fuelwood as high as possible.

Leucaena fuelwood planting, Pondicherry. (Kirk R. Smith)

Contour fuelwood planting and erosion control, Philippines. (Napoleon Vergara)

Hedgerow Plantings. Little is known of the potential of hedgerow yields and management even though this is perhaps the method of tree planting used most often by small-scale landholders. Fast-growing, low stature trees are often used as windbreaks, boundary markers, and living fences in traditional systems and may be more acceptable to small farmers than block plantings. That hedgerow plantings can be extremely productive was shown by a windbreak planting at the University of Hawaii experimental farm in Waimanalo. A 50-meter *Leucaena leucocephala* hedgerow consisting of three rows spaced one meter apart has been coppiced annually for the past three years and has been producing about 0.3 ton of wood for every 10 meter per year. (For a similar hedgerow surrounding a 1-ha plot, this would translate into about 13 tons annually!)

Even though that appears to be a huge amount of firewood, the same 10 meters of hedgerow averages about 13 meters shading width, because coppiced stems tend to grow outward toward the light. Therefore, if it were contained solely on the farmer's land, the entire hedgerow would shade nearly half of the hectare directly and would reduce crop yields substantially. More realistically, hedgerows should be confined to one or two rows and pruned heavily for forage and wood throughout the year. A project manager could develop a management system for this and reliably predict its yield.

Integrated Agroforestry Plantings. Much is being learned about mixing trees and agricultural crops on the same piece of land. In terms of fuelwood production, agroforestry applications can range from an almost casual fashion, as with the practice of pruning of branches from trees in home gardens of Java and Bangladesh, to the highly evolved forage/ fuelwood plantations I observed in Thailand.

Little hard data exist on the productivity of home gardens for fuel, but wood fuel production is likely to be low and heavily supplemented with agricultural wastes, cow dung, or miscellaneous materials like twigs and leaves (Islam 1980, Soesastro 1980). Utilizing home gardens as a production base for fuelwood has advantages. As a primary source of fuelwood in many areas, it is likely that farmers will welcome the opportunity to increase production of fuelwood from this source. Home gardens, as a familiar production factor in the life of farm families, provide a "painless" method of introducing firewood farming. However, as these gardens are often heavily shaded, multilayered, and unfenced, it is unlikely that fuelwood production from planted stands of fast-growing trees within them will ever be optimum, remaining instead a part of an integrated food production scheme. This arrangement may be satisfactory depending on the location.

Home gardens can best be utilized as a stepping stone to more formal fuelwood farming on either marginal agriculture lands or actually integrated with crop production. The Thailand example mentioned above (located near Kanchanaburi) is located on sugarcane lands. Production had decreased due to drought, declining soil fertility, and poor market conditions. The farmer sought advice from friends and was told of the possibility of growing a fuelwood/forage mix. He direct seeded *Leucaena leucocephala* over an 8 ha area. A local feed mill buys the forage fraction and harvests the trees every three to six months, depending on growth, at no cost to the farmer. The wood averages about 3 cm in diameter, a size preferred by the local ceramic and earthenware industry, which purchases all the farmer can produce. One could successfully argue that this is indeed "household firewood production," as it has become the farmer's sole means of livelihood.

It is the generation of coproducts such as forage, food, and green manure that make tree plantings particularly useful. Depending on household demand and market conditions, the manager can juggle rotation age, spacing, and species selection to maximize either fuelwood

TABLE 5.
Leaf and Wood Production of Leucaena leucocephala at Three Ages, Waimanalo, Hawaii (Dry tons/ha)

	Plant Population		
	40,000/ha	20,000/ha	10,000/ha
—6 months—			
Leaf	7.3	4.9	4.3
Wood	8.1	5.8	4.9
—12 months—			
Leaf	6.7	5.3	4.5
Wood	29.1	21.8	17.7
—24 months—			
Leaf	9.1	5.9	4.4
Wood	119.9	74.0	56.1

Source: Van Den Beldt 1983.

or coproduct production or even balance the two so that both are optimally produced. A good example of this is found in experimental planting of *Leucaena leucocephala* in Hawaii. High populations produce both the highest amount of wood and forage over a two-year period. But, as shown in Table 5, leaf production essentially stops at six months at all spacings. Thus, if forage production is emphasized, it would not pay the manager to allow trees to grow beyond this time, but would be possibly advantageous to space stems more closely together and harvest more frequently. If firewood production is desired, the trees would be wider spaced and left to grow for more than two years.

Experimentation

Justification for experimentation should by now be apparent. Long lists of suitable firewood species may be available from research reports or local forestry experts, but those sources usually report growth of a species at a given site and for different uses. Little hard data exist that systematically compare species over diverse environments. This is a routine procedure for new varieties of agricultural crops. Since the basic goal of fuelwood and agronomic production are the same (to maximize the net value of production per unit area of land after meeting costs), testing should be a basic part of any fuelwood program.

An example of the variation of performance as a function of site has been shown in Hawaii. Identical experiments involving five core fuelwood species were introduced to four sites and data taken every six months. First year data for all sites are presented in Table 6. Clearly, each species has an "optimum" location in the series of trials. Studying the table, one

TABLE 6.
Volume at One Year of Five Nitrogen-fixing
Fuelwood Species (dry tons/ha)

Species	Waimanalo	Molokai	Waipio	Niulii
Leucaena leucocephala	26.7	36.6	13.2	2.1
Leucaena diversifolia	19.3	22.7	17.6	7.5
Sesbania grandiflora	10.2	23.9	8.2	2.3
Calliandra calothyrsus	7.7	8.1	2.0	5.3
Acacia auriculiformis	10.2	4.6	1.7	0.5

Source: MacDicken 1983.

would not recommend that *Leucaena leucocephala* be grown at Niulii. Yet, the only way to be sure of this was to test it there.

Experimentation does not have to be as time consuming, expensive, and theoretical as is often thought. Preliminary data from the Hawaiian trials of fuelwood species indicate that, in most cases, six-month growth performance can be used to predict how a species will perform at a particular site (MacDicken 1983).

The experimentation phase is most effective in or near villages where programs will be implemented. A research project by the Center for Policy Research in Bangladesh identified which species to use, how to introduce them into communities most effectively, and which management methods to use. Six areas in Bangladesh were identified as pilot sites, four of them in villages. Identical experiments were installed at all sites. The trial plan consisted of eight species at two spacings in non-replicated blocks (Fig. 1). Additionally, the trials were surrounded by three-row direct seeded borders of three species. Notably, two of the species (*Samanea saman* and *Sesbania grandiflora*) were distributed to villagers for homelot plantings as trials were installed.

CPR's trials were advantageous because they

1. Provided useful information on species growth at the actual sites, including two locally preferred species;
2. Assisted in determining genotype x environment interaction for these species under Bangladesh conditions;
3. Allowed experimentation on fuelwood establishment and management, such as direct seeding, spacing, hedgerow yield and silviculture, and forage production; and
4. Provided demonstration to villages in the trial areas.

The experiments did have flaws. For example, since the blocks were

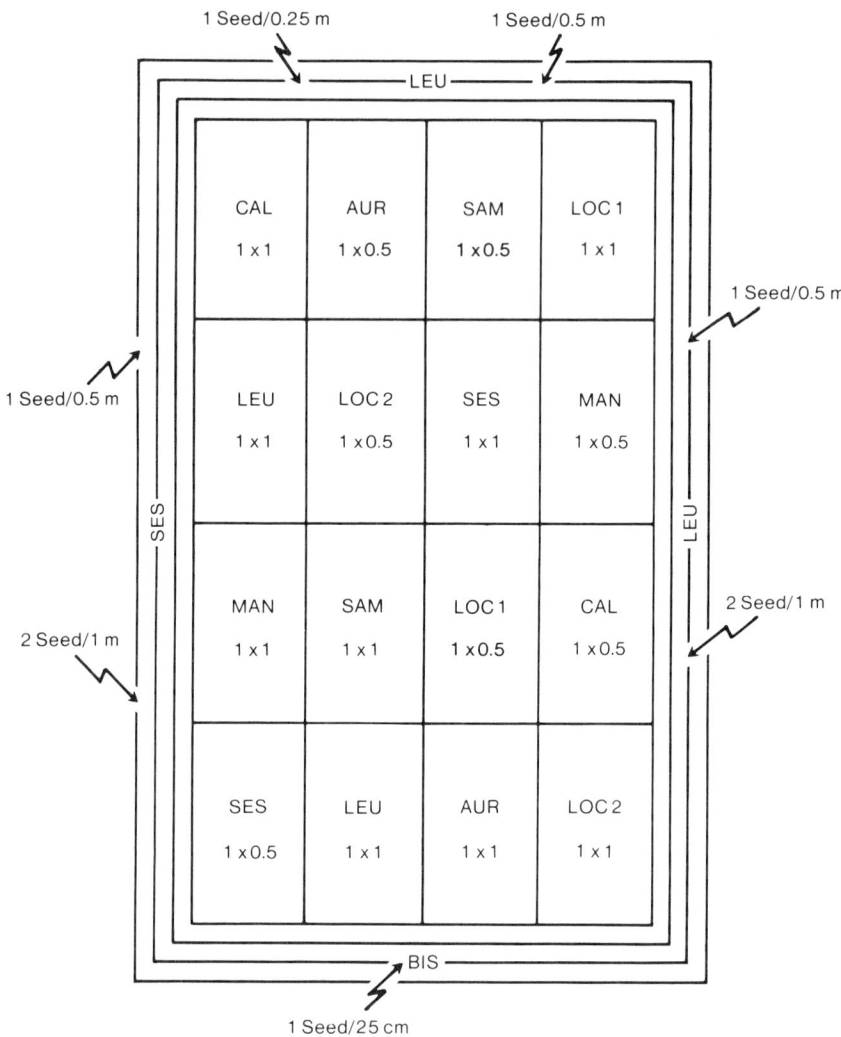

Figure 1. Proposed Species Trials with Direct Seeded Border

not replicated at a given site, systematic combination of analysis between sites was difficult. Also, since they were located in open, fenced agricultural lands, they gave little indication of how trees would fare in shaded home garden plantings or in situations where there would be grazing pressure from livestock. Still they well served their purpose of demonstration, environmental verification of species choice, and verification of silvicultural options.

Research components of fuelwood programs can be much more sophisticated, and research institutions can be tapped to perform research that supports household firewood production schemes. Much information can be gleaned from ongoing research programs that complement these schemes. Still it is best to develop a degree of autonomy in the area of silvicultural research and not to rely solely on outside information sources. Failure to do so may lead to management blunders that may be difficult to correct in the future.

CASE STUDIES

Center for Policy Research: A Bangladesh Example of Firewood Research

The Center for Policy Research of the University of Dhaka was established in 1978. In 1980, the Center, in collaboration with the Universities Field Staff International, was funded by USAID under a Technological Research Grant to study four broad areas in energy policy of Bangladesh. These were: (1) Alternative energy technologies, (2) Hydrocarbon potential of the Bengal Basin, (3) Small hydropower resources, and (4) Biomass energy.

The latter study, biomass energy, is presented here as an example of an action research program that is providing valuable information on fuelwood use in Bangladesh, while testing theories of tree dispersal and people's participation and systematically studying growth rates of fast-growing trees. The case study is derived in part from CPR's annual report of the Biomass Energy Study, which I assisted in preparing.

The role of fuelwood in rural Bangladesh and its context in the rural energy picture has been previously discussed in this volume (Chapter 2). To summarize, household use of firewood from the main trunk of a tree in Bangladesh ranks behind use of other parts of the tree, cut branches, and agricultural and animal residues (jute sticks, rice straw, rice hulls, cow dung, and others). Firewood is a valuable commodity in Bangladesh, valued both by households as an efficient energy source and in rural markets as a good profit source. Rural industries also are substantial users of firewood in Bangladesh, representing potential income for the poor farmer.

Traditionally, the major source of household wood fuels in Bangladesh is from home gardens planted in and around the homestead. Distribution of fuels from these plantings is often not equitable. Wealthier households usually retain the right to harvest branches and wood from trees they own, thus forcing rural poor to do without or to purchase firewood, both at great personal expense. Household plantings are usually in the form of slow-growing fruit species and thus may not be able to add a yearly wood increment equal to that which is taken for fuel. Although parts of rural Bangladesh may now experience an equilibrium between production and consumption of woody fuels, it seems clear that pockets of extreme fuelwood shortages are developing and can only expand in the face of increasing population and growing demand for these fuels. When one considers the substantial fuelwood demand of rural and cottage industries, this argument may be extended to include remaining forested areas—the Sundarban and the Chittagong Hill Tracts, for example.

Thus, there is an urgent need for fuelwood planting in rural Bangladesh. Given the density of human population of the delta, perhaps nowhere else in the world is people's participation in fuelwood planting more crucial for success than in Bangladesh. Recently, the Asian Development Bank (ADB) drafted a US $11 million project proposal for community-based fuelwood production in seven northern districts. There is as yet little idea of how best to proceed, however, and specific expertise in fuelwood farming is lacking among forestry and agricultural agencies.

Given the scenario, the CPR defined four areas of study in the context of biomass production.

Selection of species. A survey was taken of households in 16 districts of Bangladesh to determine people's preference of tree species for fuelwood and to measure the relative growth rates of the preferred species. On the basis of this study, two native nitrogen-fixing tree species (Table 3) and one exotic nitrogen-fixing species, *Leucaena leucocephala*, were selected. Since then, more species have been added (Fig. 1).

Silviculture and Management. Four sites in different parts of Bangladesh (two in villages) were prepared and planted to the three original species. Although initial emphasis was placed on correct nursery and establishment procedures, subsequent studies involved effects of spacing and environment on tree growth, direct seeding as an establishment option, alternative planting methods along fencelines, boundaries, and roadsides, intercropping with food crops, and management for increased coproduct generation (primarily fodder and green manure).

Extension. At first, extension was limited to informal contact between project staff, including the locally hired field caretakers. Village interest was high. As a result, all experiments were moved to villages, fulfilling the dual role of experiment and demonstration. Seedlings were dispersed to farmers to determine response.

Perhaps the strongest point of CPR's approach was its sensitivity and basic commitment to the concept of incorporation of the needs of rural people in the project. An early decision was made to utilize indigenous species first in fuelwood plantings, with the idea that there would be less resistance in the village to establishing these species. The fact that the initially selected species are used for many other purposes by villagers made the approach seem even more logical.

Locally hired caretakers, in all cases landless peasants of the poorest classes, were employed not only to scare off cattle and pull weeds but also to identify neighbors as participants and act as supervisors of seedling dispersal programs. Caretakers ran local nurseries for these programs and provided an excellent source of feedback to project staff.

Eventually, as Private Voluntary Organizations are brought into the program, it is hoped that the seedling dispersal project can be extended progressively in other areas of Bangladesh. By this time, less reliance will be placed on indigenous species and homeyard plantings, with more emphasis on formal fuelwood agronomy.

The emphasis on experimentation with species and silviculture is another strong point of the project. The staff often commented that they were not foresters and therefore at an advantage, because they had fewer preconceived notions of how fuelwood plantings are done. First efforts were based on guidelines from government foresters (for example, slow-growing species, wide spacings) that have since been altered and built upon.

A third emphasis was on establishment of links with outside agencies. Not only do these links provide advice, but often services are made available that would normally not be provided. Project staff were recruited from the National Herbarium and the Botany Department of the University of Dhaka, providing a wealth of seed sources and laboratory facilities for soil analysis and fuel character determinations. International linkages assured not only monetary support but a good access to training facilities, expertise, and seed sources.

The project was designed in a way that it may ultimately be absorbed by any of several institutions. It could well become part of the larger ADB-sponsored fuelwood program or be built into similar programs under consideration by development agencies in Bangladesh. Better still, it may well provide the needed spark to stimulate self-sustaining fuelwood production efforts by villages. Although the project overall appears

TABLE 7.
Wood Sales of Perum Perhutani, 1979 (1,000 x m^3)

	Domestic Consumption	Export Sales
Teak logs	445.6	2.5
Teak products	10.9	17.5
Teak "squares"	–	1.4
Other logs	89.3	–
TOTAL	545.7	21.4

Source: Perum Perhutani (1980).

sound, in the final analysis success will be based on how well it will be integrated with future projects or whether it can develop into a popular movement after funding ends. It is this goal that should be the ultimate objective of all fuelwood-related research done in the future.

Perum Perhutani: Meeting the People to Save the Forest

Perum Perhutani (abbreviation for Perusahaan Umum Kehutanan Negara, or State Forest Enterprise) is a government corporation responsible for the management of 3 million ha of forest on the island of Java. Its mandate is threefold: (1) to manage, protect, and exploit forest lands; (2) to provide employment to rural people; and (3) to carry out, with approval of the Ministry of Agriculture, other activities that promote the first two goals (Hartono Wirjodarmojo, pers. comm., 1982).

To understand the motivation of Perum Perhutani, one must first accept the fact that it is a profit-oriented entity. Its main source of income is from the teak forest, as shown in Table 7. More than 95 percent of Perhutani's harvest is sold on the domestic market. Other products include parquet flooring (11,423 m² in 1979); veneer (425,500 m²); wall paneling, eucalypt oil (120,000 liters); pine resin, silk, turpentine (402,000 tons); and shellac. Significantly, for the purpose of this chapter, 235,000 stacked m³ of firewood were sold as well. (Perum Perhutani 1979). This firewood was sold primarily to rural industries, and as such, cannot be tagged as "household use." Yet, the fact that residents of lands adjacent to forests are involved in the production of the firewood presents a good example of the marketing aspect of fuelwood production. As mentioned earlier, in areas where strong firewood demand exists in the market, it is likely that firewood farmers will sell first to this sector.

TABLE 8.
Land Area Under Jurisdiction of Perhutani (1,000 ha)

	Teak Forest	Non-Teak Forest	Total	Percent of Total Land Area
Unit I (Central Java)	269	287	656	19
Unit II (East Java)	476	847	1,323	21
Unit III (West Java)	167	801	968	23
TOTAL	1,012	1,035	2,947	21

Source: Perum Perhutani (1980).

About 21 percent of the land area in Java is under Perhutani's control (Table 8). The area Perhutani controls is by no means contiguous. Rather, it resembles a series of dots and patches across the map of Java — "a forest between the people," in the words of Hartono Wirjodarmojo, Perhutani's director. On an island with more than 600 persons per km^2, population pressure on these scattered clumps of forested lands is inevitable.

According to Wirjodarmojo (1981), there are 2,372 villages with 6.7 million people surrounding the forested areas in Central Java alone. Nearly 75 percent of these people are farmers or seasonal agricultural workers. Dryland farming in these areas is twice as prevalent as paddy rice farming. The familiar pattern of slash-and-burn agriculture followed by imperata grass takeover and annual burning is prevalent in these upland areas. Another severe pressure on forests is grazing, both inside the forest and outside. Finally, firewood gathering is seen as a severe problem, particularly in areas that service rural industries (mostly tile and limestone reduction) with wood fuel. Teak charcoal is especially valued in the blacksmithy industry (Raintree 1980).

A frequent reaction of administrative forestry groups in the tropics is either to do nothing about population pressures or to beef up enforcement and institutional capability in an attempt to control those pressures through law. Perhutani's reaction to the people's use of the forest is based on the second part of its mandate — to provide employment — and, in my opinion, is rare in its institutional acceptance of the concept of forestry with a human component.

Perum Perhutani community charcoal and lumber production. (R. Van Den Beldt)

A keystone to Perhutani's approach is the "MALU" system of seeking the people's cooperation and participation. MALU is derived by abbreviating *mantri* (forest guard, a junior-level employee of Perhutani) and *lurah* (government-appointed head of village). Perhutani has established a 14-package series that can be delivered to a village after consultation under the MALU aegis. Of these packages, the most notable include

— Agroforestry   Intercropping food with newly planted forest plantations.
— Agriculture   Improving production of home gardens with credit, seeds, and technical advice.
— Pasture   Planting of suitable grass species for sale as forage.
— Beekeeping   Apiary projects using native bees (*Apis indica*) and flowering forest trees.

— Water Catchment/Watershed Rehabilitation	Providing drinking water and sustained water yields for irrigation.
— Housing	Providing housing in small community plots to itinerant forest workers.
— Firewood	Planting of fast-growing trees in plantations for sale to industry.

Villages approached by Perhutani are viewed more as sources of contract labor than as rural folk in need of upliftment. For the most part, effort is made by the forest guards to determine the uses of the forest by the people and the markets these uses serve. Perhutani then invests capital and labor to increase and stabilize production of forest products, assuming the role of the marketing agent and using the people as contractors. Extension is used to ensure the success of projects by removing population pressures on the forest. People's interest in the projects is insured by involving them in the enterprise as early as possible and then maintaining that involvement in succeeding years. The procedure is well illustrated in Perhutani's firewood enterprise. At present, Perhutani's firewood plantations cover about 60,000 ha (Wirjodarmojo 1981), mostly planted to *Gliricidia sepium* and *Calliandra calothyrsus*.

Cutting of trees by rural residents for sale as firewood occurs mostly in areas already facing severe deforestation pressure. Moreover, it is done mostly by those with little economic alternative — a "last resort" for procuring such income. Thus, there is little incentive for these persons to respect replanting efforts — in fact when seedlings reach the sapling stage, they are ideal as harvestable firewood, as their small diameters require minimal handling.

The response of Perhutani to this situation is to plant belts of fast-growing firewood species around those areas most plagued with firewood gathering. This is accomplished by contracting local labor to see the trees through establishment. First harvest begins at four years, with the same village contractors harvesting the plantations for salary. Payment also is made on the basis of piecework (number of stacked meters per day), depending on local arrangements. Perhutani collects the wood at roadside and sells it directly to industry. Private cutting for sale is not allowed, although villagers are allowed to gather twigs and dead branches in the plantations, a recognized traditional right of the landless in Java.

According to the Perhutani director, some new directions are being tested. The first is to involve the people directly in raising their own firewood on private land and selling the wood to Perhutani. This involves training in seed collection, nursery operations, and planting. Another is closer links with other governmental groups (e.g., Land Rehabilitation and the Regreening Movement of the Ministry of Agriculture) to increase extension efforts. Third, there are continuing dialogues between village leasers and Perhutani junior staff to bring in other forestry-related enterprises that cement the presence and value of forests in the people's day-to-day lives. Finally, several village cooperatives are being set up on a trial basis to handle firewood sales from Perhutani forests.

This brief account is not meant to be read as a blanket endorsement of Perhutani's firewood scheme. Critics argue that the actual economic benefits to people living in the area are not as great as they would be if more control were given to the people in the management of the plantations and marketing of the firewood. Also, it is not clear whether the contract arrangements — particularly piecework harvesting and stacking — are financially equivalent to work of those individuals who allegedly cut from the forest and sell directly to middlemen. Finally, there is the question of whether or not the best species and silvicultural techniques have been chosen for a given site.

Perhutani's approach to "people's participation" defines a procedure that was dictated out of necessity rather than as an end in itself. It is, in effect, the "shortest distance between two points." If the system proves economical and widely adaptable, it will undoubtedly slow down the trend toward forest destruction on Java. Perhutani predicts that as much as 44 million stacked meters of fuelwood could be recovered annually on a sustainable basis from its fuelwood plantations, providing steady employment to more than 3 million persons in rural areas (Wirjodarmojo 1981). If these predictions are even partially fulfilled, it would be a notable achievement.

ACKNOWLEDGEMENTS

The assistance and guidance of James Brewbaker in producing data of *Leucaena leucocophala* presented here is gratefully acknowledged. The contributions of Richard Morse, East-West Center; M. Hadi Soesastro, Center for Strategic and International Studies, Jakarta; and Sunarjo Darsono and Boen M. Purnama, Central Research Institute for Forestry, Bogor, in planning and coordinating aspects of the research conducted in preparation of this chapter are acknowledged.

REFERENCES

Anonymous.
    1973   Pertanian: Jilid VII Sensus Pertanian 1973. Biro Pusat Statistik, Jakarta.

Anonymous.
  1983 Summaries of Selected Wood Fuel Surveys. Annex II in FAO, *Wood Fuel Surveys*, 131-146.
Arnold, J.E.M.
  1982 Community Forestry Development in Korea. Also Community Participation in Forestry Projects. Conference on Forestry and Development in Asia, April 19–23, Bangalore. The Asia Society.
Arnold, J.E.M., and J. Jongma.
  1978 Fuelwood and Charcoal in Developing Countries. Paper presented at 8th World Forestry Congress, September, Jakarta.
Brewbaker, J.L.
  1983 Fodder and Fuelwood N_2-Fixing Trees for Nepal. Report to International Agricultural Development Service on Consultation, March, New York.
Brewbaker, J.L., R. Van Den Beldt, and K.G. MacDicken.
  1982 Nitrogen-Fixing Tree Resources — Potentials and Limitations. In *BNF Technology for Tropical Agriculture*, D.H. Graham and S.C. Harris, eds., CIAT AA67–13. Cali, Colombia.
Cecelski, E., J. Dunkerly, and W. Ramsey.
  1980 Household Energy and the Poor in the Third World. Research Paper R–15. Resources for the Future, Washington, D.C.
Center for Policy Research (CPR).
  1982 Bangladesh Biomass Energy Study. Annual report. USAID TRD#388–0027, Dhaka.
deLucia, Russell.
  1983 Defining the Scope of Wood Fuel Surveys. In FAO, *Wood Fuel Surveys*, 5–28.
deLucia, Russell.
  1983 Different Categories of Wood Fuel Surveys. Annex I in FAO, *Wood Fuel Surveys*, 125–130.
Dick, H.
  1980 The Oil Price Subsidy, Deforestation and Equity. *Bulletin of Indonesian Economic Studies* 26(3):32–60.
Donovan, D.G.
  1981 Fuelwood: How Much Do We Need? DGD–14. Institute of Current World Affairs, Hanover, N.H.
Douglas, J.J.
  1982 Traditional Fuel Usage and the Rural Poor in Bangladesh. *World Development* 10(8):669–676.

Food and Agriculture Organization (FAO).
   1981   *Report of the Technical Panel on Fuelwood and Charcoal.* United Nations Conference on New and Renewable Sources of Energy, August, Nairobi.

FAO.
   1982   World Forest Products — Demand and Supply 1990–2000. FAO Forestry Paper. Rome.

FAO.
   1983   *Wood Fuel Surveys.* Rome.

Gamser, M.S.
   1980   The Forest Resource and Rural Energy Development. *World Development* 8(10):769–80.

Haeruman, H.
   1979   Energy Consumption in Rural Areas of West Java. Paper presented at 7th General Assembly World Federation of Engineering Organizations, November, Jakarta.

Hariyatno, H. Prahastro, and B.M. Purnama.
   1982   Case Study on Community Fuelwood Farming at Toyomerto. Report no. 160/1982. Forest Products Research Institute, Bogor.

Harker, A.P., A. Sandels, and J. Burley.
   1982   Calorific Values for Wood and Bark and a Bibliography for Fuelwood. G162, Tropical Products Institute, London.

Hoskins, M.W.
   1979   Women in Forestry for Local Community Development. USAID OTR –147–79–83, Washington, D.C.

Islam, M.N.
   1980   *Village Resources Survey for the Assessment of Alternative Energy Technology.* Prepared for International Development Research Centre, Canada. Dhaka: Bangladesh University of Engineering and Technology, Department of Chemical Engineering.

Ki-Zerbo, J.
   1981   Women and the Energy Crisis in the Sahel. *Unasylva* 33(133): 5–10.

MacDicken, K.G.
   1983   Studies on the Early Growth Rates of Selected Nitrogen-Fixing Trees. MS Thesis, University of Hawaii.

Morgan, W.B.
   1983   Urban Demand: Studying the Commercial Organization of Wood Fuel Supplies. In FAO, *Wood Fuel Surveys*, 53–74.

National Academy of Sciences (NAS).
   1980   Firewood Crops — Shrub and Tree Species for Energy Production. National Academy of Sciences, Washington, D.C.

Noronha, R.
   1981   Why is it so Difficult to Grow Fuelwood? *Unasylva* 33(131): 4–12.

Openshaw, Keith
   1983   Measuring Fuelwood and Charcoal. Annex III (b) in FAO, *Wood Fuel Surveys*, 173–178.

Panday, K.K.
   1981   Trees and Conservation of Cultivated Lands in Nepal. Southeast Asian regional symposium on problems of soil erosion and sedimentation. Bangkok.

Parera, V.
   1983   Wider Use of the Girdling System in Sikka. *Leucaena Research Reports* 4:45.

Park, W., L.C. Newman, and K. Ford.
   1982   Fuelwood Supply for Managua, Nicaragua—Sustainable Alternatives for the Las Maderas Fuelwood Supply Region. Mitre Corporation, MTR–81W285.

Pasca, T.M.
   1981   Concerning Wood Energy. *Unasylva* 33(131):2–3.

Perum Perhutani.
   1979   A Glimpse of Perum Perhatani. Perum Perhatani, Djakarta.

Petheram, R.J., L. Liem, and L. Paat.
   1983   Survival and Growth of Leucaena Stump Cuttings by Ruminant Rearers in Villages, West Java. *Leucaena Research Reports* 4:46–47.

Raintree, J.B.
   1980   Meeting the Energy Requirements of Small-Scale Rural Industry in North-Central Java. Development Alternatives, Inc., Jakarta.

Saouma, E.
   1981   The Urgency of Food and Energy Problems. *Unasylva* 33(133): 2–4.

Soesastro, M. Hadi.
   1980   Energy and Rural Development in Indonesia: Structuring of the Baseline Data. Paper presented at Energy and Rural Development Implementation Workshop, February. Chiang Mai.

Van Den Beldt, R.J.
   1981   Fuelwood Production of Leucaena with Short Rotations. Paper presented at workshop on Environmentally Sustainable Agroforestry and Fuelwood Production. November, Environment and Policy Institute, East-West Center, Honolulu.

Van Den Beldt, R.J.
- 1982 Litterfall as a Function of Population in a 1-year-old *Leucaena* Planting. *Leucaena Research Reports* 3:95.

Van Den Beldt, R.J.
- 1983 *Leucaena Leucocephala* (Lam.) deWit for Wood Production. Ph.D. dissertation, University of Hawaii.

Weatherly, W.P.
- 1980 Environmental Assessment of the Rural Electrification I Project in Indonesia. USAID, Jakarta.

Wirjodarmojo, H.
- 1981 Some Socioeconomic Aspects of Fuelwood Plantations in Java. In *Proceedings of Energy for Rural Development Activity Set III: Fixing Criteria for Selection of Fuelwood Species*, March 4–7, University of the Philippines at Los Baños, Philippines.

Wood, D.H., D. Brokensha, A.P. Castro, M.S. Gamser,
B.A. Jackson, B.W. Riley, and D.M. Schraf.
- 1980 The Socioeconomic Context of Fuelwood Use in Small Rural Communities. USAID Evaluation Special Study no. 1. USAID, Washington, D.C.

10
Financial and Resource Analyses of Anaerobic Digestion (Biogas) Systems

Michael T. Santerre

INTRODUCTION

Anaerobic digesters, also known as biogas plants or gobar gas plants, are versatile energy and waste management systems that interest the energy and rural development community because of four possible main benefits.

1. Anaerobic digesters provide a clean-burning fuel (biogas) that can provide energy for various domestic, agricultural, and other rural tasks.
2. Anaerobic digesters offer a means to conserve the fertilizer value of the nitrogen present in domestic and agricultural residues, since burning, storing, or composting wastes could result in higher nitrogen losses.
3. The physiochemical environment inside an anaerobic digester can partially or completely destroy the human or animal parasites and pathogens in fecal material as well as seeds of weeds in animal feces.
4. Widespread use of biogas plants in rural areas of developing countries could lessen deforestation by reducing demand for firewood and could promote an improved balance of payments in some countries by reducing the demand for petroleum fuels.

Biogas plants are based on the principle of anaerobic digestion (or anaerobic fermentation). By a bacterial process in an oxygen-free (anaerobic) environment, organic substances (e.g., carbohydrates and other materials in dung and straw) decompose into biogas and a mixture of residual end products. Biogas plants act as air-tight containers for fermenting dung or other organic materials and water that are fed into the fermentation pit through a feedstock inlet (Fig. 1). Anaerobic digestion by bacteria occurs in the pit. The biogas bubbles up through the watery slurry and is collected in a gas holder. The gas holder operates like an inverted cup floating in a basin of water: biogas collects in the holder, which floats on top of the slurry.

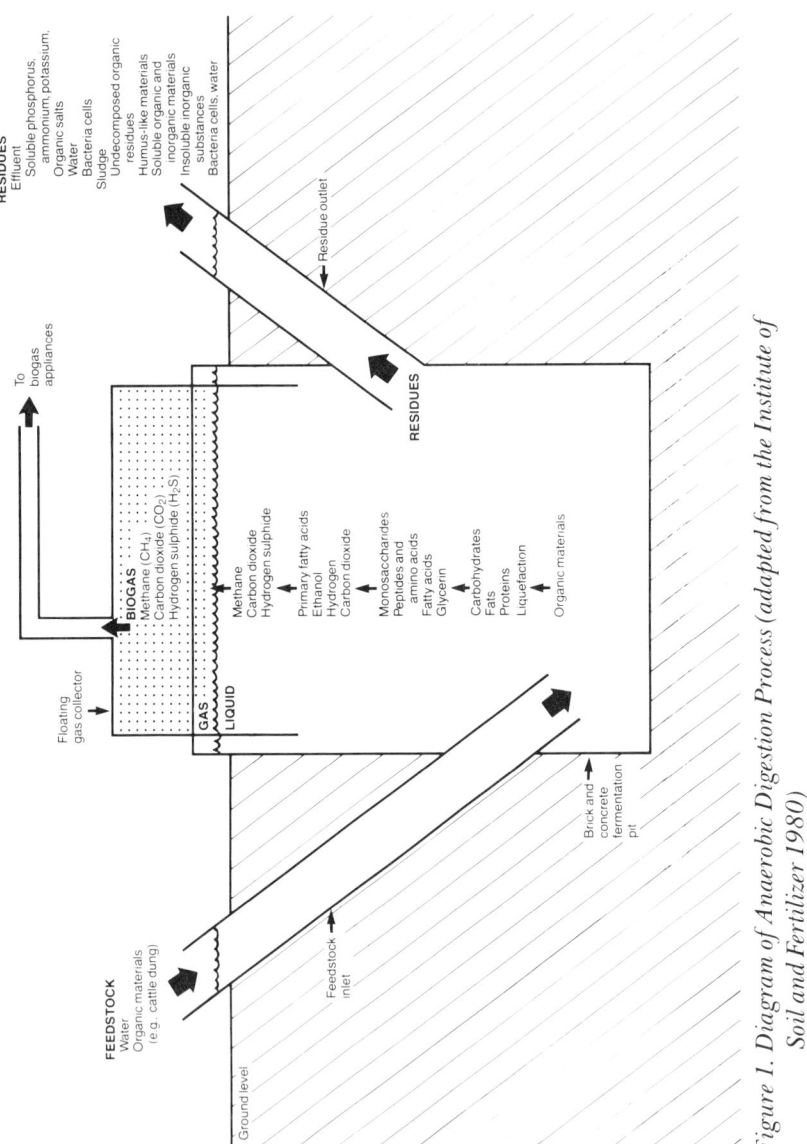

Figure 1. Diagram of Anaerobic Digestion Process (adapted from the Institute of Soil and Fertilizer 1980)

Anaerobic Digestion (Biogas) Systems   417

In this chapter, I examine the possible role of biogas plants in meeting rural energy needs. First is a brief overview of the products and potential benefits of anaerobic digesters. Second is a technique for analyzing and comparing the resource requirements for installing and operating biogas systems in rural areas. Then comes an analysis of factors affecting the viability and benefits of biogas systems, documented by brief reviews of recent biogas developments in China and India. This discussion establishes the background for the central effort of the chapter, a systematic approach to assessing the viability and benefits of biogas plants by taking account of each key factor relevant to an investment decision at a particular place and time. A framework is presented for financial analysis of household-scale biogas systems for domestic energy needs of various types of rural households. Then the analysis is extended to community biogas systems for domestic energy uses and for energy to run community-based food processing enterprises. (Detailed technical and other information about anaerobic digesters is available in P.J. Meynell 1976, the National Academy of Sciences 1977, Andrew Barnett and others 1978, and ESCAP 1980 and 1981.)

PRODUCTS AND BENEFITS

The principal beneficial product of anaerobic digestion from the standpoint of an energy technology is biogas: a combustible mixture of methane, carbon dioxide, and other gases. The methane content of biogas is usually around 50 to 60 percent, depending mainly on the type and chemical composition of the organic feedstock materials, the input rate, and the environmental conditions in the fermentation pit.

As already mentioned, biogas is a relatively versatile and very suitable energy source for performing various rural tasks. Because biogas is not readily liquified for convenient storage and transport, however, its use in powering vehicles has required cumbersome low-pressure storage bags or costly compressors and high-pressure gas cylinders. Thus, biogas seems more suited to providing energy for stationary tasks. Some basic appliances and machines can be powered by biogas, as shown with their biogas consumption rates in Table 1.

In Asia, biogas usually is used for cooking and lighting. A typical household of about five persons requires about 1.0 to 1.5 m^3 of biogas per day for cooking purposes. Operating one or two biogas lamps for about five hours each day could bring the total biogas requirement to about 1.5 to 2.5 m^3 of gas per day. A household operating a biogas plant can realize savings in purchased or collected biomass fuels and in kerosene — the most widely used fuels for cooking and lighting in rural areas. The

TABLE 1.
Biogas Utilization by Various Appliances and Machines

Appliance or Machine	Biogas Consumption Rate
Biogas Lamps	0.07 to 0.17 m^3 per hour for equivalent of 40-60 watt bulb
Electric Lamp (powered by biogas operated electric generator)	0.03 m^3 per hour for a single 40 watt electric bulb
Biogas Stoves	0.17 to 1.7 m^3 of biogas per day required for cooking for one person
Biogas Refrigerators (flame operated)	0.4 to 1.2 m^3 per hour per m^3 of refrigerated volume
Electric Generators	0.45 to 2.0 m^3 per hour per kilowatt of electric power
Engines	0.4 to 0.5 m^3 per hour per horsepower
Vehicles	
3 ton lorry	0.35 m^3 per hour
0.5 ton pick-up truck	0.19 m^3 per hour

Source: See Santerre and Smith 1980.

attractiveness of biogas systems will depend very much on the magnitude of these savings, as discussed later.

Besides potential benefits from not having to purchase or collect biomass or other fuels, biogas is a clean-burning fuel that can improve the kitchen environment significantly by reducing smoke. The smoke from burning firewood or other types of biomass is thought to be a major health problem for women in developing countries (Kirk R. Smith and others 1983). Investigations are presently underway to attempt to quantify the extent of this problem and to quantify the benefits that could result if smoke-free cooking devices such as biogas stoves are adopted.

The other important product of anaerobic digestion is the residue produced from the bacterial degradation of the organic feedstock materials. This residue, alternatively called digester residue, sludge, slurry, effluent, or digested manure contains various soluble and insoluble inorganic substances, undecomposed organic materials, humus-like substances, bacteria cells, and water (Fig. 1). The composition and quantity of residues differ from the original feedstock materials (Table 2). During the digestion process, the total solids concentration decreases

TABLE 2.
The Mass Balance of Materials in a Biogas Plant

Input	Composition
100 kg cattle dung 100 kg water	16.99 kg total solids (8.5%) 14.16 kg volatile solids (7.08%) 4.59 kg carbon (27%) 0.29 kg nitrogen (1.7%)
Outputs	
3.55 kg biogas (3.4 m^3)	60-70% methane 30-40% carbon dioxide
196.45 kg sludge	13.34 kg total solids (6.67%) 8.89 kg volatile solids (4.45%) 3.20 kg carbon (24%) 0.29 kg nitrogen (2.2%)

Source: P. Rajabapaiah and others. 1979.

TABLE 3.
Changes in the Composition of Nitrogen Species During Anaerobic Digestion

Material	Percentage of Nitrogen in Form of:	
	Organic Nitrogen	Ammonia Nitrogen
Cattle Dung	97	3
Digested Slurry	82	18
Digested Slurry (dried)	99	1

Source: M. A. Idnani and S. Varadarajan 1974.

significantly because some of the feedstock is converted into biogas. Notably, however, the quantity of nitrogen remains constant during digestion, although nitrogen concentration increases compared to total solids because some solids turn into biogas.

An important change is the conversion of a portion of the nitrogen present in the organic form into inorganic ammonia nitrogen (Table 3). Although ammonia nitrogen is regarded as being more readily available for uptake by plants, ammonia is a very volatile form of nitrogen and could be lost rather rapidly if the digester residues are stored in open air or dried prior to use on the fields. Thus, the organic nitrogen probably is the more important form of nitrogen present in digester residues, since those residues will decompose further after being applied to the soil and thereby will be available to plants in the same way as a time-release fertilizer.

Even with possible losses of organic matter and nitrogen during anaerobic digestion and handling of digester residues, these losses might be less severe than if other waste treatment measures are used. For example, studies at the Institute of Soil and Fertilizer (1980) in Sichuan have noted that, although the losses in organic matter from biogas manure are between 30 and 50 percent and nitrogen losses are about 5 percent, storing human and animal excreta in an open air pool results in a 20 percent nitrogen loss. Rapid composting methods are even less conservative: more than 50 percent of the organic matter, and 20 to 40 percent of the nitrogen are lost. However, if digester residues are stored in open air, dried, or composted (as is usual) for application to the soil once or twice each year (as in common in rural areas), their fertilizer value could also decline markedly.

Although use of dried digester residues often means better crop yields compared with use of farmyard manure or with control (no manure) treatments, those crop yields from residue treatments can be considerably lower than those from an equivalent application of commercial fertilizer nitrogen. The information presented by M.A. Idnani and S. Varadarajan (1974) suggests that, at least for some crops, the fertilizer value of nitrogen in digester residues might be only 25 percent as effective in stimulating crop yields as the nitrogen present in ammonium sulphate in the first year (Table 4). On the other hand, other fertilizer elements present in both the organic and inorganic manures also could have affected the outcome of those experiments. Considerably more research is needed in this area to better quantify the value of watery or dried digester residues for different crops under various environmental conditions.

Another important benefit from biogas plants is the possible reduction or elimination of pathogens, parasites, and weed seeds during the treatment of animal wastes. The extent to which these organisms are eliminated has been demonstrated to be dependent on the design of the biogas plant, the residence time of the organisms in the fermentation pit, and the temperature and other physiochemical conditions present in the fermenting medium (M.G. McGarry and J. Stainforth 1978, Ariane van Buren 1980). Whether such sanitation benefits are actually realized or not depends very much on the local situation. For example, if human or pig feces are used, the reduction of pathogens and parasites could be a significant benefit. Waste management practices, however, also influence how far the potential health benefits from treating animal and human wastes are realized. If fields were receiving human or pig feces treated in a biogas plant as well as *untreated* human or pig wastes, health benefits would be few.

TABLE 4.
Effects on Crop Yields of Digester Residues
and Other Manures or Commercial Fertilizers

	Grain Yield/Straw Yield (grams dry matter per pot)				
Experimental Treatment Crop	No manure	Wet digested slurry (@ 124 kg N/ha)	Sun-dried digested (@ 124 kg N/ha)	Farmyard manure (@ 124 kg N/ha)	Ammonium sulphate (@ 124 kg N/ha)
Wheat	9/13	10/15	11/17	10/16	14/20
Marua	10/31	12/34	14/37	12/35	15/41
Sannhemp	/93	/107	/117	/104	/121

Source: From Idnani and Varadarajan 1974.

CONSTRUCTION AND OPERATION RESOURCE REQUIREMENTS

The purpose of recent emphasis on the development of indicators of appropriateness for alternative energy systems has been to ensure a better match of such systems to the needs and resources of persons living in rural areas. Although economic and financial viability are very important criteria of suitability of energy systems, other goals such as alleviating deforestation or improving living conditions of rural people are also important criteria. At the same time, they are often difficult to apply in conventional economic analyses (Smith and Michael T. Santerre 1980).

One often cited indicator of appropriateness of renewable energy systems is that the technology

> ... exploits locally available energy, human and material resources in the rural areas as far as possible, so that the rural income remains within the rural areas. (UNIDO 1978)

Such technologies, R.H. Brown (1977) argues,

> ... would be fairly simple and therefore used, maintained, and repaired on the spot without great dependence on trained mechanics and imported parts. ... People could be more easily trained, supervision and organization would be simpler, and the operation in general would be geared to the capabilities of those whom it seeks to employ and serve.

In this section, I will present some of the results of recent work in examining the use of resources for constructing and operating different types of anaerobic digesters in developing countries. Such information is useful for various purposes, including (a) comparing renewable energy systems to determine which devices are the most appropriate for a specific rural situation based on locally available resources, and (b) assisting planners in allocating development resources to utilize the resources of a country most satisfactorily.

Resource accounting techniques do have certain limitations, however. Selecting a biogas plant or other energy system would not be justified just because it relies on locally available materials or uses less of a particular resource. Recent information from China underscores this point. Although Chinese fixed-dome digesters often rely heavily on local materials and conserve scarce supplies of cement and steel, Vaclov Smil (1982) points out that many of the digesters constructed in China are very unreliable and often fail in their second year of operation. Thus, measures of appropriateness must include reliability as well as resource and other criteria.

In order to compare the resource requirements for different types of energy systems, the quantity of resources should be indexed to a standard measure of energy production capacity or output over a specified time interval. Development of a consistent systems boundary also is necessary to ensure that space and time elements are equivalent in resource accounting for each energy system. With the FLERT (Fuel-Linked Energy Resources and Tasks) approach, for example, the quantities of resources needed to construct a biogas plant are indexed to a fixed energy output: 1 m^3 of biogas per day (Smith and Santerre 1980, Santerre and Smith 1982). The system boundaries used in the FLERT approach are defined spatially by the village boundary, although this boundary is not defined in rigorous geographic terms. For example, although the quantity of cement needed to construct a biogas plant is included, the quantity of energy and other resources needed to manufacture the cement is not accounted for if the cement is manufactured outside of the village area. The temporal system boundary defined in the FLERT approach begins with collection of local construction materials and ends when the lifetime of the energy system ends. It does not include resources needed for research and development or credit for the salvage value of materials at the end of the system's lifetime.

Four examples of the use of construction materials and labor for the installation of representative types of biogas plants used in Asia are enumerated in Table 5. The systems boundaries in this FLERT analysis are assumed to be equivalent, though certain inconsistencies may exist

TABLE 5.
Use of Construction Materials and Labor by Four Anaerobic Digesters

	Fixed-Dome		Floating-Dome	
	Circular Pit of Brick, (large volume/small mouth)	Janata	Khadi and Village Industries Commission	
Rated Capacity (m^3/day)	2.0	2.8	2.8	85
Quantity of Resource per m^3/day biogas production				
Bricks (#)	400	890	1,000	560
Cement, lime, plaster (kg)	150	180	270	140
Iron, steel (kg)	0	0	55	32
Sand, cinders, gravel stone (kg)	930	3,000	1,900	800
Labor (person-days)	18	15	not given	not given

Sources: Santerre and Smith 1980; van Buren 1979; Singh and Singh 1978; Sathianathan 1975.

Note: The materials and labor for installing the digesters are reported in units per m^3/day of installed biogas production capacity. The capacities given here are either as reported in the literature, or based on average biogas output. For calculations and other assumptions, see: Santerre and Smith 1980.

because the data used are secondary. Most existent sources do not refer specifically to which physical parts of the biogas system are included in the resource data or whether labor, for instance, includes the time required to procure local materials such as sand and crushed rock. The analysis, nonetheless, illustrates the varying degrees of dependency of different types of digesters on local and nonlocal resources. Floating-dome digesters require a significant quantity of steel compared to fixed-dome plants. The table also indicates the larger floating-dome plant enjoying a certain economy of scale in the use of materials. The table does not include resources needed for gas and slurry distribution. Economies of scale of community systems could be affected by the spacing of dwellings or the distances to sources of dung or fields receiving the residues. Comparisons of biogas systems using resource accounting methods such as the FLERT analysis should be made during investment assessment, since shortages of steel and cement reportedly have been hampering biogas development programs in some countries (KVIC 1976, A.B. Karki 1980).

The resources needed for operating and maintaining a biogas system include an organic feedstock material such as dung or straw, an approximately equal quantity of water (depending on the water content of the feedstock), paint and cement or other materials for maintaining and repairing the biogas system, and labor. Infrastructural support also is vital and includes, for example, repair shops and consultative services. Often when such support is overlooked, this nontraditional energy technology is abandoned long before its intended lifetime because the owner of the biogas system is not able to maintain it properly. For example, up to 50 to 67 percent of the biogas plants installed in India may be inoperative at present because what may be relatively simple technical problems in some areas are much more complex in remote rural areas when infrastructural support is inadequate (T.K. Moulik and others 1978; Moulik, pers. comm., 1982; D. Sharma 1981; Rita Bhatia 1981).

Table 6 provides estimates of the quantities of organic materials needed to produce 1 m^3 of biogas. Considerable variation exists in the literature concerning these quantities. Most of this variation can be attributed to quality of materials being used, although descriptions of the experimental conditions are usually inadequate to permit meaningful generalizations. For example, from these data it might be deduced that a household requiring 2 m^3 of biogas per day for cooking and lighting would need to have access to between 24 and 64 kg of fresh cattle dung, and about 24 to 64 liters of water. I suspect, however, that the higher estimates are more reliable. The number of cattle required to support a household biogas plant usually is reported to be between three and six head, but the actual number will depend on how big the cattle are and what their feeding regime is; whether they are free roaming (thus making dung collection more difficult) or confined; whether the dung is added in a fresh or partially dried state to the digester; and whether cattle urine or other feedstock materials are also used.

Many other types of organic feedstock materials can also be used to produce biogas. The use of plant materials and human wastes, however, occurs more commonly in China than in other parts of Asia. One major obstacle to the more widespread use of organic materials other than cattle dung appears to be aversion to handling human and pig wastes in some areas. Others are that plant materials often already have important uses (e.g., construction materials) and that their use as feedstock will usually mean additional work to reduce their size and thereby facilitate bacterial degradation and reduce clogging in the digester.

TABLE 6.
Conversions of Organic Feedstock Materials to Biogas

Organic Material	Kilograms of organic material needed to produce one cubic meter of biogas per day
Animal Manures	
Cattle dung	12 to 32
Horse dung, dried	3.3 to 5.0
Pig manure	13 to 20
Swine manure, dried	1.8
Poultry manure	12 to 16
Human Wastes	
Night soil	14
Sewage wastes, dried	1.6
Plant Materials	
Fresh grass, dried	1.6
Maize straw, dried	1.2
Flax stalk, dried	2.7
Wheat straw, dried	2.3
Rice husks, dried	1.6
Tree leaves, dried	3.4 to 4.8
Weeds, dried	1.6

Sources: Chengdu Institute of Biology, Academia Sinica 1980; ESCAP 1980; Idnani and Varadarajan 1974; Rajabapaiah and others 1979; Srinivasan 1978; van Buren 1979; Singh 1974.

The labor required to operate a household biogas plant varies considerably according to the local situation. The labor inputs are (a) collecting organic feedstock materials and water; (b) preparing these materials and adding them into the fermentation pit; (c) removing digester residues from the plant and other activities associated with storing, composting, or transporting the residues to the field; and (d) monitoring the biogas production process, stirring the digester, and other activities. An important consideration in evaluating labor inputs is distinguishing between gross and net requirements. The gross labor requirements are simply the sum of all labor needed to perform activities (a) to (d), plus the labor associated with using the biogas. The net labor requirements are the gross requirements minus the labor used for activities that are no longer performed after a biogas system is adopted. An example of such an activity might be firewood collection for cooking. However, whether the activity is actually displaced by biogas-related activities depends on local conditions. If firewood is being collected while other functions are performed, such as cattle grazing, then not having to collect firewood might have relatively little impact on time spent (i.e., time will still be devoted to tending cattle).

Other factors also must be considered when analyzing labor needs. The labor involved in handling and transporting digester residues might differ little in quantity if the household were previously handling and transporting compost made from cattle dung. For a household considering replacing inorganic fertilizers with digester residues, some important tradeoffs must be considered. Digester residues are particularly bulky forms of fertilizer. For instance, consider that if a farmer wants to apply 1 kg of nitrogen he needs to transport and apply only 2 kg of urea fertilizer. On the other hand, the application of 1 kg of nitrogen in the form of watery or dried digester residues could involve the handling, transport, and application of about 680 kg of slurry or about 80 kg of dried residues! (See Table 2.) Clearly the labor requirements for these alternatives will differ considerably and the rural household should consider carefully the labor implications of this course for meeting and energy fertilizer needs.

FACTORS INFLUENCING THE VIABILITY AND BENEFITS OF BIOGAS SYSTEMS

The quantification of the benefits and costs of biogas plants can be a particularly difficult task, especially when trying to capture the essence of factors that influence the investment decisions of rural households. The principal problem remains the assignment of monetary values to the products and benefits of biogas plants and to the resources for operating the system.

Although a number of financial and economic analyses have been made of biogas plants in developing countries (Jyoti K. Parikh and Kirit S. Parikh 1977, A.K. Sanghi and D. Day 1977, Barnett and others 1978, D. French 1979), very few of these studies have approached their particular analysis from the perspective of the rural household or community, while considering the diverse range of energy needs or other conditions existing in rural areas. Neither have they attempted to include the wide range of possible biogas system costs, digester residue values, discount rates, or practices for financing the acquisition of a biogas system.

Insights on the interplay of household and community factors that influence the valuation of biogas inputs and outputs as well as the success or failure of biogas systems are available from the experiences of countries with active biogas programs, such as China and India. To bring these complex factors into focus, I present a brief overview of biogas development in these two countries. To show how lack of attention to one or more key factors or values could lead to misguided investment decisions, I then present a framework for incorporating these factors in assessments of household and community biogas systems.

The most ambitious biogas development efforts, at least in terms of numbers, have taken place in China and India, where about 8 million, and 100,000 biogas plants have been installed, respectively. Most of these are household-scale units supplying gas for cooking and lighting purposes. There are, however, about 36,000 larger village-scale digesters in China (Smil 1982), and reportedly in 1979 about 560 digesters were powering generators producing about 1.6 MW of power (Robert Taylor 1980). India has at least three community plants in operation, and another 20 will be completed shortly (ESCAP 1981). For more information about digester programs in South and Southeast Asia, see Jamuna Ramakrishna (1980).

Biogas Development in China

The development of anaerobic digestion technology in China began in the late 1950s, although it was not until the 1970s that the efforts began to gather speed. The majority of the 8 million plants were built between 1974 and 1978, and, although the official goal was to construct 70 million units by 1985, the pace of construction has slowed considerably, and it is evident that this target will not be met (Taylor 1980, Smil 1981 and 1982).

Despite recent problems, the biogas development effort in China during the 1970s was an impressive feat, especially in Sichuan Province. About 70 percent of the biogas plants in China are located here; about 30 to 35 percent of the rural households have units. Biogas systems are highly concentrated in this one province because its relatively warm climate allows for biogas production for eight to ten months of the year, because dung is available from a well-established hog industry, and because high population density has resulted in severe fuel shortages in the past (FAO 1978, Nurul Islam 1979, Taylor 1980, van Buren 1980).

The first digesters that marked the onset of the rapid dissemination of biogas plants in Sichuan were built in 1970 by a group of peasants in Zhongjian County, whose success attracted the government's attention and support (van Buren 1980). Development of such rural technologies in China is based on the concept of local self-reliance: local team-brigades, communes, or counties mobilize their locally available resources for projects designed to increase the standard of living of persons in those areas. Resources from outside the area are imported where necessary to catalyze project development, but, wherever possible, local materials such as indigenous "triple-concrete," labor, and technical and management expertise are utilized (Taylor 1980). This local development emphasis, often referred to as the doctrine of "walking on two legs," is intended to balance centralized industries (e.g., natural gas industries) with decentralized small-scale projects under local control (Taylor 1980).

The infrastructure for supporting the development of renewable energy technologies in China is multilayered (John Hawkins 1978). National institutions and capabilities for theoretical research and for applied research, training, planning, and coordination focus on relatively advanced technologies and long-term development. Local technological infrastructure includes institutions and capabilities for applied research and adapting technologies to immediate local conditions, technology popularization, implementation, project management, maintenance, training, and local planning and coordination (Taylor 1980).

On the local scale, close meshing of the benefits of biogas systems to the individual family and to the production team or collective has played a large part in rapid development of the systems. To a great extent, such local organization accomplished the valuation of resources and outputs of the system that are internal and external to the family. The family assists in construction of the system, supplies animal and human wastes directly from the household, and performs the various tasks of operating the unit. The household can obtain loans for construction materials at rates as low as 0.18 percent per annum (van Buren 1980). The benefits to the family are biogas for cooking and lighting and improvements in comfort and health through reduced smoke during cooking, as well as in sanitary management of human and animal excreta.

The production team or collective is responsible for the allocation, remuneration, and accounting for labor for constructing the digester, transporting construction materials to the site, and supplying feedstock materials. The team or collective receives benefits such as digester residues for fertilizer, environmental benefits from the alleviation of deforestation, and improved sanitation (van Buren 1980).

Most likely because of these environmental benefits, the principal promoter of biogas at the national level has been the Environmental Protection Office of the State Council (EPOSC). Under EPOSC, biogas offices have been established from the provincial to the local levels.

Community-scale biogas digester, fixed dome, Pondicherry. (Kirk R. Smith)

Sichuan Province, for example, is said to have biogas offices in 85 percent of its 100 counties (Islam 1979). Provincial biogas offices are responsible for coordinating research programs, publishing technical and training manuals, communicating with other biogas offices, and mobilizing resources such as loans or construction materials. For example, they coordinate specialists in the areas of soil and fertilizer, architecture, microbiology, and other disciplines in order to ensure a thoroughly integrated effort (Islam 1979).

Based on the rather limited information available in the past, rural energy specialists in other countries often have tended to regard the Chinese biogas effort more or less as a paradigm of successful energy development. More recent accounts, however, suggest that biogas development is slowing down markedly from the earlier near geometric growth rates and that severe problems might be occurring in the reliability and longevity of many of China's digesters.[1]

Among the explanations offered for the slowdown is the general relaxation since 1978 of "command farming" (Smil 1982), with fewer mass construction campaigns. Rural households also might be gaining more autonomy in making decisions about the allocation of their time and other resources and might be tending toward investments that provide more attractive monetary returns than does a biogas plant. Such investments are the production of marketable handicrafts, for instance, and high value agricultural items such as meat and eggs. Other problems have arisen regarding the management of the digester. The collective exercises control over the digester's residues and requires large amounts of this material as fertilizer about twice a year. The digester must be opened at these times, interrupting the supply of gas to the household (other fuels must then be used). Attempts are now being made, however, to store quantities of sludge during these periods in order to reintroduce some anaerobic bacteria to the digester to begin biogas production more quickly after the digester is closed up.

Relevant analysis of official Chinese statistics from 1980 shows that, of all the biogas units constructed until 1979, only about 55 percent were functioning normally (Smil 1982). Even among these functioning plants, "not too many can often be used to cook rice three times a day, still less every day for four seasons" (Smil 1982). Smil suggests that a more likely figure of reliable biogas plants in China would be around one-third of the total number built. He goes on to say,

> Difficulty with the existing digesters is hardly an encouragement to build more. Indigenous "triple concrete" digesters have proven largely useless. Operating units commonly have water and gas leaks, are inconvenient to use and clean, and the generated gas is burned with such low efficiencies that at least half is wasted. As not much attention is paid to proper C:N ratios or pH with frequent loading and removal, many digesters gradually just turn into waste pits and are abandoned. Looking for quick success and instant benefits meant poorly maintained units which usually failed in their second year of operation. The simple truth may be that most of the digesters have been a burden to their builders and a costly loss . . .

Officials seem to be realizing these problems, however, and various measures are being taken to remedy the program's weak points (Taylor 1981, FAO 1983). For example, maintenance of the digester was earlier the responsibility of the family, but now is becoming the responsibility of trained professionals. Programs also are underway to have trained persons inspect digesters for cracks when the digesters are empty, with the state paying for rebuilding damaged units. In addition, now a collective must meet certain criteria for training personnel and for financing the costs of a digester before even a single unit can be installed (R. Wagner, pers. comm., 1982).

Two-burner biogas stove and kerosene stove, Pondicherry. (Kirk R. Smith)

Biogas Development in India

In some respects, the development of biogas techniques in India resembles the experience of China's program. India also experienced a rapid rate of growth in numbers of biogas units in the 1970s (although not nearly as spectacular as in China) and has encountered a similar leveling off of annual growth. About 25,000 units were installed in 1981–82 (RERIC News 1983), bringing the total estimated number of plants to around 100,000. Despite attractive subsidies of from 20 to 100 percent of the installation cost to farmers, communities, and institutions (DST 1981), the numbers of biogas plants installed to August 1982 were substantially below the target rate of 75,000 units for 1982–83 (RERIC News 1983).

India's biogas plants also are experiencing problems of reliability and longevity, as discussed earlier. A number of reasons might be offered both for the slowdown of adoption as well as for the reliability problems encountered in the Indian biogas development efforts.

1. The initial costs of installing a floating-dome plant are high,

especially because steel used in the gas dome accounts for about one-half of the total costs (DST 1981).
2. The plants that are promoted are often of inappropriate size for local environmental conditions, and retention times are not adequately tailored to the specific situation (DST 1981).
3. The steel gas holder is subject to corrosion and eventual leakage of gas if improperly maintained (DST 1981), and repairs are expensive.
4. Infrastructure for supporting the maintenance and repair of faulty units is poorly developed, and expert advice or repair services are difficult to obtain at the site of the digester (DST 1981, Santerre and Smith 1982).
5. Shortages of steel and high-grade cement in India have slowed development efforts because inadequate attention has been given to adapting locally available materials for use in biogas plants (Santerre and Smith 1982).
6. Some financial analyses of biogas plants in India have tended to be overly optimistic and to be based on unrealistic assumptions of life expectancy, of the value of benefits and costs in relation to rural energy consumption patterns, or of other factors. (See also the discussion in the next section.)

In response to some of these problems, greater emphasis is being placed on using fixed-dome digesters, on finding alternative materials for the floating gas dome, and on improving the efficiency of biogas plants. These efforts are still in the laboratory and pilot stages, although some of the results appear encouraging (see DST 1981). The fixed-dome plants, known in India as Janata biogas plants, numbered about 2,700 at the beginning of 1980 (DST 1981) and are about 40 percent less expensive than the floating-dome digesters of similar capacity (ESCAP 1981).

People in India are becoming increasingly interested in community-scale plants to provide a more equitable access to cooking fuels, recognizing the uneven distribution of cattle and buffalo among rural households. A number of interesting management and social problems have arisen in the operation of such systems, and the reader might consult these presentations for additional information: Prabhakar Ghate (1979a and 1979b), Bhatia and Niamir (1979), Shahzad Bahadur and Sushil C. Agarwal (1980), and Moulik (1982).

FINANCIAL ANALYSES OF HOUSEHOLD AND COMMUNITY BIOGAS SYSTEMS

Here I present an approach to assessing the financial viability of household and community biogas projects, keeping in mind the uncertainties in valuing the benefits and costs of this technology and the diversity among households living in rural areas. First is a case example of a financial analysis of a biogas system to illustrate a fairly typical set of assumptions and analytical methods that often contribute to the basis and rationale for large-scale developement efforts to promote this technology. Then I establish a baseline case of five hypothetical rural households, each with different levels of monetary expenditures on fuels for cooking and lighting. In order to provide consistent comparisons among the different households, financial analysis tables are presented in terms of net present value (NPV).[2] Next, I test the sensitivity of the financial outlook to variations in key factors. The financial attractiveness of biogas plants to these five households is examined while varying the following factors: (a) the market prices of kerosene and firewood; (b) the installation costs of the biogas system; (c) the method used for financing the system; (d) the net change in daily household labor and the perceived opportunity cost of that labor; (e) the net increase in the value of fertilizer from treating organic materials by anaerobic digestion; and (f) the discount rate of money, or the minimum attractive rate of return on the investment to the household. Finally, I will examine some financial aspects of investments in community biogas systems involved in community decisions on whether to use the energy from biogas for cooking and lighting purposes or for economic activities.

Case Study of a Financial Analysis of Biogas Systems

Most financial studies of biogas plants have tended to use only a single set of assumptions regarding the value of benefits and costs of this technology, while also assuming that the energy needs and available resources of the households in rural areas generally are homogeneous. I believe that the earlier chapters of this book have demonstrated sufficiently that this assumption is usually erroneous. The following case study is based on a published report evaluating the Janata fixed-dome biogas plant currently being popularized in India (PRAD 1980). These plants are seen as lower cost alternatives to floating-dome designs and have been promoted principally by the Planning Research and Action Division (PRAD) of the State Planning Institute in Lucknow, Uttar Pradesh.

The costs given in the cited report were based on interviews with 74 households operating Janata biogas systems during 1979–80. Estimates of the monetary benefits of biogas through fuel savings were based on estimates of the amounts of firewood and kerosene consumed by the households for cooking and lighting before installation of the biogas units. Expenditures on firewood ranged from Rs. 50 to Rs. 100 per month (about US $6 to $12 per month), averaging Rs. 62.50 per month. Purchases of kerosene ranged from Rs. 10 to Rs. 30 per month, averaging Rs. 14.20 per month. The actual weights or volumes of these fuels and their market value were not reported in the PRAD study. The survey revealed that the prebiogas requirements for cooking fuel were not met entirely by purchased firewood, and that one-third of the total cooking fuel was in the form of dung cakes (generally available from the household's own cattle).

The survey also found that the majority of the households were not meeting their total cooking and lighting needs with biogas after installing their digesters. Current expenditures by the households on kerosene for lighting averaged Rs. 7 per month. The percentage of households purchasing firewood to supplement biogas and their average monthly expenditures were as follows:

Percentage of Households	Quantity of Firewood Purchased per Month (Rs.)
30	15
20	16–30
30	31–50
10	50–75

Why the percentage of households column does not total to 100 was not explained. (The remaining 10 percent may be those who found biogas to be adequate in quantity for their entire cooking fuel needs. According to the report, however, 36 percent of the families were meeting their entire cooking fuel requirements from the gas.)

PRAD's financial analysis of a 3 m^3/day biogas plant assumes a savings of Rs. 40 per month in firewood purchases per household, and Rs. 16.55 per month in kerosene. Some discrepancy exists, however, between the prebiogas expenditures on firewood of Rs. 65.50 (average), and the Rs. 40 value used for savings on firewood. Another problem exists with the value assumed for kerosene savings, which is higher than the average prebiogas expenditure on kerosene of Rs. 14.20 per month. Also the value used was not put at Rs. 14.20 − Rs. 7.00 = Rs. 7.20 to incorporate both pre- and postbiogas expenditures on kerosene. Again, the report was not clear in explaining this convention.

In relation to benefits derived from the digester residues, no mention was made in the report of whether any households were using commercial fertilizers. Prior to installing the digesters, reportedly all the families were using some dung as manure.

Percentage of Households	Percentage of Available Dung Used as Fertilizer
15	33
31	50
27	67
15	75
12	100

PRAD valued the additional fertilizer benefits on the basis of having 40 kg of additional manure per day (assuming that of the 80 kg of dung needed to operate the biogas unit, 40 kg were previously burned as fuel and were not available as fertilizer). That remainder would produce 2.74 tons of compost annually. This compost was valued at Rs. 240 per year, although the values of chemical fertilizers and other assumptions determining this rupee value were not reported.

The interview of biogas users showed that the people perceived other benefits, although only fuel savings and fertilizer benefits were valued for the purposes of financial analyses. Interestingly, although 92 percent of the survey respondents felt that they enjoyed the benefits from reduced smoke, only 43 percent indicated that they enjoyed benefits from savings in money.

A 3 m^3/day Janata plant costs Rs. 2,250, but after receiving an Rs. 725 subsidy grant, the cost to the household is Rs. 1,525. PRAD's financial analysis assumes that this amount is financed by a 10.5 percent loan repaid over the 20-year lifetime of the unit, an annual cost of Rs. 160. The cost of repairs and replacement of accessories is assumed to be Rs. 38 per year, of maintenance Rs. 45 per year. The net annual cash flow and NPV calculations are given in Table 7. I have converted the PRAD analyses from the beginning-of-year format to the simpler and easier-to-use, end-of-year convention to avoid confusion in later comparisons with the five hypothetical households. The NPV from this revised analysis is the sum of the present values: Rs. 0.00 + Rs. 5,755 = Rs. 5,755. This differs somewhat from the NPV given in the original study because the end-of-year convention is used. Nonetheless, under the *assumed* benefits and costs of the study, the Janata biogas system does appear to be economically attractive. Their findings and conclusions are summarized below.

TABLE 7.
Biogas Digester Costs, Janata Plant

Year	Item	Annual benefits (Rs)	Annual costs (Rs)	Annual net benefit (Rs)	Uniform series x present value factor (@ 10%; 20 yr)	Present value (Rs)
0	Biogas system		2,250			
	Subsidy	725				
	Loan	1,525	___			
		2,250	2,250	0.00		0.00
1–20	Loan repayment		160*			
	Repairs		38			
	Maintenance		45			
	Firewood savings	480				
	Kerosene savings	199				
	Digester residue	240	___			
		say, 920	243	+677	8.514	5,755

Source: After PRAD (1980).

*Their loan repayment value would have been Rs 168 using compound interest tables and their original beginning-of-year convention. Using the end-of-year convention, the annual loan repayment amount would be Rs 185. I chose to retain the value of Rs 160 provided in the original study.

A three m³ Janata biogas plant operating efficiently will generate savings in costs equivalent to net savings of Rs. 677 annually. . . . It is realized that this benefit-cost analysis is not rigorous . . . however, [it] established fairly well the economic feasibility of the installation of a biogas plant. (PRAD 1980)

Analyses of Five Hypothetical Rural Households

Five hypothetical rural households were created to illustrate the financial attractiveness of household-scale biogas systems from the perspectives of households with different energy consumption patterns. I did this because the principal energy benefit from anaerobic digesters,

TABLE 8.
Weekly Domestic Fuel Consumption for Cooking and
Lighting, Five Hypothetical Rural Households

	Households				
	A	B	C	D	E
Weekly Domestic Fuel Consumption					
Kerosene, liters	20	5	1	0.5	0
Firewood, kilograms	0	50	50	50	50

biogas, generally is not traded in the marketplace. Its value to the household must be determined in part from having to purchase less kerosene, firewood, or other fuels sold in the marketplace. Unfortunately, most financial analyses of biogas systems to date have based these benefits on an *idealized* replacement of purchased fuels. In other words, they have accounted for the value of biogas either in direct terms (i.e., biogas being equivalent to x-kilograms of a given fuel consumed at a given efficiency), or in terms of an idealized or average rural household (as in the case study given above). As mentioned earlier in this book, considerable diversity exists in the fuel use patterns of rural households in a given community, district, region, or country. The financial attractiveness of even a specific type and size of biogas plant will be determined (from the perspective of fuel substitution) by the *actual* monetary savings of the particular household, not by the *ideal* or *average* savings. Consequently, because household expenditures on fuels for cooking and lighting differ widely, so must the benefits of using biogas. A household consuming large amounts of purchased fuels that adopts a biogas plant will enjoy greater monetary benefits than a household consuming relatively little or no purchased fuels. Although seemingly intuitively obvious, and at once an absolutely vital assumption, this principle has been almost entirely *overlooked* in earlier financial analyses of biogas systems, often leading to inappropriate conclusions as to whether biogas systems are suitable for promotion in rural areas.

The five hypothetical households are categorized as Households A to E in Table 8. For the sake of simplicity, they are assumed to consist of similarly sized families and are assumed to have equal access to the resources needed to install, operate, and maintain a household biogas system, such as land, cattle dung, water, and construction materials. They are arranged in the table in order of decreasing dependence on kerosene. Household A is assumed to perform all of its cooking and lighting using kerosene appliances, which is rather uncommon in rural areas and is

TABLE 9.
Annual Household Expenditures on Fuels for Cooking and
Lighting at Various Fuel Prices, Households A to E.

Kerosene Price (US$/liter)	Firewood Price (US$/tonne)	Households				
		A	B	C	D	E
0.10	0*	104	26	5	2	0
	10	104	52	31	29	26
	20	104	78	57	55	52
0.25	0*	260	65	13	6	0
	10	260	91	39	32	26
	20	260	117	65	58	52
0.50	0*	520	130	26	13	0
	10	520	156	52	39	26
	20	520	182	78	65	52

Note: Expenditures are rounded to nearest dollar.

*Assumes firewood is collected at zero cost.

representative of only the relatively affluent rural dweller. Households B to E are assumed to be cooking entirely with firewood (or an equivalent biomass fuel), and to consume different quantities of kerosene for lighting purposes. These are more typical rural patterns. The firewood consumption level of 50 kg per week was chosen because it represents an approximate midpoint of firewood consumption rates found in rural areas. The analyses incorporate a wide range of annual monetary and time expenditures on obtaining these two fuels, and the results and methods could also be applied to households consuming fuels other than kerosene and firewood.

The annual household expenditures on domestic fuels of the five hypothetical households are provided in Table 9. Kerosene prices between US $0.10 and US $0.50 per liter and firewood prices between US $0.00 and US $20.00 per tonne are used to illustrate the range of fuel prices existing in rural areas in South and Southeast Asia.

Assumptions of Analyses. This analysis of the financial attractiveness of household biogas plants incorporates these assumptions:

1. No one particular type of biogas plant or one particular size of a biogas plant is assumed. The biogas plant is assumed to be of appropriate size and design to meet the household's needs of fuel for cooking and lighting.

2. The biogas produced by the anaerobic digester is assumed to be substituting for 75 percent of the kerosene and 75 percent of the firewood previously consumed by the households. I chose not to assume a 100 percent substitution rate because probably some kerosene still would be needed for portable lighting (biogas lamps are stationary); presumably some firewood (or kerosene in Household A) still would be needed for cooking certain foods at certain times; and probably the biogas system would not be 100 percent operational throughout the year due to environmental conditions (e.g., low temperatures inhibiting anaerobic digestion) or interruptions in service for cleaning, repairs, and maintenance.

3. For illustrative purposes, I assume that the biogas is not used for economically productive activities, or for space heating, or for other activities not directly associated with household lighting and cooking.

4. Except for the presentation of sensitivity analyses on biogas system installation costs, the costs of a completely installed biogas system (including land, labor, materials, appliances) are assumed to be US $400. Each of the five households is assumed to install the same design and size of biogas system that is assumed to have a 20-year lifetime with no salvage value assumed.

5. Annual operating costs are assumed to include (a) loan repayment (if applicable); and (b) costs for repair, maintenance, and replacement, assumed to be a uniform annual cost of US $20.

6. Annual benefits are assumed to include (a) 75 percent of the monetary value of previously consumed firewood and kerosene; and (b) US $6.50 for the increased nitrogen content of the digester residues (see ESCAP 1980 for calculations), except in the sensitivity analyses on the value of digester residues. Because of a lack of adequate data, potential health benefits from using clean-burning biogas or from anaerobic digestion of cattle dung are not included. Likewise, possible costs associated with smoke reduction, such as more rapid deterioration of house construction materials or increases in the incidence of insect-borne diseases (e.g., malaria), are not valued due to lack of adequate data.

7. Possible changes in household labor associated with the adoption of a biogas system and the opportunity costs of labor are treated in a separate sensitivity analysis.

8. In the baseline financial analysis that follows, the financial attractiveness of biogas plants is examined from the perspective of five hypothetical households, while varying these factors.

(a) *Fuel prices.* Kerosene prices are assumed to be US $0.10, $0.25, and $0.50 per liter. Firewood prices are assumed to be US $0.00, $10.00, and $20.00 per tonne.
(b) *Financing arrangements.* Four financing schemes are examined. In the first scheme, the household is assumed to be financing the biogas system from its own savings, without a loan. In the three remaining schemes, the household is assumed to be obtaining a loan towards the total cost of installing the biogas system with a 20-year repayment period at annual interest rates of 10, 20, or 30 percent. Although banks often require the borrower to finance a portion of the total costs with his own money, to simplify the presentation, the loan is assumed to be for the entire cost of the system.
(c) *Discount rates.* Three rates of return on investment are considered: 10, 20, and 30 percent. NPVs are determined from net annual cash flows for a period of 20 years.

Results of Baseline Financial Analyses. The results of the financial analyses of the attractiveness of household biogas systems to five hypothetical rural households are presented in Table 10.[3] This table includes both the net annual cash flows and NPVs related to the installation and operation of a biogas system under the assumptions just listed.

From the results of these analyses, some generalizations can be made. First, rural households purchasing large quantities of kerosene (A type) would find biogas systems fairly attractive investments unless kerosene is available at a very low subsidized price. Second, households having relatively high rates of kerosene consumption and principally depending on firewood for cooking (B type) would find biogas systems financially attractive only at relatively high prices of kerosene and firewood, and generally only with returns on their investments anticipated at less than 30 percent. Third, households with relatively moderate to low rates of kerosene consumption (C to E types) would find biogas systems (under the previous assumptions) unattractive investments at the different rates of kerosene and firewood prices, financing schemes, and returns on investment used in these analyses.

The results of these baseline analyses with five hypothetical households demonstrate that financial analyses of biogas systems are especially sensitive to the actual levels of household expenditures on fuels for

TABLE 10.
Effects of Fuel Prices, Financing Arrangements, and Discount Rates on the
Financial Attractiveness of Household Biogas Systems to Households A-E

		Fuel Prices (US$)										
Kerosene (per liter)		0	0.10		0.25				0.50		
Firewood (per tonne)		0	0	10	20	0	10	20	0	10	20

Household A

Net Annual Cash Flows (US$)

Loan Scheme	Period											
No loan	Year 0	-400	-400	-400	-400	-400	-400	-400	-400	-400	-400
	Years 1-20	64	64	64	182	182	182	376	376	376
10%; 20 years	Years 1-20	18	18	18	135	135	135	330	330	330
20%; 20 years	Years 1-20	-18	-18	-18	99	99	99	294	294	294
30%; 20 years	Years 1-20	-56	-56	-56	61	61	61	256	256	256

Net Present Values in 20 Years (US$)

Loan Scheme	Discount Rate (%)											
No loan	10	145	145	145	1,150	1,150	1,150	2,801	2,801	2,801
	20	-88	-88	-88	486	486	486	1,431	1,431	1,431
	30	-188	-188	-188	204	204	204	847	847	847
10%; 20 years	10	153	153	153	1,149	1,149	1,149	2,810	2,810	2,810
	20	88	88	88	657	657	657	1,607	1,607	1,607
	30	60	60	60	448	448	448	1,094	1,094	1,094
20%; 20 years	10	-153	-153	-153	843	843	843	2,503	2,503	2,503
	20	-88	-88	-88	482	482	482	1,432	1,432	1,432
	30	-60	-60	-60	328	328	328	975	975	975
30%; 20 years	10	-477	-477	-477	519	519	519	2,180	2,180	2,180
	20	-273	-273	-273	297	297	297	1,247	1,247	1,247
	30	-186	-186	-186	202	202	202	849	849	849

TABLE 10, continued.

Household B

Net Annual Cash Flows (US$)

Loan Scheme	Period	Kerosene (per liter) Firewood (per tonne)				Fuel Prices (US$)					
		0	0.10			0.25			0.50		
			0	10	20	0	10	20	0	10	20
No loan	Year 0	-400	-400	-400	-400	-400	-400	-400	-400	-400	-400
	Years 1-20	6	26	45	35	55	74	84	104	123
10%; 20 years	Years 1-20	-41	-22	-2	-12	8	27	37	57	76
20%; 20 years	Years 1-20	-76	-57	-37	-47	-27	-8	2	21	41
30%; 20 years	Years 1-20	-115	-95	-76	-85	-66	-46	-37	-17	2

Net Present Value in 20 Years (US$)

Loan Scheme	Discount Rate (%)										
No loan	10	-349	-179	-17	-102	68	230	315	485	647
	20	-371	-273	-181	-230	-132	-40	9	106	199
	30	-380	-314	-251	-284	-218	-155	-121	-55	8
10%; 20 years	10	-349	-187	-17	-102	68	230	315	485	647
	20	-200	-107	-10	-58	39	131	180	278	370
	30	-136	-73	-7	-40	27	90	123	189	252
20%; 20 years	10	-647	-485	-315	-400	-230	-68	17	179	349
	20	-370	-278	-180	-229	-131	-39	10	102	200
	30	-252	-189	-123	-156	-90	-27	7	70	136
30%; 20 years	10	-979	-809	-647	-724	-562	-392	-315	-145	17
	20	-561	-463	-370	-414	-321	-224	-180	-83	10
	30	-381	-315	-252	-282	-219	-152	-123	-56	7

TABLE 10, continued.

		Fuel Prices (US$)									
Kerosene (per liter)		0.10			0.25			0.50			
Firewood (per tonne)		0	10	20	0	10	20	0	10	20	

	Period									
Household C										
Net Annual Cash Flows (US$)										
Loan Scheme										
No loan	Year 0	-400	-400	-400	-400	-400	-400	-400	-400	-400
	Years 1-20	-10	10	29	-4	16	35	6	26	45
10%; 20 years	Years 1-20	-57	-37	-18	-51	-31	-12	-41	-21	-2
20%; 20 years	Years 1-20	-92	-72	-52	-86	-66	-47	-76	-57	-37
30%; 20 years	Years 1-20	-130	-111	-91	-124	-105	-85	-115	-95	-76
Net Present Values in 20 Years (US$)										
Loan Scheme	Discount Rate (%)									
No loan	10	-485	-391	-153	-434	-264	-102	-349	-179	-17
	20	-449	-395	-259	-419	-322	-230	-371	-273	-181
	30	-433	-397	-304	-413	-347	-284	-380	-314	-251
10%; 20 years	10	-485	-315	-153	-434	-264	-102	-349	-179	-17
	20	-278	-180	-88	-248	-151	-58	-200	-102	-10
	30	-189	-123	-60	-169	-103	-40	-136	-70	-7
20%; 20 years	10	-783	-613	-443	-732	-562	-400	-647	-485	-315
	20	-448	-351	-253	-419	-321	-229	-370	-278	-180
	30	-305	-239	-172	-285	-219	-156	-252	-189	-123
30%; 20 years	10	-1,107	-945	-775	-1,056	-894	-724	-979	-809	-647
	20	-633	-541	-443	-604	-511	-414	-560	-463	-370
	30	-431	-368	-302	-411	-348	-282	-381	-315	-252

TABLE 10, continued.

		Fuel Prices (US$)								
	Kerosene (per liter)		0.10			0.25			0.50	
	Firewood (per tonne)	0	10	20	0	10	20	0	10	20
Household D										
Net Annual Cash Flows (US$)										
Loan Scheme	Period									
No loan	Year 0	-400	-400	-400	-400	-400	-400	-400	-400	-400
	Years 1-20	-12	8	27	-9	11	30	-4	16	35
10%; 20 years	Years 1-20	-59	-39	-20	-56	-36	-17	-51	-31	-12
20%; 20 years	Years 1-20	-94	-74	-55	-91	-71	-52	-86	-66	-47
30%; 20 years	Years 1-20	-132	-113	-93	-129	-110	-90	-124	-105	-85
Net Present Values in 20 Years (US$)										
Loan Scheme	Discount Rate (%)									
No loan	10	-502	-332	-170	-477	-306	-145	-434	-264	-102
	20	-458	-361	-269	-444	-346	-254	-419	-322	-230
	30	-440	-373	-310	-430	-364	-301	-413	-347	-284
10%; 20 years	10	-502	-332	-170	-477	-307	-145	-434	-264	-102
	20	-287	-190	-97	-273	-175	-83	-248	-151	-58
	30	-196	-129	-66	-186	-119	-56	-169	-103	-40
20%; 20 years	10	-800	-630	-468	-775	-604	-443	-732	-562	-400
	20	-458	-360	-268	-443	-346	-253	-419	-321	-229
	30	-312	-245	-182	-302	-235	-172	-285	-219	-156
30%; 20 years	10	-1,124	-962	-792	-1,098	-937	-766	-1,056	-894	-724
	20	-643	-550	-453	-628	-536	-438	-604	-511	-414
	30	-438	-375	-308	-428	-365	-298	-411	-348	-282

TABLE 10, continued.

			Fuel Prices (US$)								
Kerosene (per liter)			0.10			0.25			0.50		
Firewood (per tonne)		0	10	20	0	10	20	0	10	20

Household E

Net Annual Cash Flows (US$)

Loan Scheme	Period									
No loan	Year 0	-400	-400	-400	-400	-400	-400	-400	-400	-400
	Years 1-20	-14	6	26	-14	6	26	-14	6	26
10%; 20 years	Years 1-20	-60	-41	-21	-60	-41	-21	-60	-41	-21
20%; 20 years	Years 1-20	-96	-76	-57	-96	-76	-57	-96	-76	-57
30%; 20 years	Years 1-20	-134	-115	-95	-134	-115	-95	-134	-115	-95

Net Present Values in 20 Years (US$)

Loan Scheme	Discount Rate (%)									
No loan	10	-519	-349	-179	-519	-349	-179	-519	-349	-179
	20	-468	-371	-273	-468	-371	-273	-468	-371	-273
	30	-446	-380	-314	-446	-380	-314	-446	-380	-314
10%; 20 years	10	-511	-349	-179	-511	-349	-179	-511	-349	-179
	20	-292	-200	-102	-292	-200	-102	-292	-200	-102
	30	-199	-136	-70	-199	-136	-70	-199	-136	-70
20%; 20 years	10	-817	-647	-485	-817	-647	-485	-817	-647	-485
	20	-468	-370	-278	-468	-370	-278	-468	-370	-278
	30	-318	-252	-189	-318	-252	-189	-318	-252	-189
30%; 20 years	10	-1,141	-979	-809	-1,141	-979	-809	-1,141	-979	-809
	20	-653	-560	-463	-653	-560	-463	-653	-560	-463
	30	-444	-381	-315	-444	-381	-315	-444	-381	-315

Note: Results are presented as annual cash flows (in US$), and in net present values in 20 years (in US$). Assumptions of analyses are stated in the text.

cooking and lighting. In the previously cited case study, although PRAD presented some information (1980) about household expenditures on firewood and kerosene before and after installation of the biogas unit, the data provided are insufficient to categorize the households by expenditure levels as was done for these five hypothetical households. It should be mentioned also that the PRAD (1980) study was based on households that *already owned* Janata biogas systems, so the information presented on household expenditures on fuels should *not* be considered representative of the whole rural population. The installation costs of this energy system are high enough to exclude adoption by lower income groups who could also experience difficulty in obtaining loans towards the costs of the system.

If I were to recalculate the NPV of this case study using the hypothetical values of annual expenditures on fuel of Household E (which consumed no purchased fuel) while holding all other assumptions of the PRAD (1980) study constant, the biogas system would appear financially unattractive (Rs. 240 − Rs. 243 = −Rs. 3 × 8.514 = − Rs. 25.50 NPV). Although Household E was a rather extreme case of zero expenditure on fuel, negative NPVs might be experienced by our other hypothetical households if the assumed benefits from digester residues were made lower. There is reason to believe that the Rs. 240 value assumed by PRAD as the annual benefit from digester sludge is overly optimistic (see information provided below).

Sensitivity Analyses: Biogas System Installation Costs. The preceding analyses examined the financial attractiveness of a biogas system with an installation cost (complete) of US $400. Table 11 presents the results of a sensitivity analysis varying the installation cost of a biogas system from US $0.00 for a totally subsidized system to US $800. According to the results of this sensitivity analyses, the installation cost of a biogas system is quite important in determining its financial attractiveness to rural households. Although Households C to E might not find a US $400 biogas system attractive under most conditions, biogas systems costing US $200 or less demonstrate a greater than 30 percent return on investment under our assumed conditions. This information points to the possible important role of plant cost reduction (through design or material changes) if household biogas systems are to become more widely acceptable.

The analysis in Table 11 could also facilitate assessment by policymakers of the effects of various possible subsidy regimes on the attractiveness of family-scale biogas systems to different types of households. Relatively well-off households consuming large amounts of kerosene, such as

TABLE 11.
Effects of Different Installation Costs on the Financial
Attractiveness/of Household Biogas Systems to Households A-E

Household	Discount Rate (%)	Biogas System Installation Costs (US$)				
		0	100	200	400	800
A	10	1,545	1,445	1,345	1,149	745
	20	884	827	770	657	426
	30	602	563	524	448	290
B	10	632	532	432	230	-168
	20	362	304	247	131	-96
	30	246	207	168	90	-65
C	10	300	200	101	-102	-500
	20	172	114	57	-58	-286
	30	117	78	39	-40	-195
D	10	259	159	59	-145	-541
	20	148	91	34	-83	-310
	30	101	62	23	-56	-211
E	10	217	117	17	-179	-583
	20	124	67	10	-102	-333
	30	85	46	7	-70	-227

Note: Results are NPVs for 20 years (in US$). Assumptions: kerosene price $0.25/l; fuelwood price US $20/tonne; 20-year loan at 10 percent interest; other assumptions as stated for the baseline analyses in the text.

Household A, should be able to afford relatively expensive systems without the need for government subsidy. A subsidy grant to bring the installation cost down to US $400 (from whatever cost the unsubsidized plant costs) would be useful under certain conditions to make this energy system attractive to the types of rural households exemplified by Household B, while reducing the price to US $200 could make the systems attractive to types like Households C to E. My case study of a Janata fixed-dome digester in India, for example, assumes that the biogas system costs Rs. 2,250 (about US $250). The installation cost to the household is Rs. 1,525 (about US $170) after a subsidy of Rs. 725 (about US $80) is subtracted (PRAD 1980). The presubsidy installation cost of a floating-dome KVIC biogas system of similar size in India is Rs. 3,000 (about US $333) (ESCAP 1981). Note, however, that the 10 percent per annum loan used to finance the plants in this analysis is a relatively *low* cost for borrowing money. Higher interest rates would make even the lower priced digesters less attractive to these households.

Sensitivity Analyses: Labor. Because great uncertainty presently exists about the effects of adoption of a household biogas system on family labor activities and about the nature of opportunity costs for that labor, I was very hesitant about incorporating a labor cost or labor benefit into the baseline financial analyses presented in Table 10. Although accounting for labor is a particularly crucial factor in performing a good financial analysis of an energy system, very little information exists about even the gross labor quantity needed to operate a biogas plant. Bhatia (1977) reported two Indian sources with estimates of labor needed ranging from 0.6 to 4 hours per day. French (1979) assessed the economics of floating-dome systems in India and assumed that two additional (net) hours daily would be needed to operate a household-scale system.

My sensitivity analysis of the effects of labor changes and labor costs on the financial attractiveness of household biogas systems examines the possible effects of a net increase in household (or hired) labor of one hour per day and of a net decrease of one hour per day (see Table 12). These changes are examined for three perceived or actual costs of this labor, US $0.05, $0.10, and $0.20 per hour. In the NPV analysis, an increase in labor (labor is defined here as physical activity) is treated as an additional cost, while a decrease in labor is treated as an additional benefit from operating a biogas system.

My baseline analyses (Table 10) assume that labor needed to operate the household system did not result in an additional cost or benefit. This assumption represents either or both of two conditions: (a) the net change in daily labor after adopting a biogas plant would be zero hours per day; and (b) the possible change in labor would have zero opportunity cost, that is, the household would not place any value on either the additional labor required or on the labor saved from adopting this energy system.

The information presented in Table 12 shows that the quantity of net labor required to operate a biogas system and the opportunity costs of that labor can be a significant factor in the financial attractiveness of this energy technology. Especially interesting are the possible effects of labor savings associated with adopting a biogas system. If the household were able to save one hour per day and were able to utilize this labor to pursue an economic activity, then even households with moderate or little

TABLE 12.
Effects of Labor Changes and Labor Opportunity Costs on the
Financial Attractiveness of Biogas Systems to Households A-E

Household	Discount rate (%)	Net Change in and Cost of Daily Labor (+ = more; - = less)						
		+1 h/d			0	-1 h/d		
		0.05 US$/h	0.10	0.20	0.00	0.05	0.10	0.20
A	10	992	830	524	1,149	1,298	1,460	1,767
	20	567	475	300	657	743	835	1,011
	30	386	323	204	448	506	569	688
B	10	79	-83	-390	230	385	547	854
	20	45	-47	-223	131	220	313	488
	30	31	-32	-152	90	150	213	332
C	10	-253	-415	-722	-102	53	215	521
	20	-145	-237	-413	-58	30	123	298
	30	-99	-162	-281	-40	21	84	203
D	10	-295	-457	-763	-145	12	173	480
	20	-169	-261	-436	-83	7	99	275
	30	-115	-178	-297	-56	5	68	187
E	10	-336	-498	-805	-179	-30	132	438
	20	-192	-285	-460	-102	-17	75	251
	30	-131	-194	-313	-70	-12	51	171

Note: Results are reported in NPVs in 20 years (in US$). Assumptions as stated in Table 11 Note.

expenditure on domestic fuels might view biogas systems as favorable. However, to assume that this labor would indeed be used for economic purposes would be spurious unless the specific household situation was properly assessed. Also, I have greatly simplified my analysis of labor changes and costs, since considering seasonal variations in household labor activities and in the opportunity cost of labor is also very important. I have also not made any assumptions about whether a net increase or decrease in labor would affect the activities of men, women, or children, or whether household or hired labor would change if a biogas system were adopted.

Because of uncertainties in incorporating monetary values for labor in financial analyses of biogas plants, I have not performed further sensitivity analyses on my case study. PRAD (1980) also chose not to include labor as a factor in their analyses. I would like to mention, however, that a number of studies have dealt with the labor costs of operating biogas systems (e.g., Bhatia 1977, Moulik and others 1978, A. Pang 1978, French 1979, ESCAP 1980 and 1981), and the reader might wish to examine them in light of the previous discussion.

Sensitivity Analyses: Digester Residue Value. In the baseline financial analysis, the assumed value of the digester residues is US $6.50 per year. This figure was borrowed from the *Guidebook on Biogas Development* (ESCAP 1980) and is based on a possible net increase in nitrogen from digesting cattle dung in a biogas plant (household scale) instead of storing or composting by traditional techniques. The nitrogen is assumed to be equivalent in fertilizer value to the nitrogen present in urea fertilizer at a market price of US $0.44 per kg of nitrogen (ESCAP 1980).

Because disagreement exists in the literature about the actual worth of digester residues (see discussion earlier in this chapter), I performed a sensitivity analysis using values ranging from US $0.00 to US $26.00 per year. If the information from Idnani and Varadarajan (1974) presented in Table 4 is generally applicable to a wide variety of crops and environmental conditions in the region, then values lower than ESCAP's US $6.50 per year might be appropriate. On the other hand, digester residues contain other plant nutrients, such as phosphate and potassium, and soil conditioners that are also valuable. Considerably more study is obviously needed in this area.

The results of these sensitivity analyses (Table 13) suggest that the value of digester residues used in a financial analysis would not affect significantly the financial attractiveness of a biogas system and that the benefits from energy savings (evident in the different NPVs among Households A to E) would be far more significant. Thus, given presently

TABLE 13.
Effects of Assumed Monetary Benefits from Residues on the Financial Attractiveness of Household Biogas Systems to Households A-E.

Household	Discount Rate (%)	Value of Digester Residues (US$/year)			
		0.00	6.50	13.00	26.00
A	10	1,090	1,149	1,201	1,311
	20	623	657	687	750
	30	425	448	468	511
B	10	177	230	288	398
	20	101	131	164	228
	30	69	90	112	155
C	10	-155	-102	-45	66
	20	-89	-58	-25	38
	30	-60	-40	-17	26
D	10	-197	-145	-86	25
	20	-113	-83	-49	14
	30	-77	-56	-34	10
E	10	-238	-179	-128	-17
	20	-136	-102	-73	-10
	30	-93	-70	-50	-7

Note: Results are NPVs for 20 years (in US$). Assumptions as stated in Table 11 Note.

available information about the value of digester residues, not too much emphasis on their worth is justified while undertaking financial assessments of this energy technology in specific rural situations.

In my case study, however, is an example of a relatively high annual benefit assigned to digester residues, valued at Rs. 240 (US $27) (PRAD 1980). Although reducing only this value would still result in a positive NPV for the Janata system, combining a lower value for residues benefits with lower values of benefits from fuel savings could significantly alter the outcome of PRAD's financial analyses. For certain of the PRAD households, a reduction in the assumed monetary value of the digester residues seems justified, since more than 25 percent of the households were already using 75 to 100 percent of their available cattle dung as fertilizer before adopting the biogas system. Thus, further fertilizer benefits from anaerobically digesting the dung probably would be marginal.

Financial Analysis of Community Biogas Systems

Hypothetical Rural Community. I created a hypothetical small rural community of six households to illustrate the financial attractiveness of community-scale biogas systems. I used six households in this community in order to allow for a diversity of energy requirements and available resources, but, to simplify the presentation, did not consider more than six households. The households, designated Households F to K in Table 14, are assumed to be located in a relatively compact cluster of dwellings. The households also are assumed to differ in the quantities of collected or purchased firewood consumed for cooking and in the quantity of kerosene consumed for lighting. In addition, the number of cattle owned by each household is assumed to differ (Table 14). One further assumption of these analyses is that this small community would be able to organize itself to undertake a community project.

The analyses provided here examine financial aspects of prospective investments in three different types of biogas systems: (a) household biogas systems for households with adequate dung resources, (b) a community biogas system meeting the cooking and lighting needs of the six households, and (c) a community biogas system providing biogas for food processing. The three tasks of cooking, lighting, and food processing (oil expelling, rice husking, and grain milling) were identified by members of the community as priority areas where energy technologies might be employed to improve their living conditions.

Household Biogas Systems. Because a household biogas system must supply 1.5 to 2.5 m^3 of biogas per day to meet the cooking and lighting needs of a "typical" household, and since the production of 1 m^3 of biogas under average conditions requires about 32 kg of cattle dung (conservatively), a household would require 48 to 80 kg of dung daily. Assuming that an average head of cattle can supply 10 kg of recoverable dung per day (although this should be evaluated in each household considering a biogas sytem), a household would require between five and eight head of cattle. Assuming, for simplicity's sake, that the households each require

TABLE 14.
Energy Consumption, Expenditures on Fuels, and Cattle
Owned by Six Hypothetical Community Households

	Households					
	F	G	H	I	J	K
Firewood: collected (kg/wk)	0	54	71	75	0	52
Firewood: purchased (kg/wk)	65	0	0	0	78	0
Annual cost @ US $20 per tonne (US$)	68	0	0	0	81	0
Kerosene (liters/week)	5	3	1	2	4	2
Annual cost @US $0.50 per liter (US$)	30	18	6	12	24	12
Cattle owned (number)	20	5	4	0	12	2

2.5 m³ of biogas daily, only Household F (with 20 cattle) and Household J (with 12 cattle) have adequate quantities of dung available for supporting a household-scale biogas system. Each of these two households purchases about as much kerosene per week as Household B (Table 8) and slightly more firewood. Under the same assumptions, the earlier analyses for Household B serve as close indicators of the attractiveness of household biogas systems to Households F and J. Applying Table 11, for example, these households would find a home biogas system attractive at an installation cost of US $400, but unattractive at US $800.

Community Biogas System for Cooking and Lighting. Although Households G, H, I, and K might be able to operate their own individual biogas plants by purchasing dung, they would be dependent on Households F and J or persons outside the community, so the costs of operating the system would be higher. Thus, I chose not to analyze this possibility here. The community does have sufficient cattle to consider a community-scale biogas plant that would produce about 14 m³ of biogas per day, or about 2.3 m³ per day per household. The assumptions used in this analysis are

1. A biogas system with 14 m³ per day of biogas production capacity is assumed to cost US $3,700, complete with pipelines, appliances, and other appurtenances. This cost is approximately the same for installation in a village in Nepal (Deepak Bajracharya,

pers. comm., 1982). The system is financed with a loan at 6 percent interest for seven years. It is assumed to have a life expectancy of 20 years with no salvage value.
2. Annual costs are assumed to include loan repayment (Years 1 through 7 only) and an assumed annual cost for repairs, replacement, and maintenance of US $50. Annual benefits are from fuel substitution only, with no benefits assumed for increased value of fertilizer from digester residues.
3. The system is assumed to provide sufficient biogas to substitute for 75 percent of the previously consumed firewood and kerosene.
4. Firewood is assumed to cost US $20 per tonne, and kerosene US $0.50 per liter.
5. All households provide all of their available cattle dung to the community project at zero cost.
6. Labor is assumed not to be changed by adopting the biogas system, or if it would change, it is assumed to have a zero opportunity cost.

The results of the financial analyses for an unsubsidized system and a system receiving a 50 percent subsidy grant towards the initial installation costs are presented in Table 15. The analyses indicate that the system is financially unattractive, even if heavily subsidized. The principal reason for these negative NPVs arises from the assumed low rates of monetary expenditures on fuels for domestic uses by most of the households in this hypothetical community (Table 14).

Community Biogas System for Powering Food-Processing Machinery. The third application of anaerobic digestion technology considered here is the use of a community-based business enterprise. This analysis is based, in part, on information provided by Deepak Bajracharya (pers. comm., 1982) for a rural development project in Nepal, involving the use of biogas for powering a dual-fuel engine that is connected to an oil expeller, rice husker, and grain mill. The analysis uses installation costs, wage rates, diesel fuel price, and milling charges adapted from Bajracharya's experiences. The hypothetical community of six households, however, bears no resemblance to the actual community in Nepal. The assumptions used in the analysis are

1. The biogas plant is installed at a cost of US $3,700, plus an additional US $760 for a diesel engine (including modifications for it to use biogas to substitute for 80 percent of the diesel fuel)

TABLE 15.
Attractiveness of a Community Biogas System for
Cooking and Lighting Needs of Households F to K.

	Unsubsidized	50 Percent Subsidy
Annual Benefits		
a. From 75% savings in purchased fuel by Households F to K	188	188
Annual Costs		
a. Years 1-7		
Loan costs (6%; 7 years)	663	331
Repair, maintenance, replacement	50	50
b. Years 8-20		
Repair, maintenance, replacement	50	50
Net Annual Cash Flows		
a. Years 1-7	-525	-193
b. Years 8-20	138	138
Net Present Value in 20 Years		
a. 10% discount rate	-2,053	-436
b. 20% discount rate	-1,718	-521
c. 30% discount rate	-1,400	-470

Note: Annual benefits, annual costs, net annual cash flows, and net present value of the biogas system in 20 years are given in US$. Two cases are examined: (a) a community plant with no subsidy; and (b) a community plant to which a 50 percent subsidy has been applied to the installation cost.

and US $1,900 for an oil expeller, rice huller, grain mill, and accessories. The system is assumed to be financed with a loan at 6 percent interest over seven years. It is assumed to have a life expectancy of 20 years and zero salvage value. Depreciation and income taxes are not included in the analyses.

2. The analyses are conducted for three assumed capacity factors, based on the number of days per year the system would be operated: (a) 300 days per year (full capacity), (b) 150 days per year (half capacity), and (c) 75 days per year (quarter capacity).

Diesel fuel and labor are treated as variable costs. Diesel costs are based on a consumption rate of 2 liters per day for the dual-fuel engine when at full capacity, at a cost of US $0.50 per liter. Labor costs are based on an assumed need for 24 person-months of labor at full capacity, at a wage of US $30 per person-month.
3. Annual benefits are assumed to be derived solely from the sale of food-processing services, with no benefits assumed from the digester residues. Annual costs are assumed to be loan repayment costs (Years 1 through 7 only), and annual repair, replacement, and maintenance cost of US $300.

In defining the system boundary to include both the biogas plant and the food-processing facility, a key assumption is that diesel fuel is in limited supply and would not be available except at a much higher price for amounts beyond that needed for the dual-fuel operation. (See Chapter 6 for instances of diesel scarcity and Chapter 3 for sharp differentials in petroleum fuel prices at different locales.) That is, the biogas fuel is taken to be essential in order to establish the food-processing operation.[4]

The results of these analyses (Table 16) suggest that the community would find this venture attractive financially if they were able to operate the system at full or half capacity. (The breakeven point generally falls somewhere between half and quarter capacity.) It also would be financially feasible if they operated the equipment at quarter capacity and were satisfied with a return on investment of only about 10 percent. Although the values for such benefits were not incorporated in the analyses, members of this community might also derive some benefits from labor savings by not having to perform these time-consuming tasks of food processing.

DISCUSSION AND CONCLUSIONS

Although anaerobic digesters potentially can provide a variety of energy and nonenergy benefits to persons living in rural areas, much uncertainty exists about just how important the role is that biogas plants will play in meeting the energy needs of rural households and communities. This uncertainty stems partly from the lack of adequate information about the performance of existing biogas plants in the Asian region, as well as from some evidence (albeit not well documented) that the biogas development efforts in some countries are not proceeding as well as initially expected.

This chapter has referred to an apparent slowdown in China's biogas development activities and a leveling off in India's biogas program. In

TABLE 16.
Attractiveness of a Community Biogas System for
Food Processing Using a Dual-Fuel Engine

	Capacity Utilization	
	Full Capacity (300 d/y)	Half Capacity (150 d/y)
Annual Benefits (receipts)		
a. Husking rice (3 h/d 210 kg/h; US $1.00 per 180 kg)	1,050	525
b. Expelling oil from mustard seed (2 h/d @ 20 kg/h; US $1.00 per 15 kg)	800	400
c. Milling grain (3 h/d @ 50 kg/h; US $1.00 per 50 kg)	900	450
	2,750	1,375
Annual Costs		
a. Years 1-7		
Loan costs (6%; 7 years)	1,080	1,080
Repair, maintenance, replacement	300	300
Diesel fuel (2 1/d @ US $0.50/1)	300	150
Labor (base at 300 d/y; 24 person-months @ US $30/month)	720	360
b. Years 8-20		
Repair, maintenance, replacement	300	300
Diesel fuel (2 1/d @ US $0.50/1)	300	150
Labor (base at 300 d/y; 24 person-months @ US $30/month)	720	360
Net Annual Cash Flows		
a. Years 1-7	350	-515
b. Years 8-20	1,430	565
Net Present Values in 20 Years		
a. 10% discount rate	6,917	-446
b. 20% discount rate	3,070	-1,141
c. 30% discount rate	1,716	-1,153

Note: Annual benefits and costs, net annual cash flows, and net present value of the biogas system in 20 years are given in US$. Two cases are considered: (a) the food processing equipment operates at full capacity (300 d/y; 8 h/d); and (b) the equipment operates at half capacity (150 d/y).

addition, biogas systems seem to have high frequencies of inoperative or abandoned plants or of dissatisfied owners (T.B.S. Prakasam 1979, Karki and others 1980, S.N. Ratasuk and others 1979, Bhatia 1981, D. Sharma 1981). Many of these problems have been attributed to technical problems such as gas leaks but, if so, appropriate technical solutions should exist. Thus, the biogas-promoting agencies might not have devoted enough attention to ensuring that rural households were acquiring the most appropriate energy system considering each particular household's needs and available resources, or to ensuring that post-installation services (such as consultative services and service centers) were readily available and at a reasonable cost.

An Indian Institute of Management study (Moulik and others 1978), for example, observed that 55 percent of the digesters surveyed in four states in India did not have repair centers available in the village, while about 45 percent of the repair centers were more than 10 km from the village. In Sri Lanka, I interviewed several household users of biogas systems and found that relatively simple technical problems with the biogas plant or gas distribution system can cause a great frustration to the biogas plant owners, who often are provided with only very general information about the operation of their plants, and who seldom have the opportunity to discuss their problems with experts from the institutions promoting this technology. These situations are certainly not unique and often appear with other energy technologies.

These problems exist when technology-promoting institutions measure the success of their rural energy programs in terms of numbers of units installed, rather than in terms of numbers of functional units or satisfied users. Undoubtedly this will hamper rural energy development programs in the long run because from the beginning, nontraditional energy technologies (e.g., biogas plants, improved stoves, solar crop driers) probably are regarded rather skeptically by farmers who must conserve capital. That skepticism probably deepens after a system's reputation for poor performance and reliability develops.

One point that I have tried to underscore here is the importance of considering the perspective of the rural household or farmer in weighing the merits and demerits of an energy system. Even though the techniques of resource and financial analyses presented here may not correspond with the way a rural household or farmer views the situation, this type of analysis should be useful to rural development personnel in *assisting* rural families and communities in evaluating the appropriateness and financial attractiveness or unattractiveness of biogas systems. It is equally important to compare biogas systems with other energy system alternatives as well as with other investment options besides energy systems, such as irrigation pumps or fertilizer (since the farmer often might perceive his nonenergy needs as more urgent).

The financial analyses, accordingly, are from the standpoint of the household or community investors most directly concerned with viability and benefits. How lack of attention to one or more key factors or unrealistic assumptions can lead to misguided decisions has been shown. Wider economic criteria have not been incorporated in the quantitative analytic framework. Instead, steps for resource assessment and for considering benefit categories that are not readily quantifiable have been presented. Policy analysts in specific environments could use data from the several sensitivity tests, in conjunction with relevant resource and environmental factors, to compare financial returns with economic benefit-costs for candidate biogas systems. Unless these systems meet the viability and benefit tests of households or communities whose resources are to be committed, extra resources of society will be needed to initiate the systems. It will be this increment of social resources that must meet economic benefit-cost criteria.

The nature and extent of the role biogas energy systems can play in rural areas, of course, will vary with the specific rural situation, and with improvements to the technology itself. Nevertheless, characteristics of rural households or communities can be identified that might be conducive or restrictive to the adoption and successful use of biogas plants based on experience and analyses to date.

Because the principal financial benefit derived from operating a household biogas system for domestic energy needs is fuel savings, one of the more important characteristics of the household is the level of monetary expenditure on fuels to be replaced by biogas. My financial analyses suggest that biogas systems are considerably more attractive to households consuming large quantities of purchased fuels and are relatively more attractive at higher fuel prices. If cooking fuels are collected by household labor and this labor is valued at zero or low opportunity cost, and if relatively small amounts of kerosene or other commercial fuels are used for lighting, then a biogas system would probably not provide a suitable return on an investment. The possible exception to this would be if the cost of the biogas system and the cost of credit were relatively low, perhaps by means of government subsidies.

My analyses also suggest that community (and also perhaps household) biogas systems could be financially attractive if the biogas were used for powering food-processing machinery or other type of economic activity, where other sources of power were limited, uncertain, or entailed high costs. However, relatively few studies exist on the use of biogas in this manner yet a number of commercial operations in Asia (such as the Maya

Farms piggery in the Philippines) have successfully integrated biogas into their operations (F.D. Maramba 1978).

Another characteristic important in determining the potential of a biogas system is the availability of feedstock materials, including organic materials and water. In the Indian subcontinent, cattle and buffalo dung are the most widely used form of organic materials, although some biogas plants on the subcontinent use human wastes and experiments are proceeding with use of crop residues and other vegetative materials. The availability of organic materials will be an important limiting factor in many rural areas, since a typical household might require 50 to 80 kg of cattle dung each day (see O. Odeyami 1982, J.J. Huang and others 1982). Equally important will be access to an approximately equal quantity of water. This quantity of water might be unavailable for use in a biogas plant in many locations, or the time required to haul the water from its source to the plant might place further restrictions on the applicability of this technology.

Whether or not the operation of a biogas system will alter greatly existing patterns of household or hired labor should also be considered. It is impossible to generalize without prior study, whether adoption of a biogas system by a particular household or community will result either in net increases or labor savings, or whether a possible change in labor will be valued at any given opportunity cost. Seasonal patterns of labor use in rural areas influence the availability of labor for the various operations of a biogas plant and should be considered too.

Social factors play a significant role in whether or not biogas systems are adopted successfully by a rural household or community. Presently available information suggests that social factors are more important considerations for community-scale biogas development than for household biogas plants. In both cases, understanding the factors affecting the potential demand for the technology beyond financial considerations is important (such as convenience, prestige, awareness of the need for conservation), but community organizational characteristics will be particularly crucial for community-scale biogas systems. These are discussed elsewhere in this book.

Finally, the level of infrastructure development of a rural area will influence the potential of biogas systems, particularly regarding the availability of local and nonlocal materials and labor, together with facilities for post-installation consultative and repair services. The availability of credit will also be an important factor.

The potential of biogas plants in rural areas will be dependent on many factors; some have been discussed in this chapter. Because biogas energy systems are relatively complex technologies and can affect a number of aspects of rural life, the introduction of biogas plants should be preceded by a careful examination of energy use patterns and expenditures (time and money), resource requirements of the biogas system and resource availability, and socioeconomic factors relating to the rural household or community. Equally important is weighing the merits of biogas systems against the merits of other potential energy and nonenergy investments by the household or community. To a large extent, the contribution of biogas systems in rural areas will depend on just how appropriately they were applied in the context of the diverse conditions in the rural area.

NOTES

1. The brief discussion here is based on information obtained from Taylor 1980; Smil 1981 and 1982; personal communications with T.K. Moulik of the Indian Institute of Management, Ahmedabad, and T. Lumpkin of the University of Hawaii; and personal communication from R. Wagner of Cornell University as relayed to me by M. Wolfe of East-West Center.

2. A variety of microeconomic methods exist for evaluating the financial attractiveness of investments in alternative energy technologies: the net present value (NPV) technique, the benefit-cost ratio technique, the internal rate of return technique, and the payback period technique. Each of these methods has its own strengths and weaknesses but, when properly applied, each generally provides the same conclusions about whether a prospective investment is attractive or unattractive. This brief discussion about the NPV technique should acquaint noneconomists reading this chapter with some of the salient features of the method. For a more detailed discussion of NPV methods, please consult these presentations: E.L. Grant and others 1976, French 1979, and J.F. Kreidor and F. Kreith 1981.

The NPV analyses presented in this chapter are simplified greatly for the purposes of this book. I do not present, for example, methods for including depreciation accounting, methods for dealing with rising costs (such as inflation effects on kerosene prices), or methods for accounting for salvage value of the components of the energy system after it has completed its useful lifetime. The five principal steps in performing these simplified NPV analyses follow.

Determination of Annual Costs. Two categories of costs are associated with an investment in a biogas system. First, often an initial investment cost in the energy system and its accessories, including the costs of land, materials, labor, or other expenses. These initial investment costs are assigned here to "Year 0," as is the practice in many textbooks on financial analysis (e.g., Grant and others 1976). A project may entail no Year 0 costs, however, if it is financed entirely through bank loans or other sources of credit. The second category of annual costs is incurred during the operation of the system and includes labor, materials, loan repayment costs, and other expenses. Annual costs typically vary from year to year, as illustrated by Moulik (ESCAP 1981). Cash flow tables (see example) are used to depict annual costs and annual benefits and are helpful in NPV analyses, especially if information is available from prior experiences with biogas plant models on when different component parts will need servicing, repair, or replacement.

In my analyses of biogas plants it is assumed that the annual costs of operating a biogas plant, except for Year 0, are uniform over the lifetime of the plant. I do this for two reasons: first, to simplify the presentation for this generalized analysis of the financial aspects of biogas systems; and second, because I am not addressing these analyses to any one type of biogas plant and thus, to assume any detailed year-to-year cash flow would be inappropriate. The use of year-to-year cash flow tables will be desirable, however, for development workers analyzing a specific prospective energy system in actual rural situations.

Determination of Annual Benefits. A discussion of the value of benefits from a biogas plant is presented in detail elsewhere in this chapter. As with annual costs, annual benefits could also vary from year to year, due to changes in the performance of the biogas plant arising from changing environmental conditions or inflationary effects on the price of fuels that were substituted for by biogas. For simplicity, my analysis assumes that the annual benefits derived from the operation of a biogas system are uniform during the lifetime of the system.

Determination of Net Annual Benefits. These values are calculated by subtracting annual costs from annual benefits. If, for any year, annual costs exceed annual benefits, then the net annual benefit is expressed as a negative value.

Determination of Net Present Values. Two alternative procedures are followed, depending on whether year-to-year cash flow tables are used or whether uniform annual costs and benefits are assumed. With year-to-year annual costs and benefits, the individual net annual benefit values are multiplied by the appropriate "single payment present worth factor," which is readily found in compound interest tables such as those provided in Grant and others 1976 or in Kreidor and Kreith 1981. In the case of uniform annual costs and benefits, the uniform net annual benefit is multiplied by the appropriate "uniform series present worth factor," as provided in the references mentioned above.

The particular compound interest table used will depend on the applicable discount rate. The discount rate represents the time value of money and is used to convert net annual benefits to equivalent amounts at a specified point in time. Another way of looking at discount rates is that they represent the minimum rate of return on an investment attractive to the investor. For example, a prospective investor considering installing an energy system at an initial cost of US $100 might weigh his decision in terms of a 20 percent interest rate he might receive in investing his money in a savings account in a bank. This investor might thus choose a minimum attractive return on investment in an energy system at 20 percent, that is, the return he would be getting from investing his money elsewhere.

The example provided below will illustrate the two procedures mentioned above. The first table represents NPV calculations for an initial investment of US $100 in an energy system in which varying annual costs and benefits are assumed. The discount (minimum attractive rate of return) rate chosen here is 20 percent. The energy system is assumed to have a life expectancy of five years. Values are given in US dollars.

Year	Annual benefits	−	Annual costs	=	Net annual benefits	×	Single payment present worth factor (@20%)	=	Present value of benefits (@20%)
0	0		100		−100		1		−100.00
1	120		96		+ 24		0.8333		+ 20.00
2	112		48		+ 64		0.6944		+ 44.44
3	100		64		+ 36		0.5787		+ 20.83
4	106		76		+ 30		0.4823		+ 14.47
5	98		80		+ 18		0.4019		+ 7.23

The next table assumes uniform annual benefits and uniform annual costs. The initial investment costs, discount rate, and system's life expectancy are as given above.

Year	Annual benefits	−	Annual costs	=	Net annual benefits	×	Uniform series present worth factor (@20%, 5 years)	=	Present value of benefits (@20%)
0	0		100		−100		1		−100
1-5	107.20		72.80		+ 34.40		2.991		+102.89

Calculation of Net Present Value. Once the present value of benefits has been determined, the attractiveness or unattractiveness of the investment project at the minimum attractive rate of return can be determined. The procedure for this is simply to total the last column, the present

464 Rural Energy to Meet Development Needs

value of benefits. In the case of year-to-year cash flow analysis presented in the first table, NPV is equal to +US $6.97; in the second table, using uniform annual benefits and costs, the NPV is +US $2.89. That the investment meets the criterion of having a minimum return on investment of 20 percent is evident from the positive sign of the NPV. An investment that does not return 20 percent will have a negative NPV. The slight discrepancy between the NPVs of the two procedures is due to having used average values to determine the NPV in the latter case.

3. The discussion here briefly illustrates the procedures used to calculate the net cash flows and NPVs for my five hypothetical households and the sensitivity analyses that follow. I illustrate the case of Household B, which consumes 5 liters of kerosene and 50 kg of firewood per week (Table 8). I assume below the biogas system to be financed by a 10 percent loan over 20 years and firewood costs to be US $20 per tonne and kerosene costs to be US $0.50 per liter.

Annual Costs. Annual costs for Year 0 include the US $400 installation costs for the biogas system. Because I am using the end-of-year convention for illustrating annual costs, no loan repayment costs or maintenance costs are given for Year 0. Years 1 through 20 include loan repayment costs and an assumed annual maintenance cost of US $20. The loan repayment costs were determined using compound interest tables, from which was taken the uniform series capital recovery factor of 0.11746 for an interest rate of 10 percent and a 20-year repayment period. The calculation of annual loan repayment costs is as follows:

Amount of loan	×	Uniform series capital recovery factor	=	Annual loan repayment costs
US $400		0.11746		US $46.98

Annual Benefits. Household B enjoys a loan for US $400 in Year 0 to finance the installation costs of the biogas system. The annual benefits from Years 1 through 20 include US $6.50 from additional fertilizer value from anaerobically digesting cattle dung, and savings from purchasing less firewood and kerosene. Household B consumed US $52 of firewood per year and US $130 of kerosene. Because biogas is assumed to substitute for only 75 percent of the prebiogas fuels for lighting and cooking, the benefits used in the financial analysis are US $39.00 for firewood and US $97.50 for kerosene.

Net Annual Cash Flows. Net annual cash flows are determined by subtracting annual costs from annual benefits. These are illustrated in the table in this note.

Net Present Value. Present values are determined by multiplying the annual cash flows by the appropriate uniform series present worth factor (as given in compound interest tables). In the next table, I illustrate the calculation of present values for an assumed discount rate (i.e., desired rate-of-return on investment) of 30 percent over the 20-year lifetime of the biogas system. The NPV is determined by summing the present values for Year 0 and Years 1 through 20. (Note: If the biogas system were not financed by a loan, then −US $400 would be the present value for Year 0, and annual costs for Years 1 through 20 would not include loan repayment costs.)

Year	Item	Annual benefit ($US)	−	Annual cost ($US)	=	Net annual cash flow ($US)	×	Uniform series present value factor (for 30%, 20 years)	=	Present value ($US)
0	Biogas System			400.00						
	Loan	400.00								
		400.00		400.00		0.00				0.00
1-20	Loan Repayment			46.98						
	Maintenance			20.00						
	Fuel Savings									
	a. Firewood	39.00								
	b. Kerosene	97.50								
	Digester Residue	6.50								
		143.00		66.98		+76.02		3.316		+252.08

NPV = +US $252.08

4. As pointed out by Ramesh Bhatia in his review of this chapter, if diesel fuel is available at a given price for full operation of a diesel engine (100 percent) to run the food-processing plant, credit for the biogas plant should be taken only for the 80 percent diesel fuel saving, not for the combined system. The general point is that benefits from investment in the renewable energy system (the value of diesel fuel saved) should be distinguished from benefits from the use of biogas (i.e., the milling service that could be powered by technologies that do not use biogas).

REFERENCES

Bahadur, S., and S.C. Agarwal.
    1980   *Community Biogas Plant at Fateh Singh ka Purwa — An Evaluation.* Planning Research and Action Division, State Planning Institute. Lucknow, India: PRAD.

Bajracharya, D.
- 1982 Personal communication, Honolulu, Resource Systems Institute, East-West Center.

Barnett, A., L. Pyle, and S.K. Subramanian.
- 1978 *Biogas Technology in the Third World: A multidisciplinary Review.* Ottawa: International Development Research Centre.

Bhatia, R.
- 1977 Economic Appraisal of Bio-gas Units in India: Framework for Social Benefit Cost Analysis. *Economic and Political Weekly* 12 (33,34):1503–1517.

Bhatia, R.
- 1981 Gobar Gas Plant. *Indian Express*, February.

Bhatia, R., and M. Niamir.
- 1979 *Renewable Energy Sources: The Community Bio-Gas Plant.* Presented in the Seminar on November 2, Department of Applied Sciences, Harvard University. Cambridge, Massachusetts: Harvard University.

Brown, R.H.
- 1977 Appropriate Technology and the Grass Roots: Towards a Development Strategy from the Bottom Up. *The Developing Economies* 15(3): 253–279.

Chengdu Institute of Biology, Academia Sinica.
- 1980 The Process and Mechanism of Biogas Digestion. In *Biogas Training in China: A First Exchange with Developing Countries*, ed. E.A. van Buren. Nairobi: United Nations Environment Programme.

ESCAP.
- 1980 *Guidebook on Biogas Development.* Energy Development Series no. 21. Bangkok.

ESCAP.
- 1981 *Renewable Sources of Energy, Volume II: Biogas.* ECDC–TCDC: ST/ESCAP/160. Bangkok.

FAO.
- 1978 *China: Azolla Propagation and Small-Scale Biogas Technology.* FAO Soils Bulletin 41. Rome.

FAO.
- 1983 *Proceedings of the Workshop on Rural Energy Planning in the Developing Countries of Asia*, April 11–29, Beijing, People's Republic of China. Rome.

French, D.
- 1979 *The Economics of Renewable Energy Systems for Developing Countries.* Washington, D.C.: USAID.

Fulford, D.J., and G. Devkota.
   1981   *Economies of Gobar Gas — Domestic versus Commercial.* Butwal, Nepal: Development and Consulting Services, United Mission to Nepal.

Ghate, P.B.
   1979a *Action Research in the Community Biogas Programme.* Prepared for the Research Planning Workshop on Energy for Rural Development, Chiang Mai, Thailand, February 5–14, Honolulu: Resource Systems Institute, East-West Center.

Ghate, P.B.
   1979b *Biogas: A Pilot Project to Investigate a Decentralized Energy System.* Lucknow, India: State Planning Institute, Planning Research and Action Division.

Government of India, Department of Science and Technology (DST).
   1981   *Biogas Technology and Utilization.* New Delhi: Department of Science and Technology.

Grant, E.L., W.G. Ireson, and R.S. Leavenworth.
   1976. *Principles of Engineering Economy.* New York: John Wiley & Sons.

Hawkins, J.N.
   1978   Rural Education and Technique Transformation in the People's Republic of China. *Technological Forecasting and Social Change* 11:315–333.

Huang, J.J.H., J.C.H. Shih, and S.C. Steinsberger.
   1982   Poultry Waste Digest: From the Laboratory to the Farm. In *Energy Conservation and Use of Renewable Energies in the Bio-Industries 2*, ed. F. Vogt, 376–383. Oxford, England: Pergamon Press Ltd.

Idnani, M.A., and S. Varadarajan.
   1974   *Preparation of Fuel Gas and Manure By Anaerobic Fermentation of Organic Materials.* ICAR Technical Bulletin (Agric.) no. 46. New Delhi: Indian Council of Agricultural Research.

Institute of Soil and Fertilizer.
   1980   *The Utilization of Biogas Fermentation Residue-Sludge and Effluent.* Academy of Agricultural Science Research, Sichuan. UNIDO ID/WG.321/10.

Islam, M.N.
   1979   *A Report on Biogas Programme of China.* Dhaka: Bangladesh University of Engineering and Technology, Department of Chemical Engineering.

Karki, A.B.
    1980   *Biogas in Nepal: The Prospects and Problems.* Presented at the Research Planning Workshop on Energy for Rural Development, Chiang Mai, Thailand, February 5–14. Honolulu: Resource Systems Institute, East-West Center.

Karki, A.B., K.N. Pyakural, and N. Axinn.
    1980   *Techno-Socio-Economic Study of Bio-Gas Plants in the Chitwan District, Nepal.* Paper presented at the Research Planning Workshop on Energy for Rural Development, Chiang Mai, Thailand, February 5–14. Honolulu: Resource Systems Institute, East-West Center.

Kreidor, J.F., and F. Kreith.
    1981   *Solar Energy Handbook.* New York: McGraw-Hill Book Company.

Khadi and Village Industries Commission (KVIC).
    1976   *Gobar Gas on the March.* Bombay: Khadi and Village Industries Commission, Directorate of Gobar Gas Scheme.

Lumpkin, T.
    1982   Personal communication, Honolulu, University of Hawaii at Manoa.

Maramba, F.D.
    1978   *Biogas and Waste Recycling: The Philippine Experience.* Manila: Liberty Mills, Inc.

McGarry, M.G., and J. Stainforth.
    1978   *Compost, Fertilizer, and Biogas Production from Human and Farm Wastes in the People's Republic of China.* Ottawa: IDRC.

Meynell, P.J.
    1976   *Methane: Planning a Digester.* Dorchester: Prism Press.

Moulik, T.K.
    1982   *Biogas Energy in India.* Ahmedabad: Academic Book Centre.

Moulik, T.K.
    1982   Personal communication, Ahmedabad, Indian Institute of Management.

Moulik, T.K., U.K. Srivastava, and P.M. Shingi.
    1978   *Bio-Gas System in India: A Socio-Economic Evaluation,* Ahmedabad: Indian Institute of Management.

National Academy of Sciences (NAS).
    1977   *Methane Generation from Human, Animal, and Agricultural Wastes.* Washington, D.C.

Odeyemi, O.
    1982   Biogas Generation from Cassava Leaves Compared with Two Animal Manures and Sewage Sludge. In *Energy Conservation and Use of Renewable Energies in the Bio-Industries 2*, ed. F. Vogt, 554–558. Oxford, England: Pergamon Press Ltd.

Pang, A.
    1978   *Economics of Gobar Gas*. Butwal, Nepal: Development and Consulting Services, United Mission to Nepal.

Parikh, J.K., and K.S. Parikh.
    1977   Mobilization and Impacts of Biogas Technologies. *Energy* 2:441–455.

Planning Research and Action Division, (PRAD).
    1980   *Janata Biogas System: An Evaluation*. Lucknow, India: State Planning Institute. PRAD.

Prakasam, T.B.S.
    1979   Application of Biogas Technology in India. In *Biogas and Alcohol Fuels Production: Proceedings of a Seminar on Biogas Energy for City, Farm, and Industry*. Staff of Compost Science/Land Utilization, eds. Dorchester, Dorset/Emmaus. Pennsylvania: JG Press.

Rajabapaiah, P., K.V. Ramanayya, S.R. Mohan, and A.K.N. Reddy.
    1979   Performance of a Conventional Biogas Plant. *Proceedings of the Indian Academy of Science* C2(3):357–363.

Ramakrishna, J.
    1980   *Bibliography on Anaerobic Digestion*. ERD Research Materials RM-80-2: Honolulu: Resource Systems Institute, East-West Center.

Ratasuk, S., N. Chantramonklasri, R. Srimuni, P. Ploypataropinyo, S. Chavedej, S. Sailamai, and W. Sunthonsan.
    1979   *Pre-feasibility Study of the Biogas Technology Application in Rural Areas of Thailand*. Thailand: Applied Scientific Research Corporation of Thailand.

Renewable Energy Resources Information Center (RERIC) News.
    1983   India's Biogas Not on Target. *RERIC News* 6(1):6.

Sanghi, A.K., and D. Day.
    1977   A Cost-Benefit Analysis of Biogas Production in Rural India: Some Policy Issues. In *Agriculture and Energy*, ed. W. Lockeretz. New York: Academic Press.

Santerre, M.T., and K.R. Smith.
    1980   *Application of the FLERT Approach to Rural Household and Community Anaerobic Digestion Systems*. PR-80-5. Honolulu: Resource Systems Institute, East-West Center.

Santerre M.T., and K.R. Smith.
  1982 Measures of Appropriateness: The Resource Requirements of Anaerobic Digestion (Biogas) Systems. *World Development* 10(3):239–261.

Sathianathan, M.A.
  1975 *Biogas: Achievements and Challenges.* New Delhi: Association of Voluntary Agencies for Development.

Sharma, D.
  1981 Gobar Gas Plants: Cost of Neglect. *Indian Express*, February 4.

Shian, S.T., M.R. Ching, Y.T. Ye, and W. Chang.
  1979 The Construction of Simple Biogas Digesters in the Province of Szechwan, China. *Agricultural Wastes* (1979):247–258.

Singh, R.B.
  1974 *Biogas Plant: Generating Methane from Organic Wastes.* Ajitmal, India: Gobar Gas Research Station.

Smil, V.
  1981 Energy Development in China: The Need for a Coherent Policy. *Energy Policy* 9(2):113–126.

Smil, V.
  1982 Chinese Biogas Program Sputters. *Soft Energy Notes* (July/August 1982):88–90.

Smith, K.R., and M.T. Santerre.
  1980 *Criteria for Evaluating Small-Scale Rural Energy Technologies: The FLERT Approach (Fuel-Linked Energy Resources and Tasks).* PR-80-4. Honolulu: Resource Systems Institute, East-West Center.

Smith, K.R., A.L. Aggarwal, and R.M. Dave.
  1983 Air Pollution and Rural Biomass Fuels in Developing Countries: A Pilot Study in India and Implications for Research and Policy. *Atmospheric Environment* 17(11):2343–2362.

Srinivasan, H.R.
  1978 Biogas Systems and Sanitation. In *Sanitation in Developing Countries.* Chichester: John Wiley and Sons.

Taylor, R.P.
  1980 *The People's Republic of China: Experience and Capabilities in Renewable Energy Resource Development.* Science and Technology Report Series no. 35. Washington D.C.: The World Bank.

Taylor, R.P.
  1981 *Rural Energy Development in China.* Washington D.C.: Resources for the Future.

UNIDO
    1978   *International Forum on Appropriate Technology.* Report of the Technical/Official Meeting to the Ministerial Level Meeting, Anand, India. Vienna.

van Buren, E.A. (ed.).
    1979   *A Chinese Biogas Manual: Popularising Technology in the Countryside.* London: Intermediate Technology Publications, Ltd.

van Buren, E.A. (ed.).
    1980   *Biogas Training in China: A First Exchange with Developing Countries.* Nairobi: UNEP.

11
Organizing Current Information for Rural Energy and Development Planning

Richard Morse, S.C. Agrawal, Carol J. Pierce Colfer, Elizabeth Foster, Ramabhadran Govindarajan, and Jamuna Ramakrishna

A:	"In fact, we have calculated already how much the income will be if the mill runs 8 hours, 6 hours and 4 hours. The question is whether it will run 8 hours or not."
C:	"Isn't it 12 hours per day?"
E:	"No. We have to subtract at least 4 hours."
A:	"We have to be careful and have to consider whether grain will be available for milling. . . ."
F:	". . . . In Moranj, they have a lot of grains and they have many things to mill, whereas we have a very limited production. Therefore, someday we may feel it good enough if the mill runs only about 6 hours a day."

F:	"We need 15 tins of dung a day. Isn't that true?"
Deepak:	"Yes. That has to be fresh dung."
A:	"If you ask me to bring one or two tins of dung, I will bring for sure. We should bring the dung until the tank is filled."
B:	"Yesterday, I promised to provide a tin of dung everyday. This morning when I went to see it, I could not believe that there wasn't as much as a tin."
A:	"We will measure the dung."

—Chorkate, Nepal. February 1982.
(Deepak Bajracharya and Chandra Gurung 1984)

The Chorkate experience demonstrates neighborhood planning for investment in a new technology. The farm families and consumers themselves at their face-to-face planning sessions provided most of the information required to calculate a threshold volume of milling (a break-even point for the community), to determine the effect of added volume on profitability, and to estimate and negotiate the supply of adequate dung. The *lami* facilitated the analysis by providing knowledge of how to organize this information to assess viability, by adding data mainly about external factors, and by doing some basic computations.

In such local planning, one fundamental question is: what information from outside must be added to local knowledge, to build the total picture the community needs? Another is: how can local residents define and fill such knowledge requirements while retaining the initiative in decision making and in negotiations with outsiders to gain the most advantageous terms in the investment?

This chapter first identifies some approaches for rural residents to use to define their energy-related information needs, and for researchers to use in facilitating such local learning processes. The cooking sector provides one entry to understanding food-fuel links in the local farming system. These links affect efficiencies in resource use and are therefore relevant dimensions in needs assessment. Needs posed by different village groups for improving food and fuel productivity as well as access to resources provide a framework, then, for organizing data from dependable microstudies on food-fuel relations and on use of, and substitution among, different fuels.

How families allocate their labor time to the different tasks that tie the cooking sphere to the crop-animal-forest sphere is a crucial decision factor in resource assessment. Men's and women's time either as limiting or surplus resource is examined in this context. Energy scarcity and potential in farming systems can only be assessed in this interactive light. A basis is thereby set for adapting such local planning knowledge, through comparative analysis, to regional and national policy planning purposes.

RURAL RESIDENTS' INFORMATION NEEDS FOR ENERGY PLANNING AND DEVELOPMENT

Land, work, and time: these are key resource units in the language and decision criteria of farmer-landholders and of landless families who work as members of farming communities. Time and energy available for work are critical constraints during peak periods. If farm family members have two plots but only enough bullock-days to plow one in time for sowing, they must decide on the basis of available information which plot to plow, which crop or variety to sow, and how to allocate tasks. If the weather changes, or if signs of change are perceived, they may change their decisions. Among the many factors that affect these choices are how much (if any) grain the family holds in storage, which crop is regarded as most resistant to weather changes through the harvest, and whether the family has resources to borrow bullock time or lease tractor services. The factor of time enters the decision both as a point (today) and as a flow or sequence (the harvest season). Units of land, per unit of work, per unit of time are the differential or marginal units of allocation for men and women who farm and who engage in farm work and food and fuel gathering.

Diversity in choices for combining these inputs has been emphasized as "the critical factor in the efficient use of scarce resources" in intensive farming systems (Richard Harwood 1979). "As any resource becomes scarce enough to limit production, diversity becomes increasingly important to its optimum utilization." Earlier chapters have established that the knowledge held by rural residents is a crucial base for discovering and utilizing diverse opportunities. The complexity of identifying and obtaining pertinent external information that will add to their knowledge in useful ways is evident.

Decision Criteria of Farming Community Members

The number, sex, age, and dependency status of family members are central factors in their decisions both on consumption needs and on production activities. These aspects in their many permutations result in considerable local diversity. Decision roles and decision needs of men and women, although influenced by culture and society, are also family particular. The quality of specificity in knowledge needs is reinforced.

Problem identification through dialogue between researchers and rural citizens, as exemplified in Chhoprak panchayat, is an essential procedure for establishing communication paths to use in defining and meeting these information needs. Such face-to-face learning enables the researcher as well as the rural householder to broaden their knowledge base and its utility for action.

Defining energy quality and energy efficiency in terms that fit the knowledge map of people who live close to nature is a step in this process. Residents' interpretation of their own experience and information in the light of scientific formulations is a test of how far folk knowledge and modern terminology can converge. Stimuli to innovation arise in this "hazy" realm of new words and meanings. Room is open for the innovator's function of creating new values by connecting previously unseen dimensions of value in realms both inside and outside (Fredrik Barth 1966, Vijay Chebbi and others 1977).

The utility of knowledge for wider policy and planning purposes depends on such local tests of validity and on how, when, and for which questions local information can safely be aggregated. A function of this chapter is to illustrate these issues empirically.

Indicators for Analyzing Rural Energy Needs and Potential

We have established the location-specific, heterogeneous nature of rural energy systems in earlier chapters. In the effort to understand a particular rural environment, therefore, great caution must be exercised in transferring or interpreting data from another situation. Suitable terms are needed to denote the situation-specificity and quality of data, as a guard against "displaced concreteness." We will use here as a lexicon to define data quality the terms "need indicator," "value indicator," and "diagnostic indicator."

Need Indicator. Need indicator is defined as a number or ratio that establishes quantity relationships with sufficient reliability to be used in local project evaluation and resource allocation. To be regarded as reliable, either of two criteria must be met.

Criterion 1: Experimental or other physical evidence involving variables and controls that reasonably simulate the particular household and village condition in question.

Criterion 2: In-depth village or microscale studies having high sampling and data collection reliability, with statistical characteristics indicating reasonable stability in mean values or revealing clear association with specified reproducible variables.

Examples are relations between food cooked and fuel consumed, and between crop variety and water requirements. No coefficients will be termed need indicators in this volume unless they meet criterion 1 or 2. People's assessments of nonquantifiable project aspects are of course centrally important, in using need indicators to evaluate project and allocation choices.

Value Indicator. A value indicator is defined as a proxy measure of scarcity or exchange value in nonmonetized or partly monetized economies where labor and goods largely serve or are produced for the household's own use, or if exchanged, are on an in-kind or barter basis. Examples are Elizabeth Foster's estimation of the value of dung as manure (1983) and the firewood/dung cake energy valuation considered for a north Indian case later in this chapter.

Diagnostic Indicator. Ratios such as persons per sq km, cattle per person, percentage of net cropped area irrigated, and percentage of fuelwood collected from own or common sources are used herein to gain understanding of connections between energy demand and farming system characteristics. In local planning, study of variances among such diagnostic indicators helps to identify key limiting or surplus factors in the farming system, and to assess priorities. Relationships among several indicators must be studied, however, to identify points of stress and opportunity.

Significant differences between mean values of diagnostic indicators are guides for identifying the rough boundaries of farming systems and, on a wider spatial scale, of ecosystems. Maps delineating such distinctions are critically needed. Otherwise, national policy planners may assign different regions and subregions "average" values that incorrectly imply deficits or surpluses in food, fuel, and related inputs.

A research progression from "diagnostic" to "need" and "value" indicators is envisaged. When a set of data of sufficient reliability on a particular diagnostic indicator is obtained to meet criterion 2, it would qualify as a need or value indicator. This provides a test of the adequacy and quality of data at hand to meet energy and rural development planning purposes. In using such indicators, we specify which definition — need, value, or diagnostic — is intended in each case.

THE COOKING SECTOR

Cooking consumes 60 to 70 percent of the energy used in rural areas of developing countries. This major end use motivated our emphasis on domestic cooking fuels and technologies in Chapters 8 to 10. Links between cooking fuel and fulfillment of food and nutrition needs are at the root of this motivation and offer insights into rural families' outlooks in resource allocation. In this section we undertake to develop need indicators as a tool in assessing energy requirements for nutrition.

Nutrition and Cooking Fuel as Basic Needs

One observed response to fuel scarcity, as already noted, is to reduce the number of hot meals per day, to increase consumption of uncooked food, and to heat water for tea or coffee to below the boiling point. The health implications of these adjustments are inadequately understood but initially suggest a possible increase in the incidence of food-related disease. Nutrition may also suffer from decreased availability of protein and sometimes calories in uncooked or partly cooked foods. Unavailable protein in gram (cowpea or chickpea) that is parched, for example, is 40 percent, compared to 7 percent when gram is well cooked (Charlotte

Wiser 1955). Similar effects are found in soybean, if it is roasted rather than ground as flour for mixture with wheat flour (Noboru Iwamura, pers. comm., 1982). In some instances, on the other hand, local methods of preparing food without cooking can enhance nutritive values. Soybeans processed with fungus as a cake, a household industry in Indonesia, are a good protein source without cooking (Barbara Chapman, pers. comm., 1982).

The implications of these responses to fuel shortages appear to be more severe for particular groups, including the small landholder and landless families. Nutritional surveys show these groups to be especially prone to deficient diets in terms of calories, protein, and vitamins (Institute of Nutrition and Food Science 1977). Energy microstudies cited in Chapter 2 and below have shown that the same groups often have limited access to biomass cooking fuels. Thus these groups are particularly vulnerable to any reductions in the availability of food *or* fuel; both can affect their nutritional status.

Investigation of the fuel requirements for a given level of nutrition can provide the framework for a more thorough understanding of the fuel/food economy. This economy is grounded in both domestic and agricultural spheres because food and fuel are interdependent in the cooking sphere. They are also related in an often competitive manner if we consider land use.

Fuel requirements of a given nutritional level are dependent on several parameters specific to the local situation, including food and fuel types, stove efficiencies, and other factors examined below. The practical limit of data now available on these factors is quickly reached, since testing has been limited largely to simulations in laboratories or simple water boiling tests. But for the household user, whether or not the most efficient stove, fuel, and pan mix according to laboratory conditions is the best suited for efficiently cooking traditional meals and local staples remains a question. What are the fuel requirements for cooking food for a family of a given size for one day at current levels of food/fuel consumption? What is the per capita increment of fuel that would be required at improved levels of nutrition meeting the local minimum daily requirement? These numbers can contribute to an improved understanding of fuel needs derived from food needs. This relation must be understood if sustainable and adequate quantities of food and fuel are going to be produced from a given land base. Fuel links to nutrition thus have important implications for the food policy planner as well as for the village researcher and development specialist concerned with either food or energy.

Estimating Cooking Fuel Requirements and Efficiencies

Women and men who cook observe directly the relationship between quantities of fuel consumed and food prepared. They often can give quite close approximation of these ratios, in local units, and can explain several of the conditions that influence their values. Fuel/food ratios therefore provide a good starting point for discussing fuel efficiencies and requirements, even though reliable measurement of these quantities in actual use conditions is difficult.

Fuel/food ratios are sensitive to

1. fuel type and moisture content
2. type of food and its physical state
3. stove efficiencies
4. pan shape and material
5. cooking mode and practices (e.g., frying, baking, fire-tending practices, excess water use)
6. scale efficiencies.

With controls on variables (2) through (6), the heating values of different fuels could be measured. For a particular food and fuel, if three of the other factors are held constant, the effects of the final factor on efficiency will be shown by fuel/food ratios. Or, if variables (3) through (6) are held constant, the fuel requirements for cooking different foods can be determined. But all these are widely variable conditions that are rarely specified completely enough to permit firm comparisons between studies (see C.S. Martinsen and J.G. Ostrander 1982). We recommend learning from the cook's own knowledge of these conditions, as a guide in structuring local data collection and development efforts.

Efficiency and Access to Efficiency. Units of measure for recording and reporting fuel/food ratios differ importantly in meaning. Weight (kg/kg) and energy (kcal/kcal or kJ/kJ) ratios are the principal examples. For consideration of efficiency and requirements, these ratios serve different purposes.

The kg fuel/kg food ratio labeled "specific fuel consumption" by Howard Geller (1982), is shown in his studies to be highly correlated with fuel efficiency ($r = 0.94$). Geller therefore says this weight ratio makes a good field proxy for efficiency as long as the food and other cooking factors are reasonably consistent from meal to meal, or from observation to observation. This measure is really only valid for a single fuel, not for a fuel mix. If used to record fuel and food use, food quantities in the diet should be noted so that the resulting information can lend itself to a nutrition/energy analysis as well.

User discussion of efficiency factors can be facilitated through use of the fuel/food weight ratio. Users of wood fuel would immediately be struck by the comparative ratios in Table 1 for two types of aluminum pans used in cooking rice, which indicate that spherical-bottom pans (kg/kg = 1.6) are more efficient than flat-bottom pans (kg/kg = 2.3). The evidence of efficiencies of scale is also of interest. A kg/kg ratio of 0.45 is observed for cooking potatoes in a batch of 4.1 kg, compared to a ratio of 0.96 for a batch of 1.36 kg. Through such ratios, controlling for all but one variable, efficiency effects in other cooking tasks can be examined.

Such a ratio may also offer insight into factors associated with differential access to efficiency. Though Nurul Islam's study cited in Table 1 did not show a difference between clay and aluminum pans of the same shape, Geller's study found an average efficiency of 4.1 percent for clay pots and 7.2 percent for aluminum pots (1982). The kg/kg ratio for cooking with clay pots was 1.52 (mean of five household observations) and for aluminum pots 0.95 (mean of eight observations), statistically different at the 1 percent significance level. Since clay pots are typically made by local artisans, often on a barter or in-kind basis, while aluminum pots must be bought in the market, we may ask: are subsistence households dependent on the less efficient clay pots, while families with surplus cash income save fuel by using more efficient aluminum pots?

The question as posed is relevant to welfare effects of the two types of pots and to estimation of local fuel requirements. To estimate wider effects, it might be necessary to compare the employment-generating role of local firing of clay pots with that of manufacture and distribution of aluminum pots. In terms of overall energy efficiency, it might be necessary to extend the system boundary to compare the energy content of locally fired clay with that of aluminum pots from the bauxite mine to the village. We consider these secondary effects to be beyond the attained state of the art in rural energy analysis at this stage.

In adjusting to scarcity, in other circumstances, poor families have been observed to use fuel with greater technical efficiency than higher-income families. In the Nabagram study, relatively well-to-do families cook two or three times a day, while poorer families cook once a day for a longer single period (Nurul Islam 1980). The poorer families experience higher average fuel efficiency resulting from only one cold start.

Interpreted with care, and accounting for the specific conditions shown, data of the type presented in Table 1 may be useful in illuminat-

TABLE 1.
Wood Fuel Consumption in Cooking Specific Foods, Bangladesh

Food*	Utensil Type	Amount Cooked (kg)	kg Water Per kg Food	kg Wood** Per kg Food	kJ Energy Per kg Food	kJ Energy Per kJ Food	Remarks+
Rice‡	Flat pan, aluminum	1.14	4.2	2.3	35,890	2.4	
Rice	Spherical pan, aluminum	1.36	4.3	1.6	25,110	1.7	Fuel saving, spherical pan
Rice	Spherical pan, clay	1.36	4.3	1.6	25,110	1.7	
Wheat: chapati++	Flat plate	1.0	0.5	1.1	16,330	1.2	12 chapatis from 1 kg flour
Pulse: masur	Spherical pan, aluminum	0.45	3.0	1.9	29,570	2.0	
Pulse: khesari	Spherical pan, aluminum	0.45	3.0	2.5	37,630	2.6	
Potato	Spherical pan, aluminum	1.36	1.0	0.96	14,450	3.6	
Potato	Spherical pan, aluminum	4.1	1.0	0.45	6,790	1.7	Clear economies of scale

Source: Islam 1980. ++Islam pers. comm. 1983.

*Food caloric values:
Rice 3,537 kcal/kg; Pulse: khesari 3,450 kcal/kg; Whole wheat flour: 3,410 kcal/kg; Potato 970 kcal/kg;
Pulse: masur 3,430 kcal/kg.

**Firewood type sundari, good quality, air-day. Conversion norm: 15,120 kJ/kg.

+One-mouth stove, unimproved, 0.43 m depth of hearth. Tests from a cold start, using excess water ("wet" method).

‡Parboiled aman rice, moisture 16 percent.

ing or pointing to significant facets of fuel use in other conditions. In dialogue with users, knowledge of and assessment of such adjustments can be expanded.

Factors for Estimating Fuel Requirements in Relation to Nutrition. For a given fuel, diet, and set of efficiency factors, the kg/kg ratio can be converted to kJ/kJ or kcal/kcal ratios, the coefficients that will be most useful in defining energy requirements for a given diet. We have some indications of these coefficients for individual *foods* cooked with firewood (Table 1). The "food" numbers are useful because they can give an indication about the relative fuel intensities for different foodstuffs, especially staples that commonly provide 60 to 80 percent of the food energy of a population. For example, the preliminary data in Table 1 indicate that fuel requirements (1.7 kcal in spherical pans) per kcal of rice are about 40 percent more than fuel requirements (1.2 kcal) per kcal of wheat cooked as *chapatis* under the stated conditions. Such information will be helpful in discerning differences in the fuel requirements of different diets by understanding the requirements of the component parts.

We also have some indications of these ratios for specified *meals* cooked with firewood, both in actual use and in experimental cooking tests (Table 2). Difficulties in interpreting such ratios and relating them to stove design needs were emphasized in Chapter 8. It is important to gain familiarity with such data, however, as building blocks for "whole family" requirements.

An important methodological concern in fuel estimation is exemplified by Suliana Siwatibau's tests, reported in Fiji data in Table 2. She explicitly includes boiling water for tea. Including water tends to reduce the fuel/food weight ratio due to the relative ease of heating water compared to cooking food (thus the numerator increases less than the denominator). Including water tends to increase the energy ratio due to the extra work required to heat the water (which has zero caloric value). Thus adding tea water in the menu acts to make the two ratios diverge in value. Water is an extreme example of how food mix can affect these ratios, but it also nicely exemplifies how the menu affects these numbers.

Economies of Scale. Only limited evidence on scale efficiencies is available. Islam's data on potatoes have already been cited. Similar data on rice cooking have been presented in Chapter 8, Tables 4 and 5. They illustrate the importance of examining scale effects for quantities actually in use. In the households monitored by Geller, poor correlation was found between efficiency and quantity of food cooked (1982). In households in four Nabagram villages, however, per capita fuel consumption decreased as household size increased (Chapter 2, Table 3). Estimates of the economy of scale coefficient b, calculated for the power function $y = ax^b$, where y = per capita fuel consumption and x = number of family members, are -0.41, -0.42, -0.43, and -0.65 in these four villages respectively. Taking the four villages together, mean per capita fuel use by households with 7 to 9 members was 36 percent below the mean per capita use by households with 2 to 4 members. Further research on economies of cooking scale is evidently required in order to clarify the present ambiguous findings. The scale factor must be understood (a) for its implications in stove design, (b) as a factor in estimating household fuel requirements, and (c) in order to control or adjust for scale when evaluating other efficiency or use variables.

The ratios and estimating factors presented in Tables 1 and 2 meet our criteria as need indicators. Though they require further refinement and development, they offer groundwork for estimating effects on fuel use of increased food consumption necessary to meet minimum nutrition standards.

Analyzing Fuel Consumption by Source and Access

Apart from the fuel requirement and efficiency factors just discussed, determinants of the amounts and types of cooking fuels consumed per person include

1. forest, animal, and crop mix as fuel resources
2. nearness of source and collection times
3. distribution of land, trees, and animals, by household
4. number, sex, and age of family members
5. occupation
6. per capita household income (including nonmonetized income)
7. user preferences
8. nature of rights to public land: open access, common (specified), restricted
9. proportions of subsistence to commercial farming and noncommercial to commercial fuels
10. road transport and related infrastructure development
11. competing urban and industrial demands for fuels
12. fuel prices and interfuel substitution.

Rarely is it possible to sort out the separate influence of each determinant on cooking fuel consumption. Linkages and weights among these factors are integrative, and quite distinctive in each village or farming

TABLE 2.
Indicative Fuel to Food Ratios

Location and Combustion Conditions	Food Consumption	Fuel to Food Ratio		Diet	Reference
	kcal/cap	kg/kg	kcal/kcal		
Field Data					
India	2,492	1.93	2.2	Wheat, rice, pulses, sugarcane, meat, gram, maize, potatoes	Agrawal, 1981*
	2,759	1.77	2.0	Wheat, rice, jawar, milk, pulses, bajra, barley, potatoes	"
	1,948	1.96	2.0	Wheat, rice, pulses, milk	"
	2,734	2.39	2.1	Wheat, pulses, milk, rice, gram	"
	3,329	2.51	2.7	Wheat, jawar, gram, bejhar, pulses, rice, milk, meat	"
Bangladesh	1,600		3 (ave.)	0.41 kg/grain/cap/day; fish	Briscoe, 1979
Experimental Data	kcal/meal				
Smokeless stove (wood)	2,491	0.98	1.16	Rice, avare, ragi flour	Dutt, 1978
Smokeless stove (wood)	2,842	1.13	2.13	Rice, cowpea, ragi flour, vegetables	"
Smokeless stove (wood)	2,763	1.5	2.0	Rice, cowpea, ragi flour, greens	"
Smokeless stove (bamboo)	3,427	2.8	3.5	Rice, cowpea, ragi flour, vegetables	"
Smokeless stove (bamboo)	3,563	2.7	3.9	Rice, toor dal, cowpea, ragi flour, eggplant	"
Smokeless stove (casurina)	2,956	2.3	4.2	Rice, cowpea, potato, onion, ragi flour	"
Open stove (wood)	1,184	1.6	1.9	Rice, avare, ragi flour	"
Open stove (bamboo)	2,898	2.3	2.7	Rice, cowpea, ragi flour	"
Open stove (bamboo)	3,444	1.1	2.3	Rice, toor dal, cowpea, ragi flour, greens	"
Open stove (casurina)	3,194	1.4	2.5	Rice, cowpea, ragi flour, greens	"

TABLE 2, continued.

Location and Combustion Conditions	Food Consumption	Fuel to Food Ratios		Diet	Reference
FIJI					
Open fire (dogo wood) Siwatibau, 1981**	2,207	.52	2.56	Fijian+	"
	1,983	.58	2.9	Indian++	"
	2,696	.87	1.6	Indian‡	"
	3,568	.38	1.3	Chinese‡‡	"
	3,935	.26	0.81	Chinese#	"

*Excludes making tea and water for preparation of rice and curry.

**Includes making tea; excludes water for preparation of rice and curry

+Tea, coconut cream, fish and cassava or cassava, fish, tea.

++Tea, rice potatoes and fish or rice, fish curry, and tea.

‡Flour, tea, potatoes and beans or roti, fish curry, tea.

‡‡Tea, meat, rice, carrots and beans, or rice and beef chop suey.

#Chicken, beef, carrots, beans and cabbage or pot roast chicken and beef chop suey.

TABLE 3.
Annual Firewood Consumption as Cooking Fuel, Selected Villages

Country and Village	Firewood Consumption for Cooking (GJ/Person)	Reference
INDIA		
PURA	9.7	Ravindranath and others 1978
PHILIPPINES		
VERDE ISLAND	8.3	Manibog 1979
FIJI		
NATIA	8.4	Siwatibau 1981
YAROI	8.0	"
NACAMAKI	7.5	"

system. Rates of change differ among factors, sometimes substantially. Moreover, many of these variables are not defined in standard terms yet, nor are firm data available. Therefore these determinants must be considered in varying combinations, in the attempt to understand their influence.

Accordingly, the data presented in Tables 3 and 4 on village cooking fuel consumption are viewed as diagnostic indicators. With the numbers, we attempt to associate the factors and circumstances cited in the reference studies. This is not possible in each case, because of the aggregate nature of the data. As a standard practice we recommend: diagnose the numbers before using them to diagnose the situation. The data in these tables are from eight village or microregion studies in which the design and size of sample as well as care in household data collection resulted, in our judgment, in reliable estimates.

Table 3 covers five villages or rural settlements where cooking fuel consumption is almost entirely confined to firewood. Table 4 covers 33 villages or areas using from three to five principal types of domestic cooking fuels. The energy units used in the tables are calculated from the heat values of fuels reported in the cited studies, keeping each set of data as close as possible to its context. These values were in some instances established by bomb calorimetry tests. In others, the study adopted the conversion rate considered most suitable considering species variation and typical moisture content at the place and time of study.

For annual cooking fuel consumption a central tendency of 7.5 to 9.7 GJ/person is observable in Tables 3 and 4. This central range includes nine villages and areas in nations as diverse as Bangladesh, India, the Philippines, and Fiji. The annual mean for seven of these villages is about 8 GJ/person. The villages in this range are not highly commercialized and are principally dependent on biomass fuels. Pending future data

TABLE 4.
Annual Cooking Fuel Consumption by Source, Selected Villages

Country and Village or Village Area	Fuel Source						Reference
	Firewood	Dung	Crop Residue	Green Plants	Other	Total	
	(GJ/Person)						
BANGLADESH							
ULIPUR	1.4	0.2	4.1	0.5	0.8	6.9	Briscoe 1979
NABAGRAM (23 villages)	3.0	0.2	1.2		0.5	4.9	Islam 1980
KULAGHAT	4.2	0.2	3.1		0.6	8.1	Quader, Omar 1982
INDIA							
NARAICH	2.4	6.1	3.9	0.8		13.4	Agrawal 1981
PINDARI	0.1	4.9	3.1	0.6		8.7	"
PATHARHAT	4.8	2.2	0.3	0.9	0.1	8.3	"
HARIHARPUR	3.2	4.2	0.7	0.2		8.3	"
HAZIPUR	1.2	2.8		1.5		5.5	"

development, this range centering on 8 GJ/person for annual biomass cooking fuel consumption may be taken as a point of departure for diagnosing situations with much higher or lower average cooking fuel use.

A comparative picture for villages in West Java is shown in Table 5. For households using biomass or mixed fuels (biomass and kerosene), a central range from 7 to 9 GJ/person per year is observed. The annual mean for households in all districts, 8.8 for biomass users and 11.9 for mixed fuel users, is influenced upward by high use in Bandung district associated with higher per capita village income and larger number of households in modern and transitional villages (Chapter 3).

For whom are these diagnostic numbers useful? The procedures demonstrated in Chhoprak suggest that, in the first instance, they should be used in a participatory process of diagnosis and planning with local residents. The kg/kg ratio may better serve this discussion than GJ, but not without reference to heat values of different fuels and to the concept of standard "units" for measuring heat. Different terminology may be necessary in the different contexts. The complexity of factors involved in the types and quantities of cooking fuel consumed per person in each local area make this process indispensable for valid assessments and effective planning. Open discussion among community members often can reveal the structure of weights and linkages among these factors.

In examining such interactive factors, we present situations and numbers in a manner intended to be useful to rural residents and to researchers for local as well as regional assessments.

TABLE 5.
Annual Cooking Fuel Consumption by Source, West Java Villages

District	Biomass	Biomass and Kerosene (GJ/Person)	Kerosene
Bandung	12.7	15.3	4.6
Ciamis	9.3	9.3	3.6
Garut	8.4	7.1	3.9
Serang	7.4	7.0	5.5
Cirebon	4.9	7.0	5.2
All households	8.8	11.9	4.6

Source: Chapter 3, Tables 12-14.

TABLE 6.
Ownership of Fuel-Producing Assets Per Family, Ulipur

	Hindu Fishermen	Muslims Landless	Muslims Poor	Muslims Medium	Muslims Rich
Number of Families	8	14	11	8	8
LAND (decimels*)					
Median	0	8.5	66	127	242
Mean	0.7	9.5	65	135	296
TREES					
Median	0	6	8	16	182
Mean	2	11	12	18	209
CATTLE					
Median	0	0	0	1	4
Mean	0	0.3	1.3	1.3	2.6

Source: Briscoe 1979.

*One decimel = .01 acre = .004 hectare

Land Distribution and Cropping System. Ulipur in Bangladesh and Hazipur in Uttar Pradesh, India (two of the villages with low average cooking fuel use in Table 4), have extremely low firewood consumption and are heavily dependent on crop residues (Ulipur) and other vegetation (Hazipur). Hazipur has much lower cooking fuel consumption than two other villages in the same agroclimatic zone, Patharhat and Hariharpur. An examination of land, tree, and cattle distribution together with cropping patterns and use of crop, animal, and tree products can provide an adequate explanation of such diversity.

Terraced farming and mixed trees, Gorkha District, Nepal. (D. Bajracharya)

The ownership of fuel-producing assets by family in Ulipur is shown in Table 6. In this village, the land farmed provides only 46 percent of the food eaten but provides 75 percent of the fuel used and 100 percent of the fodder consumed by the small numbers of cattle (John Briscoe 1979). Some 53 percent of the fuel used by rich families and 37 percent used by middle-income families are agricultural by-products in overall short supply but stocked by landowners for use in the wet season before harvest of the main (monsoon) rice crop. During the wet season, landless and poor families face an acute fuel shortage that they attempt to meet by collecting twigs from beneath the trees of the rich and breaking off small branches from trees planted along paths (Briscoe 1979).

Low energy consumption in Hazipur is associated with an extremely limited land resource base. No families with large landholdings were reported. Landless workers make up 34 percent of sample families. Marginal holdings (as defined in Chapter 6) account for 49 percent of agricultural land, small holdings the rest. Animal distribution is less unequal than land distribution, however. Laborers and farmers with less than 2 ha hold 67 percent of animals compared to 49 percent of farm-

Folk and place, northern Nepal slope. (D. Bajracharya)

land. Their dung cake consumption for fuel, 45 percent of the sample total, is substantially less than their share of animals. An interesting pattern emerges. To gain cash income, 32 percent of the dung produced is sold for compost and 11 percent for dung cakes. Gathering of other vegetation (green plants) accounts for 29 percent of total fuel consumption (Sushil Agrawal 1981). At this low level of fuel use per person, green plants take on an implicit positive value, equivalent to the cash value of dung sold less the difference in whatever value is ascribed to collection time for the two biomass materials.

Related Resource Constraints and Potential. Even among villages with the average per capita cooking fuel consumption in Table 4, the mix of biomass fuels varies widely. The quantities of different bioresources available in the village are major factors influencing this mix. Naraich and Pindari villages, in an agroclimatic zone characterized by extensive farming, have more land, cropped area, and animals per person than the other three, intensively farmed villages in Uttar Pradesh. High dung cake and crop residue use in the former two villages is attributable to this resource factor. Higher numbers of trees per person in Patharhat and Hariharpur permit higher firewood use. Competition among crops, animals, and trees for land and water use is critical in farm families' decisions about these resource factors (see below).

Only a few studies have attempted to assess, for a particular area, the pressure of fuelwood collection on land quality by comparing estimates of wood consumption and net forest growth. Bajracharya (1980) places this issue in the context of food needs by sorting out different settlement and land use patterns in a Nepal hill village panchayat. He maps proximity of residences and farmland to different forest areas to begin to analyze food and fuel demand and supply in each area. In this manner, he is able to identify household groups and locations with severe food problems and to examine their associated fuel regimes. He concludes that for the panchayat as a whole, the most pressing need is food, not fuel. Scattered wooded areas (distinct from forests) are under slight fuelwood stress, and demand for fuelwood in the poorest sections of the village is likely to heighten the deforestation process. The detailed texture of geographic and land use analysis needed to establish such patterns is evident. Also evident is the need for systematic compiling and analysis of diagnostic indicators, as demonstrated below, to permit a start on generalizing the findings of particular studies.

In Pura (Table 3), where cooking is done almost entirely with firewood, average two-way distance walked to collect the 10 kg of wood used per family per day is 4.8 km. Fuelwood collecting here does not appear to contribute to deforestation. The real problem is the enormous time spent collecting (Ravindranath and others 1978).

Preferences, Income, and Fuel Availability

> "Smoke from these *babool* twigs and branches bothers me. If we had 2 or 3 *bighas*, I'd be happy."*
>
> > — Babuti, mother of landless family of eight members, Garhi Kherwa, Uttar Pradesh, India. March 1982. (Varun Vidyarthi, pers. comm., 1982)

The cynicism in Babuti's voice said, "If we had land, what couldn't we have?" . . . including better fuel. Attentively and cumulatively recorded from many people, statements such as hers add up to patterns of need and preference.

Babool: *Acacia nilotica*. *Bigha*: approximately 850 m^2.

Work, Gujarat. (Kirk R. Smith)

Fuel preferences are hard to identify and evaluate, nonetheless. Even so, a Nile delta study recorded three clear preference levels among biomass fuels. Considered prime in this area is animal dung in a mixture locally termed *gella*, made with wheat or rice chaff, straw, and ashes. This gives the highest heat and least smoke. Dung is being increasingly regarded as more valuable for fertilizer, however. Next in desirability are dried cotton stalks. They give good heat with little smoke or ash but are clumsy to store and handle. Least desirable fuels are rice and corn wastes, twisted into knots. They give average heat, much unpleasant smoke, and much useless ash (Salah Arafa, Cynthia Nelson, and Samira Amin Megally 1980).

User preferences in Bangladesh for different fuelwood species are reflected in price differentials. Tamarind, the most favored fuelwood, costs 20 percent more than the next good quality wood (*sundari*) and is priced 50 to 70 percent more than ordinary grade firewood (Islam 1980). When researchers are able to develop a relationship with fuel users over time, they are able to assess preferences accurately; the grades in Chapter 2, Table 5, were evolved this way. The preference order in Uttar Pradesh is firewood, dung cake, crop residue, and green plants, excepting dung cake's preferred place in the slow boiling of milk (Agrawal 1981).

The degree of disutility or penalty experienced by families or groups who are forced to use inferior fuels is not easily indexed. Use of low-quality fuels by landless and small farm holders was reported in Chapter 2 (Tables 4 and 5). That use indicates suppressed demand for fuelwood, which however is not readily estimable without an indicator of strength of preferences. Compilation of local quality attributes of fuels would be worthwhile. From these, a lexicon could be prepared and also a scale of user preferences in different areas, as a basis for more informative consumer assessments (see David Brokensha and Bernard Riley 1983).

West Java villages are the only areas covered in these microstudies that use substantial quantities of kerosene for cooking. Chapter 3 (Tables 8 and 12) shows kerosene use for cooking in West Java increasing with per capita income. Substantial increases in per capita firewood consumption in West Java with increases in income were also recorded.

Such data on the effect of per capita income on per capita consumption of different fuels are limited. This is partly because researchers often tabulate fuel consumption by household income classes. This obscures per capita relationships. The number of family members is often positively associated with family income as shown in several studies (Agrawal 1981, James Douglas 1982). Economies of scale, as we have seen, may tend to reduce per capita cooking fuel use in larger families. Income effects might tend to pull up per capita consumption of preferred fuel or pull down an inferior fuel's per capita consumption. These factors are impossible to sort out when numbers of family members are hidden in household income classes. Future studies should cross-tabulate per capita use of cooking fuels by number of family members, on one hand, and by household per capita income on the other. An underlying essential step in data handling is to preserve each set of data in its original units, so that follow-up analysis of anomalies or interesting relations can be done.

Subsistence, Transition, and Commercial Areas. Location and degree of commercialization also affect types and amounts of cooking fuel consumed. This is brought out most clearly in the West Java data discussed in Chapter 3. In more remote, forested, and less densely populated *kabupatens* (in Ciamis, Garut, and Serang districts) most households use only firewood for cooking. Significantly lower amounts of kerosene are used, and by lower proportions of households. Crop residues in these areas tend to substitute for or supplement firewood and are used more often in households that do not cook with kerosene (Raymond Atje and James Tarrant 1980).

The greater use of kerosene in densely populated districts with good roads and closer access to cities (Cirebon, Bandung) is associated with

"Except as a person handled an ax," Gujarat. *(Kirk R. Smith)*

lower kerosene prices. While mean kerosene prices in Cirebon and Bandung were 52 and 60 percent above the official price, respectively, kerosene prices in Garut were 76 percent and in Ciamis 104 percent above the official rate. Centrality and complexity of villages interact with these regional factors, creating subpatterns within regions that are analogs — but require specific identification — of these broad subsistence/commercial contrasts between regions (Bruce Koppel 1980).

Energy densities of biomass fuels are low, though varying. Proximity to supply is therefore important. Transport cost is high relative to energy value. Within a rural area, availability and allocation of work time of family members to collect these fuels can be a limiting factor in their use. Values placed by family members on work time are part of the value built into different quantities and forms of biomass, whether collected for own use or for exchange through nonprice or price mechanisms. Experiences of scarcity and associated perceptions of value lead to the question, "In a particular place or season, which fuel or biomass resource sets the opportunity cost against which a new or substitute fuel or technology must compete to be viable?"

WOMEN'S AND MEN'S VALUATION AND ALLOCATION OF TIME

> We women have this kind of work — collect fuelwood, do wage labor, return late in afternoon (one o'clock), then cook food, clean utensils, go for wage labor again, return by evening. Water has to be pulled up. This is very troublesome. Water is very deep in the well. It pulls out our heart. . . . We get up at four in the morning and start grinding.
>
> — Kodya, landless agricultural worker, Garhi Kherwa, Uttar Pradesh, India. December 1982.

Varun Vidyarthi, in quoting Kodya and her neighbor Meena, observes:

> Pulling water from a well is a back breaking job as there is no proper method for lifting water. A log is put over the well and women literally risk their lives standing on it and pulling the bucket by a rope directly. It is interesting to note that pulling water for domestic purposes is a task performed by women exclusively. To quote a male villager, "Men will die of thirst but shall not pull water." This may not be entirely correct. But it does indicate the trend. (Vidyarthi, pers. comm., 1982).

Cooking and fuel gathering, though central, are only part of the wider set of essential food preparation activities, including processing, storage, and water gathering. Food preparation tasks are largely outside of the formal economic sphere and are usually done by women. Thus, these vital activities and their articulation within the family production unit have received inadequate attention relative to their importance in the survival of millions of rural poor.

To better understand rural energy systems and opportunities for improved rural energy technologies, obtaining improved knowledge of the food preparation sphere is crucial. Problems of cooking fuel availability and implications for the nutrition of rural people and for their time and work allocation are often compelling by themselves. Improved stoves, village forestry, and biogas systems have been examined in this context. But cooking is only one part of the entire food production (agricultural) and food preparation (domestic) chain of activities that form the backbone of the rural economy. As such, analysis of cooking alone is incomplete and may be misleading unless delineated within the more encompassing framework of the food preparation sphere.

Widening the frame of reference beyond cooking and fuel collection illuminates other constraints that are, in some cases, more important and more limiting to the user concerned. In food processing and storage, for example, constraints of time, weather, and energy can cause food losses, a significant hazard to subsistence and marginal families. The time and energy necessary to collect water have implications for sanitation and spread of disease through food. In many locations, the time spent gathering water overshadows time spent gathering fuel. Finally, these activities and their sum comprise a time expenditure schedule with notable implications for the availability of women's time for child care (Sue Schofield 1979, J.C. Waterlow 1982), for alternative production activities, for education, and for possible labor allocation to experiment with new technologies. The labor demands and existing technologies of these subsistence activities are principal determinants of female productivity. Because this labor is used in essential survival activities, the social strictures limiting the availability of alternative work opportunities for women are reinforced.

Time is the most important resource available to poor people, according to Julie Da Vanzo and Donald Lye Poh Lee (1978). Therefore, naturally, a positive valuation should be ascribed to work outside of the formal market sector. We have suggested, in a basic sense, that knowledge is an even more important resource. These are important considerations in any analysis of nonmonetized or partly monetized economies, either on the scale of the village or of the household. Recognizing nonformal market activity as "productive" and worthy of economic examination has become more acceptable in recent years, but methods of determining values of work remain elusive. The joint production of human capital (childbearing and rearing) with household or agricultural production obscures

Sugarcane crushing, Gorkha District, Nepal. (D. Bajracharya)

efforts at proxy valuation for any single activity. The limitation of women's access to activities with monetary remuneration, both due to their commitment to child care activities and to strictures on their outside productive activities, results in understatement of the value of women's productivity. A more complete understanding of the allocation of women's time including child care provides a means by which the food preparation sphere may be delineated in terms of energy, productivity, drudgery, employment — in sum, the opportunity costs of a woman's time.

Understatement of the value of women's labor time is an understatement or reduction in the value of total labor time of men and women in the village. If a woman's labor is regarded as free or approaching zero, the valuation of all labor is undercut. Opportunity costs of women's time therefore become a central factor in the valuing of men's and women's employment, and in the choice and management of alternative technologies or resource combinations in which labor inputs are important. Women's time commitments also condition their ability to engage in entrepreneurial activities and devise more productive opportunities.

On the other hand, if severe unemployment or underemployment is characteristic of the rural area, release of women's work time would add to the labor supply in a manner which could, in a relatively stagnant economy, put further downward pressure on wages. In either case,

TABLE 7.
Women's Time Allocation for Food Preparation Activities (Hours/Day)

Village or Region	Food			Fuelwood Collection	Water	Source
	Cooking	Processing	Cooking and Processing			
W. Sumatra, Indonesia	2.69	n.a.	n.a.	n.a.	n.a.	Wulfe, 1982
Pakistan	1.75	3.5	5.25	n.a.	0.50	Anwar and Bilquees, 1976 (in Dixon 1978)
Bangladesh	4.0	4.85	8.85	0.25	0.50	von Harder 1977 (in Whyte and Whyte 1978)
Nepal	2.05	0.97	3.02	0.38	0.67	Acharya and Bennett 1981 (in Tinker 1981)
Pura, South India	3.0	n.a.	n.a.	1.2	1.2	Ravindranath et al. 1978
South India		1.82*	3.65	.66	1.24	Reddy et al. 1980
Uttar Pradesh, India						
Patharhat			1.5	1.2		Agrawal 1981
Hariharpur			4.3	5.2		"
Hazipur			5.1	1.0		"
Pindari			6.4	3.8		"
Naraich			3.3	3.9		"

N.A. = information not available
*Includes carrying food to fields, adjusted to remove other chores.

Cultivating tobacco, Gujarat. (Kirk R. Smith)

knowledge of time allocation is essential to an assessment of family and village entrepreneurial and employment prospects (see Robert and Pauline Whyte 1982).

Farm Household and Labor Household Time Economy

The time required for cooking, food processing, and for fuel and water gathering varies by region, class, and caste (Ruth Dixon 1978). The type and amount of staple food consumed, methods used for processing and cooking, as well as proximity and accessibility of fuel and water supply, access to child and servant labor, and work performed for others or in agricultural operations, contribute to a particular household's time use pattern. Other considerations are allocation between men and women of farm tasks, the nature of other traditional activities, as well as commercial opportunities for female household members. The variation between households and regions can be significant, but Table 7 looks beyond this divergence and speaks to an underlying similarity, namely, the preponderant amount of time spent in food preparation activities.

Together, cooking and food processing consume from about 3 to 6 hours of a woman's day in most of the villages cited in the table. To these we add the collection time for water and fuel, which gives us a range from 4.5 to more than 9 hours per day on food preparation activities. Significantly, food-processing time rivals cooking time in the studies cited, and the collection of water outweighs the collection time for fuel. This implies a need to address the labor requirements of food processing and of water supplies in addition to those of cooking and fuel gathering.

A wide range of time expenditure in different food preparation activities can be seen in Table 7. Toward the higher end of this range, the opportunity cost of labor probably increases. As more time is spent in food preparation, less time is available for alternate productive activities. In an Andean community in Peru, an upper limit apparently exists to the time available for food preparation activities as a whole. As the time allotted to one of these activities, namely, firewood collection, increased, time available for cooking decreased. This appeared to have observable negative effects on nutrition (Sarah Lund Skar, Nelida Arias Samanez, and Saturno Garcia Cotarma 1982).

Long working hours in food preparation not only limit the alternative productive activities that women may benefit from but could limit effective participation in tasks possibly prerequisite to such innovations as biogas, social forestry, high-yield grain varieties, education, and others. These innovations all require labor inputs not possible for already overburdened women, or may simply add to their work day. Thus, it is vitally important that the time use requirements of innovations be understood and coordinated with the existing time use constraints of people whose inputs they require. Innovations that adequately address the time-intensive labor of women, including perhaps the food preparation activities just discussed, can potentially reduce drudgery and free time for other activities. This "liberated" time could be used productively in improved child care; in livestock management, agricultural production, fish rearing, and gardening to augment the family's food and nutrition supply; or in home industry or craft production, trading, and wage labor opportunities to improve the family cash income. For these potential increases in productivity to be realized, women must be permitted access to resources and organizational as well as informational inputs.

Technological advances that address some of these labor-intensive activities are cited frequently as having had adverse effects on the earnings of village women who pounded rice or pressed oil for sale to their neighbors (Dixon 1978; Kathleen Cloud and Catherine Overholt 1982). Notably, these women are often among the very poor. Ruth Dixon suggests that these grain-processing technologies be under the control of

Making ghur, Chorkate, Nepal. (D. Bajracharya)

women and on a village industry scale. They could take advantage of economies of scale, buy in bulk or at harvest, manage inventories to gain advantage of price changes, and preserve rural female employment while upgrading the quality of work.

Seasonal factors are specially important, here. Peak labor-days for three crops — *aus* rice, *aman* rice, and jute — coincide in May and June as ascertained in a recent Bangladesh rural energy study (A.K.M.A. Quader and K.I. Omar 1982). Peak labor requirements — for men and women — do not exceed available person-days of farm family and hired labor for any individual crop. But the sum of person-days required for the entire crop mix exceeds available labor by more than 30 percent in May and by 15 percent in June. Work allocation data for one crop alone would not reveal this issue. Only in the context of the cropping system that actually obtains in the village is this stress revealed.

Women in landless or marginal farm households most probably have work and earning regimes sharply different from those of women in farm families on small and viable landholdings. More diverse, productive earning opportunities are a central concern for that former group. The time allocation window is thus likely to open some central resource allocation and access issues. Near-subsistence farm families who farm part-time and seek other earning opportunities part-time would be particularly motivated to discuss these issues. Such examination of work in the informal sector can lead to effective dialogue in which the knowledge and expressed needs of rural women help in identifying development alternatives.

IDENTIFYING FARMING SYSTEM ENERGY OPPORTUNITIES AND CONSTRAINTS

> Husband, cultivating tomatoes: "We use much manure on these beds near the house, but not much on the wheat and paddy fields outside the village."
>
> Wife, lifting water for plot with rope bucket: "We have two bullocks and a buffalo, but burn almost half their dung for cooking, even though we get wood chips from the carpenters nearby."
>
> Researcher: "How many cycle rickshaw loads of vegetables do you send to the market each week?"
>
> Grower: "Usually a couple a week."
>
> Researcher: "What do you use the waste leaves for?"
>
> Grower: "Mostly we put them on the compost pile for later use on the vegetable plots."

Decision questions: Would adding the waste leaves to the dung provide enough inputs for a family-size biogas plant? Would dung and vegetable wastes from the 20 growers nearby be enough for a neighborhood digester?

—Patharhat, Uttar Pradesh, India.
April 1981.
Richard Morse (field notes).

Diagnostic indicators from which such local resource users and decision makers can gain new information and perspectives make up the focus of this section. An important input to these decisions is better knowledge of potential surplus elements in local resources and of critical limiting factors. Energy as a key limiting factor may block increased farm output if (a) its quantity is inadequate; (b) its form or quality is inappropriate, requiring a technical transformation not possible utilizing local knowledge and resources; (c) it is not available on a timely basis; or (d) it is held by certain families or groups but could be used more productively by others. Potential energy surpluses exist when (a) little or no private or social cost is entailed in collecting or processing abundantly available biomass resources, (b) energy use is inefficient, through either management or technical factors, (c) energy is consumed wastefully, or (d) it is used for purposes that a significant number of local residents regard as having low priority. Stimulated by new information, introduction of new ways of thinking about these end-use, resource, organizational, and allocative aspects can help local groups establish priorities and evaluate new resource and technology strategies.

Local Systems Assessment: Indicators of Scarcity and Potential

A resource assessment tool must be in a form usable by, and with data relevant to, the decision makers directly concerned — here, rural residents. If the tool and resulting data also are to have wider utility, they must be adaptable to situations at least somewhat different from the local one. This implies as well that the tool should possess the content needed to define the conditions of such differences. Borrowing from the business world, "ratios of manufacturing" are an example. "Return on sales" and "sales turnover" can be used by a small or large industrial company to analyze its own marketing and distribution policies; they also can be compared with those of firms in the same or different performance ranks in the specified line of industry to see what factors may be responsible for better or worse performance.

To illustrate development and use of such diagnostic indicators, fuel

TABLE 8.
Patterns of Domestic Energy Consumption, By Fuel Type,
Five Uttar Pradesh Villages

Village	Population (sample)	Domestic Energy Consumption per Person, Monthly (Kcal x 10³)	Consumption by Type of Fuel (percentage)				
			Dung Cakes	Crop Residues	Firewood	Spring Plants	Leaves and Twigs
Pattern 1: Dung Cakes, Crop Residues							
Naraich	289	267	46	29	19	6	1
Pindari	863	174	56	35	1	7	0
Pattern 2: Firewood, Dung Cakes							
Patharhat	1,394	166	26	3	57	10	3
Pattern 3: Dung Cakes, Firewood							
Hariharpur	648	165	50	8	38	3	0
Pattern 4: Dung Cakes, Spring plants, Firewood							
Hazipur	332	113	49	0	22	25	4

Source: Agrawal (1981).

use and farming system characteristics in five north Indian villages are assessed. Contrasts between the two subregions in which these villages are located are revealed by their profiles of domestic fuel consumption (Table 8). The two villages Naraich and Pindari are in the Trans Jamuna agroclimatic zone characterized as "alluvial veneer," with hill outcroppings and ravines. Significant use of crop residues is a distinctive feature of their high cooking-fuel consumption. The other three villages are in the alluvial Ganga-Ghagara agroclimatic zone. Each of the three exhibits a distinctive pattern both in amount and types of fuel used, having its own mix of dung cakes, firewood, and green plants (Agrawal 1981).

Village-scale indicators are used to analyze connections between these fuel use profiles and the local farming system (Table 9). Unfortunately, separate indicators for each landholding and income group are not available. However, data on the distribution of land and animals reveal that different occupational and income groups combine farming and fuel resources differently in each village.

Medium firewood and high dung cake consumption by a divided community, Hariharpur, are analyzed to exemplify use of diagnostic indicators for planning and policy analysis. Even though firewood provides 38 percent of village cooking fuel, scarcity of this fuel is shown by scrutiny of village indicators. The weak firewood supply position is evident in numbers of trees owned per family (1.2) and per hectare (0.4). To identify the nature of pressure on firewood and other resources, however, distributive and geographic aspects of landholding and use must be examined.

Land distribution in Hariharpur is highly unequal. Sixty percent of farm families hold 19 percent of the land in holdings of less than 2 ha. Thirteen percent of the farmers hold 50 percent of land in holdings of 5 ha and larger. Landless labor make up 48 percent of the occupational distribution (Agrawal 1981).

Shifting stream beds inhibit sustained land management in sections of the village land. Low cropping intensity (index 0.99 in Table 9) and a small irrigated area (38 percent) are resulting features. High wheat and paddy yields are evidently associated with high fertilizer and manure use on cropped land, but the low cropping intensity and relatively small number of families in agriculture in Hariharpur (only 34 percent) result in very low grain output per person per year (133 kg) and a correspondingly low share of grain marketed (11 percent).

The high numbers of cattle in Hariharpur per farm family (5.1) and per landless family (2.4) are more equally distributed than land, however. Landless families own about 34 percent and small-scale farmers 30 percent of the bovine animals. High annual fodder consumption per bovine

TABLE 9.
Village Farming System and Domestic Fuel Consumption Indicators, Five Uttar Pradesh Villages

Village	Population (sample)	Population below Rs. 2,400 annual family income (%)	Share of farm assets, families below 2 ha		Cooking fuel consumption per person (10^3 kcal/mo)	Gross cropped area per person (ha)	Irrigated to gross cropped area (%)	Multiple cropping index	Grain production per person (kg/yr)	Crop residue output per person (kg/yr)
			Land (%)	Animals (%)						
Profile 1: Dung Cakes, Crop Residue										
Naraich	289	22	8	32	267	.48	30	1.04	516	546
Pindari	863	21	12	31	174	.38	65	1.08	356	413
Profile 2: Firewood, Dung Cakes										
Patharhat	1,394	27	41	71	166	.20	100	1.64	263	170
Profile 3: Dung Cakes, Firewood										
Hariharpur	648	32	19	64	165	.15	38	.99	133	158
Profile 4: Dung Cakes, Spring Plants, Firewood										
Hazipur	332	40	49	67	113	.17	93	1.79	241	218

Village	Dung cake fuel use per person (10^3 kcal/mo)	Grazing area per bovine animal (ha)	Cattle per person (No.)	Buffaloes in total livestock (%)	Fodder consumption per bovine (100 kg/yr)	Milk output per buffalo (kg/yr)	Dung recovery per animal person (100 kg/yr)		Share of dung used as:		Manure per gross cropped hectare (100 kg/yr)
									Manure (%)	Fuel (%)	
Profile 1: Dung Cakes, Crop Residues											
Naraich	122	.51	.53	21.2	14.4	208	10.0	5.3	24	75	2.8
Pindari	98	.70	.31	20.2	20.8	218	16.9	5.2	29	71	4.0
Profile 2: Firewood, Dung Cakes											
Patharhat	43	.02	.22	17.8	32.8	865	17.5	3.9	56	43	10.8
Profile 3: Dung Cakes, Firewood											
Hariharpur	83	.07	.33	22.7	39.0	1,060	21.0	6.9	42	40[1]	19.8
Profile 4: Dung Cakes, Spring Plants, Firewood											
Hazipur	56	.00	.34	30.0	44.8	1,465	23.3	7.9	29	28[2]	13.4

TABLE 9, continued.

	Fuel Consumption Per Person		Fuelwood consumption per person (10^3 kcal/mo)	Trees Owned		Percentage of production marketed		Dung Cakes purchased (%)	Families purchasing fuelwood (%)	Price Ratios	
	Crop residues (10^3 kcal/mo)	Spring plants (10^3 kcal/mo)		Per Farm family (No.)	per hectare (No.)	Grain (%)	Milk (%)			dung cake/ manure (P_d/P_m)	fuelwood/ dung cake (P_f/P_d)

Profile 1: Dung Cakes, Crop Residues

Naraich	78	16	48	2.5	0.5	42	0	17	3	1.33	1.69
Pindari	61	12	2	1.8	0.4	39	45	3	0	1.50	1.79

Profile 2: Firewood, Dung Cakes

Patharhat	6	.17	96	4.9	4.4	26	58	7	3	1.25	2.00

Profile 3: Dung Cakes, Firewood

Hariharpur	14	4	63	1.2	0.4	11	52	4	19	1.29	2.44

Profile 4: Dung Cakes, Spring Plants, Firewood

Hazipur	0	29	24	5.4	4.0	19	75	18	2	1.10	2.00

Source: ERD based on Agrawal (1981).

Note: At assumed efficiencies, technical rate of substitution fuelwood for dung = 1.64.

[1] Nine percent used in biogas plant, 7 percent sold as compost, 2 percent sold as dung cake.

[2] Thirty-two percent sold as compost, 11 percent sold as dung cake.

animal (3,900 kg), dung recovery per animal (2,100 kg), and dung recovery per person (690 kg) sustain the role of dung cake as the most used cooking fuel in Hariharpur (50 percent). Relatively high application of manure per gross cropped hectare (1,980 kg) suggests how intensely this resource is used by marginal and small-scale farmers.

The shares of dung cake consumed as fuel by poor and very poor families, however, are far less than their shares of animals. Conversely, shares consumed by upper income families exceed their share of animals. The high 24 percent of dung cakes purchased is evidence of substantial dung cake sales by landless families to upper income families. Collection of firewood from open access sources is common (28 percent) and evidently helps small-scale farmers retain significant amounts of dung for use as manure.

These actions by different resource holders in Hariharpur offer a basis for reexamining the firewood situation. Firewood's substantial share in fuel consumed by upper income groups, its high market price (Rs. 22 per 100 kg), and the high percentage of families purchasing firewood (19 percent) indicate demand pressure on this limited resource. One value or scarcity indicator, the 2.4:1 firewood/dung cake price ratio, is far above the 1.6:1 approximate energy equivalent value of 1 kg of firewood compared to 1 kg of dung cake. That wood as a fuel is extremely scarce is clear. These figures may additionally reflect a preference of upper income groups for firewood, which in turn reinforces its overall scarcity.

This diagnosis has these implications:

1. Possibilities for afforestation on underutilized and flood-prone land should be assessed. An interdisciplinary team made up of forest, soil, water management, social science, and animal husbandry specialists might work with the community to find out if such land could be utilized successfully.

2. Through discussion with landless and small farm families the team would need to determine whether such land is currently used for grazing their numerous animals. If so, land use rights would have to be confirmed or adapted to encourage selection and management of fodder and fuelwood species suited to these lands.

3. Hariharpur agriculturalists, especially those with small holdings, place high value on the use of dung as manure. They appear to have incentive, therefore, to replace dung as fuel by installing appropriately sized biogas systems, using the residual slurry as manure.

4. Assessment of new means of organizing energy technologies must take into account the large family size (12 to 16 members) of upper income groups in Hariharpur. Like other groups, this segment of the

population evidently needs resource and employment development opportunities. The team should assist in their search for new activities that reduce, not widen, existing segmented structures in the village.

Valuing Agricultural and Animal Residues as Energy Resources

In several village situations described, cattle are held more equitably than land. In those cases, as V.S. Vyas (1981) has observed, distribution of a mobile farm resource (livestock) tends to be more equal than that of land. Mobile resources apparently offer development opportunities that warrant special attention. Thus energy input and output characteristics of livestock are particularly important.

Increasing the ability of people and communities to assign values to bartered or in-kind commodities and services, as a basis for developing strategies for coping with market mechanisms, would be an important organizational resource. This is one purpose for developing value indicators. Also of obvious relevance is the utility of such indicators in conventional project appraisals. In spite of the importance of animal dung, crop residues, and even natural vegetation in rural energy systems, few formal markets exist for buying and selling these materials. Their valuation in terms of a market price is thus difficult to ascertain. Other methods of ascribing value to animal and vegetable residues are necessary to understand the optimal allocation of these resources within a rural energy and agricultural system.

The problems of identifying and controlling the factors that form a basis for valuing dung used as manure have been addressed by Elizabeth Foster (1983). In her analysis, the average physical product of manure is the proxy measure most relevant to the farmer's actual situation. Average physical product was estimated from the few crop experiments that have compared crop outputs over a period of several years on plots with specified applications of manure to control plots without manure. The difference in crop yield was then valued at market price. Value of marginal physical product is the relevant measure for a farmer's decision, but marginal data for the relevant rates of manure application are not available.

From experiments conducted over a 13-year period in West Bengal, India, where rice was grown on laterite sandy-loam alkalite soils, a proxy value for manure of approximately Rs. 20 per 100 kg (1980 prices) was estimated at an application rate of 6,500 kg/ha. This value cannot be strictly compared to manure prices (Agrawal 1981) used to derive the ratios shown in Table 9 because (a) the effect of soil and rice varietal differences cannot be ascertained and (b) the experimental manure application rate was more than three times the average of the village (Hariharpur) with the highest average manure use per hectare. Nevertheless, the proxy value at twice the market price of the limited proportion of manure purchased indicates strongly that this basic resource is being undervalued. On-farm research, while difficult to organize, appears essential to obtain a more reliable set of values for dung in its alternative uses.

Estimation and valuation of crop residues in their competing uses as fodder, fuel, pulp, and building material are also very area-specific. Energy values for a variety of these materials were obtained in a Thailand study (John Arnold and Russell deLucia 1982). Assigning proxy values was not attempted, however. Zero opportunity costs for crop residues have been assumed in a number of studies to illustrate methods of resource and project appraisal (Michael Halse 1979, Meta Systems 1980). However, unless these residues are indeed found to be surplus in a particular area or community, the feasibility of their proposed alternate fodder or energy use cannot be assumed. If they are not in surplus, methods of field assessment incorporating the judgments of farmers and workers who produce and transport these materials and of families who consume them will be required in order to evaluate their opportunity or displacement costs.

In sum, in order to estimate reasonable values for animal and crop residues as energy resources, close interaction is required between specialists in the physical properties and market potential of these materials, on one hand, and their local producers and users on the other.

POLICY IMPLICATIONS OF SUBREGIONAL AND REGIONAL VARIATIONS

What policy and planning issues are affected by local and subregional variations in rural energy sources and use? What levels of variation are significant in connection with different issues? At what points in the planning process is it necessary to take these variations into account? These questions are important in themselves and, anticipating Chapter 12, in attempts to design productive sequences of local and wider scale research. To examine these questions, we review aspects of the energy studies reported in the Bangladesh and Thailand chapters.

The Nabagram study in Bangladesh (1980) covered all households in 23 villages within a compact, relatively homogeneous area of 9 square miles. Variations in biomass fuels consumption among these villages are indicated in Table 10 in ascending order of per capita consumption. The ordering provides evidence of apparent fuel deficits in several villages.

TABLE 10.
Annual Biomass Fuel Consumption, by Village, Nabagram

Village Number	Fuel Use Consumption (GJ/Person)	Village Number	Fuel Use Consumption (GJ/Person)
15	.9	11	4.6
14	1.7	12	4.8
10	2.3	18	4.9
16	3.0	23	4.9
	(2.3)*	20	5.0
21	3.5		(4.7)
8	3.5	19	
5	3.7	4	5.7
22	3.8	9	5.9
1	4.0	17	5.9
	(3.8)	3	6.7
2	4.4		(6.6)
7	4.5		
6	4.5		

Source: Calculated from Islam (1980), Tables 3.1.1 and 3.3.3.

*Figures in brackets denote means for the preceding group of villages.

Mean annual fuel consumption for the entire 23-village population was 4.9 GJ/person, and in a central range of 9 villages was 4.4 to 5.0 GJ/person. We might infer from this narrow central pattern that villages in this range on the whole meet their fuel requirements, at their present levels of food consumption. Since the report did not classify fuel consumption by income, however, it is not possible to determine whether individual households or groups of households in this central range of villages experienced deficits.

Analyzing villages with annual fuel consumption below the assumed threshold requirement range of 4.4 to 5.0 GJ/person throws some light on probable overall village fuel deficits. Those villages are grouped in two classes in Table 10, the first four having apparent severe fuel shortages, the next five with less acute conditions. If the 23-village mean is assumed as a basis for estimating deficits, the shortfall in these two groups of villages is about 350 tons of biomass fuel per year. Compared to total 23-village consumption of 5,600 tons, this is about a 6 percent deficit. For the area as a whole, this suggests that the aggregate fuel shortage is not large. For the deficit 9 villages, however, the required increase in fuel supply would amount to 22 percent of their present consumption. Difficult land use and resource mobilization decisions would be faced by people in these 9 villages to achieve an increase of biomass fuel production of this magnitude.

This analysis, while serving as a guide for specific planning in

TABLE 11.
Annual Biomass Fuel Purchase and Consumption, Nabagram

Number of Villages	Fuelwood Purchased (Percentage)	Biomass Fuel Consumption (GJ/person)
5	0 - 10	3.3
10	20 - 29	4.7
8	40 - 100	5.7

Source: Calculated from Islam (1980).

Nabagram union, illustrates a general issue. Fuelwood shortages in rural situations can sometimes appear to be minimal when examined on an areawide basis. But within these areas there may be pockets of extreme fuel shortages. The potential for rapid exacerbation of these "micro" fuel shortages in the face of increasing population and concomitant pressures on fuel resources may be dramatic. This is especially the case in areas where the average consumption level may already be somewhat in deficit.

Patterns of fuel purchases in Nabagram villages suggest two more features relevant to planning. There appears to be a trend of increasing village per capita fuel consumption with greater percentages of fuel purchased (Table 11). If we associate low consumption and low percentage of purchase with poverty, are we noticing in the first group of villages a suppressed demand for fuel due to poverty? What are the implications with regard to fuelwood plantations? Is purchasing an indication of village fuel deficit? If it is, the prospect for firewood plantations in the latter group of villages may be good since the incentive is evidently present.

The Nabagram statistics are instructive for understanding sources of variation. The coefficient of variation among the 23 villages for total biomass fuel use per capita is 37 percent. Firewood, accounting for 56 percent of biomass fuels by weight, has a coefficient of variation of 46 percent; about two-thirds of the villages consume from 100 to 250 kg per person per year of firewood. Coefficients of variation for twigs and leaves and rice straw — 22 percent of biomass fuel by weight — are 59 and 69 percent, respectively. The remaining one-fifth of biomass fuel is made up of cow dung, jute sticks, rice husks, bagasse, and other agricultural residues. Coefficients of variation for these fuels are 80 percent or more.

These intervillage differences are entry points for further local assessment and planning together with village residents. Such assessments would test, among other factors, whether annual fuel consumption of about 5 GJ/person (in local fuel equivalent) accords with village perceptions of adequacy. A probe of such issues with local residents by forestry, community organization, land use, and stove design specialists is a logical next step.

These differences among 23 villages in a relatively homogeneous area are relevant to policy issues such as the incidence of need; allocation of land and other resources to enhance fuel production; and choices between homestead or village woodlots and commercially oriented plantations. Anticipating later discussion we conclude: "No village is a region."

The experience and data acquired in an in-depth study such as that in Nabagram are a foundation not only for direct use in local planning but also for research training that could provide guidance on how to establish these crucial variations—as well as the observed regularities—by less time-consuming methods. Learning how to make qualitative assessments of such factors in dialogue with village residents and learning how to determine what specific data are needed to evaluate alternatives are perhaps "new" Nabagram tasks in applied policy research.

Intervillage differences in Nabagram pose new questions bearing on the findings of the wood and bamboo consumption study (FAO 1981a). Data in the latter study were not reported by village, or even by *thana* but as averages for six major "regions" (Chapter 2, Table 6). The lowest annual average fuel consumption — 2.95 GJ/person — was in western Region 2. This is close to the level just assessed, among Nabagram villages, as severely deficit. The FAO report breaks down the Region 2 figure into below-average supply of both agricultural residues and fuelwood (FAO 1981a). No policy conclusion is presented, however, on whether this very low figure signifies a serious deficit, and whether Region 2 should have high priority in fuelwood investment programs.

Nabagram is in Region 4. The FAO report compares the Nabagram 23-village mean, 4.9 GJ/person, with the Region 4 survey estimate, 3.8 GJ/person, and finds the former "well within the error range" about the latter (FAO 1981a). Since the former is the mean for a universe of all households and the latter for a sample, perhaps the phrasing should be reversed. From a policy perspective, the question to be raised is whether the 3.8 GJ/person estimate is within a deficit range.

With major emphasis on estimating forest wood supply, the principal policy issue addressed by the FAO study was the Bangladesh supply-demand balance for wood. The report concludes by assigning high priority to firewood investments, regarding tree fuel as "a basic ingredient in the nutritional cycle of the rural population," and finding that in the absence of programs to increase traditional fuel supplies the consequences "are most likely to impact disproportionately on the rural poor" (FAO 1981a).

A more informative, disaggregate presentation of regional variations for firewood and charcoal is provided in the Thailand NEA study (1980) described in Chapter 5 (Tables 19 to 24). Here the analyses of free and commercial sources, within one's own compound, from nearby or distant forests, or from salespersons, are important guides to local planning.

A reverse use has been made of NEA regional data, "scaling down" aggregate data from Thailand's Northeast Region to simulate a "model" village (Meta Systems 1980). This may be regarded as the analytic analog of the "representative firm" of an industry (Alfred Marshall 1890). The purpose was to simulate parameters of village size, farm size (small, medium, and large), and water requirements as a basis for assessing alternative irrigation scenarios. With simplifying assumptions on crop mix, crop residue availability, and other input and output factors, these scenarios served as guides to technology design and appraisal. For project planning and evaluation, it would of course be necessary to replace these assumptions and statistical constructs by collecting and analyzing local data. Inverting our earlier conclusion, "No region is a village."

Experiences in this chapter reveal many instances of synergy between rural people, knowing a particular place and time, and "You, the researcher," knowing the national. The challenge is to bring these perspectives together in ways that clarify and connect local and national policy choices, in terms that will be most advantageous to those people who have greatest need.

REFERENCES

Agrawal, S.C.
   1981   *Rural Energy Systems in Two Regions of Uttar Pradesh: First Phase Report.* Honolulu: Resource Systems Institute, East-West Center.

Arafa, S., C. Nelson, and S.A. Megally.
   1980   Energy Consuming Activities and Traditional Energy Resources in the Village of Basaisa: A Preliminary Report. For project report, Utilization of Solar Energy and the Development of an Egyptian Village: An Integrated Field Project.

Arnold, J., and deLucia, R.J.
   1982   *Rural Energy Surveys: The Thailand Experience.* Cambridge, Massachusetts: Meta Systems, Inc.

Atje, R., and J. Tarrant.
1980 Aspects of Rural Energy Use and Resources in West Java: A Preliminary Analysis of the Household Sector. In *Energy Analysis in Rural Regions: Studies in Indonesia, Nepal, and the Philippines* by Atje and others. ERD Program Report I-80-2; Honolulu: Resource Systems Institute, East-West Center, September.

Bajracharya, D.
1980 Fuelwood and Food Needs versus Deforestation: An Energy Study of a Hill Village *Panchayat* in Eastern Nepal. In *Energy Analysis in Rural Regions: Studies in Indonesia, Nepal, and the Philippines* by Atje and others. Honolulu: Resource Systems Institute, East-West Center.

Bajracharya, D., and C. Gurung.
1984 Dialogue as a Method for Village Level Energy Planning: An Approach to Action Research in a Nepali Village. Honolulu: Resource Systems Institute, East-West Center. Forthcoming, 1984.

Barth, F.
1966. *Models of Social Organization.* London: Royal Anthropological Institute of Great Britain and Ireland.

Briscoe, J.
1979 Energy Use and Social Structure in a Bangladesh Village. *Population and Development Review* (December): 616–641.

Brokensha, D., and B. Riley.
1983 Wood Fuels in a Marginal Area of Kenya. In Food and Agriculture Organization, *Wood Fuel Surveys*, Annex II:145.

Chebbi, V.K., A. Abdullah, R.M. Conti, A.O. Mangabat, R. Morse, and Suparno.
1977 *Initiating Rural Nonfarm Projects: A Working Guide.* Honolulu: Technology and Development Institute, East-West Center.

Cloud, K., and C. Overholt.
1982 Women's Productivity in Agricultural Systems: An Overview. Prepared for the International Agriculture Economics Meetings, August 31, Jakarta.

DaVanzo, J., and D.L.P. Lee.
1978 *Compatibility of Child Care with Labor Force Participation and Non-Market Activities: Preliminary Evidence from Malaysian Time Budget Data.* Santa Monica: Rand Corp.

deLucia, R.J., H.D. Jacoby, J.D. Gavan, J.C. Houghton, M.C. Lesser,
J.J. Stern, R.D. Tabors, and R. Tyers.
    1982   *Energy Planning for Developing Countries: A Study of Bangladesh.* Baltimore: The Johns Hopkins University Press.

Dixon, R.
    1978   Women's Cooperatives and Rural Developments: A Policy Proposal. Baltimore: The Johns Hopkins University Press.

Douglas, J.
    1982   Traditional Fuel Usages and the Rural Poor in Bangladesh. *World Development* 10(8):669–676.

Foster, E.
    1983   *Proxy Valuation of Dung.* Working Paper no. WP-83-13. Honolulu: Resource Systems Institute, East-West Center.

Geller, H.S.
    1982   Cooking in the Ungra Area: Fuel Efficiency, Energy Losses, and Opportunities for Reducing Firewood Consumption. *Biomass,* reprint, eds. J. Coombs and D.O. Hall, 2:83–101. London: Applied Science Publishers Ltd.

Gill, J.
    1982   Fuelwood and Stoves in Zimbabwe: A System in Change. Presented at the 2nd E.C. Conference on Energy from Biomass, September 20–21, West Berlin.

FAO.
    1981a   *Supply and Demand of Forest Products and Future Development Strategies: Consumption and Supply of Wood and Bamboo in Bangladesh.* Field Document 2. Project of UNDP/FAO/Planning Commission. (J.J. Douglas, project coordinator.)

Harwood, R.R.
    1979   *Small Farm Development: Understanding and Improving Farming Systems in the Humid Tropics.* Boulder, Colorado: Westview Press.

Halse, M.
    1979   Producing an Adequate National Diet in India: Issues Relating to Conversion Efficiency and Dairying. *Agricultural Systems* 4(4):239–278.

Institute of Nutrition and Food Science, University of Dhaka.
    1977   *Nutrition Survey of Rural Bangladesh, 1975–76.* Dhaka.

Islam, M. Nurul.
    1980   *Village Resources Survey for the Assessment of Alternative Energy Technology.* Prepared for the International Development Research Centre, Canada. Dhaka: Bangladesh University of Engineering and Technology.

Koppel, B.
　1980　A Preliminary Analysis of Fuelwood Consumption in the Bicol River Basin. In *Energy Analysis in Rural Regions: Studies in Indonesia, Nepal, and the Philippines* by Atje and others. Honolulu: Resource Systems Institute, East-West Center.

Manibog, F.R.
　1979　Patterns of Energy Utilization in a Philippine Village: Sources, End-Uses and Correlation Analyses. A draft report presented to the International Energy Agency and Organization for Economic Cooperation and Development, December, Paris, France, and to the Rockefeller Foundation, New York.

Marshall, A.
　1890　*Principles of Economics*. London and New York: Macmillan and Co.

Martinsen, C.S., and J.G. Ostrander.
　1982　Waterless Cooking: Influence on Energy Consumption and Nutrient Retention. In *Energy Conservation and Use of Renewable Energies in the Bio-Industries 2*, ed. F. Vogt, 525–530. Oxford, England: Pergamon Press Ltd.

Meta Systems, Inc.
　1980　State-of-the-Art Review of Economic Evaluation of Non-Conventional Energy Alternatives. U.S. Department of Agriculture Forest Service.

Quader, A.K.M.A., and K.I. Omar.
　1982　*Resources and Energy Potentials in Rural Bangladesh: A Case Study of Four Villages*. Prepared for Commonwealth Science Council, London.

Ramakrishna, J., and K.R. Smith.
　1982　Smoke from Cooking Fires: A Case for Participation of Rural Women in Development Planning. Working Paper no. WP-82-20. Honolulu: Resource Systems Institute, East-West Center.

Ravindranath, N. H., H. I. Somashekar, R. Ramesh, A. Reddy, K. Venkatram, and A. K. N. Reddy.
　1978　The Design of a Rural Energy Centre for Pura Village. Draft. Bangalore: ASTRA.

Reddy, A.K.N.
　1983　Rural Fuelwood: Significant Relationships. In *Wood Fuel Surveys*. Rome: FAO.

Schofield, S.S.
　1974　Seasonal Factors Affecting Nutrition in Different Age Groups and Especially Preschool Children. *Journal of Development Studies* 2(1):22–40.

Siwatibau, S.
　1978　*A Survey of Domestic Rural Energy Use and Potential in Fiji.* A report to the Fiji Government and to the International Development Research Centre, Canada. Suva: The Centre.

Skar, S.L., N.A. Samanez, and S.G. Cotarma.
　1982　Rural Employment Policy Research Programme: Fuel Availability, Nutrition and Women's Work in Highland Peru — Three Case Studies from Contrasting Andean Communities. Working Paper for World Employment Programme Research. Geneva: ILO.

Smith, K.R., A.L. Aggarwal, and R.M. Dave.
　1983　Air Pollution and Rural Biomass Fuels in Developing Countries: A Pilot Village Study in India and Implications for Research and Policy. *Atmospheric Environment* 17 (11):2343–2362.

Tinker, I.
　1981　Energy for Essential Household Activities. Dames and Moore. Center for International Development and Technology. Washington, D.C.

Vyas, V.S.
　1981　Structural Changes in South Asia Agriculture: Implications for Food Grains Production. Outline of a research proposal. Ahmedabad: Indian Institute of Management.

Waterlow, J.C.
　1982　Nutrient Needs for Man in Different Environments. In *Food, Nutrition and Climate*, eds. Sir Kenneth Blaxter and Leslie Fowden, 271–283. London and New Jersey: Applied Science Publishers.

Whyte, R.O., and P. Whyte.
　1978　Rural Asian Women: Status and Environment. Research notes and discussion paper no. 9. Singapore, Institute of Southeast Asian Studies.

Whyte, R.O., and P. Whyte.
　1982　*The Women of Rural Asia.* Boulder, Colorado: Westview Press.

Wiser, C.V.
　1955　The Foods of a Hindu Village of North India. *Annals of the Missouri Botanical Garden* 42(4):301–407.

Wulfe, M.
　1982　Household Energy in West Sumatra, Indonesia. Unpublished Working Paper. Honolulu: Resource Systems Institute, East-West Center.

12
Converting Rural Energy Needs to Opportunities

Richard Morse, Deepak Bajracharya, Carol J. Pierce Colfer, Barry Gills, and Martin Wulfe

RESEARCH FINDINGS AND LIMITATIONS

The previous chapters demonstrate that the new research community in energy for rural development has created and validated new research methods bringing attention to important issues and perspectives. These include

- diversity of rural energy uses and needs, of biomass and other sources, and of monetized, nonmonetized, and partly monetized forms;
- cultural, social, political, and economic diversity among rural communities and areas affecting rural energy patterns and changes;
- importance of biomass energy in terms of indigenous supply, use, familiarity, and potential for new productivity;
- village interactions needed to shape more productive and equitable energy systems;
- rural area needs for technology information, skills, and related inputs to facilitate local energy innovations;
- action research potential through a rural facilitator working in individual villages to foster local organization of innovations and to provide linkages with external research and development agencies; and
- need for integration of regional and national energy and development policies with rural energy planning and management.

The chapters also show important gaps in the knowledge and understanding required to transform rural energy systems to meet development needs. Principal limitations include

- inadequate recognition of rural cultures, informal institutions, and community dynamics and their effects on energy needs and development;

- nonexistent or rudimentary knowledge of local variations in the efficiency of biomass and fossil fuel energy use;
- consequent wide and unexplained variations in estimates of rural energy consumption;
- failure to link rural energy baseline data collection with analysis of local farming systems and related development priorities;
- inadequate economic and environmental impact assessment of existing and potential biomass energy technologies;
- need for more comparative studies of rural electrification in the context of local and regional bioresource systems.

The book also demonstrates that rural residents do succeed in identifying realistically their needs for energy and in organizing means to fulfill these needs. We bring together in this chapter integrated sets of research methods to construct a new agenda dealing with energy for rural development and founded on the principles of people's participation in cooperative research.

PEOPLE'S PARTICIPATION IN COOPERATIVE RESEARCH

Confidence gained by rural residents in agreeing on their priority needs and organizing new resource combinations and technologies to meet those needs was a critical feature of the Nepal experience reported in Chapter 7. The Kumhal community realized new self-esteem in establishing the mill and biogas plants. Families in Chhoprak who opted for improved stoves asserted their individual autonomy but also built new paths of community cooperation by mobilizing skills and inputs required to implement the project. Aided by the *lami's* effective interactions with external agencies (technology suppliers, the development bank, and the land record office), these village groups also gained a new degree of influence on their external environment. Their achievement supports Samuel Paul's Proposition 11.1:

> Program performance is facilitated when beneficiary participation, negotiation, and internal autonomy of implementors vary in proportion to the complexity (uncertainty, diversity, and scope) of the program and the environment. (Paul 1982)

Confidence Through Local Control

Building on their own knowledge base, these various community groups acquired detailed new knowledge on the scale factors and construction techniques for biogas digesters and improved *chulos,* on the workings of fermentation pits and chimney drafts, and on the timing and organization of work and inputs. They also expanded their understanding of the motivations and resilience of neighbors in negotiating benefits and costs, and grew in appreciation of the singular dedication of the lami and his family. Undoubtedly the people of Chorkate secured the most enabling knowledge: of renewal as a community. In the biogas enterprise, five neighborhoods cooperatively discovered a new community resource and gained a new sense of efficacy.

This cooperative learning experience manifests four principles:

1. User participation in actively assessing priorities, needs, and new resource combinations is a critical ingredient to implementing and sustaining new energy technologies.

2. In interaction with a skilled facilitator, communities can innovate and adapt their existing institutions to provide for fair sharing of new opportunities. In the biogas case, 65 families were able to agree on differentially structured resource commitments and obligations representing a fair distribution of benefits and costs.

3. Individual and group capabilities, enhanced through experiential learning and well-focused training, can be applied to the management and operation of newly emerging energy systems.

4. Analogous accomplishments are possible in other rural areas using patterns of continuing action research that contribute to further experiential training of facilitators and local researchers.

Making Research Relevant to Rural Reality

> There is a short limit on the steps in deductive reasoning that can be taken without correcting for, or checking with, the facts.
> — Lawrence J. Henderson (Chester Barnard 1949)

This pointer by a physician-sociologist to the inductive bases of knowledge accords with our observations on the importance of knowledge held by rural users and producers of energy. The perception of the actor or doer is essential in identifying facts based in action. In turn, deductive reasoning suggests wider significance of facts. If research is to be effectively developmental, cooperative methods are essential in order to encourage village residents to join with researchers in interpreting and synthesizing relevant facts and in testing proposed new systems.

Other chapters have established a foundation of methods and information to enhance the skills of rural residents in quantifying and assessing

Comments by Dr. Vijaya Shrestha, Social Action Research Program, Chorkate. (D. Bajracharya)

energy development needs and opportunities. From this experience, we draw together four principles for a development process to advance rural initiatives, productivity, and equity through better use of energy and related resources:

1. Increase food and fuel productivity by improved land and work arrangements, with continuing attention to the cooking sector as an integrating dimension of the farming system. This procedure involves strengthening local capabilities for measuring and evaluating the efficiency of energy production and use in the bioresource and cooking sectors, as a basis for formulating efficiency improvements.

2. Reduce and overcome constraints on crop, animal, and fuelwood production in local farming systems by attending to priority energy needs for irrigation, farm power, agroprocessing, rural industry, and transport.

3. Facilitate community organization and management of energy systems that achieve scale economies within farming areas and present new negotiating advantages for local production and marketing of farm products and energy services.

4. Sustain stimuli of new added value, employment, and income within the rural area to transform land, work, and earning relationships, thereby deepening the base for improved income distribution and productivity.

These elements offer an organizing framework for assessments and planning by rural residents, cooperatively with interdisciplinary research teams, to identify and meet local, regional, and national information needs bearing on these transformations.

Research and Training Commitment

A program to translate the action research lessons exemplified at Chhoprak into curricula development and field training has been initiated by Chulalongkorn University and the East-West Center in consultation with institutions in other countries. This experiential learning program will prepare men and women, motivated to live in rural areas, to serve as facilitators and bridges between villages and the organized research system. To strengthen two-way flows of information, participants' skills for ascertaining local preferences and building on the knowledge of rural residents will be emphasized. The program will assist in orienting technologists and district planners to user-based energy systems development.

Paul's Proposition 11.2 is pertinent here:

Program performance is facilitated by the program management's flexibility in staff selection, and a strong emphasis on training and commitment creation. (Paul 1982)

A New Development Policy Research Agenda

The framework of energy research for rural development starts with needs and priorities in countless village homes and farms. Development research grounded in the decision needs of men and women in these local centers is required. Research must be set within this decision context, but it must also move outward to the decision environments of regional and national policymakers who need information bearing on their particular planning and decision roles. Research priorities therefore are not neatly definable in linear or sectoral terms: they must be woven together operationally in time and space (see Andrew Vayda, Carol Colfer, and Mohamad Brotokusumo 1980). For such a design, we draw on analysis and planning experience in other domains where local individuals and groups face the need to organize a diversified range of social and technological innovations and where national policies to foster these changes are sought.[1]

The new energy for rural development agenda that we are advocating encourages social, allocative, and technological learning on dimensions outlined in Figure 1. In the next sections, we discuss each of these research endeavors, the variables involved, the kinds of research methods suited to each aspect, and how these can be connected.

RURAL ENERGY AND FARMING SYSTEMS RESEARCH AND DEVELOPMENT (1.0)

We see the necessity of integrating rural energy research with farming systems research and development and recommend two criteria for consideration by sponsoring agencies in selecting areas for program initiation: (1) the presence of a clearly recognized energy constraint or opportunity, and (2) existence or potential of cooperative relations with local community groups. In applying the first criterion, energy problems in relation to farm production, poverty, and population pressure should, for example, provide the subject focus. In addressing the second, institutions initiating the program would explore ways of achieving representative villager participation, recognizing the diverse and perhaps divergent interests and capabilities of local communities and individuals.

Identification of Energy Problems and Opportunities in Local Farming Systems (1.1)

The problem identification and research planning phase, including an alert search for potential solutions and promising opportunities, is critical to the program's effective formulation and eventual success. We illustrate its primary steps in the next paragraphs.

Team Formation and Orientation to Scope of Problem (1.1a). First, a core research group comprising at a minimum an agronomist, an anthropologist or rural sociologist, an agricultural engineer or water management specialist, an agricultural economist, and one or more rural energy specialists assemble to develop an initial statement of the area's pressing problems. Added members might include a forester and livestock specialist. Two or more women team members should be sought. The team has or develops skills in integrated, problem-solving rural research and in styles of participatory research. Representatives of the research area are invited to orient the team on approach and scope. From existing secondary data, the team organizes relevant diagnostic indicators

RURAL ENERGY AND FARMING SYSTEMS RESEARCH AND DEVELOPMENT (1.0)

    Identification of Energy Problems and Opportunities in Local Farming Systems (1.1)
        Team Formation and Orientation to Scope of Problem (1.1a)
        Problem Diagnosis and Opportunity Identification (1.1b)
        Organization of the Work Plan (1.1c)
    Technology Appraisals and Organization (1.2)
    Institutional and Resource Adaptations (1.3)
    Policy Innovations in Energy and Farming Systems (1.4)
    Evaluation, Iteration, and Generalization (1.5)

REGIONALIZING RURAL ENERGY TRANSFORMATIONS (2.0)

    Identification of Geographic Areas for Extending Rural Energy Successes (2.1)
        Mapping Crop and Related Bioresource Systems (2.1a)
        Regional Population and Infrastructure (2.1b)
    Converting Local Information Advantages to Asset Creation Through New Energy Systems (2.2)
    Comparative Assessment and Generalization of Renewable Energy Systems (2.3)

RESEARCH ON NATIONAL AND INTERNATIONAL POLICY ISSUES (3.0)

    Data Development Guidelines (3.1)

                    Priority Issues (3.2)

    Innovations in Development Loans for Rural Energy Systems (3.2a)
    Nutrition Policy Implications of Food and Fuel Deficits, and Procedures for Their Elimination (3.2b)

                    Future Research Issues (3.3)

    Rural Transport Policy and Energy Development Alternatives (3.3a)
    Comparative Regional Trends in Substitution of Renewable for Petroleum-Based Fuels (3.3b)
    Comparative Regional Structures for Renewable Energy and Electricity (3.3c)

Figure 1. Energy for Rural Development Research Agenda

of the kinds defined in Chapter 11 (Table 9). Through analysis of these indicators, working propositions are formulated as guides to inquiry.

Problem Diagnosis and Opportunity Identification (1.1b). By establishing residence in the area or through the sondeo or *Gaun Sallah* method exemplified in Chapter 7, the team undertakes a comprehensive appraisal of the area's problems and potential. The quality of dialogue between representative local groups and the researchers is crucial at this stage. From whose standpoint are the severity of problems to be judged and potential solutions tested? It is vital that not only the most vocal citizens be heard. As the complexity of issues unfolds, people whose problems are pressing may themselves voice the criteria highest on their agendas — equity, food adequacy, resource access, community cohesion, specific energy needs — and identify the problem categories to be studied. Although it will by no means be simple to rank or reconcile the priorities of different local groups, their voices will bring to the research frontier the motivations and needs that require some measure of resolution, if solutions are to be acceptable.

Interviews with individual family members, and with groups who articulate particular concerns or knowledge of resources, are conducted. Open-ended inquiry and dialogues are the prevalent modes. Three key techniques help link perspectives: team members alternate in pairs to share disciplinary outlook and skills and to avoid unwieldy numbers in local interviews; each pair writes a daily assessment note including significant statements, observations, and data; and, on completing the diagnosis phase, the team promptly writes the opportunity and problem identification report. Travel and work schedules are planned to ensure that the report is jointly written while impressions and facts are vividly in mind.

Organization of the Work Plan (1.1c). During the initial appraisal, area residents and researchers identify farm, home, or community energy technologies that are of immediate interest to families or groups either as trials or demonstrations or as operational units. An implementation plan is formulated taking into account the organization required for mobilizing internal resources and, when necessary, for negotiating external inputs. This process clearly involves delicate negotiations among village residents and between village residents and outside resource people. Clear lines of responsibilities should be established. Realistic assessment of internal and external resources is important so that commitments made by village residents and external resource agencies can in fact be met. The identified technologies would be implemented promptly as agreed by those taking responsibility for their costs, organization, use, and maintenance. Criteria and plans for monitoring, evaluation, iteration, and extension of successful technologies would be adopted.

If the steps in this first phase are bypassed or given inadequate resources and time, the program's benefits will "most likely be sharply reduced" (Willis Shaner and others 1982). Farming system research specialists typically schedule about three months for research area selection and problem identification. We consider four to six months more realistic phasing for energy problem identification and research planning, since inquiries into the social and technical complexities of rural energy systems are at a much younger stage than are similar farming system inquiries.

Technology Appraisals and Organization (1.2)

Other promising technologies, more complex or uncertain, require structured feasibility assessments as a basis for investment decisions. The fit of a particular design or scale of technology to local needs and resources is the central criterion. These projects might include

1. Family or community biogas plants for cooking and lighting, or for neighborhood-scale parboiling, milling, or other agroprocessing.
2. Biogas, producer gas, coconut, physic nut (*jatropha*) or other biofuels for dual-fuel diesel engines to be used in irrigation pumping, agroprocessing, saw milling, or other stationary shaft power.
3. Improved stoves for cooking.
4. Improved charcoal conversion for cooking and for local industry.
5. Solar kilns for timber and plywood drying.
6. Solar refrigeration for food storage and preservation.
7. Crop and wood residue compaction, drying, and pelletizing for thermal power generation or industrial heat applications.
8. Producer gas for electricity generation.

A common feature of such single-process technologies is that they convert nonuniform and often scattered materials to a relatively more standard and compact — though not necessarily clean — fuel. Among several factors that make on-site project appraisal essential is the need for close attention to the physical and management characteristics of such nonhomogeneous inputs.

Community or individual sponsors of such candidate projects would be equipped by researchers with standards of data and methods of data collection to determine local project viability. The research team would organize data on performance and benefits of the technology elsewhere, to define key elements of necessary local appraisal data. Appraisal data would be collected mostly by local people. Obtaining and organizing the required data equip the sponsors for most aspects of production and cash flow management. Marketing and asset management are likely to be more complex. Counseling by researchers on these aspects would in turn provide data and capabilities for generalizing the project performance record and for adapting it to other areas.

Such energy conversion technologies, like more complex processes, increase the density and change the form, utility, and often the quality of the original energy source. The weight- and bulk-reduction of materials and the potential economies of scale in these processes require locational analysis to assess the competitive viability of local, small-scale systems. Appraisals by local sponsors must determine the attractiveness of plants of varying scale, labor/capital intensity, and product mix to meet village or neighborhood needs and for possible wider markets. These analyses will, in turn, contribute to defining data categories and levels of accuracy necessary for comparative analysis and extension of rural energy systems (phases 2.1 and 2.3).

Institutional and Resource Adaptations (1.3)

Core research opportunities relate to changes in social institutions that can emerge with changes in use practices, both temporal and spatial, of land, water, and sunlight. Research can facilitate transitions from present (perhaps unjust) user rights to rights (possibly more equitable) to new added value. Also important is the prospect of taking advantage of integrated, more efficient, and enhanced use of limited resources.

Instances of such potential transformations include

1. *Integrating fuelwood into local farming systems.* Land use assessments, single or multiple species use, species adaptability to the environment, seasonal and multiyear labor balancing, loan terms to fit harvest schedules, and negotiation of benefit rights for disadvantaged participants make up elements of enterprise evaluation and creation (see D.A. Hoekstra 1983a and 1983b).

2. *Combined firewood and stove improvement programs.* These entail determining stove efficiencies with current fuel mixes of typical moisture content (VITA 1982, FAO 1983a); obtaining women's critiques and concepts on candidate systems from fuel source to ash, including cooking practices and related family care; assessing firewood expansion and improved air drying of firewood; and evaluating and testing improved local stove designs to burn preferred wood fuels. They also entail evaluating smoke pollution hazards (Kirk Smith and others 1983); valuing possible replaced vegetative fuels for alternative fuel, fodder, fiber, or pulp uses, and for soil loss avoidance (William Lockeretz, G. Shearer, and D. Kohl 1981, Gerald Marten and others 1981, NRC 1981); and devising local conventions and instruments balancing family and community investments and benefits.

3. *Mixed fodder and fuelwood species with animal husbandry development.* Included here are fodder quantities and quality; other intercropping; animal species and feed improvement; grazing and feed management;

Animals in the farming system, Gujarat. (Kirk R. Smith)

dairy and other product development, processing, refrigeration, transport, and marketing; and cooperative dairy management. (For elements of analysis see M.A. Altieri and others 1983, S.S. Brar and others 1982, S.P. Carruthers and M. Jones 1982, and L.P. Walker 1982.)

4. *Animal power efficiency improvements.* Small farm options with improved yokes, harnesses, tool bars, tools, water lifts, and pumps. Improved fodder provision and energy intake for draft animals are pivotal features.

5. *Manure and slurry use practices and valuation.* Perceived crop responses to dung and biogas residues; monitored multiyear assessments; favored crop systems; and organizational design.

6. *Yield variability of new fuel and fodder crops.* Components include productivity gains and variability in more intensive cropping systems; and insurance procedures for short-term (one year) and medium-term risks (two to five years).

Research by farmers and interdisciplinary teams, in some instances supplemented by structured surveys within a given agroclimatic zone, will be essential in analyzing these new systems and in identifying and valuing particular limiting factors and opportunities. The discussion by

Amulya K.N. Reddy (1983) on relationships between fuelwood and other productive components of an agroecosystem provides a framework for assessing competition and complementarity among these components. Each inquiry of this type centrally involves local use practices and rights. Many of these rights are socially and culturally established, though not necessarily immutable. For new uses and discoveries they are negotiable. Major labor tasks are required to establish these new systems. Retaining values from one's own group labor and transforming these values into capital in the form of renewable energy make up a potential system transformation, to which the research may contribute.

While project information and appraisal reports on individual technologies (phases 1.2 and 1.3) will merit dissemination for interregion and intercountry adaptation, reports on new resource mixes and institutional adaptations will also have wider theoretical and policy interest. The primary audience for key research findings, however, and indeed their coauthors, are rural organizers who initiate new systems.

Policy Innovations in Energy and Farming Systems (1.4)

In local planning situations, communities are originators of new policies. Solutions reached and actions taken by rural groups on the kinds of organizational and resource allocation issues just considered shape new policies that can cumulate beyond a single system solution. In Chorkate, the respective abilities of individuals and groups to provide resources — cattle dung, land at risk as collateral, and labor—evolved through negotiation as the basis for investments in the biogas digesters and mill. With the prospect of benefits for each member, ability to pay evolved as the guiding criterion for project inputs. Such institutional adaptations are resource policy in the making.

Local policy innovations may extend spatially through nearby environments — for example, within relatively similar farming systems — until they influence or are blocked or distorted by price or tax policies or by laws shaped in regional or national centers. At such intersections, interesting new policy research issues are revealed. Probably the most interesting policy initiatives involving a new research challenge will be those raised by disadvantaged groups who often have not succeeded in communicating coherent policy concerns in the past. Research flexibility to pick up these new issues is sought.

Certain critical issues can be identified in the research we have reported here. Instances are

1. *Nutrition, energy, and work.* Objectives would include testing relationships between food and nutrition deficiency and fuel deficits, estimating

Prototype 3 kW gasifier, scrubber, and irrigation pump based on wood chips, Jyoti Solar Energy Institute, Gujarat. (Kirk R. Smith)

effects of substandard energy intake on men's and women's work activities and related agricultural production, and assessing incidence of these effects among landless families and small farmholders. Other objectives are estimating added quantities of food and fuel needed to overcome these conditions; and, through counsel by representative families concerned, identifying possible policy alternatives for resource allocation, institutional adaptation, and productivity improvement to overcome these effects.

2. *Effects of farm size and crop systems on energy expansion paths.* The objective would be to determine suitable energy sources and technologies for enterprise expansion in subsistence, small-scale, and medium-scale farms in the local farming system. Account would be taken of choices and mixes of draft animal and mechanical power for irrigation, field, and postharvest operations, of potential substitution or mixes of biofuels and fossil fuels, and of potential contributions of biomass development for fuel (see R. Wijewardene 1982, R.D. Bell and T.J. Willcocks 1982).

3. *Employment-generating potentials (related to nutrition and work)*. With energy-related farming system changes, these include job creation through intensified cropping systems, agroprocessing, small-scale industry, and road transport development, with a research focus on retaining economic multiplier effects within the area.

4. *Innovative use of traditional common property rights*. As in Chorkate, adaptations are sought to enhance the usage of limited resources in ways that ensure equitable and just distribution of costs and benefits.

Participatory research would establish priorities among these issues in the local area, refine the aims and scope of work, and establish the basis for analyses to provide comparability with studies in other areas.

Evaluation, Iteration, Generalization (1.5)

Each of the phases of local research just outlined has its own time scale and measures of effectiveness and benefits related to costs. Wide adoption of practices and technologies is the first evidence of success for phases 1.1 to 1.4. Perhaps of greatest importance is formation by area residents of their own criteria and evaluation procedures. These will guide researchers in structuring more systemic reviews, in studying underlying success factors, and in assessing their implications for wider use and adaptation.

REGIONALIZING RURAL ENERGY TRANSFORMATIONS (2.0)

We have proceeded in building a design based on people's initiatives, under conditions of scarcity, for integrated research centered on energy, food, and related resource organization. Rural communities' capacities to realize wider energy potential may be enhanced by the steps outlined next to regionalize research on renewable energy development.

Identification of Geographic Areas for Extending Rural Energy Successes (2.1)

Where can successful innovations in supplying energy to increase farming system productivity be effectively extended or adapted? How can performance measures on individual projects and diagnostic, need, or value indicators of particular resource configurations be used to guide such adaptations and locational choice?

In farming systems research, several techniques are used to identify recommendation domains in which "roughly homogeneous" groups of farmers may be advised to use "more or less" the same new practices, based on farmer-managed trials (Shaner and others 1982). Six components in a stratification approach are enumerated by Shaner (1983): agroclimatic zones, soil and land classification, biological and cropping environment, farm practices, economic conditions, and sociocultural conditions including land tenure. Noting that each international crop research institute analyzes these factors somewhat differently, Shaner cautions against working with too many combinations: "A futile exercise in orderliness."

To establish a direct, comparative approach to identifying recommendation domains for rural energy systems, we suggest a focus on two principal dimensions: (1) the cropping system and associated distribution of livestock, forests, and trees; and (2) regional population, sociocultural, and infrastructure characteristics.

Mapping Crop and Related Bioresource Systems (2.1a). The prevailing crop, livestock, and forest system sets the most immediate context for rural energy potential and constraints. Underlying endowments of climate, land, and biology influencing cropping systems may be considered secondary factors for purposes of regional energy assessment. A starting point for mapping energy and bioresource systems, for example, is in the methods and information used to develop maps of rice-cropping regions, that is, shallow rainfed, wet season irrigated, intermediate rainfed, dry upland, dry season irrigated, and deepwater (Robert Huke 1981, 1982). These maps, published by the International Rice Research Institute, are scaled to identify areas having 3,000 ha in a particular rice-cropping system (see also NATMO 1980). Rural energy researchers could combine these with maps and diagnostic indicators for other principal crop systems and for livestock, forests, and tree groves to establish indicative energy and farming system domains. Soil taxonomies as utilized in the Benchmark program (Benchmark Soils Project 1982) and the International Benchmark Sites Network for Agrotechnology Transfer (IBSNAT) could progressively supplement these analyses. This procedure would permit substantial advances in defining sample areas for structured studies and in generalizing their findings.

This recommendation is consistent with the conclusion in Chapter 3 that ecological rather than administrative areas should be the sampling basis for rural energy studies. The crop-livestock-forest system is a manifestation of the ecosystem and is measurable in energy-relevant terms. Bioresource system maps and diagnostic indicators would be useful to district administrators and planners, as well as to national policymakers, since they provide location-specific guidelines for different parts of administrative districts.

Regional Population and Infrastructure (2.1b). Energy recommendation domains identified by studying bioresource configurations are subject to revalidation on other dimensions: (1) geographic or spatial scale; (2) sociocultural characteristics; (3) population density and patterns of settlement; (4) proportions of subsistence to commercial agriculture, and of noncommercial to commercial energy forms; (5) road density and related communications; and (6) types and scale of industry and commerce.

The geographic extent or diversity of farming systems influence the physical means and costs of transport, communications, liquid fuel distribution, rural electrification, marketing, banking, and regulatory structures. Population size and income distribution, market and other central place functions, and size distribution of villages influence the size and location of regional demands for local and imported energy services. Road connections to interior villages and accessibility factors such as transport distances and costs from villages to central places further influence supply-demand relations. Proportions of nonmonetized and monetized production and trade in food and energy materials are at once a function and measure of these interacting factors and a principal determinant of their evolution.

We suggest that subsistence to commercial proportions of agricultural and energy production be mapped and compared with crop-livestock-forest maps, forming a grid on two analytic dimensions to identify recommendation domains for energy-farming system innovations. Consistent use of these two dimensions for comparative analyses among different regions and countries would refine these instruments and expand their applicability.

These two indicator sets together provide a framework for analyzing specific input, output, and transport factors influencing the viability of energy technologies and mixed energy systems in various farming systems. In this framework also, other selected variables could be used to structure comparative analyses on specified policy issues and to study changes over time. Maps of cultural variations would be important for several issues. By those means, progress toward generalizing the understanding of technology, management, and policy aspects of renewable energy systems could be advanced.

Converting Local Information Advantages to Asset Creation Through New Energy Systems (2.2)

Close knowledge of renewable energy linkages with farming systems should prove to be an important advantage of local communities and entrepreneurs in their discovery of new energy development opportunities. Information advantages will have accrued through experience of seasonal and annual fluctuations in crop and fuel output related to marketing, cash flow, and asset management. Adequacy of new fodder supplies to sustain larger cattle populations, for example, as a basis for dairy processing, refrigeration, and distribution chains, would have been validated through experiences particular to each energy-farming system.

Information advantages on changes in energy density, quality, and scale of conversion (phase 1.2) are essential elements in enabling energy developers in rural areas to assume leadership in establishing viable new systems. Typically, advantages of local information can be expected to be greatest for technology changes directly linked to rural end uses, for which local users can directly influence design innovations. Changing fuel production technologies are more problematic for rural locations. If conversion to greater energy density implies economies of scale and materials reduction, then, in many instances, location factors may give advantages to installations in central or even metropolitan places. To reverse such a trend, rural areas will need to mobilize their countervailing information advantages and adjust system scale and performance to rural endowments and needs.

End-use changes also may have negative local effects, often on particular groups. Portable chain saws and diesel-powered boats in Kalimantan, Indonesia, have enhanced the efficiency of men's work, making women's work relatively less productive (Colfer 1983). The prospect of displacing some petroleum-based fuel with biofuels such as producer gas requires assessment in such areas, linked with possible opportunities for the planting and tending of biomass seedlings or other employment to restore status and earnings in forms acceptable to women. Links of this type between rural end-use innovations and new fuel conversion systems based on local resources (phase 1.3) could offer important locational advantages to rural areas. Rural energy organizers who perceive such advantages can combine their ideas with those of researchers in technology development to retain these benefits in the local area.

The research aim here is to build on these advantages of local system knowledge to discover and develop new energy system opportunities. Sources of added value through this process could be converted to asset formation by and for local enterprises and communities. These include efficiency gains in end use practices and conversion technologies. They include producer's surplus realized in scale economies or higher energy quality, if retained in the area. With institutional innovations to ensure that end users and producers share in these efficiency rents and in transport or transfer savings, rural residents' income and asset base could be strengthened with multiplier benefits, enlarging local markets and stimulating new employment.

Comparative Assessment and Generalization of Renewable Energy Systems (2.3)

To realize such potentials for rural asset creation on a wide scale, bridges are necessary between new capabilities in need and market assessment, on the one hand, and the programs of technology development agencies, on the other. Technology breakthroughs typically require concentration of a high order of scientific and engineering capabilities. This is so because each technology package integrates a host of physical properties and design parameters. Allocation of research and development funds and personnel fosters specialization built around major energy resources and conversion systems. Given the need for specialization, this is a necessary stage in scientific and technological discovery.

The agenda we advocate for bringing the perspectives and ideas of rural residents into technology development would provide a strong complement to this technological infrastructure. This is essential if the thrust of technology development is to ensure appropriate solutions for improving the quality of rural life. Design parameters based on rural end-use experience and on research trials of local energy applications are required. Comparative assessments across larger regions are recommended to validate rural energy needs and priorities by taking into account different agricultural, social, and cultural factors.

Such assessments can be made with the aid of systemic technology attributes on a grid that would include

1. *Physical characteristics.* Size, divisibility, substitutability, biological and environmental interactions, precision and tolerances.
2. *Locational and spatial characteristics.* Energy density, weight and volume characteristics, input and output mobility, diffuse or concentrated distribution patterns.
3. *Temporal factors.* Periodic, random, or peak supply and use patterns; daily and seasonal variability; maturation period; life cycle and duration.
4. *Infrastructure requirements.* Service and repair industries, specialized skills, distribution.
5. *Organization attributes and fit to user conditions.* Management complexity, maintenance, flexibility for change, performance monitoring, benefit distribution, form and control of asset creation, reinvestment of proceeds.

Using such an assessment grid, for example, the capital intensity of energy technologies would be assessed to determine the potential for high capacity utilization when the system under consideration involves high fixed costs compared to variable or operating costs. Renewable energy resources — wind, many biomass forms, and solar energy — would be assessed for possible variability that might interrupt or sharply reduce facility operation. The implications of potential supply interruptions when biomass resources in a particular area are controlled by a few sellers or buyers would be assessed. Systemic study of such regional influences can significantly advance knowledge of how to fit energy technologies and organization to local need and resource conditions. Funds alloted by development agencies to such ground-based assessments can thereby be expected to yield high incremental returns to technology-specific research and development.

RESEARCH ON NATIONAL AND INTERNATIONAL POLICY ISSUES (3.0)

In the context of emerging policy issues in each country, experience of studies on local policy innovations (phase 1.4) would help define the future objectives and scope of national energy and rural development policy studies. Through better definition of aims and scope, future research including comparative international studies can find answers to specific policy issues in more cost-effective ways.

Data Development Guidelines (3.1)

The research experiences reviewed in this book furnish certain guidelines for developing rural energy data and for determining the objectives and scope of major studies.

1. The objectives of energy and rural development studies invariably must bring together the perspectives of the social and policy sciences with the perspectives of the physical sciences and technology. Hundreds of millions of people in rural areas are engaged in complex efforts to assure and advance their future well-being. Energy, though central to these transformations, is only part of them. Ability and readiness to build new knowledge consistent with the actions, experience, and beliefs of rural people is essential in research to understand, for example, the interactions of energy with high-yield crop varieties or with widening rural transport systems.

2. To meet the purposes of baseline data collection, the regular inclusion of rural energy in censuses and sample surveys periodically spon-

sored by governments would be more instructive and cost-effective than studies conducted in isolation. One instance is the Philippines' energy needs survey planned as a refined effort in cooperation with the National Census and Statistics Office (Chapter 4). Standardized, carefully structured questions on energy are being introduced into agricultural censuses, household expenditure surveys, and socioeconomic surveys in other countries also. Sustained use of these standard statistical systems is necessary for consistent development of rural energy time series data.

3. These standardized surveys deal in a significantly limited way, however, with noncommercial energy sources and use. Reliance on "standard" or "consistently defined" energy data based only on priced transactions distorts the national picture and is value-laden against more traditional parts of the society. In partly monetized areas, therefore, complementary studies should periodically be undertaken to fill this major gap. Participant observation is essential in these complementary studies to obtain understanding of people's behavior in using and developing energy resources, to understand aspects of scarcity and valuation, and to define characteristics for structured questionnaires. Rural residents' perceptions of constraints and opportunities for energy development should be sought to help focus data collection. Studies of this kind in the nonformal sector would in no sense substitute for the development and action research outlined earlier. Indeed, geographic classifications and indicators established in phase 2.1 would help define sample strata for periodic studies, and thereby gradually would portray the more complete rural energy scene.

Priority Issues (3.2)

In each country, national agencies face pressing policy issues that shape their research agendas and priorities. An important source of research priorities is evident in this book: the policy initiatives of rural groups and communities who are finding new ways of harmonizing their interests in order to choose, organize, and benefit from renewable energy alternatives. Institutional adaptations and innovations as exemplified in Chhoprak offer rich new learning ground for national policy formation. We recommend that national bodies seek ways of being alert to these rural policy initiatives and sponsor studies actively involving rural groups in order to fashion national policy innovations.

Two priority concerns that merit early consideration are outlined next.

Innovations in Development Loans for Rural Energy Systems (3.2a). In determining allocation levels and conditions of development loans for energy projects linked with crop, forest, and livestock systems, three aspects of financial structure should be recognized: (1) wide seasonal and annual variations in project maturity and yield, (2) wide variations in capital to operating ratios, and (3) consequent need for adaptable equity and debt profiles and sharing of risk. The Chorkate community adapted traditional institutions to structure obligations and benefits differentially, thereby evolving an acceptable schedule of benefits over the expected project life. Structuring of development loan terms to meet such adaptive capacities is critical to building local assets. Development finance assumes a crucial policy role in light of the locational variability of rural energy systems. Whereas national price and tax policies may not reach their intended beneficiaries where regional barriers or contrasts are high, development loan managers can tailor financial packages to the structure and development objectives of particular project profiles. A systemic, comparative study of rural energy projects in different farming systems, then, would contribute to the evolution of loan policy guidelines to reinforce rural asset commitment and creation.

Nutrition Policy Implications of Food and Fuel Deficits, and Procedures for their Elimination (3.2b). Assessments indicate that nutritional deficiency associated with inadequate food consumption reduces human energy for work, often with delayed time effects, and may also reduce efforts to find work (David Seckler 1978). Studies cited in this volume have identified instances where lack of access to fuel, usually associated with inadequate food supply as well, has reduced cooking periods with sometimes inferred reduction in nutrition and other potentially negative health effects. Research in specific rural energy and farming systems (1.4) is designed to test these factors and provide evidence of their effects on men's and women's productive work, health, and agricultural production activities especially in peak periods. Assessment of effects on food output and on system capacity for more intensive production would result.

Sustained residence in villages involved in this research will be required to establish voluntary cooperation by the families concerned, to observe representative food and fuel situations as the basis for purposive sampling, and to measure and evaluate these sensitive factors (Elizabeth Cecelski 1982 and 1983, Subachari Dasgupta and Govinda Joshi 1983, Myra Gunawan and others 1983). In turn, the active participation of rural families is expected to bring out ideas for practical shaping of policies and programs to alleviate nutrition and energy deficits. Sponsorship of carefully conducted studies of this type in different farming regions would provide an estimation of the severity and incidence of fuel

Energy of the future, Gujarat. (Kirk R. Smith)

and food constraints, thereby leading to suggestions for policy design and implementation tailored to varying local requirements.

Future Research Issues (3.3)

Three other policy areas of major significance for the future of energy in rural development are inherent in the structure of this agenda.

Rural Transport Policy and Energy Development Alternatives (3.3a). In many regions, two generic transport policy issues are whether road network development within the local energy and farming system offers greater rural advantages than arterial highways connecting different areas, and whether road development can be accomplished in ways that avoid or minimize ecological damage. Policies favoring internal transport development could potentially foster economies of scale in local conversion of energy resources and transport savings through local assembly of products serving rural end uses. Policies favoring arterial transport development, conversely, could give scale and location advantages to metropolitan centers. Transport of fuelwood to Dhaka and Khulna at low marginal cost in trucks otherwise returning empty from commercial deliveries (as portrayed in Chapter 2) exemplifies another effect of road transport policy. To what extent does this exacerbate rural fuelwood shortages? Could suitable transport pricing policies coupled with rural development loans modify or reverse such effects?

Biofuel potentials in local farming systems introduce an important new variable in these calculations. System variables include the potential for new design of limited-distance, limited-volume trucks, tricycles, and other vehicles for use within the farming system (FAO 1983b); design criteria on road width, materials, capital/labor construction costs, and life expectancy; and environmental impacts from alternative road systems. Selective research on these rural transport issues merits support, especially in areas where studies of regional characteristics (phase 2.1 and 2.2) reveal significant potential for asset creation in rural energy systems by residents of the area.

Comparative Regional Trends in Substitution of Renewable for Petroleum-Derived Fuels (3.3a). Only when comparative studies on energy choices for farms of different size in various farming systems are well advanced (phase 1.4) can adequate analysis be made of the overall advantages of specific renewable energy systems relative to petroleum-based fuels. Assessments of rural transport alternatives just mentioned will also significantly shape such substitution potential. Iterative studies of these factors will be necessary to establish a foundation for comparative evaluations among regions and countries.

Comparative Regional Structures for Renewable Energy and Electricity (3.3c). Analogously, studies on energy constraints experienced by different size farms with varying crop, livestock, and forestry mixes bear directly on the cost-effectiveness and benefits of rural electrification programs. These situations in different farming areas also, of course, are interdependent with liquid and gaseous fuel developments and with local potential for electric power generation. Fairly extensive evidence from such studies in diverse regions is probably necessary as a basis for comprehensive assessment of structural relations among renewables, electric power, and electrical grid expansion.

POLICY EMPHASES

The social and physical basis of the outlined agenda, clearly, is in the innovative new efforts to enable rural people, where they live, to engage with specialized professionals in assessing promising opportunities — as well as limiting factors — for using energy more productively, efficiently, and equitably on farms, in cooking, and for community tasks. Highest priority for resource mobilization and program scheduling is assigned to informed dialogues for diagnosis of problems and priorities in energy and farming systems (1.1); to local appraisal and organizing of energy technology systems (1.2); to institutional and resource innovations for more intensive crop and biomass production, linking energy with cultural practices to assure more effective protection of the environment (1.3); and to policy research with active participation of farm and labor families on urgent issues of nutritional adequacy in relation to food and fuel access and on energy requirements to increase productivity on small and transitional farms (1.4).

These development research foundations will strengthen rural people's ability to build on their own knowledge advantages, to organize more productive and equitable farming and energy systems, and to retain the benefits of these institutional changes to improve rural well-being (2.2). Priority policy research on development loans (3.2a) and on nutrition, energy, and work (3.2b) should be closely linked with these local system efforts.

Research and development institutions located within farming regions, in most instances, will require budget enhancement and flexibility to build local outreach and to meet the interdisciplinary requirements of such a program. Numerous fields of specialization are involved, the particular combinations varying among regions. Disciplines centrally involved include the agricultural, animal husbandry, and forestry sciences; social sciences, including anthropology, home and farm economics, farm management, nonformal education, women in development, and rural sociology; and mechanical, chemical, and electrical engineering, as well as agricultural engineering and water resource management.

Clearly, few institutions located in rural regions enjoy such arrays of talent. Interinstitution cooperation, especially in forming core research teams, will usually require at least five-year commitments. Organizational flexibility in fielding and supporting development research teams will be at a premium. National policy support for generating these energies in rural development perhaps emerges as a principal issue on the agenda. Developmental research undertaken where most rural people live, deriving insights from their beliefs and experiential heritage, and guided by the severity of constraints they express — this is the policy challenge.

NOTE

1. Tested research designs that have fostered local development initiatives by linking these with national policy formulation are found, as examples, in such diverse spheres as small industry (Nanjundan, Robison, and Staley 1962, Staley and Morse 1965, ILO 1977, Anderson 1982); community health (Taylor 1978); farming systems (Shaner, Philipp, and Schmehl 1982, Harwood 1979, Zandstra 1983, Mark and Lucas 1983); social education (Roughan, forthcoming, 1985); and development administration (Paul 1982, Korten 1980, Johnston and Clark 1982).

REFERENCES

Anderson, D.
  1982 Small Industry in Developing Countries: A Discussion of Issues. *World Development* 10(11):913–948.
Altieri, M.A., D.K. Letourneau, and J.R. Davis.
  1983 Developing Sustainable Agroecosystems. *BioScience* 33(1): 45–49.
Barnard, C.
  1949 *Organization and Management*. Cambridge: Harvard University Press.
Bell, R.D., and T.J. Wilcocks.
  1982 Energy Conservation in Mechanisation of Agriculture in Developing Countries. In *Energy Conservation and Use of Renewable Energies in the Bio-Industries 2*, ed. F. Vogt, 88–89. Oxford, England: Pergamon Press Ltd.
Benchmark Soils Project.
  1982 *Assessment of Agrotechnology Transfer in a Network of Tropical Soil Families*. Honolulu, Hawaii: Department of Agronomy and Soil Science, College of Tropical Agriculture and Human Resources, University of Hawaii, and Department of Agronomy and Soils, College of Agricultural Sciences, University of Puerto Rico. Progress Report 3, July 1979–September 1982.
Brar, S.S., and others.
  1982 Continuous Maize and Wheat Production Under No-Tillage and Conventional Tillage System: A Ten Year Study. In *Energy Conservation and Use of Renewable Energies in the Bio-Industries 2*, ed. F. Vogt, 100–107. Oxford, England: Pergamon Press Ltd.

Carruthers, S.P. and M. Jones.
:   1982   The Role of Agriculture in British Biofuel Production. In *Energy Conservation and Use of Renewable Energies in the Bio-Industries 2*, ed. F. Vogt, 162–179. Oxford, England: Pergamon Press Ltd.

Cecelski, E.
:   1982   Household Fuel Availabilities, Rural Women's Work and Family Nutrition. Draft. Geneva: Rural Employment and Development Department, International Labour Organization (ILO).

Cecelski, E.
:   1983   Report of Technical Consultation: ILO/Netherlands Energy and Rural Women's Work Project 21–25 March. Geneva: ILO.

Colfer, C.J.P.
:   1983   Change and Indigenous Agroforestry in East Kalimantan. *Borneo Research Bulletin*, April.

Dasgupta, S., and G. Joshi.
:   1983   Energy and Rural Women's Work in Rural India. Background paper for ILO/Netherlands Energy and Rural Women's Work Project, Technical Consultation, Geneva.

Food and Agriculture Organization (FAO).
:   1983a   *Wood Fuel Surveys*. Rome.

FAO.
:   1983b   Proceedings of the Workshop on Rural Energy Planning in the Developing Countries of Asia, April 11–19, Beijing, People's Republic of China.

Gunawan, M., P. Arbianto, and P. Sajogyo.
:   1983   Energy and Rural Women's Work in Indonesia. Background paper for ILO/Netherlands Energy and Rural Women's Work Project, Technical Consultation, Geneva.

Harwood, R.R.
:   1979   *Small Farm Development: Understanding and Improving Farming Systems in the Humid Tropics*. Boulder, Colorado: Westview Press.

Hoekstra, D.A.
:   1983a   *Leucaena Leucocephala* Hedgerows Intercropped with Maize and Beans: An Ex Ante Economic Analysis of a Candidate Agroforestry Land Use System for the Semi Arid Areas in Machakos District, Kenya. Nairobi: International Council for Research in Agroforestry.

Hoekstra, D.A.
:   1983b   The Use of Economics in Agroforestry. Nairobi: International Council for Research in Agroforestry.

Huke, R.E.
  1981   Agroclimatic and Dry Season Maps of South, Southeast, and East Asia. Philippines: International Rice Research Institute.
Huke, R.E.
  1982   Rice Areas by Type of Culture: South, Southeast, and East Asia. Philippines: International Rice Research Institute.
International Labour Office (ILO).
  1977   *Small Enterprise Development: Policies and Programmes.* Edited by Philip A. Neck, Management Development Series no. 14. Geneva: International Labour Organisation.
Johnston, B.F., and W.C. Clark.
  1982   *Redesigning Rural Development: A Strategic Perspective.* Baltimore: The Johns Hopkins University Press.
Korten, D.C.
  1980   Community Organization and Rural Development: A Learning Process Approach. *Public Administration Review* (September/October):480–511.
Lockeretz, W., G. Shearer, and D.H. Kohl
  1981   Organic Farming in the Corn Belt. *Science* 211 (February):540–547.
Mark, S.M., and R.L. Lucas.
  1983   *Agricultural Research and Planning: Some Lessons from the Hawaii Experience.* Honolulu: University of Hawaii, Hawaii Institute of Tropical Agriculture and Human Resources.
Marten, G.G., D. Babor, P. Kasturi, D.A. Lewis, C. Mulcock, L.C. Widagda, and I.P. Willington.
  1981   *Environmental Considerations for Biomass Energy Development: Hawaii Case Study.* Research Report no. 9. Honolulu: Environment and Policy Institute, East-West Center.
Nanjundan, S., H.E. Robison, and E. Staley.
  1962   *Economic Research for Small Industry Development.* Bombay: Asia Publishing House.
National Atlas and Thematic Mapping Organization (NATMO).
  1980   *Atlas of Agricultural Resources of India.* Calcutta: Department of Science and Technology.
National Research Council (NRC).
  1981   Board on Science and Technology for International Development. Advisory Committee on Technology Innovation. *Food, Fuel, and Fertilizer from Organic Wastes.* Washington: National Academy Press.

Paul, S.
  1982 *Managing Development Programs: The Lessons of Success*. Boulder, Colorado: Westview Press.
Reddy, A.K.N.
  1983 Rural Fuelwood: Significant Relationships. In *Wood Fuel Surveys*. Rome: FAO.
Roughan, J.J.
       Organization for Village Development. Ph.D. diss., University of Hawaii, forthcoming 1985.
Seckler, D.
  1978 The Role of Nutrition in Agricultural Development: Variations on an Indian Theme by Dandekar and Rath. Mimeo.
Shaner, W.W.
  1983 Stratification: An Approach to Cost-Effectiveness for Farming Systems Research and Development. In *Proceedings of Kansas State University's 1982 Farming Systems Research Symposium — Farming Systems in the Field*, ed. Cornelia Butler Flora, 162–181, Paper no. 5, Kansas State University, Manhattan.
Shaner, W.W., P.F. Philipp, and W.R. Schmehl.
  1982 *Farming Systems Research and Development: Guidelines for Developing Countries*. Boulder, Colorado: Westview Press.
Smith, K.R., A.L. Aggarwal, and R.M. Dave.
  1983 Air Pollution and Rural Biomass Fuels in Developing Countries: A Pilot Study in India and Implications for Research and Policy. *Atmospheric Environment* 17 (11):2343–2362.
Staley, E., and R. Morse.
  1965 *Modern Small Industry for Developing Countries*. New York: McGraw-Hill, Inc.
Taylor, C.E.
  1978 Charting Health Status — Food and Population. Speech presented to American Public Health Association annual meeting, October, The Johns Hopkins University, Baltimore.
Vayda, A.P., C.J.P. Colfer, and M. Brotokusumo.
  1980 Interactions Between People and Forests in East Kalimantan. *Impact of Science on Society* 30(3):179–190. UNESCO.
Volunteers in Technical Assistance (VITA).
  1982 *Testing the Efficiency of Wood-Burning Cookstoves*. Provisional International Standards. VITA, 1815 North Lynn Street, Suite 200, Arlington, Virginia.

Walker, L. P.
  1982 Energy Integrated Dairy Systems. In *Energy Conservation and Use of Renewable Energies in the Bio-Industries 2*, ed. F. Vogt, 689–705. Oxford, England: Pergamon Press Ltd.

Wijewardene, R.
  1982 Low Energy Farming Systems and Appropriate Tools for Farming in the Humid Tropics. In *Energy Conservation and Use of Renewable Energies in the Bio-Industries 2*, ed. F. Vogt, 708–724. Oxford, England: Pergamon Press Ltd.

Zandstra, H.G.
  1983 An Overview of On-Going Applied Farming Systems Development Projects: What are Farming Systems and How Do They Relate to Development. In *Proceedings of Kansas State University's 1982 Farming Systems Research Symposium — Farming Systems in the Field*, ed. Cornelia Butler Flora, 33–48, Paper no. 5. Kansas State University, Manhattan.

INDEX

Information in tables, figures, and photographs is listed by subject and page number. Country references are for pertinent subjects only, in most instances outside the country chapter.

Abilities. *See* Capabilities, local
Action research, 2–3, 85, 519–524
  definition of, 25, 281–285
  example of, in Nepal, 30, 33, 279–336
  facilitator, 5, 24, 285, 288–320, 325–330, 474, 519–523
  incentives, 325, 327
  interactions in, 243, 267, 272, 280–285, 332, 337, 401, 405, 510, 519
  See also Participatory research
Afforestation, 11, 93, 95, 173, 182, 269, 271, 508. *See also* Fuelwood plantings; Silviculture; Tree species
Agricultural development
  rate of, 169–173
  strategy for, 44, 181–184
  See also Rural development
Agroforestry, 395, 399, 404, 408
Agroprocessing mills, 4–5, 15, 522, 527
  costs and benefits of installing biogas plants for, 454–457
  costs and benefits of installing, in Nepal, 289–290, 294–299, 306–307, 315
  rice, 139, 220–223
  and rural employment, 45, 53
Alternative technologies
  dual-fuel engine, 291, 304, 454–457, 527
  gasifiers, producer gas, 527, 531, 535
  pedal power, 307–308
  steam, 267
  See also Animal power; Biogas; Solar energy; Stoves; Wind energy
Anaerobic digestion. *See* Biogas
Analysis
  economic, 185, 217, 459, 465

  financial, 432–465
  statistical, 144, 153, 216–217
  See also Data
Animal power, 15, 474–475, 531
  for agricultural production, 46, 172–173, 184
  efficiency improvements in, 529
  for irrigation, 260–262, 267, 270
  for transportation, 264–267
  use of, by landholding size, Thailand, 218, 220–222
  use of, by region and crop, 210–216
Animal residues. *See* Dung
Animal resources, 14, 20, 55, 388, 403, 528–529, 533
  distribution of, 50–51, 58–59, 66, 81–82, 139, 156, 159, 249–251, 254–258, 488–491, 505–510
  for milk production, 268, 271
  See also Animal power; Dung
Appliances
  used with biogas, 418
  use of electrical, 157–158, 206, 208
  use related to family income, 150, 155
  See also Rural electrification
Appropriate technology
  and alternative energy systems, 421
  in Nepal, 279, 319
  for rural development, 90
Arid regions, firewood shortage
  in Africa, 381
  in Indonesia, 93
Artisans
  information source for rural energy development, 14
  use of skills, 309, 344, 372
  See also Knowledge, of rural people; Occupations

Benefits
  distribution of, 22, 44, 177, 271–272

549

to primary and secondary interest groups, 337–341, 370, 388, 410
probabilistic, 270–271
in relation to risks, 289
sharing of, and group cohesion, 290, 304–305, 318, 530
See also Biogas
Biogas
assessment factors and methods, 10, 30, 185, 317, 415–471
benefits and costs, 81, 83, 294–295, 316, 415–421, 433–455
in China, 427–430
community organization of, in Nepal, 30, 290–300, 306–307, 316–318, 332
community-scale, 81–82, 452–456
dung requirements for, 81–82, 291, 317, 424–425
family-size, 81, 433–452
in India, 431–432, 458
limitations, 81
organic feedstock, 424–425, 460
and pathogen reduction, 415, 420
potential for irrigation, 81–82, 527
sensitivity analysis of financial viability, 31, 433–452
technological research on, 82–83, 333
and waste management, 415, 420
Biomass
dependence on, 16, 486–487, 519
depletion of, 11, 91
production of, 531, 542
in trees, estimate of, 55
variability of, 17–18
Biomass fuel
composition of, 15, 96, 188–196, 487, 504–508, 512–514
consumption of, 94, 102–121, 128–131, 488, 510–512
See also Smoke from biomass fuels
Biomass fuel, sources of
bamboo, 58, 65, 68, 120, 513
cassava, 172, 223
coconut, 119–120, 138, 145, 154, 156, 159, 172, 189–196, 527
sugarcane, 172, 183
vegetation, 11, 14, 269, 381, 391, 487–493, 504–510, 528
woodwaste, 139, 157, 159
See also Charcoal; Crop residues; Dung; Firewood

Caloric value
relevance for energy analysis, 70
of rice and wheat, 66, 363, 366–367, 481
See also Heat value
Capabilities
enhancement of, 285, 322, 521–524, 527–528
local, 1, 31–33, 421
See also Knowledge
Capacity utilization
of biogas digesters, 422–423, 455–456, 473
as project appraisal factor, 456, 537
Case study
of biogas, 433–436
of fuelwood, Bangladesh, 403–406
of fuelwood, Indonesia, 406–410
Ceramics, fuelwood used for, 197, 199, 224
Charcoal
definition of, 188
efficiency of, 199–204, 353
heat value of, 102, 162, 227, 353
markets for, 156, 159
production of, 220, 223, 353, 527
residual, in stove tests, 366, 368, 374
Charcoal, consumption of
in cooking, 29, 69, 189–204
income effects on, 220
in industry, 189–196
for ironing, 119, 145, 189–196
regional, 148–149
See also Fuel substitution
Children
demographic characteristics of, 251, 253
as fuel collectors, 259–260
future of, 5, 289, 540
nutrition of, 177
and stove hazards, 357
Climate
drought, 80, 248–249, 267–268, 339
and farming system decisions, 474–475
floods, 58, 248, 269, 271, 339
rainfall, 60, 136–137, 183, 248, 290, 390
suitability to biogas production, 427
temperature, 60–61, 390
and tree species selection, 390
Commercial energy, 17, 23
coal as, 46, 53, 271
definition of, 167
policy issues and, 53–54, 70
sources of, in Bangladesh, 46–47
supply planning for, in India, 275
use of, in rural areas, 171
See also Energy consumption
Communications, 150, 188, 206, 280, 288, 318, 321, 324, 370

Community characteristics, 282–284, 286, 307, 315, 318
conflict, 245, 272, 290, 308
cooperation, 245, 294–296, 318, 521
initiatives, 18, 32, 35, 89–90, 474, 522, 530, 538
leadership, 6, 99, 188, 244, 307
organization, 82, 270–273, 280, 293–307, 332, 460, 522, 524
Community negotiations, 32, 281, 283, 290–307, 316–318, 474, 520–524, 526, 528–532
Community-scale system, 81–83, 290–307, 423, 427, 429, 432, 452–457, 459–460
Conservation
 energy, 12, 181, 197, 226, 340–341, 373
 forest, 53
 project in Nepal, 287–288, 331
 soil and water, 268
Constraints, 6, 524
 on construction materials, 320
 on energy, 10, 19, 115–119, 226, 241, 503–510, 522, 524
 on food, 20
 on housing and space, 345
 indicators of, 31–32
 on time, 474–475, 499–502
 See also Food deficit; Fuel deficit
Construction materials, 59, 83, 298–301, 317, 338–339, 341, 343–344, 354–355, 360, 422–424, 432
Cooking
 experiments in, 360–370
 practices and fuel consumption in, 310–312, 340–349
 predominance of, in rural energy use, 16, 46, 102, 145–146, 189–190, 226, 477
 process of, 337–343
 in relation to nutrition, 477–478, 530–531, 539–540
 utensils used in, 341
 See also Cooking fuels; Cooking time
Cooking fuels, 1, 337
 access to, 483–496
 biogas, 81, 417–418, 424, 427–428, 430, 434, 437–446, 452–455
 biomass, 349–352
 composition of, 46, 52, 67–70, 91–92, 102–111, 128–131, 148, 191–196, 197–203, 210, 258–260
 consumption of, per capita, 56, 58–61, 63–70, 75, 102–103, 108–111, 128–131, 191–196, 199–203, 486–488
 estimation of requirements for, 477–485
 preferences in, 105–106, 491–493
 sources of, 483–396
 See also Landholding size and land tenure; Socioeconomic factors
Cooking sector, 31, 477–502, 522–523
Cooking time
 and fuel saving, 21, 343, 348
 and stove design, 311, 314, 348–349
 valued for other uses, 5
Cooperatives, 24, 45, 139, 267–269, 295, 385, 410
Costs, 31
 of biogas, 81–83, 291, 295, 333, 339, 430–436, 453–457
 of cooking fuels, 52, 81–83, 189
 of fuelwood, 52, 387, 400
 of lighting fuels, 189
 of windmills, 270
Crop residues
 access to, in relation to land ownership, 53, 56, 75, 84, 487–493, 504–510
 consumption of, in cooking, 52–53, 60, 66–67, 69, 71, 189–196, 202, 337, 487–493
 consumption of, by farm size, 66, 68
 effects of depletion of, on soil quality, 46, 528
 effects of, on stove efficiency, 345, 351–352, 357
 heat value of, 86, 162, 227
 local knowledge of, 14–15, 492
 multiple uses of, 24, 80, 510
 rural reliance on, 1, 351, 354
 substitution of, for firewood, 46, 55–56, 80, 351, 384, 487–493
 supply effects of crop variation, 55, 61, 74, 513–514
 valuation of, 509–510
Culture, 10, 18, 35, 75, 274, 321, 330, 360, 371, 488, 519, 534

Data
 analysis of, 77, 85, 143, 209, 246–247, 368, 493
 base, energy, 12, 24, 34, 96, 185, 537–538
 collection of, 9, 16, 61–63, 66, 76–77, 99, 141–143, 245–247, 273–275, 368, 527
Decision makers
 district, 6–8
 external, 279–281, 321
 family, 30, 523
 international, 8
 national, 8, 73, 337, 370

552   Index

researchers, 9
village, 3–6, 30, 288, 307, 321
Deforestation, 11–12, 53, 81, 93, 97, 176, 248, 339, 373, 381–383, 409, 415, 491
Development bank, 8, 24, 292–293, 296–298, 301–303, 316, 332, 538–539
Development research, 1, 32, 184, 226, 542
Dialogue, 3, 243, 273, 285, 288, 307, 315, 321, 327, 330, 371, 475, 525, 542. *See also* People's participation
Diesel engine
  costs of, 148, 159
  ownership of, by farm size, 255, 258
Diesel fuel
  heat value of, 162, 227
  price of, 1, 180
  price advantage of, 169, 216
Diesel fuel, consumption of
  for agricultural machines, 218–220, 228–237
  in crop production, 210–216, 241
  in fishing, 218
  in industry, 223–224
  in irrigation, 190–196, 208, 260–263, 267
  for lighting, 190–196, 204–205
  in milling, 221, 290, 306, 454–457, 465
  for transport, 148, 225
Digester residues, value as soil conditioners, 418, 426, 430, 439, 450–452, 529
Districts, 48, 50–54, 59–65, 77, 97–98, 100, 104–105, 115–118, 120, 122, 184, 326
Dung
  for biogas, 81–82, 291–292, 298, 306, 316–317, 332, 415–416, 419, 424–425, 427, 452–454
  collection of, 14, 298, 332, 473
  as cooking fuel, 52–53, 60, 68–71, 258–260, 337, 351–352, 487–493, 504–508
  environmental effects of burning, 11
  heat value of, 86, 507–508
  left untouched, 189
  use of, as manure, 509–510
  valuation of, 509–510

Economic development, 89, 176–177, 183, 532
Ecosystems, 285–286
  and development planning, 184, 243
  and energy efficiency, 18–19
  energy resources in, 18–19, 477, 533

  environmental damage in, 11
  and geographic characteristics, 97–98, 122
  sample design, basis for, 28, 30, 33, 65–66, 68, 97–98, 122, 125
  See also Farming systems; Regional analysis; Zones
Education, 12, 44, 84, 90, 251, 253, 269–270, 289, 296, 496, 543
Electricity
  distribution of, 47, 145, 158, 262–263, 541
  felt need for, 3
  generation of, 47, 159, 262–263, 527, 541
  prices of, 47, 155, 180
  rural households and, 139
  supply of, 47–48
Electricity, consumption of, 145, 149, 158–159
  for appliances, 190–196, 206, 208
  in cooking, 91, 190–196, 199, 201, 210, 343, 353, 361–365, 374
  in industry, 223–224
  in irrigation and farm production, 53–54, 81, 190–196, 207–208, 241, 248, 258, 260–263, 267
  for lighting, 103, 107–108, 112, 130, 190–196, 204–205, 263–264
Employment, 1, 24, 45, 80, 91, 175, 184, 268, 271, 322, 407, 410, 509, 523, 532, 536
Energy, agricultural, 29, 45, 522
  for crop production, 1, 45, 160, 182, 210–218, 524
  projected demand for, 182
  use of, by landholding size, 228–237
  See also Diesel fuel, consumption of; Farm mechanization; Fertilizer; Work
Energy consumption, 55, 151, 179
  balance of payments effect, 176
  behavioral aspects, 13, 538
  commercial, 55, 91, 96, 145, 147, 149, 153, 163–165, 171, 208–210, 216
  in cooking experiments, 363–368
  effect of, on environment, 1, 123, 203–204, 370, 381
  and farming system indicators, 504–514
  for farm tractors, 218–222
  noncommercial, 55, 145, 147, 149, 152, 479–490
  in personal transportation, 225
  structure of, 92, 104, 106, 209
  by type of village, 104–105, 110–112, 115–118, 128–129

Index 553

See also Biogas; Biomass fuel;
  Charcoal, consumption of;
  Cooking fuels; Crop residues;
  Diesel fuel, consumption of; Dung;
  Electricity, consumption of;
  Firewood, consumption of;
  Gasoline, consumption of;
  Kerosene, consumption of;
  Petroleum fuels; Solar energy;
  Wind energy
Energy conversion
  changes in system boundaries and,
    21–22, 34
  indexing issues and, 22, 102, 422
  intermediate stages of, 12, 22, 124,
    145–146, 160, 190
  location of, 22, 34, 535–536
  losses in, 145–146
  scale of, 22, 34, 220, 527–528,
    535–536
  using local resources for, 534–537
  variable characteristics of, 17, 527
Energy demand and supply, 2, 6,
  10–11, 13, 29, 55–56, 73, 75, 83,
  96, 124, 146, 160, 190
Energy efficiency
  combustion factors in, 311, 348–349,
    353–357
  in cooking, 56, 61, 199–204, 226,
    339–343, 346, 348–349, 353–354,
    479–483
  definition of, 353
  effect on energy needs, 96, 102,
    119–121, 131
  experiments in, 27, 353–372,
    374–375
  heat utilization factor in, 353–358
  measurement of, 22, 31
  policies for improving, 131, 203–204,
    226
  seasonal variation of, 61–62
  social benefits of improvement of,
    22, 337, 535–536, 542
  and user acceptance, 337, 373, 478
  users' knowledge of, 475, 478–483
  value added by improvement of, 22,
    475, 535–536
  variation of, among ecosystems,
    18–19, 21, 520
  variation of, among income groups,
    18, 21, 67, 107–110, 478–483
  See also Energy quality
Energy flows. See Energy conversion;
  Energy demand and supply
Energy needs, 11, 13, 31, 34, 70, 73,
  75, 83, 102, 370, 526
Energy planning, 43, 73, 79, 170, 226

Energy quality, 22, 34, 45, 225, 353,
  475, 527, 535
Energy resources, 2, 10, 13, 146,
  155–159, 169–171, 241, 264–266
Energy supply
  decentralization of, 16–17, 112–113,
    156, 159, 395
  distribution system of, 17, 113,
    115–117, 130, 263–264
  policy factors of, 338
  uncertainty of, 1, 115–117, 262–263
  See also Energy demand and supply
Energy technologies
  assessment of, 1, 526–528, 536–537
  design of, 10, 21–22, 226, 280,
    282–284, 312, 317–318, 328, 333,
    337, 424, 458, 527
  development of, 8–10, 27, 79, 85
  diffusion of, 57, 85, 280, 324, 328,
    355, 370–372, 536
  identification of, 13, 265–270,
    281–282, 289–292, 307–308,
    314–316, 328
  integrated systems of, 82, 528–530
  management of, by local control,
    303–305, 410, 520
  operation and maintenance of, 292,
    296, 305–306, 318, 430–432, 458
  options for, 265–270, 281, 287
  reliability of, 422, 424, 429–433,
    458
  See also User-directed technology
    design
Equity, 13, 28, 32, 43–45, 91, 102,
  170, 432, 522, 526, 528, 532, 542
Erosion, 20, 176, 269, 271, 339, 381,
  390, 398
Evaluation, monitoring, 83, 85, 281,
  284, 306, 310, 319, 337–339, 370,
  372, 458, 526, 532
Exchange, 19, 112, 115, 345
  monetized, 90, 534
  nonmonetized, 15, 18, 34, 437, 476,
    496, 534, 538
Experiments
  basis for need indicators, 476
  field, 30, 58, 342–349, 370–375
  on fuel consumption, 58, 121,
    362–363, 479–483
  on fuelwood species, 400–406
  laboratory, 121
  pumps, 267
  stove, 359–370
  See also Energy efficiency
Extension, 54, 288, 324–329, 374,
  404, 409–410

Farmers and farm families, 6, 14–15, 18, 32–33, 80, 177, 385, 395, 398–399, 406–407, 428, 473
Farming systems
  animals in, 21, 249–251, 502, 505–509
  crops and cropping intensity in, 46, 80, 286, 420, 488–493, 533, 542
  definition of, 19
  forests in, 113, 248–249, 502–509
  indicators for, 477, 502–510
  integrated use of farm and homestead land, 79–80
  in relation to energy systems, 21, 31–32, 80, 502–510, 522, 538–543
  research and development, 33–34, 524–534
  social organization in, 18, 21, 251–258
  transitions in, 18
Farm mechanization, 15, 169, 217–222, 228–237, 254–255, 307–308, 531
Farm size, 29, 34, 63–68, 218, 228–237, 252, 293, 297, 505–509, 531, 542
Feasibility assessment, 270–272, 281–283, 288, 294–295, 338, 527
Fertilizer, 1, 45, 69, 172, 210–217, 241, 256, 258, 415, 419–420, 426, 450, 505–508
Financing, 35, 73, 75, 257, 281, 290, 292, 372, 440, 538–539, 542
Firewood
  access to, in relation to land ownership, 53, 56, 75, 84, 387–388, 487–493, 505–509
  assessment of needs for, 31, 386–389
  and combustion in stove experiments, 359–365, 368
  and commercial development, 23–24, 95, 112, 130–131, 383–384, 387–389, 406–410
  efficiency improvements in use of, 131, 310–312, 346
  heat value of, 69, 86, 102, 119, 162, 227, 353, 359–360, 363, 507–508
  industrial demand for, 24, 74, 80, 384, 389
  influence of, on stove design, 348–354
  markets for, 156, 387, 512
  overcutting, 55
  preferences in, 68, 491–493
  prices of, 24, 95, 105, 108, 131, 260, 386, 389, 508–509
  production program for, 385–410, 528–530
  rural reliance on, 1, 130, 145, 286
  savings potential of, 81, 354, 370, 372
  scarcity of, 24, 46, 93, 115–117, 351, 381–385, 404, 487–493, 510–514, 530–531, 539–540
  substitution for, by biogas, 316, 433–454, 459
  supply and availability of, 48–53, 75, 112–114, 130–131, 286, 337, 382–383
  urban demand for, 24, 49–52, 67, 351, 383–384, 541
  use of, for charcoal production, 220, 223
Firewood, consumption of
  Bangladesh estimates, 28, 55, 71
  in cooking, 29, 60, 66, 71, 103–105, 107–111, 148, 190–204, 258–260, 286, 481–482, 484–495, 504–510
  by farm size, 66, 68, 70, 84
  in fumigation, 190–196, 207
  in heating, 61, 190–196, 206–207
  income effects on, 23, 55, 104, 107–110, 115–117, 131, 154, 345
  in industry, 190–196, 206, 223–224
  regional, 67, 69, 105, 149, 510–514
Fixed-dome digester, 83, 291, 333, 422–423, 429, 432–433, 447
Floating-dome digester, 81, 83, 416, 423, 431–433, 448
Fodder, 6, 31, 46, 80, 249, 271, 333, 396, 404, 489, 506, 528–529
Food consumption, 64–66, 177–178, 268, 477–482, 511, 526, 530–531
Food deficit, 7, 20, 44, 46, 56, 59, 249, 478, 491, 539–540
Food-fuel competition, 11, 491, 530
Food preparation, 5, 31, 55, 339–341, 495–500
Food processing, 5, 55, 339, 452, 454–457, 496–501
Food production, 1–2, 5, 45, 55, 399
  rice, 138, 156, 159, 172, 199, 342, 361–367, 481–482, 484–485
  wheat, 481–482, 484–485
Forage, 14, 396, 399–400, 408
Forest products, 173, 384, 393, 396, 406–410
Forestry, 53, 74, 84, 97, 173, 249, 268, 326, 373, 395, 407
Forests, 48, 52, 66, 68, 113, 120, 138, 156, 175, 184, 197–200, 381–384, 513, 533
Fuel collection, 14, 19, 23, 31, 108, 113–114, 127, 259–260, 339, 381, 407, 426, 438, 452, 459, 475, 491, 495–500

Fuel consumption factors, 66–68, 75, 101, 105–114, 200–203, 210, 342–343, 358–370, 479–495
Fuel deficit
  assessment of, 20–21, 24, 115–117, 203–204, 381–385, 404, 487–493, 510–514
  constraint on irrigation, 260–263
  effect on land use, 80, 203–204, 248
  effects on nutrition and work, 20, 24, 354, 477–478, 513–514, 530–531, 539–540
  and priority areas for stove improvement program, 370–372
  as stimulus to biogas development, 427
  and valuation of alternative fuels, 456, 459, 465
Fuel purchase, 434, 437–446, 452–453, 459, 507–508, 512
Fuel substitution
  biomass and fossil fuels, 1, 28–30, 104–105, 110, 197, 541
  charcoal and firewood, 29, 200–204, 225
  diesel and electricity, 53–54, 158
  diesel, gasoline, and LPG, 158–159, 171
  electricity, charcoal, and LPG, 210
  electricity and firewood, 158
  electricity and kerosene, 107, 111–112, 210
  firewood, kerosene, natural gas, and LPG, 28, 52–53, 105, 108–109
  price policy for, 12, 182
  sensitivity analysis of biogas, firewood, and kerosene, 316, 433–454, 459
  and stove efficiency, 226
Fuelwood plantings, 385, 387–389, 395, 512–514

Gasoline
  heat value of, 162, 227
  price of, 180
Gasoline, consumption of, 145, 149
  for agricultural machines, 218–220, 228–237
  in crop production, 210–216
  effect of prices on, 155
  in irrigation, 190–196, 208
Geography. *See* Ecosystems; Regional analysis; Zones
Government, 6, 45, 48, 243–247, 285, 305, 323–324, 330–331, 385, 387–389

Health, 44, 177, 184, 251, 269–270, 326, 354, 418, 420, 439, 477
Heating, 61, 189, 197, 205–207, 393
Heat value
  gross and net, 359–360
  of specific fuels, 69, 86, 102, 121, 143, 162, 227, 353, 363, 366–367
  variability of, with fuel condition, 18, 359–360
Homelot, 120, 127, 382–383, 385, 399, 401
Homesteads, 48, 52, 58–59, 79, 82, 84, 381, 482
Household characteristics, related to fuel consumption, 12, 75, 101, 188
  education and, 251, 253
  energy resource distribution and, 19, 56, 58–61, 67, 69–70, 112–114, 197–200, 251–258, 483–495, 502–509
  family size and, 12, 59, 104–106, 110–111, 122, 128–131, 153–155, 365, 368, 482–483
  food supply and, 59
  housing and, 58–59
  income and, 55, 81, 99, 104, 106–112, 115–119, 128–131, 150–152, 208–210, 266–271, 480–482
  occupation and, 14–15, 24, 254
Housing, influence on stove design, 344–345, 352, 357, 369–370
Human energy, 19–20, 46, 188–190, 210–216, 260–262, 264–269, 539
Hydroelectricity, 46–48, 179

Implementation, 85, 271–273, 279–285, 296–310, 314, 316, 387–389, 526
Income
  distribution of, 23, 95–96, 177–178
  and earning opportunities, 1, 14, 523
  and energy consumption in transportation, 224–225
  and energy purchasing power, 266–270
  and incentives for fuel saving, 339, 345
  per capita, rural, correlated with energy use, 216–218, 220
Indicators
  definition, of, 476–477
  diagnostic, 476–477, 486–493, 503–509, 524–525, 532–534, 538
  of energy program success, 84
  for energy regionalization, 275, 532–534, 536–537

of environmental depletion, 20
need for, 476–483
for resource accounting, 421–426
value of, 31, 476–477, 508–510
for village classification, 90, 183–184
Industry, 24, 53, 69, 81, 93, 169–171, 176, 179, 184, 197, 205–206, 220–224, 381, 383, 385, 387–389, 399, 403, 406, 522
Information
gaps, 2, 16–18, 28, 135, 519
needed for local energy planning, 31–34, 474–477, 519
needs of national decision makers, 2–3, 9, 78
two-way flows, 523
See also Knowledge
Infrastructure, 18, 90, 105, 125, 129, 155–159, 217, 424, 432, 460
Innovation, 5, 10, 30, 79, 279–280, 319, 330, 370, 389, 475, 500, 519, 530, 535, 542
Institutions, 5, 57, 90, 319, 325–329, 370–372, 407, 519, 528, 530–532, 535, 542
Interdisciplinary research, 2, 33, 35, 57, 76–77, 80, 372, 523–526, 537, 542
Investment, 2, 21, 305, 426–428, 430, 440, 446, 461, 513, 527–532
Ironing, 145, 148, 150, 205–206, 210
Irrigation, 7, 15, 45, 53–54, 81–83, 139, 182–184, 207, 218, 220, 248–249, 260–263, 265–268, 305, 522, 527, 531

Kerosene
efficiency of, compared to biomass, 102, 108–110, 121, 131
heat value of, 86, 102, 121, 162, 227, 363, 366–367
imports of, to Indonesia, 95
price variation of, 109, 113, 131, 132, 180
in stove experiments, 341–343, 349, 361–367, 374
subsidy policy on, in Indonesia, 28, 91, 93–95, 109, 130
substitution of, by biogas, 316, 433–454, 459
supply and availability of, 1, 113–118, 130
use of, by higher income families, 23, 107–108, 128–129
Kerosene, consumption of, 145, 149, 154–155
in cooking, 69, 103, 129–130, 148, 193–194

in lighting, 59–60, 66–67, 82, 103, 112, 129–130, 148, 189–196, 204–205, 286
Knowledge
advantages of, in asset creation, 534–536
local, 13–15, 323, 330, 358, 390, 534–536
as a resource, 496
of rural people, 1, 31–32, 127, 473–476, 514, 520–523
See also Information

Labor. *See* Work
Labor commitment, 298–300, 316–317, 530
Land commitment, 296–298, 302–303, 530
Landholding size and land tenure, 53, 56, 81, 169, 173, 184, 218, 293, 387–388, 395, 505–510
and cooking fuel use, 63–70, 488–493
and food consumption, 63–66
Landless, 6, 14, 18, 45, 54, 56, 58, 63–65, 251–260, 478, 488–493, 502, 505, 508, 531
Land use, 23, 31, 79, 98, 169, 184, 217, 226, 248–249, 269, 325, 381, 385, 391, 407, 474, 491, 505–511
Large-scale surveys, 27–29, 55, 65, 73, 75, 139–161, 171, 185–188
Learning
experiential, 281, 285, 310–315, 318–320, 327–329, 521–524, 538
mutual, 14, 475
Lighting, energy for, 46, 60, 82, 91–92, 95, 102–103, 106–107, 111–112, 128–131, 145, 150, 204–205, 210, 417, 424, 427–428, 434, 437–446, 452–455, 527
Liquefied petroleum gas (LPG)
consumption of, in cooking, 148, 189–197, 201–202, 210
price of, 180
use of, in modified gasoline engines, 181
See also Fuel substitution
Location and site selection, 34, 54, 82, 298, 332, 370, 390–391, 393–394, 400–401, 422, 432, 459, 527–528, 533, 535, 540–541
LPG. *See* Liquefied petroleum gas

Manure, 82, 292, 396, 399, 404, 505–510, 529

Mapping
  of energy recommendation domains, 532–534
  of environmental data, 11, 391
Markets, 15, 19, 21, 26, 109, 139, 154, 156, 159, 241, 264, 372, 374, 383–387, 399, 403, 406, 430, 437, 507–510, 534, 537
Measurement
  energy indexing issues and, 22
  evaluation of methods of, 126–128
  of fuel consumption, 56, 66–67, 74, 77, 99, 101, 310
  in subsistence economies, 19
  of tree resources, 127
Men
  and allocation of farm tasks, 499
  as cooks, 479
  and fuel collection, 259–260
  and needs assessment participation, 3–7
  task limitation and, 495
  and time as scarce resource, 474–475, 501–502
  training of, as stove makers, 310
  and tree species selection, 5–6
  value of work for, 497
  wage rates of, 300
Methane, 171, 181. *See also* Biogas
Microstudies, 28, 30, 43, 56–65, 74–75, 83, 144, 182, 226, 241–242, 274–275, 381, 386, 476, 486
Milk products, 249, 268, 271, 529
Moisture content, 18, 119–120, 360, 393, 479
Motivations
  toward innovations, 320
  of rural people, 1, 6, 81, 526

Natural gas, 46–48, 52–54, 70, 170, 179, 181, 183, 361–367, 374
Needs
  assessment of, 10, 13, 20–22, 27, 30, 70, 72–73, 75, 458, 526
  basic, 1–2, 46, 91, 243
  energy for rural development, 1–2, 10–11, 13, 31–32, 34, 43, 45–46, 79–83, 91
  felt, 3–6, 10, 81, 370
  See also People's participation
Networking, 320, 391, 401, 404, 415, 419, 426, 439, 450
Nitrogen, 391–394, 401, 404, 415, 419–420, 426, 439, 450
Noncommercial energy
  composition of, 145

  consumption of, by income group and region, 208–210
  definition of, 167, 208
  dominant rural role of, 379
  potential of, 161
  research requirements for, 28, 538
  See also Energy consumption
Nutrients, 292, 339, 393, 450
Nutrition, 1, 19, 34, 44, 177, 184, 354, 477–479, 482–483, 500, 513, 530–532, 539, 542

Objectives
  as determined by local people, 243, 281
  development of, 30, 226
  equity of, 43–44, 83
  external perceptions of, 279–281
  government and community, 387–389
  for reduction of poverty, 43–44
  for regional development, 43, 83
  for rural energy development, 2
  for satisfying basic needs, 325
  for village modernization, 90
Occupations, 12, 14–15, 24, 184, 244, 251, 253–254, 286
Opportunity cost, 300, 387, 448, 454, 459–460, 495, 497–502, 510
Organizations
  local, 3, 10, 14, 18, 30, 32, 54, 272–273, 279–280, 318, 428, 519, 542
  nongovernmental, 372–374
  voluntary, 295–300, 405

Paradigm
  for resource transformations and technology generation, 279–285, 320–321, 323–324, 429
  for rural study and action, 24
Parboiling, 60–64, 66, 342–343, 527
Participant observation, 6, 15, 25, 27, 34, 56–57, 74, 77, 127, 337, 538
Participatory research, 25, 30, 33, 243–247, 273, 487, 524–532
People's participation, 270–275, 279, 288
  and *Gaun Sallah*, 33, 327–329, 525
  in need assessment, 3–6, 30, 32–33, 289–292, 322–329, 520–528
  in planning and implementation, 7, 387–389, 403–405, 473–476, 539–542
  policy for, 183–184, 325
  in resource allocation decisions, 32–35, 293–301, 304–307, 528–532

558   Index

and *Sondeo*, 33, 327, 525
  See also Dialogue; Participatory research
Pesticides, 1, 256, 258
Petroleum fuels
  consumption patterns and effects of, 169–171, 179–181, 188–196
  imports of, 169–170, 180–181
  local knowledge of, 15
  supply and demand, 1, 11, 415
  See also Diesel fuel; Gasoline; Kerosene; Liquefied petroleum gas
Planning, 2, 6, 8, 43, 169, 308, 316, 387–389, 475
Policy
  on agricultural and rural development, 2, 13, 89–90, 182–185, 325–327, 519, 523–525
  on development of commercial energy sources, 47, 170
  on energy conservation, 373
  on energy price and subsidy, 12, 28, 47, 53, 91, 93–95, 109, 130, 179–182, 431, 446–448
  on energy supply-demand balance, 28, 513
  and energy variability, 54, 128–131, 510–514
  on the environment, 1, 373, 541
  on equity and income distribution, 91
  on food and nutrition, 478
  local initiation of, 11, 18, 475–476, 513, 530–532, 537–543
  on petroleum fuels, 1–2, 11
  on regional development, 54
  on renewable energy research and development, 8, 10–11, 24
  on rural electrification, 54, 132, 175–176, 541
  on stove improvement, 131, 337, 372–374
  for user-based technology development, 319, 327–329
Policy research
  on agricultural energy requirements, 34, 182, 217–220
  decision makers' perspectives on, 2–3, 8–9
  defining objectives of, 9–13, 33, 91, 96
  on development and equity, 102, 243
  on development loans for rural energy, 538–539
  on energy and nutrition, 34, 530–531, 539–540
  on energy price policy, 182, 210, 226

on energy supply, 102, 210, 226, 513
on fuel substitution, 210
for national energy policy, 70–75, 135–136
new agenda for, 523–525
on rural transport policy, 540–541
Poor households
  and access to energy resources, 18–21, 203–204, 345, 512–514
  in commercial transitions, 23–24
  dependence of, on biomass fuels, 55, 81, 384
  energy problems of, 96, 115–119
  food consumption by, 21, 56, 127–128
  fuel use by, 21, 56, 127, 154, 345, 351, 480, 488–490, 493, 508
  housing of, 345
  incentives to market fuelwood, 403
  interaction with research groups, 322
  lack of participation in planning and implementation, 244, 271
  loan program for, 332
  value of time for, 496
  See also Poverty
Population, 48, 50–51, 137, 169, 171, 184, 251, 524
Population density, 45, 60, 68, 90, 97, 105, 138, 172, 427, 534
Poverty
  definition of, by food calorie requirement, 44, 104
  persistence of, 43, 45, 241
  policies for alleviation of, 43, 183–185, 226
  and productivity, 21
  recommended action-oriented studies on, 226, 524
  in rural areas, 18–21, 177, 185, 242–243
  and social organization, 18–19
  See also Poor households
Preferences, 13, 160, 393–395, 404, 491–493, 508
Price
  of commercial cooking fuels, 52
  effect of low energy prices, 179–181
  as market mechanism, 19, 218
  of petroleum fuels, 1, 93, 180
  as scarcity indicator, 20, 507–510
Priority
  energy program, 12, 185, 242
  regional, 265
  research, 1, 31, 373, 542
  resource allocation, 2, 5
  of rural people, 279, 281

Problem feedback, 85, 306, 310–315, 318–320, 328, 372, 430, 458, 532
Problem identification, 10, 18, 85, 114–118, 244, 280, 475, 508, 513, 524–527
Productivity, 1, 32, 91, 184, 241, 522, 542
Project appraisal, 294–295, 318, 436–456, 509, 514, 527–528, 542
Pumping, 207, 228–237, 258, 260–263, 270

Regional analysis, 34, 44, 83
  for rural energy transformations, 532–537, 541
  typology for, 275, 533–534
  See also Ecosystems; Variations, regional; Zones
Regulations, 301–303, 307, 318, 422, 424, 429, 458
Renewable energy, 10, 170, 422, 537, 541
Research design elements
  household interviews, complete coverage, 27–28, 56–58, 244–247
  output design, 76
  questionnaire design, 28–30, 56–57, 66, 68, 76, 95–99, 122–127, 135–136, 141–143, 160–161, 185–188, 244–247, 538
  report design, 67, 77–78
  sampling, 56, 65, 68, 75, 96–97, 99–100, 122–125, 140, 186, 225, 273–275, 358, 386
Research, development, and extension, 319–329, 338, 370–374, 428–430, 536
Researchers, academic, 2, 83–85, 242
  commitment to implementation, 273–274, 370
  expatriate, 73
  interaction with rural residents, 25, 35, 85, 243–244, 246, 273, 474–475, 514, 524–526
  need for training, 71, 73, 78, 523–524
  in policy and planning agencies, 2
  as readers, 2–3, 34, 337
  research preferences, 73
  in scientific and technological institutes, 2, 79, 337
  and technology innovation, 355, 373
  See also Interdisciplinary research
Research methods, 1, 3, 9, 24–25, 27, 31, 83–85, 226, 275, 374, 520, 523–532, 535
Research users and sponsors, 2–3, 34, 78

Resource access, 18, 24, 50–51, 54, 56, 81, 251–260, 290, 488–495, 502, 526, 542
Resources
  external, 80, 280–281, 422–424, 428–429, 526
  internal, 83, 243, 280–281, 421–426, 428, 526
Rights
  to benefits, negotiation of, 528, 530
  to firewood collection, 92–93
  of man, 243
  to resources, 19, 532
  to tree branches, sale of, 383
Risks
  earning variability and, 14
  food and fuel availablilty and, 14, 21, 55, 262
  of innovation, 289
  insurance for, 329
  rural people's evaluation of, 21
Roads, 52, 90, 105, 113, 129, 158, 172, 221, 264, 286, 532, 534, 540–541
Rural areas, 2, 171, 177, 241, 421
Rural development, 2, 13, 31, 43–45, 85, 89, 171, 183–185, 241–243, 279, 374. See also Agricultural development
Rural electrification, 7, 171, 179, 534, 541
  alternative to reduce kerosene use, 130–132
  development plans, 47–48, 92
  for increasing economic productivity and living standards, 175–176
  potential and weaknesses of, 54
  research for, 123
  substitution of, for diesel fuel, 158
  use of appliances as an indicator of, 206, 208

Scale economies, 32, 34, 63, 65, 67, 110, 365, 423, 480, 482–483, 493, 501, 522, 527–528, 535, 540–541
Seasonal variation, 20, 49, 52–53, 57, 60–63, 68–69, 76, 82, 188, 267, 345–351, 368–369, 371, 390, 407, 450, 489, 501–502
Self-reliance, 89–90, 183, 243, 272, 288, 428
Silviculture, 31, 385, 395–401, 403–405, 410
Small-scale industry, 173–175, 532, 543
Small-scale systems, 16, 307, 322, 337, 345, 349, 351, 354, 393, 418, 428, 439, 491, 528
Smoke from biomass fuels, 345, 351, 491–493

discomfort of, 345
and housing conditions and ventilation, 352, 357, 439
and pollution and health effects, 22, 418
and preference tests, 393
reduction as stove design criterion, 337-338, 354, 428
research on, 29, 356, 372, 528
and smokeless stoves, 307-315
and stove and chimney dimensions, 349, 356-357

Social structure, 205, 226, 242-245, 251-260, 286, 288, 294, 307, 310

Socioeconomic factors
in cooking fuel use, 52-53, 56, 103-105, 107-111, 127-131, 340-341
in lighting fuel use, 106-107
as research parameters, 10, 33, 68, 74-75, 82-83, 90, 97, 99, 122, 153, 186, 244-247, 271-272, 281, 286, 314, 386, 460-461
in stove improvement programs, 337, 368-372

Solar energy, 16, 170, 268, 271, 527
Spread effects, 288, 329, 372, 532
Storage, 52, 345, 352, 474

Stoves, 17, 62-64, 308, 309, 313, 343, 355, 356, 362, 369
biogas, 418, 431
design and development of, 27, 30, 338-339, 346, 348-349, 354-358, 370-374
field testing of, 30, 370-372
influence of, on fuel requirements, 478-485
kerosene, 431
manufacture of, 344
performance tests on, 58, 358-375
programs for improvement of, 70, 80, 84, 131, 225-226, 307-315, 337-338, 370-374
types in use, 29-30, 60-64, 199-204, 286, 342-349, 358
user assessments of, 310-315, 318, 371-372, 527-528
users and use factors of, 338-342, 345-346, 348-352, 354
See also Cooking, process of

Subsistence economy
agricultural, 249, 286
development of, 18, 184
energy resources in, 19, 34, 480, 531
firewood use in, 381-384
and marginal farm families, 14, 18, 499-502, 531

research methods in, 19, 34, 127-128, 496-502, 534, 538
transition to commercial exchange, 18, 23-24, 383-384, 493-495
Survey lessons, 70-75, 121-128, 143-144, 246-247
Surveys, 2, 25, 27, 34, 57, 68, 84-85, 96, 135, 140, 160, 371, 387

Time
allocation of, 31, 430, 474-475, 495-502
consistent treatment in project appraisal, 422, 426
in fuel collection, 259-260, 339, 491
as a resource, 474-475, 496, 501-502
sensitivity analysis of, in biogas systems, 448-450
value of, 5, 7, 316, 496-499
and women's work, 5, 21, 289-290, 311, 495-502
See also Cooking time; Work

Trade
exports, 105, 169
imports, 54, 82, 93, 95, 105, 169, 180-181, 264
Training, 57, 84-85, 90, 142, 270, 272-273, 303, 308, 310, 326, 372, 374, 405, 430, 521, 523
Transformations, rural, 44, 177, 241, 270, 280, 321, 323, 330, 523, 530, 537
Transport, 34, 45, 49, 52, 145, 158-160, 179, 224, 264-265, 317, 352, 387, 426, 495, 522, 532, 534, 540-541

Tree resources
distribution of, by farm size, 66, 488-491, 505-508
estimation of, 55-71, 382
location of, 48-52, 533
research on, 101

Tree species
distribution among, 58-59, 119-120
fast-growing, 30-31, 80, 170, 184, 389
fruit, 5, 58, 119-120
heat value of, 119-120, 360
local knowledge of, 14, 382-383
multiple-use, 30-31, 48-49, 333, 388-389, 394, 399-401, 528-530
nitrogen-fixing, 391-393
rural people's choices of, 5-6, 394-395, 404
and selection criteria, 31, 269, 389-395

Unemployment, 44–45, 65, 497
Urban energy demand, 52, 66, 95, 171, 177, 197, 340–341, 345–346, 352, 374, 383
User-directed technology design, 21, 28–29, 226, 280–285, 337–354, 370–372, 478–483, 521–530, 534–538

Value
  of benefits in biogas systems, 426–428, 432–459
  of biogas digester residue, 292, 450–452, 529
  of dung, 509–510
  of food, human and animal, 20
  of land, 23
  of nonmeasurable services and burdens, 22, 493
  of nonmonetized biomass resources, 14, 490, 495, 509–510, 538
  of work time, 5, 7, 31, 474–475, 495–502
Value added, 210, 523, 528–530, 535
Variations
  household, 59–70, 122, 128–129, 433–453, 519–520
  regional, 13, 31, 67, 69, 128–129, 145, 155–158, 191–222, 241–242, 247–271, 346–349, 486–491, 505–514, 519–520
  village, 13, 15–16, 59–65, 122, 128–129, 318–319, 346–349, 386, 486–491, 505–514, 519–520
Villages, 6, 33, 43, 56, 59, 89–90, 100, 122–123, 125, 128, 183–184, 186, 188, 384, 401, 404–405, 422

Water resources, 3, 7, 172, 183–184, 286, 315, 325, 416, 424, 460, 481–482
Wind energy, 170, 267–268, 270
Women
  and changing farm work roles, 21, 499, 535
  as decision makers in fuel use, 16, 479–483, 528
  as decision makers in stove design, 30, 340–341, 370–372

  as decision makers in tree species selection, 5–6, 387, 528, 535
  and food preparation time, 495–502
  and fuel saving, 21, 340–341, 343
  participation of, in needs assessment, 4–7
  in rural energy research, 524, 542
  as stove makers and operators, 17, 310, 344, 360
  time allocation by, 21, 31, 495–502
  and time as scarce resource, 474–475, 501–502
  and time spent in fuel collection, 259–260
  wage rates of, 300
  work of, 495
Woodlots, 48, 52, 184, 197–200, 385
Work
  of agricultural laborers, 44, 64–65, 407
  allocation of, 340, 495
  dynamics of, 18, 535
  farm, 211–215, 228–237, 474–475, 495–502
  in food preparation, 495–502
  in fuel transport, 495
  in fuelwood production, 385, 387, 404–405, 409–410
  mechanical, 189, 216
  nutrition adequacy for, 530–531, 539–540, 542
  organization of, 31–32
  requirements for biogas systems, 422–426, 428, 439, 448, 454, 459–460
  and rural wage negotiation, 14
  See also Human energy

Yields
  crop, 55, 170, 172, 182, 210–217, 226, 420, 505–510, 529
  fishery, 176
  fuelwood species, 31, 389, 392, 396, 529
  milk, 506

Zones
  for agricultural planning, 44
  agroclimatic, 52, 67, 77, 84, 488–493, 505, 529–537
  agroeconomic, 171–174, 182, 226

Other Titles of Interest from Westview Press

Critical Energy Issues in Asia and the Pacific: The Next Twenty Years, Fereidun Fesharaki, Harrison Brown, Corazon M. Siddayao, Toufiq A. Siddiqi, Kirk R. Smith, and Kim Woodard

Energy in the Transition from Rural Subsistence, edited by Miguel S. Wionczek, Gerald Foley, and Ariane van Buren

Science and Technology for International Development: An Assessment of U.S. Policies and Programs, Robert P. Morgan

Technology Transfer: A Project Guide for International HRD, edited by Angus Reynolds

Renewable Energy for Industrialization and Development, David John Jhirad

Energy Planning in Developing Countries, Peter M. Meier

Rural Energy Planning in Developing Countries, Romir Chatterjee

The Economics of New Technology in Developing Countries, edited by Frances Stewart and Jeffrey James

Transferring Food Production Technology to Developing Nations: Economic and Social Dimensions, edited by Joseph J. Molnar and Howard A. Clonts

Westview Special Studies in
Social, Political, and Economic Development

*Rural Energy to Meet Development Needs:
Asian Village Approaches*
edited by M. Nurul Islam, Richard Morse,
and M. Hadi Soesastro

This book integrates policy, technology, and action research methods in providing new perspectives and tools for Asian village decision makers and planners who seek more effective uses of energy in important rural tasks. The cooperative research on which the book is based was motivated by two policy concerns: the supply instability and price uncertainty of petroleum-based fuels, fertilizers, and pesticides; and the environmental depletion associated with widespread dependency on firewood and farm residues for cooking fuel. The authors combine the voices and knowledge of women and men who produce and use rural energy with analyses and assessments by engineers, economists, agricultural scientists, and anthropologists to clarify these issues while filling serious information gaps about the use and substitution of fossil and biomass fuels.

The book focuses initially on cooking fuel required to meet food and nutrition needs. It demonstrates research methods linking energy with farming systems to increase agricultural productivity and to support other income- and employment-generating activities in rural areas. The authors thereby establish a research agenda through which rural residents, interacting with specialists and policymakers, can build upon their own experience and values in organizing socially and environmentally appropriate rural energy systems.

M. Nurul Islam is associate professor of chemical engineering and director of the Institute of Appropriate Technology at Bangladesh University of Engineering and Technology, Dhaka. **Richard Morse** is research associate and coordinator, Energy for Rural Development, East-West Center Resource Systems Institute, Honolulu. **M. Hadi Soesastro** is head of the Department of Economic Affairs at the Center for Strategic and International Studies (CSIS), Jakarta.